Wireless Information Highways

Table of Contents

Wireless Information Highways

Dimitrios Katsaros
Aristotle University of Thessaloniki, Greece

Alexandros Nanopoulos
Aristotle University of Thessaloniki, Greece

Yannis Manolopoulos
Aristotle University of Thessaloniki, Greece

IRM Press
Publisher of innovative scholarly and professional information technology titles in the cyberage
Hershey • London • Melbourne • Singapore

Acquisitions Editor: Mehdi Khosrow-Pour
Senior Managing Editor: Jan Travers
Managing Editor: Amanda Appicello
Development Editor: Michele Rossi
Copy Editor: Shanelle Ramello
Typesetter: Rachel Shepherd
Cover Design: Lisa Tosheff
Printed at: Integrated Book Technology

Published in the United States of America by
IRM Press (an imprint of Idea Group Inc.)
701 E. Chocolate Avenue, Suite 200
Hershey PA 17033-1240
Tel: 717-533-8845
Fax: 717-533-8661
E-mail: cust@idea-group.com
Web site: http://www.irm-press.com

and in the United Kingdom by
IRM Press (an imprint of Idea Group Inc.)
3 Henrietta Street
Covent Garden
London WC2E 8LU
Tel: 44 20 7240 0856
Fax: 44 20 7379 3313
Web site: http://www.eurospan.co.uk

Library of Congress Cataloging-in-Publication Data

Wireless information highways / Dimitrios Katsaros, Alexandros Nanopoulos and Yannis Manolopoulos, editors.
p. cm.
Includes bibliographical references and index.
ISBN 1-59140-568-8 (h/c) -- ISBN 1-59140-540-8 (s/c) -- ISBN 1-59140-541-6 (ebook)
1. Wireless communication systems. 2. Information superhighway. I. Katsaros, Dimitrios. II. Nanopoulos, Alexandros. III. Manolopoulos, Yannis.
TK5103.2.W57328 2004
004.6--dc22

2004022151

British Cataloguing in Publication Data
A Cataloguing in Publication record for this book is available from the British Library.

All work contributed to this book is new, previously-unpublished material. The views expressed in this book are those of the authors, but not necessarily of the publisher.

Preface

Wireless communication systems have been around for some time, and they have found usefulness in areas such as commerce, defense, and education. Wireless communications allow users carrying portable computing devices, such as mobile phones and personal digital assistants, to retain their network connectivity even when they are on the move. Wireless has been one of the most significant breakthroughs among all human achievements in the past few decades. Thirty years ago, it was very difficult to assume that telecommunication services can be provided to people irrespective of their geographical location and while they are moving around.

The proliferation of wireless networks and small, portable computing devices led to the emergence of the mobile computing paradigm, or, as it is frequently called, "*anytime, anywhere computing*." Mobile users carrying laptops or handheld computers are able to connect to databases in fixed hosts or to the Internet through publicly available wireline or wireless networks.

Universal access and management of information are driving forces in the evolution of anytime, anywhere computing toward pervasive computing, which aims at providing a platform and paradigm for "*all the time, everywhere computing*." The emergence of pervasive computing is a natural outcome of the tremendous advances in wireless networks, mobile computing, sensor networks, and distributed computing technologies.

The focus of *Wireless Information Highways* is on the impact of wireless networks and mobile computing on data organization and access. This book is based on a number of self-contained chapters and provides an opportunity for practitioners and researchers to explore the connection between various computer science techniques and develop solutions to problems that arise in the rapidly growing field of wireless networks and mobile computing. Its purpose is to provide a thorough and comprehensive overview of the techniques proposed for the establishment of a new information highway: the wireless highway.

Information Management in Wireless Mobile Computing

Mobile computing is an umbrella term used to describe technologies that enable people to access network services anytime and anywhere. Ubiquitous computing (term introduced by Mark Weiser) and nomadic computing (term introduced by Leonard Kleinrock) are synonymous with mobile computing. Information access via a mobile device is plagued by low available bandwidth, poor connection maintenance, poor security, and problems regarding the addressing mechanism. Therefore, mobile computing introduces several data management challenges related to wireless and wireline communications, mobility, and system issues.

Wireless networks usually use a fixed infrastructure as a backbone. For instance, cellular networks connect a mobile phone to the nearest base station (BS). A BS serves hundreds of mobile users in a given area (cell) by allocating frequencies and providing handoff support. BSs are linked (by wireline, fiberline, or wireless microwave links) to BS controllers that provide switching support to several neighboring BSs and serve thousands of users. The wireless link poses design challenges. The main difference between wired and wireless links is in the type of communication. Wired links normally provide one-to-one communication without interference, whereas wireless links use one-to-many communication that has a considerable noise and interference level and bandwidth limitations. Simultaneous wireless communications require channel separation, where channel may refer to time, frequency, or code. The channel capacity typically available in wireless systems is much lower than what is available in wired networks. For instance, connection admission control (CAC), which is the process that decides which connection requests are admitted to the system and decides, also, the allocated resources, is different in wireless networks from wireline networks due to mobility and scarcity of wireless resources and the physical properties of the radio channels.

Some wireless networks do not have a fixed infrastructure as a backbone. Examples are sensor networks. Wireless networks of sensors are likely to be widely deployed in the near future because they greatly extend our ability to monitor and control the physical environment from remote locations, and they improve the accuracy of information obtained via collaboration among sensor nodes and online information processing at those nodes. Sensors are normally small, cheap devices with limited computing power. One of the most challenging tasks in sensor networks is to synthesize the information requested by users from the available data measured or sensed by a large number of sensor nodes. Since there are a sheer number of nodes with stringent energy constraints in a sensor network, it may not be feasible to fetch every reading of nodes for central processing. Instead, effective data querying and aggregation techniques are needed. Data queries in sensor networks can be continuous and periodical, continuous and event driven, or snapshots, that is, onetime queries. Sensor network queries can also be categorized as aggregated or nonaggregated; they can also be complex or simple. Finally, queries for replicated data can be made. The users should be able to carry out any of these types of queries by using the data-querying scheme for sensor networks.

Mobile wireless environments are characterized by asymmetric communication. The downlink communication from BS, satellite, or other server is much greater than the uplink communication capacity. In broadcast, data are sent simultaneously to all users residing in the broadcast area. Each user in a data broadcast problem has a list of desired files or data it wants to receive from the server. The order and frequency for each file or datum that is broadcast should take access efficiency and power conservation into account. Access efficiency concerns how fast a request is satisfied, whereas power conservation concerns how to reduce a mobile client's power consumption when it is accessing the data it wants. Broadcasting poses several challenges to data management issues, such as where to keep the data, when to transmit it, whether to use caching and where in the network, how to optimize data placement, the pull-push approach, and transactional services.

Tracking mobile users and routing calls are basic functions of the wireless network system. The system needs to update and provide, upon request, information about the location of mobile users. Queries can be formulated by both the users and the network resource management facility, for example, aggregate information for adaptive channel allocation. The criterion for efficient update is low signaling cost incurred by relocation of users between cells. The cost should be kept small enough so as not to affect network performance. Providing information about user location is coupled with updating. Because of frequent relocation of mobile users, especially for small cells, it is impractical to keep track of their exact locations at all times. The more general approach is to keep the location information updated upon query, which requires extra search work. Keeping more complete information updated requires higher signaling cost, while less search work is needed upon query. Finding efficient methods or schemes can balance the effects of these factors and achieve optimal overall performance.

A modern research direction is location-aware computing, which aims at providing services to users taking into consideration the location of the users in space. Location-aware computing is a special case of context-aware computing, which describes the special capability of an information infrastructure to recognize and react to a real-world context. Context, in this sense, comprises any number of factors, including user identity; current physical location; weather conditions; time of day, date, or season; and whether the user is asleep or awake, driving or walking. The most critical aspects of context are location and identity. Location-aware computing systems respond to a user's location, either spontaneously (for instance, warning of a nearby hazard) or when activated by a user request (for example, is it going to rain in the next hour?). In order for location-aware computing to be feasible, several data management issues must be addressed. Geographical data must be available in order to determine the position of user terminal devices. Efficient techniques must be available to compute the location of a terminal device. Moreover, algorithmic techniques that take into consideration the mobility of users must exist in order to predict a future location or a trend.

Security requirements for mobile-computing environments are different from those in fixed networks. This is due to the intensity and complexity of communication between the user and the infrastructure, the mobility of the user, and dynamic sharing of limited resources. As mobile and pervasive computing makes information access and processing easily available for everyone from anywhere at anytime, the close relationship between distributed systems and mobile computing with pervasive infrastructure leads us to take a closer look at different types of vulnerabilities and attacks in such environments. Pervasive computing includes numerous, often transparent, computing devices that are frequently mobile or embedded in the environment and are connected to an increasingly ubiquitous network structure. For example, when an organization employs pervasive computing, the environment becomes more knowledgeable about the users' behavior and hence becomes more proactive with each individual user as time passes. Therefore, the user must be able to trust the environment and the environment must be confident of the user's identity. This implies security is an important concern in the success of pervasive computing environments.

Brief Outline of This Book

The broad range of topics of the present book makes it an excellent reference on data management for wireless networks and mobile computing. The book is organized so that it could cover a wide range of audiences including senior-level undergraduate university students, postgraduate students, research engineers, and system developers. Because each chapter is self-contained, readers can focus on the topics that most interest them. Most of the chapters (if not all) in this book have great practical utility. The book emphasizes computer science aspects, including implementation. Mathematical and engineering aspects are also represented in some chapters, since it is difficult to separate clearly all the issues among the three areas.

A short outline of the material presented in each of the chapters of this book follows. The purpose is to identify the contents and also to aid diverse readers in assessing just what chapters are pertinent to their pursuits and desires. Each chapter should provide the reader with the equivalent of consulting an expert in a given discipline by summarizing the state of the art and providing pointers to further reading.

It is a challenging task to clearly divide chapters into discrete areas because of overlaps. One such classification that will be attempted here is to divide the chapters into five main research areas: broadcasting (Chapters 1 to 5), location management (the complementary Chapters 6 and 7), network support (Chapters 8 to 10), location-based services (Chapters 11 to 14), and advanced topics (Chapters 15 and 16). Many chapters deal with more than one of these areas, so clear separation is difficult. We shall now elaborate in more detail on each chapter.

Chapter 1 discusses the issue of data broadcasting over asymmetric wireless environments when there are no dependencies among the broadcasted data. Communications asymmetry is due to a number of facts, the most important being equipment, network, and application asymmetry. The chapter starts with a presentation of preliminary issues and terminology for asymmetric environments for data broadcasting. It discusses broadcast schedule construction for systems employing a single broadcast channel, and schedule construction by taking into account the effect of reception errors. It then presents an algorithm that tries to provide better support for clients whose access patterns deviate a lot form the overall access pattern of the client population. It also presents algorithms for environments where item requests by clients are dropped if not served in a certain time period.

Chapter 2 introduces advanced client-side data-caching techniques to enhance the performance of mobile data access. The authors address three mobile caching issues. The first is the necessity of a cache replacement policy for realistic wireless data-broadcasting services. The authors present the Min-SAUD policy, which takes into account the cost of ensuring cache consistency before each cached item is used. Next, the authors discuss the caching issues for an emerging mobile data application, that is, location-dependent information services (LDISs). In particular, they consider data inconsistency caused by client movements and describe several location-dependent cache invalidation schemes. Then, as the spatial property of LDISs also brings new challenges for cache replacement policies, the authors present two novel cache replacement policies, called PA and PAID, for location-dependent data.

Chapter 3 discusses data management issues that are necessary for wireless data broadcasting. The major topics included in this chapter are broadcast data indexing and clustering. Mobile clients can access the wireless data in an energy-efficient way with the index on the broadcast channel, and the well-clustered broadcasted data enables mobile clients to access the wireless data in a short latency.

Chapter 4 discusses the applicability and effectiveness of data broadcasting from two viewpoints: energy and response time. Within the scope of data broadcasting, the chapter presents different data allocation schemes, indexing approaches, and data retrieval methods for both single and parallel air channels. Comparisons of different algorithms are demonstrated through simulation results.

Chapter 5 explores algorithms for placing broadcast data into multiple wireless channels so as to reduce the client access time. For reasons like heterogeneous communication capabilities and variable quality of service offerings, we may need to divide a single wireless channel into multiple physical or logical channels. The chapter discusses algorithms for placing data to multiple wireless channels, assuming that there are no dependencies among the transmitted data. It gives an algorithm for obtaining the optimal placement to the channels, and explains its limitation since it is computationally very demanding and thus unfeasible. Then, it presents heuristic schemes for obtaining suboptimal solutions to the research problem and reports on their implementation cost and their relative performance.

Chapter 6 discusses location modeling, that is, the representation of inclusive mobile objects and their relationship in space, dealing with how to describe a mobile object's location. The goal of mobility modeling, on the other hand, is to predict or statistically estimate the movement of mobile objects. With the increasing demand for multimedia applications, location-aware services, and system capacity, many recognize that modeling and management of location and mobility is becoming critical to locating mobile objects in wireless information networks. Mobility modeling and location management strongly influence the design and performance of wireless networks in many aspects, such as routing, network planning, handoff, call admission control, and so forth. The chapter presents a comprehensive survey of mobility and location models, and schemes used for location-mobility management in cellular and ad hoc networks, which are discussed along with necessary, but understandable, formulation, analysis, and related discussions.

Chapter 7 analyzes and proposes some mobility management models and schemes by taking into account their capability to reduce search and location update costs in wireless mobile networks. The first model proposed is called the built-in memory model; it is based on the architecture of the IS-41 network and aims at reducing the HLR access overhead. The performance of this model was investigated by comparing it with the IS-41 scheme for different call-to-mobility ratios. Experimental results indicate that the proposed model is potentially beneficial for large classes of users and can yield substantial reductions in total user-location management costs, particularly for users who have a low call-to-mobility ratio. The built-in memory model is also compared to the forwarding pointers' scheme. The results show that this model consistently outperforms the forwarding pointers' strategy. A second location management model to manage mobility in wireless communications systems is also proposed. The results show that significant cost savings can be obtained compared with the IS-41 standard location management scheme depending on the value of the mobile unit's call-to-mobility ratio.

Chapter 8 discusses various issues and challenges facing the design and selection of a proper service discovery mechanism. This chapter also investigates service discovery mechanisms such as Service Location Protocol (SLP), Jini, Salutation, and others, assessing their suitability for applications in wireless and mobile environments.

Chapter 9 deals with wireless sensor network design and deployment. The position of sensor network nodes need not be engineered or predetermined. This allows random deployment in inaccessible terrains, which implies that their protocols and algorithms must possess self-organizing capabilities. Sensor network nodes use their processing abilities to locally carry out simple computations and transmit only the required and partially processed data. Realization of sensor networks requires wireless ad hoc networking techniques. Although many protocols and algorithms have been proposed for traditional wireless ad hoc networks, they are not well suited for the unique features and application requirements of sensor networks. One of the most important constraints on sensor nodes is the low power-consumption requirement. Therefore, while traditional networks aim to achieve a high quality of service provisions, sensor network protocols must focus primarily on power conservation. The chapter presents a survey of protocols and algorithms proposed thus far for sensor networks, attempts an investigation into pertaining design constraints, and outlines the use of certain tools to meet the design objectives. After the introduction, the chapter discusses the factors that influence the sensor network design and provides a detailed investigation of current proposals in this area in a layered approach. Then, it explains the application, transport, and network layer protocols as addressed by various research and development efforts.

Chapter 10 discusses CAC in wireless networks and how it differs from wireline networks due to mobility and scarcity of wireless resources and the physical properties of the radio channels. In this chapter, the basic issues in CAC for wireless systems are discussed. Following the discussion of the trade-off between blocking and dropping rates and the comparison of CAC in both wireline and wireless systems, admission control in next-generation wireless systems is explained. A brief discussion of quality of service (QoS) provisioning is also provided due to its relationship with CAC.

Chapter 11 presents an overview on advances made in databases during the last few years in the area of mobile object indexing and discusses issues that remain open or, probably, are interesting for related applications. Since mobile computing emerged as a new application area, due to recent advances in communication and positioning technology and the fast and on-time access to a constantly changing data set, the indexing of moving objects appears as a very interesting, as well as crucial, research subject, especially as location-based services are becoming a part of our daily life.

Chapter 12 presents PoLoS, a novel platform for Location-Based Services (LBSs) deployment and provisioning. PoLoS is a highly scalable, component-based Java platform. It supports the most commonly used communication protocols (SMS, WAP, HTML) for interfacing with terminal mobile devices, as well as all of the currently available position-tracking techniques. It is the only LBS platform today that can be used both for indoor and outdoor applications and at the same time provides full support for service creation through the Service Creation Environment. Its architecture shows how the compliance with open interfaces and standards results in configurable modular systems with extended capabilities in adapting to future technological needs. The chapter starts with a presentation of the current trends in the area of LBSs and related work in all technological areas that are relevant to the subject. The platform's architecture is presented in detail on a per functional component basis. Apart from the platform, the basic characteristics of a new service specification language that was created to simplify the service creation process are presented in the chapter.

Chapter 13 discusses benchmarks, which define techniques that can be followed to determine the effectiveness of a given software or hardware design. Ever since the development of the Wisconsin Benchmark and subsequent Transaction Processing Performance Council (TPC) benchmarks, there has been a consensus and general acceptance of these performance comparison tools. However, these benchmarks are not sufficient to determine the performance of mobile-based applications. For example, these traditional benchmarks ignore some of the important wireless-mobile features such as location-dependent queries and movement of the mobile host. This chapter examines the issues needed for the development of such a mobile query benchmark. In particular, it focuses on queries that involve location-dependent features; it first examines the unique aspects of this mobile architecture, which impact any benchmark design, and then proposes a benchmark suitable for it.

Chapter 14 presents the infrastructure that makes it possible to create LBSs and provides an insight on key application development issues. First, an overview is given on positioning, location-information retrieval technologies, and application development issues specific to mobile devices. Following this, storage and management of location information is discussed as various examples of database queries are presented. Finally, an overview of the hottest research topics is given. The focus of this chapter is on location-information access and spatial database technology as the latest achievements of this field are presented.

Chapter 15 evaluates the suitability of existing security methods for pervasive environments. Due to the intensity and complexity of communication between the user and the infrastructure, the mobility of the user, and dynamic sharing of limited resources, security requirements for pervasive computing environments are different from those in fixed networks. Pervasive computing makes information access and processing easily available for everyone from anywhere at anytime. The close relation of distributed systems and mobile computing with pervasive computing leads us to take a closer look at different types of vulnerabilities and attacks in such environments. Pervasive computing includes numerous, secured, often-transparent computing devices which are frequently mobile or embedded in the environment and are connected to an increasingly ubiquitous network structure. Security is a critical concern in the success of pervasive computing environments.

Chapter 16 discusses transaction-processing issues for broadcast databases. It explains why broadcast transaction processing is more complicated than transaction processing in traditional database systems due to the necessary decoupling of the clients from the server in the broadcast architecture. Traditional database management systems guarantee that clients always access a current and consistent set of data that reflects the effects of all successful transactions. These guarantees are made by regulating each transaction's data access and by keeping a log of the updates made to the database state. Such control is unavailable in broadcast databases. Broadcast databases achieve their main benefit (i.e., scalability) because multiple clients can simultaneously and independently read a single data item. After reviewing briefly the broadcast model and some transaction-processing concepts, the chapter describes a natural architecture for broadcast transaction processing. Then, it discusses concurrency control, recovery, and transaction design in broadcast databases.

Recommended Reading

Each chapter in the book is accompanied by its own reference section. However, the reader should refer to journals and conference proceedings to keep up with the recent developments in the field. Important journals that publish articles in the area related to the present book are the following:

- ACM Mobile Computing and Communications Review
- IEEE Transactions on Mobile Computing
- IEEE/ACM Transactions on Networking
- ACM/Kluwer Mobile Networks and Applications
- ACM/Kluwer Wireless Networks
- Communications of the ACM
- IEEE Computer
- IEEE Internet Computing
- IEEE Communications
- IEEE Pervasive Computing
- IEEE Transactions on Knowledge and Data Engineering
- IEEE Transactions on Parallel and Distributed Systems
- IEEE Journal on Selected Areas in Communications
- IEEE Transactions on Vehicular Technology
- IEEE Transactions on Wireless Communications

- Ad Hoc Networks (Elsevier)
- Computer Communications (Elsevier)
- International Journal of Mobile Computing and Commerce (Idea Group Publishing)
- Wireless Communications and Mobile Computing (Wiley)
- International Journal of Wireless Information Networks (Kluwer)

We also encourage the reader to refer to the proceedings of some of the main conferences that cover the book topics, for example:

- ACM Conference on Mobile Computing and Networking (MOBICOM)
- ACM Symposium on Mobile and Ad Hoc Networking and Computing (MOBIHOC)
- ACM Workshop on Discrete Algorithms and Methods for Mobile Computing and Communications (DIAL M)
- ACM Workshop on Data Engineering for Wireless and Mobile Access (MOBIDE)
- ACM Symposium on Information Processing in Sensor Networks (IPSN)
- ACM Workshop on Principles of Mobile Computing (POMC)
- IEEE Conference on Computer Communications (INFOCOM)
- IEEE Conference on Data Engineering (ICDE)
- IEEE Conference on Pervasive Computing and Communications (PERCOM)
- IEEE Conference on Mobile Data Management (MDM)
- IEEE Vehicular Technology Conference (VTC)
- IEEE Conference on Distributed Computing and Systems (ICDCS)
- IEEE Conference on Computer Communications and Networks (ICCCN)

What Makes This Book Different

Although there are many books related to wireless communications and mobile computing, these books can be roughly classified into two groups. The first group focuses on readers in the radio frequency (RF) communication field, whereas the second group covers the general knowledge of data communications, both wireline and wireless, and mobile computing. The books in the first group require substantial background in RF communication and signal processing and, hence, are not appropriate for computer scientists or students and information technology (IT) managers and practitioners. The books in the second group deal with mobile computing, but mainly from a networking-level point of view, presenting topics like multiple access schemes, cellular networks, data communication, multihop networks, and other networking issues.

Many academic institutions do offer courses related to data management in mobile environments primarily for graduate students. New enterprises in the mobile sector (e.g., mobile commerce) develop advanced applications targeting geographical regions covered by wireless networks. We are aware of only a couple of books that deal with information dissemination and management in wireless mobile-computing environments. However, these books are outdated and do not cover significant recent advances that have reshaped the overall picture of mobile computing. For instance, these books do not present issues related to LBSs or sensor networks and how the traditional data management techniques are affected by their requirements. We envisioned the creation of *Wireless Information Highways* as an answer to most of the questions related to broadcast data management, location management, LBSs, sensor networks, and security in mobile and pervasive environments.

Intended Audience

Wireless Information Highways is intended for academic institutions and for working professionals, for technical and nontechnical readers. The broad range of topics in this book makes it an excellent reference on information management for wireless networks and mobile computing. The book is organized so that it could cover a wide range of audiences including senior-level undergraduate university students, postgraduate students, research engineers, and system developers.

- *Computer science instructors* could use this book to teach information management issues in pervasive computing to senior undergraduate or postgraduate students. The chapters are organized to provide a great deal of flexibility; emphasis can be given to different chapters, depending on the scope of the course and the instructor's interests. Equivalently, computer science students could use it in the context of a course or as a supplementary book for their independent study.
- *Computer science researchers* could benefit from this book because it surveys a vast body of recent research in the area of information management in mobile and pervasive computing. The research coverage is likely to benefit researchers and students from academia as well as industry.
- *IT managers* could use this book for training current or new employees during short-term training courses or they could use it to discover mobile solutions to their enterprise's computing needs.
- *General (scientific and industrial) community* interested in the issues of data management in mobile and pervasive computing could benefit from this book and acquire a basic knowledge of the issues involved.

Acknowledgments

The editors are grateful to all the authors for their contributions to the quality of this book. The assistance of chapters' reviewers is also greatly appreciated. The Aristotle University of Thessaloniki, Greece, provided an ideal working environment for the preparation of this book, including computer facilities for communication by electronic mail and for writing our own contribution.

The editors are thankful to Mehdi Khosrow-Pour, Senior Academics Editor of Idea Group Publishing, for his support and encouragement in publishing this book. The editors also appreciate the support of Michele Rossi, Development Editor, for her timely and professional cooperation, and for her decisive support of this project. A further special note of thanks goes to all the staff at Idea Group Publishing, whose contributions throughout the whole process from inception of the initial idea to the final publication have been invaluable.

Finally, we thank our families for making this effort worthwhile and for their patience and support during the numerous hours at home that we spent in front of the computer.

In this book, we have tried to provide an overview of the basic principles and practices about data management in wireless mobile environments. We hope that we have been able to achieve our goal of helping students, professionals, practitioners, and others working in this area to have a good knowledge about this exciting field. We hope that the readers will find this book informative and worth reading. Comments from readers will be greatly appreciated. Please contact us at dimitris@skyblue.csd.auth.gr, alex@skyblue.csd.auth.gr, and manolopo@skyblue.csd.auth.gr.

Dimitrios Katsaros
Alexandros Nanopoulos
Yannis Manolopoulos
Thessaloniki, Greece
May 2004

Section I

Broadcast Data Management

Chapter I

Push and Pull Systems

Petros Nicopolitidis, Aristotle University of Thessaloniki, Greece

Georgios I. Papadimitriou, Aristotle University of Thessaloniki, Greece

Andreas S. Pomportsis, Aristotle University of Thessaloniki, Greece

Abstract

Data broadcasting has emerged as an efficient way for the dissemination of information over asymmetric wireless environments where the needs of the various users of the data items are usually overlapping. In such environments, data broadcasting stands to be an efficient solution since the broadcast of a single information item is likely to satisfy a possibly large number of users. Communications asymmetry is due to a number of facts, the most important being equipment, network, and application asymmetry. This chapter starts with a discussion of preliminary issues and terminology for asymmetric environments for data broadcasting. The chapter then discusses broadcast schedule construction for systems employing a single broadcast channel, schedule construction for systems employing multiple broadcast channels, and schedule construction for systems that take into account the effect of reception errors. It then presents an algorithm that tries to provide better support for clients whose access patterns deviate a lot form the overall access pattern of the client population. It also presents algorithms for environments where item requests by clients are dropped if not served in a certain time period. Brief comments on issues that affect performance of the discussed data broadcasting methods are also made.

Introduction

Push and pull data delivery systems are members of a family of systems known as data broadcasting or information dissemination systems. Such systems have emerged as efficient ways for the dissemination of information over asymmetric wireless environments where the needs of the various users of the data items are usually overlapping. Examples of such applications are information retrieval ones, like weather and traffic information systems. For example, a traffic information system in an airport could be of much benefit to waiting passengers. A user coming to the airport will want information regarding departure of his flight (e.g., exact time of departure, possible delays, etc.). A broadcast server could deliver such data for all flights in the near future. Sometimes, demand for some flights is likely to be higher than for others (either due to more passengers for that flight or due to this flight departing in the very near future). Thus, one can see that client needs for data items are usually overlapping and, consequently, data broadcasting stands to be an efficient solution since the broadcast of a single information item is likely to satisfy a possibly large number of users.

Communications asymmetry is due to a number of facts, the most important being equipment, network, and application asymmetry. Equipment asymmetry is caused by the fact that a broadcast server usually has transceivers that are not subject to power limitations, whereas client transceivers are usually hindered due to finite battery life. Moreover, it is desirable to keep the mobile clients' cost low, which sometimes results in the lack of client transmission capability. Network asymmetry is due to the fact that, in many cases, the available bandwidth for transmission from the server to the clients (downlink transmission) is much more than that in the opposite direction (uplink transmission). Furthermore, there exist extreme network asymmetry cases where the clients have no available uplink channel (back channel). Even in the case of back-channel existence, however, the latter is subject to becoming a bottleneck in the presence of a very large client population. Application asymmetry concerns the pattern of information flow. Since most information retrieval applications are of client-server nature, the flow of traffic from the server to the clients is usually much higher than that in the opposite direction. Furthermore, application asymmetry concerns the pattern of accessing the broadcast information items. This is because, in many cases, the majority of the clients are interested only in a subset of the server's information items. Thus, some items tend to be a lot more "popular" than others and consequently, the environment is characterized by a "skewed" demand pattern.

So far, two major approaches have appeared for designing broadcast schedules. These are the pull (also known as on-demand) and the push approaches. In pull systems, the server broadcasts information after requests made by the mobile clients via the uplink channel. The server queues up the incoming requests and uses them to estimate the demand probability per data item. In push systems there is no interaction between the server and the mobile clients. The server is assumed to have an a priori estimate of the demand per information item and transmits items according to this estimate. Finally, combinations of push and pull lead to hybrid approaches. Hybrid systems divide the available downlink bandwidth into two transmission modes: the periodic broadcast mode, in which the server pushes data periodically to the clients, and the on-demand mode, which is used to broadcast data explicitly requested by the mobile clients through the uplink channel.

Chapter Scope

This chapter is organized as follows: It starts by discussing some preliminary issues and terminology that will be used later during the presentation of the various algorithms. It then discusses broadcast schedule construction for systems employing a single broadcast channel and then broadcast schedule construction for systems employing multiple broadcast channels. Next, it discusses broadcast schedule construction by taking into account the effect of reception errors. The section after that presents an algorithm that tries to provide better support for clients whose access patterns deviate a lot form the overall access pattern of the client population. Following this, the case where item requests by clients are dropped if not served in a certain time period is discussed. The next section briefly comments on issues that affect performance of the discussed data broadcasting methods. Finally, the chapter ends with a brief concluding summary.

Preliminaries

The Need for Scheduling

As will be seen later on, the major part of this chapter is concerned with the construction of efficient broadcast schedules. A broadcast schedule is a sequence of data item broadcasts by the broadcast server. It will be made clear that the primary goal of an efficient broadcast schedule is to minimize the mean waiting (response) time at the clients. An obvious thought (albeit naive after a thorough examination) would be that all a scheduling algorithm needs to do is to broadcast items according to their popularity among the clients. This subsection will explain that an efficient broadcast scheduling algorithm needs to take more into account than this fact. The following example, which resembles that which appears in Acharya, Franklin, and Zdonik (1995), is illustrative.

Consider a database of three equal-length data items, A, B, and C, and a single client. The client accesses items A, B, and C with probabilities p_A, p_B, and p_C, respectively. The broadcast server is able to broadcast different items with different frequencies so as to provide more bandwidth to the popular pages than to the unpopular ones. Furthermore, we consider three schedules:

- The *flat* one, which schedules pages cyclically. Thus, all pages are broadcast with the same frequency. The schedule will be A, B, C, A, B, C...
- The *skewed1* one, in which item A is broadcast twice as often as items B and C. Furthermore, subsequent broadcasts of page A are clustered together. Thus, the schedule will be A, A, B, C, A, A, B, C...
- The *skewed2* one, differing from the above in such a way that instances of the same item are separated by equal gaps. Thus, the schedule will be A, B, A, C, A, B, A, C..

For various values of p_A, p_B, and p_C, Figure 1 shows the mean waiting time for a client for each of the above three schedules. The mean waiting time is calculated by multiplying the probability of access for each item with the associated waiting time and summing up the results. It can be seen that for uniform item demand probabilities, the flat schedule performs best. This is to be expected since the server broadcasts items the way these are demanded by the client. However, as item demand probabilities begin to shift away from uniform distribution, the last two schedules perform better. A third remark is that the skewed program with fixed gaps between instances of the same item always performs better than the other skewed schedule. This is attributed to the fact that if gaps between instances of the same item are fixed, then the mean waiting time for an item request arriving at a random time is one half of the gap between successive instances of the item. On the other hand, if the skewed schedule with nonfixed gaps between instances of the same item is used, gaps will be of variable length. Thus, the probability of an item request being made during a large gap is greater than the probability of one being made during a short gap. Therefore, an increase in the variance of gap length between instances of the same item tends to increase the mean waiting time.

Minimization of the mean waiting time is not the only possible target for an efficient scheduling algorithm. Another possible goal is to make a system energy efficient. This would be possible if clients are able to go to a mode with lower energy consumption ("sleep" mode) and "wake up" only when the data item that interests them is about to be broadcast. Thus, the goal is to minimize the time a mobile client spends actively listening to the communications medium (tuning time). Minimization of tuning time obviously results in energy preservation at the client. Indexing methods (Imielinski, Viswanathan, & Badrinath, 1994a, 1994b; Yee, Navathe, Omiecinski, & Jermaine, 2002)

Figure 1. Mean waiting time for three different schedules

p_A	p_B	p_C	Mean waiting time (Flat)	Mean waiting time (Skewed)	Mean waiting time (Skewed /fixed gaps)
0.33	0.33	0.33	1.5	1.75	1.65
0.5	0.3	0.2	1.5	1.625	1.5
0.8	0.05	0.05	1.5	1.2	1.0

enable clients to achieve energy efficiency by multiplexing index items along with data ones. Since index items inform about the time of broadcast of subsequent data items, clients need to be in active mode only when they need to receive a data item. A naive form of indexing is to send an index item once for every period of the data broadcast. This, however, would not perform very well due to the fact that clients who miss an index have to wait the entire broadcast period to receive the next index. A better solution is (1, *m*) indexing, which is proposed by Imielinski et al. (1994a), where the index is broadcast *m* times during the period of the broadcast schedule. The optimal value for *m* is provided by the authors in Imielinski et al. (1994b). An enhancement of (1, *m*) indexing is distributed indexing. According to this method, each index item need not contain the entire index, but only information regarding the data items that succeed it.

Environment

In all the algorithms presented in this chapter, we assume an asymmetric communication environment comprising a single broadcast server and *N* mobile clients. Moreover, the broadcast server possesses a database that comprises *M* information items not necessarily of the same size. As far as transmission speed, item size, and time are concerned, we do not consider pure numbers; rather, we assume that the smallest data item is of unit size and is transmitted over the server-client link in a unit time interval. The size of an item i, $i \leq M$ is denoted by l_i.

In an asymmetric information dissemination environment, clients request information items from the broadcast server. The probability that item i is requested (demand probability for item i) is denoted as d_i. As was mentioned above, due to overlapping demand patterns among different clients, the demand pattern for data broadcasting applications is often characterized by a certain amount of skew. In order to map this fact into the demand probabilities d_i, the Zipf distribution is used. Although it shouldn't be used as the de facto workload for any wireless information system, we briefly mention it here since it is commonly used to carry out performance analysis in research. Mathematically, it is expressed as follows:

$$d_i = c\left(\frac{1}{i}\right)^{\theta}, \quad where\ c = 1/\sum_k \left(\frac{1}{i}\right)^{\theta}, \quad i,k \in [1..M] \qquad (1)$$

where θ is a parameter named access skew coefficient. For increasing values of θ the Zipf distribution produces increasingly skewed demand patterns and can thus model commonality in client demands for information items. For a sample database of 500 information items, the item demand probabilities for various values of θ are plotted in Figure 2.

Broadcast Schedules and Client Cache Management

The goal pursued in most of the proposed data delivery approaches is twofold:

1. Determination of an efficient sequence for data item transmission (broadcast schedule) so that the overall mean access time t at the clients is minimized. The overall mean access time is essentially the mean waiting time observed by a client for a requested information item. It can be computed when knowing the average waiting time t_i for every information item i. Thus:

$$t = \sum_{i=1}^{M} p_i t_i \,.$$

2. Management of client local memory (cache management) in a way that efficiently reduces performance degradation when mismatches occur between a client's demand pattern and the server's schedule. It has to be noted, however, that not all the methods presented below assume clients with cache memory.

Figure 2. Plot of the Zipf distribution for various values of θ

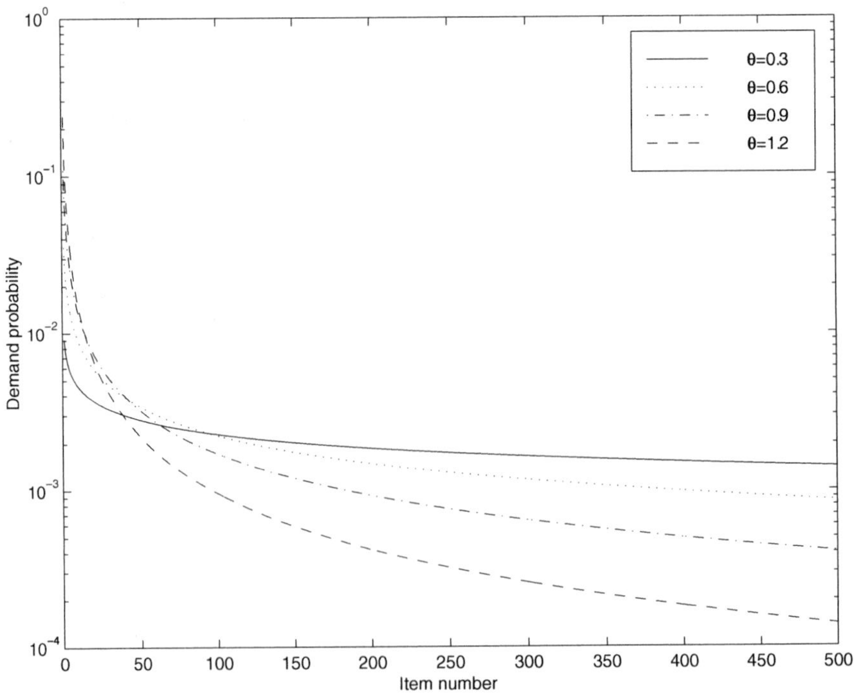

Static and Dynamic Client Demands

In terms of client demand pattern, there exist two types of environments: static and dynamic ones. Static environments are those in which the way that clients access the server's information items does not change over time. Thus, the demand probability for all information items remains the same and the server does not need to adapt to new item demand probabilities. By extending the above reasoning, one might define environments where demand probabilities change, with the nature of changes being known to the broadcast server, as static environments. This is because a fixed a priori model that dynamically estimates the timeliness of the time-sensitive demand could easily be imagined and thus always provides the server with the current item demand probabilities.

However, in many data broadcasting applications, overall client demands are likely to change with time with the nature of changes being a priori unknown. A possible example of such an application with dynamic overall client demands could be the case of a museum possessing the necessary infrastructure in order to deliver to the users information regarding the exhibits. Most museums contain several sectors, with each sector containing exhibits of a different type (e.g., Egyptian, Greek, etc.). It would be desirable for visitors within a sector to be aided in their tour by receiving information regarding the contents of the sector. Upon the arrival of a group of visitors to a specific sector, the demand for information regarding the exhibits of this sector will grow. When the group leaves the sector, this demand will lower to reflect the demand of visitors remaining in the sector. It can be easily seen that in such a situation, demands for information are characterized by a certain amount of commonality since all users inside a certain sector will demand the same information and thus the same data item (or set of data items). Moreover, overall client demands are sure to be (a) dynamic since the transition of a group of visitors to another sector will produce a change in client demands and (b) unknown a priori due to the fact that the amount of time spent by groups of visitors inside a sector is not known. Such environments with changing client demand patterns, where the nature of occurrence of these changes is unknown to the broadcast server, are hereinafter characterized as dynamic.

Assumptions

Unless stated otherwise, the following assumptions are made in the discussion of the algorithms in this chapter.

1. The communication medium is a broadcast channel, thus every transmitted information item is visible to all clients.
2. In push systems, the broadcast server has by some means acquired knowledge regarding the overall demand probabilities of the various information items. Furthermore, these probabilities do not change. These assumptions are lifted in the section discussing the adaptive push system.
3. All information items are self-identifying. This can be easily implemented via header information inclusion in the items.

4. Clients are continuously listening to the broadcast; thus no power-saving procedures are performed.
5. There are no interdependencies between information items.
6. A broadcast information item is received by clients only if wanted.
7. Values of information items do not change.
8. No back channel exists. This is true only for the pure push methods. Pull and hybrid methods use a back channel.

Schedule Construction for a Single Channel

Push Systems

Some of the early work relevant to push data broadcasting used the flat approach (Bowen, 1992; German, Gopal, Lee, & Weinrib, 1987; Gifford, 1990), which schedules all items with the same frequency. However, a very interesting paper (Ammar & Wong, 1985) revealed the following two facts, which are also used by the push systems we will describe later on. Specifically, Ammar and Wong (1985, 1987) showed that in order to minimize mean access time, schedules must be periodic and:

- The variance of spacing between consecutive instances of the same item must be reduced.
- Items should be broadcast with frequencies proportional to the square root of the probability of being demanded.

It has to be mentioned that the push algorithms presented in this section can be easily applied to pull environments by substituting vector d containing the demand probability estimates for the various items with a vector that contains the estimates of demand probabilities based on pending requests for those items.

Broadcast Disks

A method that satisfies both the above-mentioned constraints is broadcast disks (Acharya et al., 1995). According to it, schedules are periodic and the spacing between consecutive instances of the same item is the same. This method defines a disk to be a subset of the server's database that contains information items having close demand probabilities. It proposes a way of superposition of multiple disks spinning at different frequencies on a single broadcast channel. This essentially results in the server interleaving the transmissions of information items belonging on different disks so that (a) items on the faster disks are broadcast more often and (b) items on the same disk are broadcast with the same frequency. The most popular information items are placed on the faster disks and, as a result, periodic schedules are produced, with the most popular data being broadcast more frequently. Figure 3 shows a sample database of five information items, I1, I2, ..., I5, split into disks D1, D2, and D3 spinning at relative frequencies given by $f_{D1} = 2f_{D2} = 2f_{D3}$. It has to be noted that Acharya et al. (1995) assume static client demands and equilength information items.

We assume that K disks are used and that the server's database comprises M information items. Schedule construction involves the following steps:

1. Information items are ordered in ascending order of demand probabilities.
2. Items are split into the K disks so that each disk contains items with similar demand probabilities.
3. The relative frequencies f_i, $1 \leq i \leq K$ of the disks are defined with the only constraint of being integer multiples of the frequency of the slower disk.
4. Disks are split into smaller units named chunks. The split is made such that disk i is split into *max_chunks*/f_i chunks, where *max_chunks* is the least common multiple of f_i, $1 \leq i \leq K$. We denote as chunks(i) the chunks of disk i and note that chunks of different disks might be of different sizes, as in general, a chunk might comprise more than one information item.
5. The broadcast program is created by broadcasting a chunk of the first disk, followed by a chunk of the second, and so forth. In pseudocode, broadcast schedule construction is expressed as follows:

```
while(1)
        for j:=1 to max_chunks
                for i:=1 to K
                        broadcast chunk (j mod chunks(i)) of disk i.
```

Figure 4 shows the resulting schedule for a database of six information items broadcast via three disks with $f_{D1} = 2f_{D2} = 4f_{D3}$. Notice that the algorithm causes unused slots to appear in the broadcast if it is not possible to evenly divide a disk into the required number of chunks. In our example, disk D3 is divided into three chunks whereas it should be divided

Figure 3. Disks and schedule construction

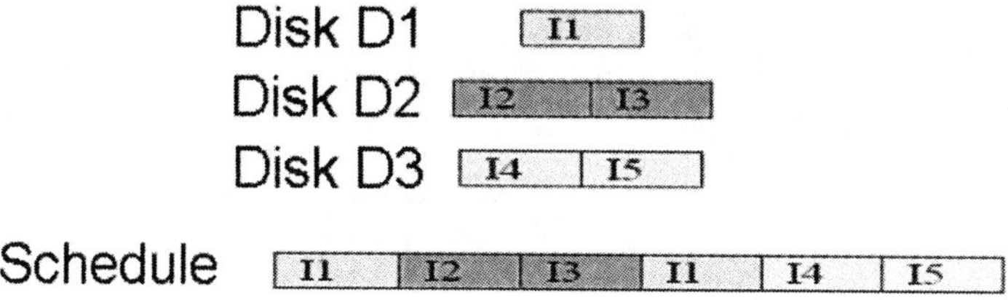

into four. Thus, at the fourth time a chunks from D3 is broadcast, the algorithm will broadcast an empty item rather than an item between I4 and I6 so as to preserve $f_{D1} = 4f_{D3}$. It can be seen that the resulting schedule period (called a major cycle, inside which every database item is transmitted at least once) comprises minor cycles, each of which contains one chunk from each disk.

Energy-Efficient Broadcast Disks

As already mentioned, broadcast disks target reduction of overall mean access time. An interesting paper that builds on the method of broadcast disks is Yee et al. (2002). It proposes techniques for scheduling data broadcasts that are favourable in terms of both response and tuning time. This means that overall, client requests will be satisfied in a low time, requiring at the same time low energy consumption by the client. The contribution of Yee et al. is threefold:

- Determination of a method to assign the M data items to be broadcast to the various disks so that overall mean access time is reduced.
- Determination of the number of disks to use, K.
- Integration of indexing in order to provide energy efficiency.

Assignment of the data items to the K disks has also been addressed in Yee, Omiecinski, and Navathe (2001), which solves the problem by using dynamic programming. However, this method has an $O(KM^2)$ computational complexity and may not be of practical use in many situations. To this end, a heuristic approximator is used in Yee et al. (2002). In this method, item assignment is made by ordering the M data items in descending order of popularity and finding the best K-1 splitting points so that the resulting overall mean access time is minimized. The algorithm tests each possible splitting point s in each partition and accepts it if $C^s_{i,j} < C_{i,j}$, where $C_{i,j}$ is the cost in terms of overall mean access

Figure 4. Broadcast schedule construction

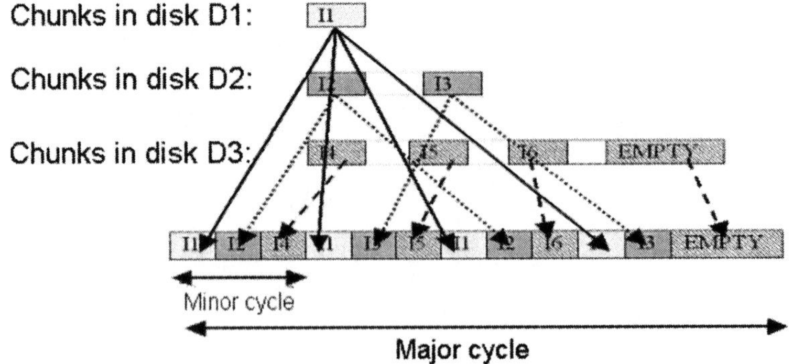

time of maintaining a partition starting from item i and ending with item j, and $C_{i,j}^{s} = C_{i,s} + C_{s+1,j}$ is the cost in terms of overall access time of splitting that partition at point s. In order to find the optimal number of disks, K, the algorithm that assigns items to partitions is run for all values of K between 1 and M, and the one that yields the lower overall mean access time determines the value of K to be used.

Integration of indexing in the method of Yee et al. (2002) takes advantage of the fact that items at each disk are broadcast in a flat manner. Thus, it applies $(1, m)$ indexing to each partition. Partitions now contain index and data items and are interleaved into a single channel. The only needed modification to $(1, m)$ indexing is that each data item contains an offset to the next index to appear, either belonging to the same disk of the data item or to another disk.

Simulation results in Yee et al. (2002) show the expected behaviour of increasing performance gain over the flat scheme for increasing skewness in item demands. When the index scheme is taken into account, the following observations occur:

- Overall mean access time is not better than that of a flat $(1, m)$ indexing scheme for low values of data skewness.
- The indexed broadcast disk method beats the flat $(1, m)$ index scheme at medium- and high-demand skewness.
- The tuning time of the indexed broadcast disk schedule is significantly lower than that of the nonindexed one, with obvious advantages for client energy efficiency.

The Vaidya-Hameed Method

Broadcast schedule construction in Vaidya and Hameed (1996, 1999b) is based on the following two arguments:

Argument 1: Broadcast schedules with minimum overall mean access time are produced when the intervals between successive instances of the same item are equal (Ammar & Wong, 1985).

Argument 2: The actual structure of the broadcast is determined by the so-called square-root rule: Under the assumption of equally spaced instances of each item, the minimum overall mean access time occurs when the server broadcasts information items with frequency proportional to the factor: $\sqrt{\frac{d_i}{l_i}}$, where d_i is the demand probability for item i and l_i is the item's length. This is a generalization of the result presented by Ammar and Wong (1985) for fixed-length items.

Thus, according to Argument 2, assuming that s_i is the constant spacing between the instances of item *i*, the overall mean access time is minimized when the following relation stands:

$$s_i \propto \sqrt{\frac{l_i}{d_i}}, \ \forall i \in [1..M] \tag{2}$$

where *M* is the number of candidate items for broadcasting. The work in Vaidya and Hameed (1996, 1999b) proposes an algorithm that is motivated by the arguments presented above. Specifically, it schedules broadcasting while trying to equalize the space between successive instances of the same item *i*, such that the following equation is achieved to the extent possible:

$$s_i^2 \frac{d_i}{l_i} = constant, \ \forall i \in [1..M]. \tag{3}$$

The algorithm operates as follows: Assuming that *T* is the current time and *R*(*i*) the time when item *i* was last broadcast, the broadcast scheduler selects to broadcast item *i* having the largest value of the cost function *G*:

$$G(i) = (T - R(i))^2 \frac{d_i}{l_i}, \ \forall i \in [1..M]. \tag{4}$$

For items that have not been previously broadcast, *R*(*i*) is initialized to -1. If the maximum value of *G*(*i*) is shared by two or more items, the algorithm selects one of them arbitrarily. Upon the broadcast of item *i* at time *T*, *R*(*i*) is updated so that *R*(*i*) = *T*. After the completion of the item *i* broadcast, the algorithm proceeds to select the next item to broadcast. The algorithm described above is of O(*M*) complexity.

Computational cost reduction via bucketing. In order to reduce the computational cost of the method, a scheme known as bucketing is proposed (Vaidya & Hameed, 1996, 1999b). According to this scheme, the server's *M* information items are split into *K* buckets, B_1, B_2, ..., B_K. Bucket *i* contains q_i items. Obviously, $\sum_{i=1}^{K} q_i = M$. We define the following:

- $d_{\text{avg}(B_i)} = \frac{\sum_{j \in B_i} d_j}{q_i}$ as the average demand probability of items in bucket B_i and

- $l_{\text{avg}(B_i)} = \frac{\sum_{j \in B_i} l_j}{q_i}$ as the average length of items in bucket B_i.

It is shown (Vaidya & Hameed, 1996, 1999b) that in order to minimize overall mean access time, for each item j in each bucket B_i it must hold that:

$$s_j^2 \frac{d_{avg(B_i)}}{l_{avg(B_i)}} = constant, \forall i \in [1..K], and\ j \in B_i \tag{5}$$

Each bucket is organized as a queue. Bucketing modifies the basic algorithm by working only with items that are in the front of the bucket. Assuming that T is the current time and $R(B_i)$ the time when an item from bucket B_i was last broadcast, the broadcast scheduler selects to use a bucket such that the cost function G is maximized:

$$G(B_i) = (T - R(B_i))^2 \frac{d_{\text{avg}(B_i)}}{l_{\text{avg}(B_i)}}, \quad \forall i \in [1..k]. \tag{6}$$

For buckets that have not been previously selected, $R(B_i)$ is initialized to -1. If two or more buckets share the maximum value of G, the algorithm selects one of them arbitrarily. Upon the broadcast of the item at the front of bucket B_i at time T, $R(B_i)$ is updated so that $R(B_i) = T$. After the completion of the item at the front of bucket B_i, that item is placed at the rear of B_i and the algorithm proceeds to select the next item to broadcast.

Using the bucketing scheme, the push method (Vaidya & Hameed, 1996, 1999b) reduces its complexity to $O(K)$. Since items in each bucket are broadcast with the same frequency, it is obvious that the members of each bucket must have close l_i/d_i values for good results.

Simple performance evaluation. In what follows, we briefly evaluate the performance of the basic method by Vaidya and Hameed (1996, 1999b). Figure 5 plots the performance of the method. It can be seen that the method is able to take advantage of increased commonality in item demands for large values of θ, and exhibit a low overall mean access time. Several parameters that are involved in the performance comparison are explained below.

- Clients are cacheless.
- Every client is initially set to access server information items in the interval [1...*Access Range*], with *Access Range* $\leq M$. All items outside this interval have a

Figure 5. Overall mean access time in unit items versus data skew coefficient θ, for single channel broadcast

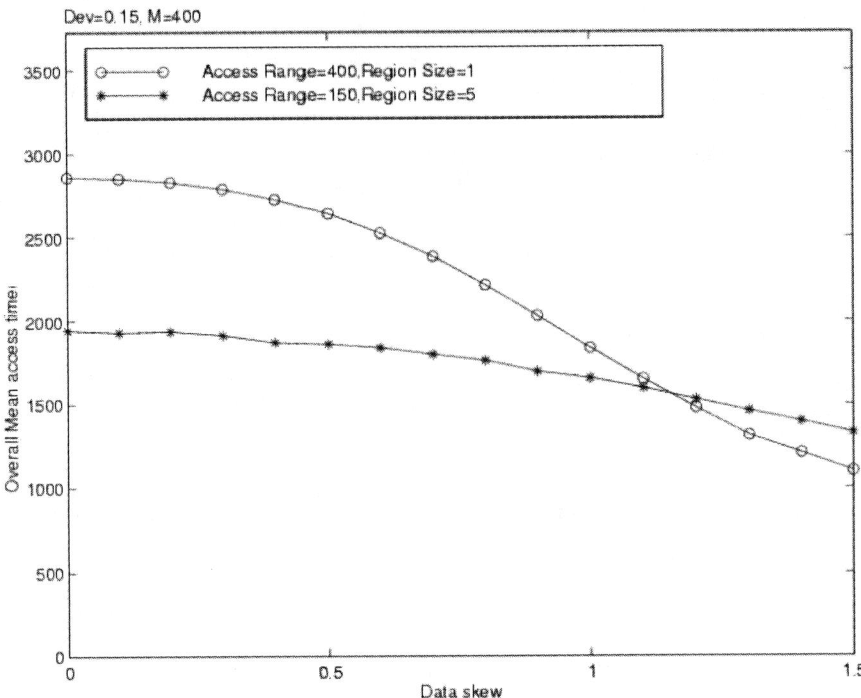

zero demand probability at the client. This item interval consists of an integral number of regions with each region containing *Region Size* items. The probabilities of regions are computed using the Zipf distribution. Items inside the same region have the same demand probability.

- To simulate differences among the demands of different clients, the parameters *Dev* and *Ns* are used. A *Dev* percentage of clients deviate from the initial overall client demand. The demand patterns for such clients are generated as follows: with probability *Ns*, the demand probability of each information item accessed by such a client is swapped with that of another item selected uniformly among all the items in the server database. In the simulation results of Figure 5, *Ns* = 0.5.

Furthermore, results in Vaidya and Hameed (1996, 1999b) show that the performance of the bucketing method for a relatively small number of buckets compared to the server database size (10 buckets are used with a database of size 500 in Vaidya & Hameed, 1999b) is quite close to that of the original algorithm, implying that bucketing can significantly reduce the computational cost of the method.

An Adaptive Push System

Both the push methods described above, and also all others that have been proposed, are only practical for static environments since no means for updating item probabilities is provided. However, data broadcasting applications operate in dynamic environments. The method in Nicopolitidis, Papadimitriou, and Pomportsis (2001, 2002) proposes a push-based system that is adaptive to dynamic client demands. The system uses a learning automaton at the broadcast server in order to provide adaptivity to the method of Vaidya and Hameed (1996, 1999b) while maintaining its computational complexity. Using simple feedback from the clients, the automaton continuously adapts to the overall client population demands in order to reflect the overall popularity of each data item. It is shown via simulation that, contrary to the Vaidya and Hameed method, the adaptive system provides superior performance in an environment where client demands change over time, with the nature of these changes being unknown to the broadcast server.

Learning Automata

Learning automata (Narendra & Thathachar, 1989) are structures that possess the ability to learn the characteristics of a system's environment. A learning automaton improves its performance by interacting with its random environment. Its target is to efficiently select among a set of A actions the optimal one, so that the average penalty given by the environment is minimized. Thus, there must exist a mechanism that notifies the automaton about the environmental feedback to a specific action. A learning automaton operates via a sequence of cycles, which eventually lead to minimization of average received penalty.

Each learning automaton typically uses a vector $p(n) = \{p_1(n), p_2(n), ..., p_A(n)\}$ that represents the probability distribution for choosing one of the actions a_1, a_2, ..., or a_A at cycle n. Obviously:

$$\sum_{i=1}^{A} p_i(n) = 1. \qquad (7)$$

The core of a learning automaton is the algorithm used to update the probability vector. Also known as the reinforcement scheme, this algorithm uses after each a_i at cycle n the environmental response $\beta(n)$ triggered by this action in order to update the probability distribution vector p. After the updating has taken place, the automaton selects the action to perform at cycle $n + 1$ according to the updated probability distribution vector $p(n + 1)$.

The Adaptive Scheduling Algorithm

In what follows, we introduce the estimation probability vector *p*, which stores the server's estimation of the actual demand probability vector *d* that contains the actual choice probabilities of the various information items averaged over the entire client population. The adaptive push system uses at the server a learning automaton whose probability distribution vector contains the server's estimate p_i of the actual demand d_i of the overall client population for each data item *i*; d_i is the overall popularity of each item at the client population, meaning that it is the probability that when an item request is made by the client population, this will be a request for item *i*. Obviously:

$$\sum_{i=1}^{M} d_i = 1,$$

where *M* is the number of information items in the database of the broadcast server.

According to the Vaidya and Hameed (1999b) method, the server selects to broadcast item *i* having the largest value of the cost function *G*. The adaptive approach extends this method: After broadcasting item *i*, the server waits for acknowledgment from all clients that were satisfied by this item broadcast. All clients that were waiting for item *i* acknowledge their reception via a short feedback pulse. The sum of the received pulse strength at the server will be used by the automaton to update the probability distribution vector *p*.

Nevertheless, the pulse strength of each client's pulse at the server depends on its distance from the server and is not constant due to the mobility of the clients. Assuming that the path loss is a $\frac{1}{x^n}$ type loss with a typical $n = 4$, the feedback pulse of clients located close to the broadcast server will be extremely stronger than those of clients further away. To prevent the clients close to the server from favouring the automaton decision toward their demands, there exists a power-control mechanism on the returning pulses. Thus, every item can be broadcast including information about the signal strength used for the item's transmission. Assuming that information on the signal strength at which the item was originally transmitted is A_0, and the signal strength measured at the item reception is *A* (obviously it will stand that $S \leq S_0$), clients will set the strength of their feedback pulse to A_0/A. Thus, this form of power control forces contribution of each client's feedback pulse at the server to be the same irrespective of the client-server distance.

The probability distribution vector at the automaton at the server defines the demand probability estimate of the server for each information item. Via this scheme, the automaton uses the amplitude of the received pulse to update the server's estimate of information item probabilities. Thus, for the upcoming broadcast, the server chooses which item to transmit by taking into account the updated values of the item probability estimates.

Probability Updating Scheme

When the broadcast of an information item i does not satisfy any client, the server probability estimates of the data items do not change. Following a useful broadcast of an item i, however, the probability estimate of item i will be increased. In general, after the server's kth broadcast, the following probability updating scheme is used:

$$p_j(k+1) = p_j(k) - L(1-\beta(k))(p_j(k)-a),\ \forall i \neq j$$

$$p_i(k+1) = p_i(k) + L(1-\beta(k))\sum_{i \neq j}(p_j(k)-a). \qquad (8)$$

L is a parameter that governs automaton convergence speed. Low values of L lead to more accurate estimations. However, low values for L also reduce convergence speed. It holds that $L, a \in (0,1)$ and $p_i(k) \in (a,1), \forall i \in [1..M]$. The role of parameter a is to prevent the probabilities of nonpopular items from taking values very close to zero in order to increase adaptivity. This is because if the probability estimate p_i of an item i becomes zero, then $G(i)$ will also take a value near zero. However, item i, even if unpopular, still needs to be transmitted since some clients may request it. Furthermore, the dynamic nature of demands is likely to make this item more popular in the future.

Upon reception of the aggregate feedback pulse, its amplitude is normalized in the interval [0,1]. $\beta(k)$ stands for the normalized environmental response after the server's kth broadcast. A value of $\beta(k) = 1$ represents the case where no client acknowledgment is received. Thus, the lower the value of $\beta(k)$, the more clients were satisfied by the server's broadcast at cycle k.

Figure 6. Convergence of automaton estimation of the demand probabilities for items 1 to 2

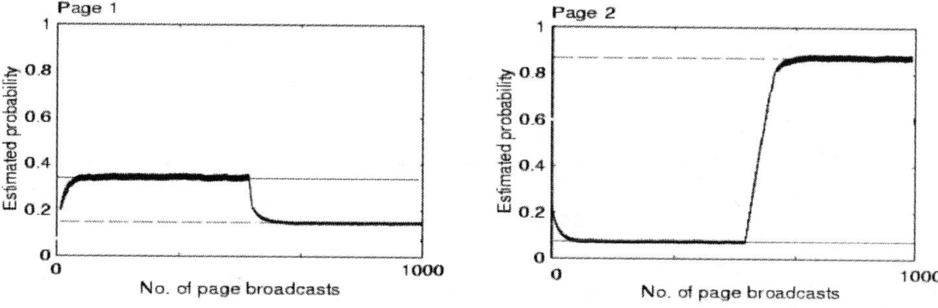

Using the reinforcement scheme of Equation 8, the data item probability estimates converge near the actual demand probabilities for each item. This makes this approach attractive for dissemination applications with dynamic client demands. This convergence is schematically shown in Figure 6, which plots the convergence of the item probability estimates toward the actual overall demand probabilities for data items 1 and 2, in a simulation of an environment with a priori unknown client demands that change after some time. It is evident that convergence of the automaton item probabilities estimates for items 1 and 2 to the overall client demand for these items is achieved.

The normalization procedure in the calculation of $\beta(k)$ suggests the existence of a procedure to enable the server to possess an estimate of the number of clients under its coverage. This is made possible by broadcasting a control packet that forces every client to respond with a power-controlled feedback pulse. The broadcast server will use the sum of the received pulses S to estimate the number of clients under its coverage. Then, upon reception of a pulse of amplitude Z after the server's kth broadcast, $\beta(k)$ is calculated as Z/S. This estimation process will take place at regular time intervals with the negligible overhead of broadcasting a unit-length item. More on the normalization procedure can be found in Nicopolitidis et al. (2002).

The probability updating scheme of Equation 8 is of $O(M)$ complexity. Therefore, the adaptive push method does not increase the computational complexity of the original nonadaptive one by Vaidya and Hameed. The bucketing described in the latter method would not further reduce computational complexity in the adaptive one, as complexity would remain of $O(M)$ due to the probability updating scheme.

Simple Performance Evaluation

Simulation results in Nicopolitidis et al. (2001, 2002) show that the adaptive push system outperforms the nonadaptive one in Vaidya and Hameed (1999b) in dynamic environments. This is due to the fact that the adaptive system is able to "learn" the changing environment and adapts its operation by taking into account the actual demand probabilities. The nonadaptive push system, however, operates "blindly" by utilizing the same set of demand probabilities all of the time. This superiority of the adaptive system for various values of the access skew coefficient θ_s is shown in Figure 7. In these comparisons, the static push system's server always broadcasts the entire range of its items assuming equiprobable demand per item. This is because such an approach is the best without knowledge of the nature of overall client demands. The simulation assumptions are the same with those mentioned in the section describing the nonadaptive method by Vaidya and Hameed .

Pull Systems

In general, pull systems have the advantage of being adaptable to dynamic client demands due to the knowledge of the demand at the server through client requests. However, they are not easily scalable to large numbers of clients. In such cases, requests carried over the back channel will either collide with each other or saturate the server.

Figure 7. Overall mean access time in unit items versus data skew coefficient θ

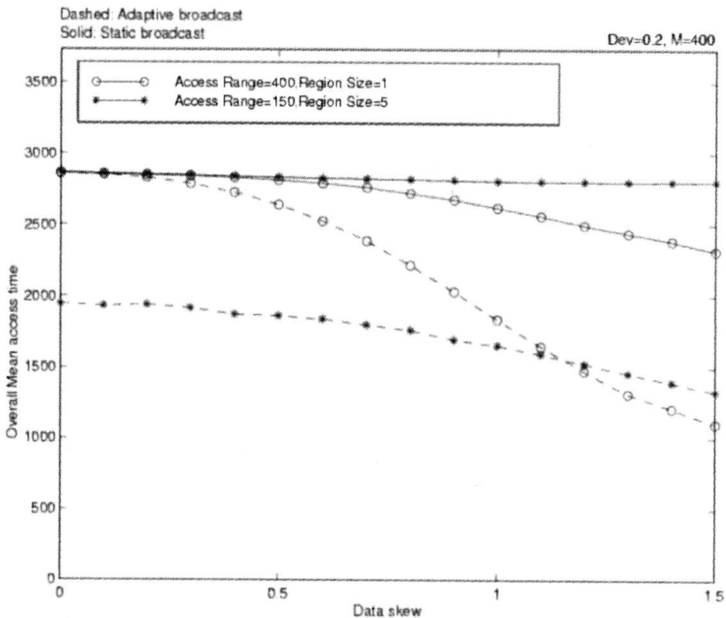

In this section we will describe the work of Aksoy and Franklin (1998, 1999) as a representative pull system. In that paper, it is reported that the main algorithms that have been proposed for pull-based data broadcasting are:

- *First Come First Served (FCFS).* This broadcasts information items in the order they were requested.
- *Most Requests First (MRF).*This schedules broadcasts by transmitting items based on the number of their pending requests in descending order.
- *Most Requests First Lowest (MRFL).* This is similar to the previous one; however, for two or more items having the same number of pending requests, the algorithm first transmits the item having the lowest access probability.
- *Longest Wait First (LWF).* This transmits the item having the largest aggregate waiting time first (the sum of waiting times for all pending requests for that item).

Simulation results in Aksoy and Franklin (1998, 1999) show that LWF is the most preferable algorithm. However, it is not practical for large systems since at each step, it needs to calculate the aggregate waiting time for every pending item.

The proposed method in Aksoy and Franklin (1998, 1999), *R*x*W*, provides a method that is shown via simulations to provide close performance to LWF while having a lower overhead. Unlike most other pull algorithms (and also push ones as well), *R*x*W* constructs the schedule based on the state of the request queue rather than the item probabilities. Specifically, the *R*x*W* algorithm dictates that the broadcast server maintains a list that contains the R_ixW_i value for each item *i* that has pending requests, where R_i is the number of pending requests for item *i* and W_i is the time that has elapsed since the oldest request for item *i* was made. In order to further reduce the overhead of *R*x*W* and to provide it with scalability to systems operating on larger databases, the following algorithms are proposed in Aksoy and Franklin (1998, 1999):

- A variant of *R*x*W* that reduces the calculations needed per item broadcast and still manages to broadcast the information item having the largest *R*x*W* value. This is achieved by reducing the item range examined by the algorithm in its search for the item to broadcast.
- A heuristic version of *R*x*W* that further reduces overhead by searching, at each step *t*, not for the item having the largest *R*x*W* value, but instead for the first one that has an *R*x*W* value greater than or equal to *a*x*Thresh*(*t*), where *Thresh*(*t*) = (*Thresh*(*t* - 1) + *PrevRxW*)/2. *PrevRxW* is the *R*x*W* of the previously broadcast item and *a* is a parameter that can be tuned to provide a trade-off between scalability to large databases and broadcast-schedule construction optimality. Specifically, as *a* approaches 0, the heuristic algorithm reduces toward a constant time approach to the selection of the item to broadcast, whereas as *a* increases to infinite, the behaviour heuristic algorithm approaches that of the original *R*x*W* one.

Hybrid Systems

In hybrid methods clients can use a back channel to submit requests to the server for information items that are not cached in the client and are not scheduled to appear in the broadcast in the near future. It can be easily seen that there must exist a policy on efficient back channel usage as a large number of client requests will either collide or saturate the server due to the latter's finite processing power and request-storage capability.

To support a mixture of push and pull delivery, the broadcast server will interleave pushed items with pulled ones (items that are broadcast in response to specific clients' requests). The percentage of bandwidth that can be allocated to pulled items is a matter of study in Acharya, Franklin, and Zdonik (1997), Lee, Hu, and Lee (1999), and Stathatos, Roussopoulos, and Baras (1997).

In Acharya et al. (1997), this bandwidth percentage is controlled by a parameter named *PullBW*, taking values in [0...1]. When *PullBW* = 0, the system is a pure push one, whereas when *PullBW* = 1, the system becomes a pure pull one. When *PullBW* takes values in (0...1), the system employs a combination of push and pull and thus becomes a hybrid one.

- A refinement of this method uses a threshold parameter, *Thresh*, that limits back channel use by clients. Specifically, clients pull items (i.e., submit explicit requests) only if the requested item will appear in the broadcast later than the duration of a 1-*Thresh* portion of the broadcast schedule's major cycle. Thus, this method constrains a client to using the back channel only for information items whose miss would cost a lot.
- A complementary method to increase performance of the system is to exclude from the push schedule those pages that are most unlikely to be demanded by clients.

> Thus, these pages can be accessed only on demand. This method has the obvious effect of reducing push bandwidth to the favour of pull bandwidth, without a significant performance penalty due to the very limited popularity of the data items excluded from the push schedule. However, if the pull bandwidth is not enough despite its increase, performance can degrade since clients will not be able to access on demand the items that were excluded from the push schedule.

Simulation results by Acharya et al. (1997) show that while pure pull and pure push perform better than the hybrid method in underutilized and overutilized systems respectively, the hybrid method provides nearly uniformly good overall mean access times in almost all cases and is thus of practical use in a wide range of system loads.

Although being able to efficiently combine push and pull, the hybrid method of Acharya et al. (1997) has the disadvantage of not being able to dynamically adapt the ratio of push to pull bandwidth to changing degrees of system utilization. A method that overcomes this problem is proposed in Lee et al. (1999). This paper proposes an algorithm, which, according to system workload, dynamically (a) allocates bandwidth to push and pull modes and (b) decides the mode (either push or pull) for the dissemination of each data item. The proposed method is compared to pure push, pure pull, and a hybrid method with a static ratio of push-pull bandwidth allocation. Simulation results regarding the first three methods are similar to those of Acharya et al. (1997) that were mentioned above. The proposed dynamic hybrid method, however, shows an optimal performance in all cases, as it is able to flexibly assign bandwidth to either push or pull mode according to system workload.

In Stathatos et al. (1997), a hybrid method that dynamically estimates the demand probability of information items is used. Items are sorted based on their demand and a line is drawn that separates the pushed popular items from the pulled nonpopular ones. The server artificially changes the demand probability estimate of items so that even a very popular item is excluded for short time periods from the push mode. This exclusion is useful since it will cause a number of explicit requests for that item which will be used by the server to obtain an estimate on the overall item demand probability. Repetition of this function obviously enables the server to keep up with changing demand and thus adapt to it. Simulation results by Stathatos et al. show that the method enjoys scalability and adaptivity to changing demands, with the rate of adaptivity being possible to tune.

Another interesting hybrid system is proposed by Hu and Chen (2002). In this paper, the authors propose a model that uses client impatience to devise an algorithm that can work online and establish knowledge of item-demands patterns in a granularity of a broadcast cycle. A client usually has limited patience for a push request it makes. Thus, after a certain period of time has elapsed and the client has yet to receive the desired item from the broadcast, it will send a pull request to explicitly demand the item. However, such requests also have the benefit of helping the server estimate the demand probability for the pushed items. To this end, the server is programmed to deliberately generate a push-broadcast miss, which will obviously force clients to explicitly demand that item. Thus, by counting the pull requests, the server can calculate the access frequency proportion of push items on the broadcast channel. Simulation results by Hu and Chen show that the mean difference between the estimated access frequency distribution and the real one is very small, proving the estimation mechanism to be very useful.

Hybrid systems are also covered in Datta, VanderMeer, Celik, and Kumar (1999). This paper addressed the problem of data broadcasting by investigating both (a) good strategies that the server can use to decide on broadcast content, given that users are highly mobile, and (b) the identification of efficient retrieval algorithms for the clients so that their energy consumption is minimized. Datta et al. consider broadcast strategies being either of constant schedule size (constant broadcast size [CBS] strategies, or periodic strategies) or of variable schedule size (variable broadcast size [VBS] strategies, or aperiodic strategies). The main characteristics of these strategies are summarized:

- In the constant schedule size strategy, items are included or excluded from the broadcast based on their popularity. Furthermore, in order to prevent client starvation, the system will support monitoring of requests for nonpopular items. Thus, after a certain period elapses after such a request, the item will be included in the schedule despite its low overall demand among the client population.
- In the variable schedule size strategy, schedule size changes according to the number of items that have been requested during the preceding broadcast period. As this strategy can possibly include all items in the database, items are dropped from the broadcast based on the expected time that a client requesting an item will stay in the cell.

Results in Datta et al. (1999) show that:

- The aperiodic approach performs better than the periodic one at low system loads.
- For moderate loads, the aperiodic approach provides lower tuning times, whereas the periodic one provides lower access times.
- For high loads, the periodic approach outperforms the aperiodic one and provides lower tuning and access times.
- When clients are interested in only a small portion of the database, the relative performance of the periodic and aperiodic approaches from the energy-efficiency point of view resembles that from the access-time point of view.
- Minor starvation of clients requesting items of low popularity starts to occur only in relatively high loads.

Scheduling Construction with Multiple Channels

The above discussion so far assumed a single broadcast channel to which all clients are tuned. However, an extension of the basic Vaidya and Hameed method for operation in environments with more than one broadcast channel is also proposed. Assume that

$H = \{1, 2, ..., c\}$ is the set of available broadcast channels and each client listens to a nonempty set $S \subseteq H$ with probability Π_S. The cost function G in Equation 4 changes to the following one, which is used to determine which item i will be broadcast on channel h, $1 \leq h \leq c$:

$$G_h(i) = \frac{d_i}{l_i}\left(\sum_{S \subseteq H, h \in S} \Pi_s \left(T - R^S(i)\right)^2 \right) \tag{9}$$

The broadcast scheduling algorithm changes accordingly and again targets minimization of the overall mean access time averages over all clients: Let T be the current time and $R_h(i)$ the time when item i was last transmitted on channel h, $1 \leq h \leq c$. For items that have not been previously broadcast, $R_h(i)$ is set to -1. The broadcast scheduler takes the following steps in order to determine the item to transmit on channel h:

- Calculates $R_s(i) = \max_{h \in S} R_h(i), \ \forall S, \ \forall i, \ S \subseteq H, \ 1 \leq i \leq M.$
- Selects item i with the largest value of $G_h(i)$. If the maximum value of $G_h(i)$ is shared by two or more items, then the algorithm selects one of them arbitrarily.

Figure 8 plots the performance of the method for two available broadcast channels. Π_1, Π_2, and $\Pi_{1,2}$ are the probabilities of a client listening either to channel 1, channel 2, or both

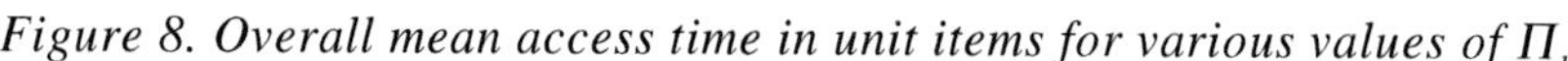

Figure 8. Overall mean access time in unit items for various values of Π_1

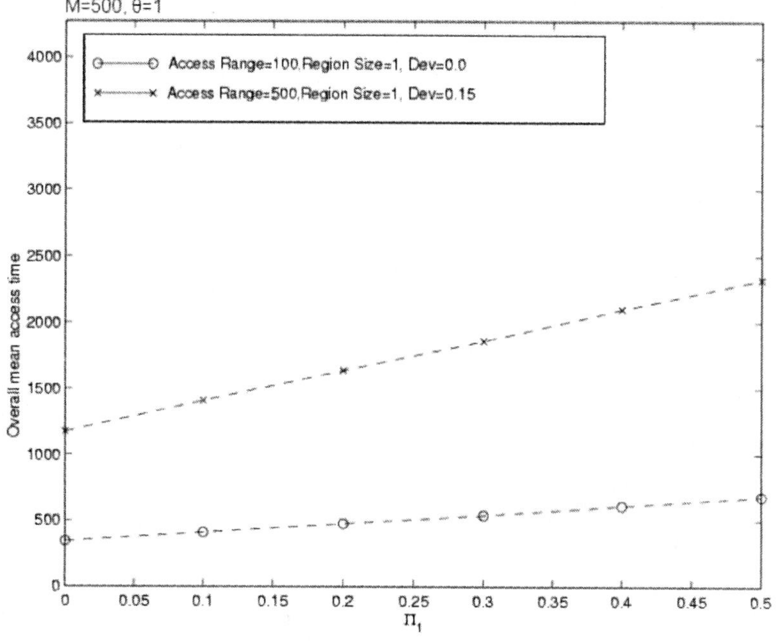

of these channels, respectively. Obviously, $\Pi_1 + \Pi_2 + \Pi_{1,2} = 1$. In this experiment we assumed that $\Pi_1 = \Pi_2 = 1 - \Pi_{1,2}$ (Vaidya & Hameed, 1999b). As in the case of the single channel, it can be seen that the method is able to take advantage of increased commonality in item demands for large values of θ, and exhibit a low overall mean access time.

Scheduling with Reception Errors

The methods presented so far do not consider the effects of reception errors, which frequently occur in wireless links, on optimum schedule construction. Such a method is discussed by Vaidya and Hameed (1999b). Specifically, Relation 1 is enhanced by modifying it in order to take into account the effect of unrecoverable reception errors (errors that cannot be corrected using error control codes). Assuming that $E(l)$ is the probability that an item of length l contains an unrecoverable error, Relation 2 changes to Relation 10 in order to take errors into account:

$$s_i \propto \sqrt{\frac{l_i(1-E(l_i))}{d_i(1+E(l_i))}} \tag{10}$$

Obviously, in cases of no unrecoverable reception errors or items of the same length, Relation 10 reduces to Relation 2 since $\frac{1-E(l_i)}{1+E(l_i)}$ reduces to unity or a constant, respectively.

Based on Relation 10, the broadcast scheduler modifies the cost function G so as to take errors into account, as follows:

$$G(i) = (T - R(i))^2 \frac{d_i}{l_i} \frac{1+E(l_i)}{1-E(l_i)}, \quad \forall i \in [1..M]. \tag{11}$$

Taking Variance of Access Time into Account

The methods presented so far primarily focus on minimization of mean access time over the entire client population. However, in several applications, it could be beneficial to reduce performance degradation of clients with demands largely deviating from the overall demand of the client population, as it is the mean access times of these clients that will largely vary from the overall mean access time. To this end, the work in Vaidya and Hameed (1996, 1999b) was augmented in Vaidya and Hameed (1998), which proposes two additional algorithms. The first one targets minimization of the variance of access time, while the second one implements a balance between minimization of overall mean access time and access time variance. It has to be noted that these two performance metrics are contradictory goals and often minimization of one of them leads to an excessive value for the other.

The variance-minimization algorithm is based on the fact that, under the assumption of equally spaced instances of the same item, the minimal variance of response time is achieved when:

$$\frac{d_i s_i^{\;2}}{l_i}\left(\frac{2}{3}s_i - \mu\right) = constant,\ \forall i,\ 1 \le i \le M, \tag{12}$$

where μ is the mean response time. The rest of the parameters have been explained in the section describing the method in Vaidya and Hameed (1996, 1999b). Based on the above relation, the cost function G changes to:

$$G(i) = (T - R(i))^2 \frac{d_i}{l_i}\left(\frac{2}{3}(T - R(i)) - \frac{1}{2}\sum_{i=1}^{M}(T - R(i))\right) \tag{13}$$

The mean-variance trade-off algorithm targets a trade-off between minimization of the overall mean and the variance of response time. To this end it tries to satisfy the following relation:

$$\frac{d_i s_i^{\;a}}{l_i} = constant,\ \forall i,\ 1 \le i \le M \tag{14}$$

and thus, the cost function G changes to the following one:

$$G(i) = (T - R(i))^a \frac{d_i}{l_i},\ 2 \le a \le 3. \tag{15}$$

For a = 2, the above cost function reduces to that of the basic method in Vaidya and Hameed (1996, 1999b). Jiang and Vaidya (1998) propose that using values of a between 2 and 3 will provide a trade-off mechanism between minimization of the variance of access time and minimization of the overall mean access time. These claims are supported by simulation results in Jiang and Vaidya (1998), which show that the proposed two algorithms are useful in medium-skewed environments, while in lightly skewed or severely skewed environments, their original push method also performs satisfactorily.

Scheduling for Requests with Deadlines

The discussion so far assumed environments where clients may wait endlessly for the desired item to arrive via the broadcast channels. This section deals with the case where item requests by clients are dropped if not served in a certain time period. To this end, the work in Vaidya and Hameed (1996, 1999b) is augmented in Jiang and Vaidya (1999), which considers the case of request deadlines. In this situation, clients that request a certain information item do not wait endlessly to receive it, but they drop the request after a time period. The motivation is to provide an algorithm that minimizes the overall mean access time while it also maximizes the percentage of served requests (service ratio).

The proposed method is based on the fact that maximum service ratio and minimum mean overall access time are achieved when the following equation stands:

$$\frac{d_i}{l_i}(\tau s_i e^{-\tau si} + e^{-\tau si} - 1) = constant, \ \forall i, \ 1 \leq i \leq M \tag{16}$$

where $1/\tau$ is the mean of the exponential distribution that represents the time lengths that a client waits for a requested item before it drops that request. The rest of the parameters have been explained in the section describing the basic Vaidya and Hameed method. Based on the above relation, the cost function G changes to

$$G(i) = \frac{d_i}{l_i}\left(\tau(T - R(i))e^{-\tau(T-R(i))} + e^{-\tau(T-R(i))} - 1\right). \tag{17}$$

Simulation results by Jiang and Vaidya (1999) show that in environments with request deadlines, this algorithm is useful when the demand skew is medium, while in lightly skewed or severely skewed cases, the basic push system also performs satisfactorily.

Critique on Performance

In this section we briefly comment on general aspects that characterize performance of the push, pull, and hybrid families of broadcasting systems and focus on several methods from each family.

- *Pull methods.* Pull methods have the advantage of being able to adapt to dynamic client demands since the server possesses knowledge regarding the demands of the clients. However, they are inefficient from the point of view of scalability. When the client population becomes too large, requests will either collide with each other or saturate the server. The latter problem is caused as the server has (a) a finite queue to store incoming requests and (b) finite computational power.
- *Push methods.* Push methods provide high scalability and client hardware simplicity since the client does not need to include data packet transmission capability. However, they are unable to operate efficiently in environments with dynamic client demands.

    1. Although the broadcast disks approach is more efficient than the flat approach, it is constrained to fixed-size data items and does not present a way of determining either the optimal number of disks to use or their relative frequencies. Those numbers are selected empirically and, as a result, the server may not broadcast data items with optimal frequencies, even in cases of static client demands. Furthermore, the rigid enforcement of the constraint for minimization of the variance of spacing between consecutive instances of the same item leads to schedules that possibly include empty and thus unused periods (holes). The broadcast disks approach is not adaptive to client demands that change over time since it is based on the server's a priori knowledge of static client demands. This results in predetermined broadcast schedules for the broadcast disks method. In environments with updates in the context of data items, for low to moderate update rates, the system can approach the performance of the read-only case.
    2. The indexed enhancement of broadcast disks in Yee et al. (2002) shows increasing energy efficiency over the flat $(1, m)$ index scheme at medium- and high-demand skewness.
    3. The method in Vaidya and Hameed (1996, 1999b) has the advantage of automatically using the optimal frequencies for item broadcasts, contrary to Acharya et al. (1995). Furthermore, the constraint of equally spaced instances of the same item is not rigidly enforced, a fact that leads to elimination of empty periods in the broadcast. Additionally, it is capable of taking into account the effect of reception errors. The method by Vaidya and Hameed works with items of different sizes, too. This assumption is obviously more realistic compared to that of fixed-length items made in the broadcast disks approach. The efficiency of the method also holds for the case of multiple broadcast channels. In environments with request deadlines, the method by Jiang and Vaidya (1999) is useful when the demand skew is medium. However, the methods by Vaidya and Hameed are also inefficient in terms of adaptivity in environments with dynamic client demands.
    4. The method by Nicopolitidis et al. (2002) combines the obvious advantages of push and pull systems' scalability and adaptivity, respectively. The incorporated adaptivity (a) helps the system adapt to overall client demands in order to reflect the overall popularity of each data item and (b) does not

increase the computational complexity of the nonadaptive method by Vaidya and Hameed.

- *Hybrid methods.* In this case, a combination of push and pull is used. Nevertheless, hybrid methods have to carefully strike a balance between push and pull and cope with a number of additional issues (determination and dynamic allocation of bandwidth available for push and pull, determination of items to be pushed and those to be pulled, etc.). Furthermore, such systems still impose the need for client packet transmission capability and existence of a back channel to carry client requests.
    1. The method by Acharya et al. (1997) shows nearly uniformly good overall mean access times in almost all cases.
    2. The method by Stathatos et al. (1997) shows scalability and adaptivity to changing demands, with the rate of adaptivity being possible to tune.
    3. The method by Lee et al. (1999) dynamically adapts the ratio of push to pull bandwidth to changing degrees of system utilization and shows an improved behaviour compared to the method in Acharya et al. (1997) as it is able to flexibly assign bandwidth to either push or pull mode according to system workload.
    4. The method by Hu and Chen (2002) is shown to provide a very useful estimation mechanism for efficient operation in environments with changing client demands.
    5. The method by Datta et al. (1999) provides two approaches: a periodic and an aperiodic one. For high loads, the periodic approach outperforms the aperiodic one and provides lower tuning and access times. For moderate loads, the aperiodic approach provides lower tuning times, whereas the periodic one provides lower access times. The aperiodic approach performs better than the periodic one at low system loads.

Conclusion

With the increasing popularity of mobile and wireless networks, data broadcasting has emerged as an efficient way of delivering data to mobile clients having a high degree of commonality in their demands. This chapter described the asymmetric environment that characterizes data broadcast. Furthermore, via representative algorithms, it covered broadcast schedule construction for systems employing a single broadcast channel, schedule construction for systems employing multiple broadcast channels, and schedule construction by taking into account the effect of reception errors. It then presented an algorithm that tries to provide better support for clients whose access patterns deviate a lot form the overall access pattern of the client population. It also presented algorithms for environments where item requests by clients are dropped if not served in a certain time period. Brief comments on issues that affect performance of the discussed data broadcasting methods were also made.

References

Acharya, S., Franklin, M., & Zdonik, S. (1995). Dissemination-based data delivery using broadcast disks. *IEEE Personal Communications, 2*(6), 50-60.

Acharya, S., Franklin, M., & Zdonik, S. (1997). Balancing push and pull for data broadcast. *Proceedings of the ACM International Conference on Management of Data (SIGMOD)*, (pp. 183-194).

Aksoy, D., & Franklin, M. (1998). Scheduling for large-scale on-demand data broadcasting. *Proceedings of the IEEE International Conference on Computer Communications (INFOCOM)*, (pp. 651-659).

Aksoy, D., & Franklin, M. (1999). RxW: A scheduling approach for large-scale on-demand data broadcast. *ACM/IEEE Transactions on Networking, 7*(6), 846-860.

Ammar, M. H., & Wong, J. W. (1985). The design of teletext broadcast cycles. *Performance Evaluation, 5*(4), 235-242.

Ammar, M. H., & Wong, J. W. (1987). On the optimality of cyclic transmission in teletext systems. *IEEE Transactions on Communications, 35*(1), 68-73.

Bowen, T. (1992). The datacycle architecture. *Communications of the ACM, 35*(2), 71-81.

Datta, A., VanderMeer, E. E., Celik, A., & Kumar, V. (1999). Broadcast protocols to support efficient retrieval from databases by mobile users. *ACM Transactions on Database Systems, 24*(1), 1-79.

Fernandez, J., & Ramamritham, K. (1999). Adaptive dissemination of data in time-critical asymmetric communication environments. *Proceedings of the 11th IEEE Euromicro Conference on Real-Time Systems*, (pp. 195-203).

German, G., Gopal, G., Lee, K., & Weinrib, A. (1987). The datacycle architecture for very high throughput database systems. *Proceedings of the ACM International Conference on Management of Data (SIGMOD)*, (pp. 97-103).

Gifford, D. (1990). Polychannel systems for mass digital communications. *Communications of the ACM, 33*(2), 141-151.

Hu, C.-L., & Chen, M.-S. (2002). Dynamic data broadcasting with traffic awareness. *Proceedings of the 22nd IEEE International Conference on Distributed Computing Systems (ICDCS)*, (pp. 112-119).

Imielinski, T., Viswanathan, S., & Badrinath, B. R. (1994a). Energy efficient indexing on air. *Proceedings of the ACM International Conference on Management of Data (SIGMOD)*, (pp. 25-36).

Imielinski, T., Viswanathan, S., & Badrinath, B. R. (1994b). Power efficient filtering of data on air. *Proceedings of the Fourth International Conference on Extending Database Technology (EDBT)*, (pp. 245-258).

Jain, R., & Werth, J. (1995). *Airdisks and airRAID: Modeling and scheduling periodic wireless data broadcast* [Extended Abstract] (Tech. Rep.95-11). Rutgers University, Piscataway, NJ.

Jiang, S., & Vaidya, N. H. (1998). Response time in data broadcast systems: Mean, variance and trade-off. *Proceedings of the Workshop on Satellite-Based Information Systems (WOSBIS)*, (pp. 52-59).

Lam, K. Y., Chan, E., & Yuen, J. C. H. (2000). Approaches for broadcasting temporal data in mobile computing systems. *Journal of Systems and Software, 51*(3), 175-189.

Lee, W.-C., Hu, Q., & Lee, D. L. (1999). A study on channel allocation for data dissemination in mobile computing environments. *ACM/Kluwer Mobile Network and Applications*, *4*(2), 117-129.

Liberatore, V. (2002). Multicast scheduling for list requests. *Proceedings of IEEE International Conference on Computer Communications (INFOCOM)*, (pp. 1129-1137).

Narendra, K. S., & Thathachar, M. A. L. (1989). *Learning automata: An introduction.* Englewood Cliffs, NJ: Prentice Hall.

Nicopolitidis, P., Papadimitriou, G. I., & Pomportsis, A. S. (2001). On the implementation of a learning automaton-based adaptive wireless push system. *Proceedings of Symposium on Performance Evaluation of Computer and Telecommunication Systems (SPECTS)*, (pp. 484-491).

Nicopolitidis, P., Papadimitriou, G. I., & Pomportsis, A. S. (2002). Using learning automata for adaptive push-based data broadcasting in asymmetric wireless environments. *IEEE Transactions on Vehicular Technology, 51*(6), 1652-1660.

Stathatos, K., Roussopoulos, N., & Baras, J. S. (1997). Adaptive data broadcast in hybrid networks. *Proceedings of International Conference on Very Large Data Bases (VLDB)*, 326-335.

Su, C. J., & Tassiulas, L. (1999). Broadcast scheduling for information distribution. *ACM/Baltzer Wireless Networks, 5*(2), 137-147.

Su, C. J., Tassiulas, L., & Tsotras, V. J. (1997). Broadcast scheduling for information distribution. *Proceedings of IEEE International Conference on Computer Communications (INFOCOM)*, 109-117.

Vaidya, N. H., & Hameed, S. (1996). Data broadcast in asymmetric wireless environments. *Proceedings of the Workshop on Satellite-Based Information Systems (WOSBIS)*, (pp. 97-108).

Vaidya, N. H., & Hameed, S. (1999a). Efficient algorithms for scheduling data broadcast. *ACM/Baltzer Wireless Networks*, *5*(3), 183-193.

Vaidya, N. H., & Hameed, S. (1999b). Scheduling data broadcast in asymmetric communication environments. *ACM/Baltzer Wireless Networks, 5*(3), 171-182.

Vaidya, N. H., & Jiang, S. (1998). Response time in data broadcast systems: Mean, variance and trade-off. *Proceedings of the Workshop on Satellite-Based Information Systems (WOSBIS).*

Yee, W. G., Navathe, S. B., Omiecinski, E., & Jermaine, C. (2002). Bridging the gap between response time and energy efficiency in broadcast schedule design. *Proceedings of International Conference on Extending Database Technology (EDBT)*, (pp. 572-589).

Yee, W. G., Omiecinski, E., & Navathe, S. B. (2001). *Efficient data allocation for broadcast disk arrays* (Tech. Rep. GIT-CC-02-20). Georgia Institute of Technology, College of Computing, Atlanta, Georgia.

Chapter II

Mobile Cache Management

Jianliang Xu, Hong Kong Baptist University, Hong Kong

Haibo Hu, Hong Kong University of Science and Technology, Hong Kong

Xueyan Tang, Nanyang Technological University, Singapore

Baihua Zheng, Singapore Management University, Singapore

Abstract

This chapter introduces advanced client-side data-caching techniques to enhance the performance of mobile data access. The authors address three mobile caching issues. The first is the necessity of a cache replacement policy for realistic wireless data-broadcasting services. The authors present the Min-SAUD policy, which takes into account the cost of ensuring cache consistency before each cached item is used. Next, the authors discuss the caching issues for an emerging mobile data application, that is, location-dependent information services (LDISs). In particular, they consider data inconsistency caused by client movements and describe several location-dependent cache invalidation schemes. Then, as the spatial property of LDISs also brings new challenges for cache replacement policies, the authors present two novel cache replacement policies, called PA and PAID, for location-dependent data.

Introduction

The past few years have seen tremendous advances in mobile computing and wireless communication technologies, including wireless high-speed networks, portable wireless devices, mobile application standards, and supporting software technologies. In particular, mobile computing concerns users who carry portable devices and need to access information anywhere and at anytime. The ubiquity of mobility has opened up new classes of data applications which promise to make our society more efficient and our lives more enjoyable. For example, people can query location-dependent information (e.g., the nearest restaurant) based on their current locations. The market drive as well as the technological advances has been flooding the commercial market with mobile data. However, various constraints of mobile computing environments, such as scarce wireless bandwidth and limited client resources, remain as barriers that must be overcome before the vision of mobile computing can be fully realized. The unique characteristics of mobile computing environments are summarized below (Barbara, 1999; Imielinski & Badrinath, 1994):

- *Constrained and unreliable wireless communications:* The radio spectrum used for wireless communications is inherently scarce. For example, Global System for Mobiles (GSM) operates only between 880 MHz and 960 MHz. The bandwidth for a single wireless channel is limited, varying from 1.2 Kbps for a slow paging channel, through 115 Kbps for General Packet Radio Service (GPRS), to about 11 Mbps for an 802.11b wireless Local Area Network (LAN). Furthermore, wireless transmission is error prone. Data might be corrupted or lost due to many factors such as signal interference and obstruction by tall trees and buildings.
- *Limited power source:* The battery power of wireless portable devices is limited, ranging from only a few hours to about half a day with continuous use (i.e., active wireless communication). Moreover, it is anticipated that only a modest improvement in battery capacity can be expected over the next few years. It is also worth noting that sending data consumes much more power than receiving data. For example, a Wavelan card consumes 1.7W when the receiver is "on" but 3.4W when the transmitter is "on."
- *Frequent disconnections:* To save energy or connection costs, mobile clients frequently disconnect themselves from the network and are kept in a weak connection status. Furthermore, mobile clients are also often disconnected due to unreliable wireless communication links.
- *Asymmetric communication:* Due to resource constraints of mobile clients, the *upstream* communication capacity from clients to servers is much less than the *downstream* communication capacity from servers to clients. Even in the case of an equal communication capacity, the data volume in the downstream direction is estimated to be much greater than that in the upstream direction (Acharya, Alonso, Franklin, & Zdonik, 1995).
- *Unrestricted mobility:* Mobile users can move from one location to another freely while retaining network connectivity, which enables their almost unrestricted mobility. Locations and movements of mobile users are therefore hard to predict.

- *Limited client capacities:* Since portable wireless devices are restricted by weight, size, and ergonomic considerations, this limits their capacities for CPU cycle, storage, and display.

In summary, the characteristics of a mobile computing environment pose many challenging problems for mobile data applications which do not exist in traditional computing environments. Thus, advanced data management and resource management techniques have been and are being developed to enhance the system performance of mobile data access. In this chapter, we attempt to address some of the performance issues by applying advanced client-side data-caching techniques.

The rest of the chapter is organized as follows: The next section introduces some background and preliminaries on the mobile computing model, data dissemination techniques, and client-side caching. Thereafter, the temporal-dependent cache invalidation issue for broadcasting environments is discussed. The next section discusses the general cache replacement issue for wireless data dissemination and presents a realistic and optimal cache replacement policy, Min-SAUD. The authors then introduce a new problem with location-dependent information services, namely, location-dependent cache invalidation. A cell-based symbolic location model is assumed and three location-dependent cache invalidation mechanisms are described. Next, location-dependent data invalidation for a geometric location model is discussed. A new performance criterion, called caching efficiency, is introduced to present a generic location-dependent cache invalidation method. The chapter then discusses two enhanced cache replacement policies for location-dependent data. Finally, we conclude this chapter with our vision on the future research trend of mobile cache management.

Figure 1. The mobile computing model

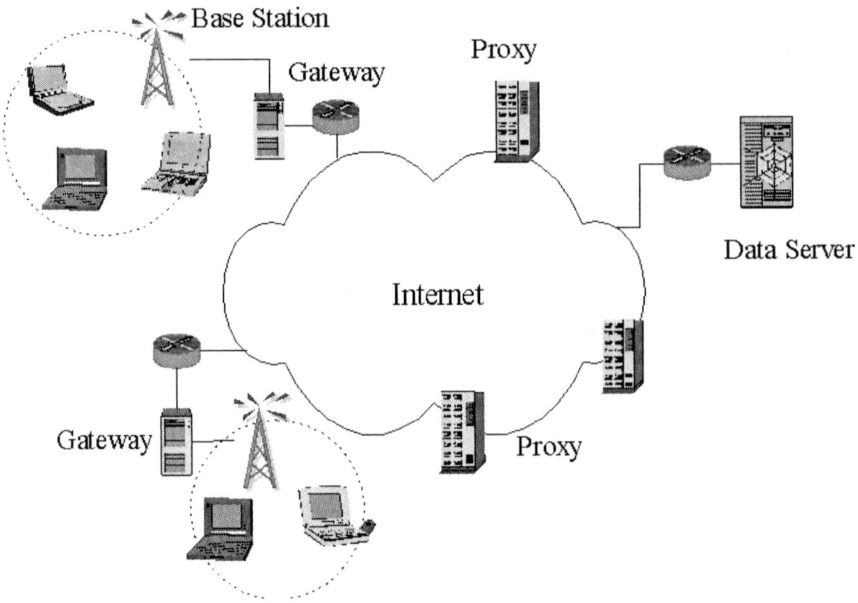

Background and Preliminary

Mobile Computing Model

Figure 1 depicts the general mobile computing system model (Barbara & Imielinski, 1994). It consists of two distinct sets of entities: mobile clients (MC) and fixed hosts. Some of the fixed hosts, called mobile support stations (MSSs), are augmented with wireless interfaces. An MSS can communicate with the MCs within its radio coverage area, called a wireless cell. An MC can communicate with a fixed host or server via an MSS over a wireless channel. The wireless channel is logically separated into two subchannels: an uplink channel and a downlink channel. The uplink channel is used by MCs to submit queries to the server via an MSS, while the downlink channel is used by MSSs to disseminate information or to forward the answers from the server to a target client. We assume that an MC contacts only one MSS at the same time. Each cell is associated with an ID (CID) for identification purposes. A CID is periodically broadcast to all the MCs residing in a corresponding cell. An MC can move from one cell to another (called handoff). After handoff, its wireless connection is switched to the new cell.

Wireless Data Delivery

There are two fundamental information delivery methods for wireless data applications: *point-to-point access* and *broadcast*. In point-to-point access, or synonymously, *unicast*, a logical channel is established between the client and the server. The channel behaves the same way as a wire; queries are submitted to the server and results are returned to the client as in a wired network. In data broadcast, data are sent simultaneously to all users residing in the broadcast area. It is up to the client to select the data it wants. Note that a wireless communication system essentially employs a broadcast component to deliver information, thus data broadcast can be implemented without introducing any system overhead. Data broadcast has elegant scalability and thus is an attractive solution for heavy-load systems.

In general, there are two data broadcast models, that is, push-based broadcast and on-demand (or called pull-based) broadcast. In push-based data broadcast, the server broadcasts data proactively to all clients according to the broadcast program generated by the data-scheduling algorithm. The program determines the order and frequencies that the data items are broadcast in. The scheduling algorithm may make use of precompiled access profiles in determining the broadcast program. In the simplest flat program, the server sequentially sends out all the objects; the duration of this period is called a broadcast cycle. Such a broadcast cycle is repeated continually at the server side. More sophisticated programs, such as broadcast disk (Acharya et al., 1995), reduce the average access latency for the clients by considering the skewness of object access frequencies.

A wireless on-demand broadcast system (Acharya & Muthukrishnan, 1998) supports both broadcast and on-demand services through a broadcast channel and a low-bandwidth uplink channel. The uplink channel can be a wired or a wireless link. When a client needs a data item, it sends to the server an on-demand request for the item through the uplink. Client requests are queued up, if necessary, at the server upon arrival. The server repeatedly chooses an item from among the outstanding requests, broadcasts it over the broadcast channel, and removes the associated request(s) from the queue. The clients monitor the broadcast channel and retrieve the item(s) they require.

Client-Side Data Caching: Overview

Caching is a commonly used technique for improving access latency and data availability in traditional operating systems, distributed systems, and Web environments (Xu, Liu, Li, & Jia, 2004). In the framework of a mobile wireless environment, client-side data caching is much more desirable due to constraints such as limited wireless bandwidth and frequent disconnections.

Figure 2 depicts a general model of cache management. The system employs either unicast, pushed broadcast, or on-demand broadcast for data dissemination. There is a cache management module in the client. Whenever an application issues a query, the local cache manager first checks whether the desired data item is in the cache. If it is a cache hit, the cache manager still needs to validate the consistency of the cached item with the master copy at the server. This process is called *cache invalidation*. In general, data inconsistency is incurred by data updates at the server (called *temporal-dependent invalidation*; see the next section for details). For location-dependent information in a mobile environment, cache inconsistency can also be caused by location change of a client (called *location-dependent invalidation*; see the section on location-dependent invalidation for details). If it is a cache hit but its content is obsolete or invalid, or it is a cache miss, the cache manager requests the data from the server. When the requested data item arrives, the cache manager returns it to the user and retains a copy in the cache. The issue of *cache replacement* arises when the free cache space is not enough to

Figure 2. A general model of cache management

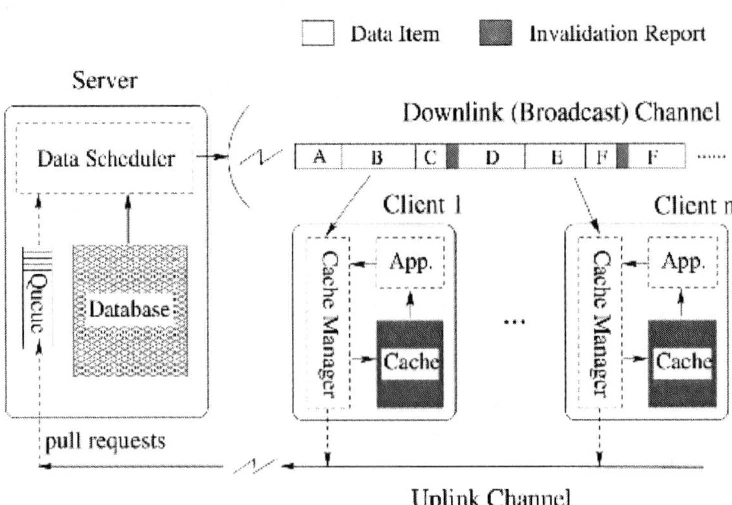

accommodate a data item to be cached. It determines the victim data item(s) to be dropped from the cache in order to allocate sufficient cache space for the incoming data item(s). The *cache prefetching*, or called cache hoarding, automatically preloads the data on the channel into the cache for the purpose of reducing cache-miss costs of future requests.

The advantages of using client-side data caching in mobile environments are as follows:

- It improves data access performance since a portion of queries, if not all, can be satisfied locally.
- It can help save energy since wireless communication is required only for cache-miss queries.
- It can reduce contention on the narrow-bandwidth wireless channels and off-load workload from the server; as such, the system throughput can be improved.
- It can improve data availability in circumstances in which clients are disconnected or weakly connected since we can use cached data to answer queries.

However, in contrast to the typical use of caching techniques in traditional environments, client-side data caching in mobile computing systems has three major differences:

- First, a mobile client may frequently go to disconnected states voluntarily (to save power and/or connection costs) or involuntarily (due to network failure) and freely move among different cells. As such, a good cache management mechanism should require as little cooperation as possible between the server and the client.
- Second, location-dependent data may show different results for different locations even with the same query, which brings new challenges for cache management.
- Finally, data retrieval delays (i.e., cache-miss penalties) might differ significantly for different data items in a wireless environment. Although the effect of different retrieval delays also occurs in distributed systems such as the Internet, the problem is amplified in a wireless system, which may employ a multitude of data-scheduling algorithms and access methods with widely different data retrieval delays.

These factors, together with scalability, variable data sizes, heterogeneous access patterns, and frequent data updates, make the design of client cache management a challenge in mobile environments. In the following few sections, we discuss caching schemes for various cache management issues including temporal-dependent cache invalidation, cache replacement for general data access, location-dependent cache invalidation, and cache replacement for location-dependent data.

Temporal-Dependent Cache Invalidation

Since data may be updated at the server from time to time (referred to as *temporal-dependent updates*), the data cached at a client should be kept consistent with those at the server. Various cache invalidation schemes have been developed to ensure the data consistency between a client and the server.

Due to the mobility and disconnection features of mobile environments, stateless approaches, in which the server is not aware of the state of a client's cache, are dominant. The server keeps track of the update history (of a reasonable length) and provides this information in the form of an *invalidation report (IR)* to the clients by periodic broadcasting or as requested by the clients. An IR is a list of IDs for the items that have changed since a past time. The mobile clients, if active, listen to the broadcast IRs and update their caches accordingly.

According to Barbara and Imielinski (1994) and Tan, Cai, and Ooi (2001), most of the existing invalidation schemes are based on the IR-based stateless approach. The basic IR-based algorithm is broadcasting time stamp (TS), described in Barbara and Imielinski (1994). It works as follows (Figure 3).

Let L be the IR broadcast interval and w the invalidation broadcast window. In the TS strategy, an IR consists of the current time stamp T and a list (d_i, t_i), where d_i is a data item ID and t_i is the most recent update time stamp of d_i such that $t_i > (T - w \times L)$. Every mobile client updates the status of its cache according to the received IR. Denote by T_l the time stamp of last validity checking. If the difference between T and T_l is larger than $w \times L$, the entire cache is dropped. Otherwise, if an item is reported to have changed at a time later than the TS stored in the cache, the mobile client purges it from the cache. The effectiveness of this scheme heavily depends on the system workload, such as client disconnection periods and data update rates.

There are some extensions based on the TS algorithm. To address the problem of long disconnections, Bukhres and Jing (1996) developed two adaptive TS algorithms, namely, basic adaptive TS (BATS) and hybrid adaptive TS (HATS). In the BATS method, the window size w is periodically evaluated according to the cache invalidation information piggybacked with the uplink queries so that the window of IR is just long enough to serve most of the clients. In the HATS method, the server partitions the database into a frequently updated data set and a rarely updated data set, and the window size for each set is computed separately.

Figure 3. Cache invalidation schemes based on IR

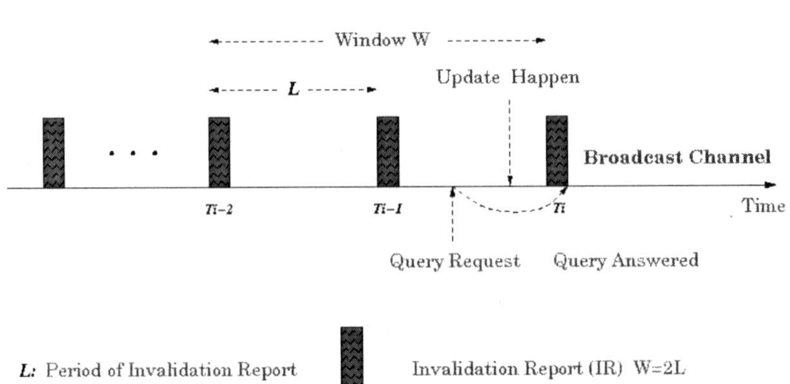

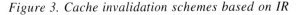

To salvage as many cache items as possible after a client disconnection, Wu, Yu, and Chen (1996) suggested a cache-checking scheme, called *group with cold update-set retention* (GCORE), based on the TS algorithm. Unlike the previous schemes, this scheme checks cache validity with the server even if the client disconnection time is longer than $w \times L^1$, and thus retains as many valid objects as possible. To reduce the uplink traffic, GCORE carries out the validity checking at the group level. In addition, GCORE excludes from a group the data items that have already been included in the most recent IR (i.e., the data items have been updated at least once within the broadcast window w since the last disconnection). In this way, it significantly improves the *false invalidation rate* within a group. However, this scheme requires a substantial amount of uplink requests, and is thus not energy efficient.

Tan and Cai (1997) improved GCORE by broadcasting group-based invalidation (BGI) reports. A BGI report consists of a pair of IRs: object invalidation report (OIR) and group invalidation report (GIR). The OIR is similar to the basic TS report, which consists of the server's update history up to w intervals, which enables the clients to salvage the most recently cached objects. The GIR report consists of the server's update history at a group level up to W ($W > w$) intervals, which aims at cutting down the probability of discarding the entire cache.

In IR-based approaches, a client must listen to the next IR to conclude if a cached copy is valid before answering a query from its local cache. Consequently, the average latency for answering a query is the sum of the actual query processing time and half of the IR interval. Based on this observation, Cao (2000) proposed a scheme that replicates, for several times, a small fraction of the essential invalidation information within an IR interval. Thus, it reduces the IR waiting time in the query latency.

General Cache Replacement and Prefetching

Overview of Existing Work

Cache replacement and prefetching issues for wireless data dissemination were first studied in the broadcast disk (Bdisk) project. Acharya et al. (1995) proposed a cache replacement policy called *PIX*. With *PIX*, the cached data item with the minimum value of p/x was evicted for cache replacement, where p is the item's access probability and x is its broadcast frequency. Thus, a cached item either has a high access probability or has a long broadcast delay. Simulation-based study showed this strategy could significantly improve the access latency over the traditional Least Recently Used (LRU) and Least Frequently Used (LFU) policies.

Acharya, Franklin, and Zdonik (1996) explored the use of prefetching to further improve client access performance in a broadcast environment. Compared with prefetching in traditional environments (such as database and file systems), this technique can be performed without introducing any additional cost in a broadcast system. When prefetching from a Bdisk, the importance of an arriving item is estimated to determine if it should replace a less valuable cache-resident item. The proposed scheme, *PT*, computes the importance value of an item by taking the product of its access probability (p) and the time (t) that will elapse before that item appears on the broadcast again. It was shown by simulation that the prefetching could provide great performance improvement by reducing cache-miss costs.

Song and Cao (2004) proposed a cache-miss initiated prefetching (CIMP) scheme. This scheme relies on two prefetch sets: the always-prefetch set and the miss-prefetch set. The always-prefetch set consists of data that should always be prefetched if possible. The miss-prefetch set consists of data that are closely related to the cache-missed data items. When a cache miss happens, instead of sending an uplink request to ask for the cache-missed data item only, the client also requests for the data items which are within the miss-prefetch set. This reduces not only future cache misses but also the number of uplink requests.

Tassiulas and Su (1997) presented a cache update policy that attempts to minimize the average access latency. In their paper, time on the broadcast channel is divided into slots with equal length of the broadcast time of a single item. Let λ_i be the request rate for item i and $\tau_i^f(n)$ the amount of time from slot n to the next transmission of item n; a time-dependent reward (latency reduction) for item i is given by $r(i,n) = \lambda_i \tau_i^f(n) + \frac{\lambda_i}{2}$. Under the assumption of constant request rates, it was claimed that an optimal caching strategy should make the cache update decision at each slot n such that the total reward kept in the cache (i.e., $\sum_{n=0}^{t-1}\sum_{i\in C(n)} r(i,n)$) until time t is maximized. However, the computational complexity of this approach is linear with respect to t, and thus actually infeasible for a real system. Instead, the proposed W-step look-ahead heuristic makes the cache update decision at slot t such that the cumulative average reward from slot n up to slot $n+W$ is maximized. The larger the window W, the better the performance but the worse the complexity. With $W=1$, this algorithm is equivalent to the *PT* prefetching scheme proposed.

Khanna and Liberatore (2000) introduced an online caching algorithm for the Bdisk system. Different from the previous work, they assumed that neither knowledge of future data requests nor knowledge of access probability distributions over items is available at the client side. Two traditional online algorithms, LRU and closest first (CF), were analyzed. In LRU, the replacement decision depends on the access history only and not waiting times at all. In contrast, in CF, which ejects the item with the least waiting time for a page fault, the replacement decision does not use the access history but waiting times only. The online gray algorithm was proposed based on the intuition that a better performance can be achieved by combining these two algorithms. Furthermore, this scheme considers prefetching. It works as follows: Initially, all items in the cache are colored gray and all other items are white. When an item is requested, it is marked black. When all items in the cache are black, the algorithm changes the color of gray items to white and the color of black items to gray, and starts a new phase. At each step, the algorithm keeps in the cache all the black items, plus the set of gray items that are farthest away along the transmission schedule. The theoretical study showed that, in terms of worst-case performance, gray outperforms LRU by a factor proportionate to $CacheSize/\log(CacheSize)$.

Min-SAUD: An Optimal Cache Replacement Policy

As discussed, existing studies on cache replacement and prefetching for wireless data dissemination are based on simplifying assumptions, thereby making the proposed schemes impractical for a realistic wireless environment. For example, in the previous studies, data items were assumed to have the same size. Furthermore, data updates and client disconnections were not considered. In this section, we introduce an optimal cache replacement policy, called **Min-SAUD**, concerning a realistic environment. We argue that a cache replacement policy should take into consideration the cache validation delay caused by the underlying cache invalidation scheme. In addition, Min-SAUD considers access probability, update frequency, retrieval delay, and data size in developing the gain function which determines the cached item(s) to be replaced. **Stretch** (Acharya & Muthukrishnan, 1998) is considered as the major performance measure since it accounts for the data service time[2] and thus is fair when items have different sizes. It is defined as the ratio of the **access latency** of a request to its **service time**, where service time is defined as the ratio of the requested item's size to the broadcast bandwidth.

To facilitate the derivation of derivation, the following notations are defined (note that these parameters are for one client only):

- N: the number of data items in the database.
- C: the size of the client cache.
- $\overline{\lambda_i}$: mean access arrival rate of data item i, $i = 1, 2, ..., N$.
- $\overline{u_i}$: mean update arrival rate of data item i, $i = 1, 2, ..., N$.
- x_i: the ratio of update rate to access rate for data item i, that is, $x_i = \overline{u_i} / \overline{\lambda_i}$, $i = 1, 2, ..., N$.
- p_i: access probability of data item i, $p_i = \overline{\lambda_i} / \sum_{k=1}^{N} \overline{\lambda_k}$, for $i = 1, 2, ..., N$.
- l_i: access latency of data item i, $i = 1, 2, ..., N$.
- b_i: retrieval delay from the server (i.e., cache-miss penalty) for data item i, $i = 1, 2, ..., N$.
- s_i: size of data item i, $i = 1, 2, ..., N$.
- l_v: cache validation delay, that is, access latency of an effective IR.
- d_k: the data item requested in the kth access[3], $d_k \in \{1, 2, ..., N\}$.

- C_k: the set of cached data items after the kth access, $C_k \in \{1, 2, ..., N\}$.
- U_k: the set of cached data items that are updated between the kth access and the (k+1)th access, $U_k \subseteq C_k$.
- V_k: the set of victims chosen to be replaced in the kth access, $V_k \subseteq (C_{k-1} - U_{k-1})$.

The key issue for cache replacement is to determine a victim item set, V_k, when the free space in the client cache is not enough to accommodate the incoming data item in the kth access. In Xu, Hu, Lee, and Lee (2000), we have observed that a cache replacement policy should choose the data items with low access probability, short data retrieval delay, high update frequency, and large data size for replacement. As mentioned, a cache replacement policy should also take into account the cost of cache validation. Thus, in Min-SAUD, a gain function incorporating these factors is defined for each cached item i:

$$gain(i) = \frac{p_i}{s_i}\left(\frac{b_i}{1+x_i} - l_v\right). \qquad (1)$$

The idea is to maximize the total gain for the data items kept in the cache. Thus, to find space for the kth accessed data item, the Min-SAUD policy identifies the optimal victim item set V_k^*, $V_k^* \subseteq (C_{k-1} - U_{k-1})$, such that:

$$V_k^* = \arg \min_{V_k \subseteq (C_{k-1} - U_{k-1})} \sum_{i \in V_k} gain(i). \qquad (2)$$

It is easy to see that Min-SAUD reduces to *PIX* when Bdisk is used and when data items have equal size and are read-only, since under that circumstance the data retrieval delay of an item is inversely proportional to its broadcast frequency. Therefore, Min-SAUD can be considered a generalization of *PIX*.

The analytical study shows that Min-SAUD achieves the optimal stretch under the standard assumptions of the independent reference model and Poisson arrivals of data accesses and updates (Xu, Hu, Lee, & Lee, 2001, 2004). In the following, we address three critical implementation issues, namely, heap management, estimate of running parameters, and maintenance of cached item attributes, for the Min-SAUD policy.

Heap Management

In Min-SAUD, the optimization problem defined by Equations 1 and 2 is essentially the 0/1 knapsack problem, which is known to be NP hard. Thus, a well-known heuristic for the knapsack problem is adopted to find a suboptimal solution for Min-SAUD:

Throw out the cached data item with the minimum value until the free cache space is sufficient to accommodate the incoming item.

This heuristic can obtain the optimal solution when the data sizes are relatively small compared to the cache size (Shim, Scheuermann, & Vingralek, 1999). A (binary) min-heap data structure is used to implement the Min-SAUD policy. The key field for the heap is the $gain(i)/s_i$ value for each cached data item i. When cache replacement occurs, the root item of the heap is deleted. This operation is repeated until sufficient space is obtained for the incoming data item. Let N_c denote the number of cached items and M the victim set size. The time complexity for every cache replacement operation is $O(M\log N_c)$, and the time complexity for every adjustment operation is $O(M\log N_c)$.

Estimate of Running Parameters

Several parameters are involved in computation of the $gain(i)$ function. Among these parameters, p_i is proportional to $\overline{\lambda_i}$, s_i can be obtained when item i arrives, and l_v is a system parameter. In most cases, $\overline{u_i}$, b_i, and $\overline{\lambda_i}$ are not available to the clients. Thus, we need methods to estimate these values.

A well-known **exponential aging** method is used to estimate $\overline{u_i}$ and b_i. Initially, $\overline{u_i}$ and b_i are set to 0. When a new update on item i arrives, $\overline{u_i}$ is updated according to the following formula:

$$\overline{u_i}^{new} = \alpha_u /(t^c - t_i^{lu}) + (1-\alpha_u) \cdot \overline{u_i}^{old}, \tag{3}$$

where t^c is the current time, t_i^{lu} is the time stamp of the last update on item i, and α_u is a factor to weight the importance of the most recent update with those of the past updates. The larger the α_u value, the more important the recent updates. Similarly, when a query for item i is answered by the server, b_i is reevaluated as follows:

$$b_i^{new} = \alpha_s \cdot (t^c - t_i^{qt}) + (1-\alpha_s) \cdot b_i^{old}, \tag{4}$$

where t_i^{qt} is the query time and α_S is a weight factor for the running b_i estimate.

The parameters of $\overline{u_i}$ and b_i, estimated at the server side, are piggybacked to the clients when data item i is delivered; t_i^{lu} is also piggybacked so that the client can continue to update $\overline{u_i}$ based on the received IRs. The client caches the data item as well as its $\overline{u_i}$, t_i^{lu}, and b_i values.

Different clients may have different access patterns while some of their data accesses are answered by the cache. It is difficult for the server to know the real access pattern that originated from each client. Consequently, the access arrival rate $\overline{\lambda_i}$ is estimated at the client side. The exponential aging method might not be accurate because it does not age the access rate for the time period since the last access. Therefore, a **sliding average** method is employed in the implementation. We keep a sliding window of k most recent access time stamps ($t_i^1, t_i^2, ..., t_i^k$) for item i in the cache. The access rate $\overline{\lambda_i}$ is updated using the following formula:

$$\overline{\lambda_i} = \frac{k}{t^c - t_i^k}, \tag{5}$$

where t^C is the current time and t_i^k is the time stamp of the oldest access to item i in the sliding window. When fewer than k access TSs are available for item i, the mean access rate $\overline{\lambda_i}$ is estimated using the maximal number of available samples.

Maintenance of Cached Item Attributes

To realize the Min-SAUD policy, a number of parameters must be maintained for each cached data item. They are s_i, $\overline{u_i}$, t_i^{lu}, b_i, $\overline{\lambda_i}$, and t_i^k. We refer to these parameters as the **cached item attributes** (or simply **attributes**). To obtain these attributes efficiently, one may store the attributes for all data items in the client cache. Obviously, this strategy does not scale up to the database size. On the other extreme, one may retain the attributes only for the cached data items. However, this will cause the so-called "starvation" problem, that is, a newly cached data item i could be selected as the first few candidates for replacement since it has only incomplete information (it may incorrectly produce a relatively smaller *gain* value). If the cached item attributes are evicted from the cache together with the data item i, upon reaccessing item i, these attributes must be collected again from scratch. Consequently, item i is likely to be evicted again. We employ a heuristic to maintain the cached item attributes. The attributes for the currently cached data items are kept in the cache. Let N_C be the number of cached items. For those data items that are not cached, we only retain the attributes for N_C items with the largest $gain(j)/s_j$ values. Since the attributes themselves can be viewed as a kind of special data, as in the management for cached data, a separate heap is employed to manage the attributes for noncached data. This heuristic is adaptive to the cache size. When the cache size is large, it can accommodate more data items and hence, attributes for more noncached data can be retained in the cache. On the other hand, when the cache size is small, fewer data items are contained and thus fewer attributes are kept.

Performance Evaluation and Summary

Min-SAUD is optimal when the independent reference model and the Poisson processes of data accesses and updates are assumed (Xu et al., 2001). We have also conducted extensive simulation-based experiments to evaluate the performance of the Min-SAUD policy under different system settings (Xu, Hu, et al., 2004). The simulation removed the assumptions made in the analysis so that we can observe the impact of the assumptions during the analysis. The simulation simulated on-demand broadcasts and compares Min-SAUD to two cache replacement policies, that is, LRU and SAIU (Xu et al., 2000). The results show that Min-SAUD achieves the best performance under various system configurations. In particular, the performance improvement of Min-SAUD over the other schemes becomes prominent when the cache validation delay is significant. This indicates that cache validation cost plays an important role in cache replacement policies.

As a follow-up work, Yin, Cao, and Cai (2003) extended the Min-SAUD to a general-cost cache replacement policy, where the cost of a cached item can be configured as the query delay or download link traffic, and so forth. Nevertheless, when the cost is defined as the stretch, their replacement policy is virtually the same as Min-SAUD.

Location-Dependent Cache Invalidation: A Symbolic Model

Location-dependent information services (LDISs) are services that answer queries based on the locations with which the queries are associated: normally, the locations where the queries are issued. The most straightforward application is the map navigation tool for mobile users. Obviously, caching and prefetching the location-dependent data along the user's route can improve his or her navigation experience. To do so, semantic caching has been applied to describe a cached result together with the valid location(s) associated with the result (Dar, Franklin, Jonsson, Srivatava, & Tan, 1996; Lee et al., 1999; Zheng et al., 2002). When the client location changes, location-dependent data cached at a mobile client may become invalid with respect to the new location. The procedure of verifying the validity of location-dependent data with respect to the current location is referred to as *location-dependent cache invalidation*. To perform location-dependent invalidation efficiently, we have proposed a number of schemes (Xu, Tang, & Lee, 2003; Zheng, Xu, & Lee, 2002). The proposed schemes can be classified into two categories according to the underlying location model employed. In this section, we focus on the location-dependent cache invalidation issue for a symbolic location model. In the next section, we consider the issue for a geometric location model.

Preliminaries with Location-Dependent Data Caching

In LDISs, a data item can show different values for different locations, and the answer to a query depends on the location where the query originates. We distinguish data item value (or data instance) from data item, that is, an item value for a data item is an instance of the item valid for a certain geographical region. For example, **nearest restaurant** is an item, and the data values for this item vary when it is queried from different locations. This section assumes that the basic location granularity is a cell. The **valid scope** of an item value is defined as the set of cells within which the item value is valid. Since an item may have different values in different cells, an item is associated with a set of valid scopes, which is called the **scope distribution** of the item. To illustrate, let's consider a four-cell system. Suppose that the nearby restaurant for cell 1 and cell 2 is value A, and the nearby restaurant for cell 3 and cell 4 is value B. Then, the valid scope of A is {1, 2}, and the valid scope of B is {3, 4}. The scope distribution of the nearby restaurant item is {{1,2},{3,4}}.

To carry out location-dependent cache invalidation, since the server generally does not know which items are cached in each client and where each client resides, it is more suitable for clients to initiate the cache validity-checking procedure. The idea is to make use of validity information of data items. Specifically, the server delivers the valid scope along with a data item value to a mobile client and the client caches the data as well as its valid scope for later validity checking. The strategy involves two issues, namely, validity-checking time and validity information organization. Since a query result only depends on the location specified with the query, we suggest doing validity checking for a cached data value until it is queried. For validity information organization, in the following we present three methods.

Location-Dependent Invalidation Methodologies

We introduce three schemes for location-dependent invalidation under a symbolic location model, that is, *Bit Vector with Compression* (BVC), *Grouped BVC* (GBVC), and *Implicit Scope Information* (ISI).

Bit Vector with Compression

In the BVC method, the complete validity information is attached to a data item value. That is, the complete set of cells in which the data value is valid is kept in the cache. As every cell is associated with a CID to distinguish it from the others, BVC uses a bit vector (BV), corresponding to all the cells, to record valid scope. Obviously, the length of a BV is equal to the number of cells in the system. A 1 in the *n*th bit indicates that the data item value is valid in the *n*th cell while 0 means it is invalid in the *n*th cell. For example, if there are 12 cells in the system, then a BV with 12 bits is constructed for each cached data item value. If the BV for a data item value is 000000111000, it means this value is valid in the seventh, eighth, and ninth cells only. The validity-checking algorithm then works as follows: Whenever a data item value is required for location-dependent validation, the client listens to the broadcast for the current cell's ID, CID_C, and uses it to examine the cached BV of that item value. If the CID_C-th bit is 1 in the BV, it is valid, and otherwise, invalid.

It is obvious that with BVC the overhead would be significant when the system is large. With the locality of a valid scope, it is possible to perform compression on a BV in a real life application.

Grouped Bit Vector with Compression

To remedy the large overhead in BVC, the GBVC method keeps track of the validity information of each data item value only for some of the adjacent cells to reduce overhead. The motivation is twofold. First, a mobile client may seldom move to a cell that is far away from its home area. Consequently, it is enough for a client to know the validity of a data value in the current and neighboring cells. Second, even if a mobile client moves to a distant cell, it takes quite some time for it to do so. During this period of time, the data may have already been updated on the data server, and therefore the complete validity information of a data item value in distant cells is useless.

With the GBVC scheme, the whole geographical area is divided into disjoint districts and all the cells within a district form a group. A CID, denoted by (group-ID, intra-group-ID), consists of a group ID and a CID within the group. Validity information attached to a cached data value is represented as a vector of the form (group-ID, BV) and includes the current group-ID and a BV which corresponds to all the cells within the current group. Note that while delivering a data value to a client, only the BV is attached since the group-ID can be inferred from the current CID.

With the same example used in the previous section, suppose that the whole geographical area is divided into two groups, such that Cells 1 to 6 form Group 0 and the rest form Group 1. With the GBVC method, one bit is used to construct group-ID and a six-bit BV is used to record the cells in each group. For the data item value mentioned earlier, in Group 0, the attached bit vector is (0, 000000); in Group 1, the attached bit vector is (1, 111000). As can be seen, compared with the BVC method, the overhead for scope information is reduced in the GBVC method.

When a mobile client checks the validity of a data item value, it listens for the current CID, that is, ($group\text{-}ID_C$, $intra\text{-}group\text{-}ID_C$), and compares $group\text{-}ID_C$ with the one associated with the cached data. If they are not the same, the data is invalid. Otherwise, the client checks the $intra\text{-}group\text{-}ID_C$-th bit in the BV to determine whether the cached data is valid.

Note that the group size is crucial to the performance of this strategy. If the group is too large, the overhead for maintaining validity information is still very significant. On the other hand, if the group is too small, the chance of mistaking a valid data as invalid might be high since a data item value is regarded as invalid if it is outside its original group. A method to analyze the optimal group size for this scheme is discussed in Xu et al. (2003).

Implicit Scope Information

In BVC, a client stores in the cache the complete validity information of each cached data in the form of a bit vector. The disadvantage of this method is that the size of the validity information could be very large, especially when the system consists of a large number of cells. Consequently, a large bandwidth and cache memory are needed. The advantage is that the validation process is very simple; only the current CID is needed. GBVC attempts to reduce the size of the validity information by only keeping partial information in the cache.

The ISI method attempts the other direction by trying to minimize the size of validity information at the expense of the validation procedure. Under this scheme, the server enumerates the scope distributions of all items and numbers them sequentially. The valid scopes within a scope distribution are also numbered sequentially. For any value of data item i, its valid scope is specified by a 2-tuple (SDN_i, SN_i), where SDN_i is the scope distribution number and SN_i denotes the scope number within this distribution. The 2-tuple is attached to a data item value as its valid scope. For example, suppose there are three different scope distributions (Figure 4) and Data Item 4 has distribution 3. If Item 4 is cached from Cell 6 (i.e., CID = 6), then $SDN_4 = 3$ and $SN_4 = 3$. This implies that Item 4's value is valid in Cells 6 and 7 only.

It can be observed that the size of the validity information for an item value is small and independent of the actual number of cells in which the value is valid. Another observation is that a set of data items may share the same scope distribution. As such, the number of scope distributions could be much smaller than the number of items in the database.

At the server side, a **location-dependent IR** is periodically broadcast. It consists of the ordered valid scope numbers (SN) in a cell for each scope distribution. For example, in Cell 8, the server broadcasts {8, 3, 4} to mobile clients, where the three numbers are the SN values in Cell 8 for Scope Distributions 1, 2, and 3, respectively (Figure 4).

The validity-checking algorithm works as follows: After retrieving a location-dependent IR, for item i, the client compares the cached SN_i with the SDN_i-th SN in the location-dependent IR received. If they are the same, the cached item value is valid. Otherwise, the data item value is invalid. For example, in Cell 8, the client checks for the cached Data Item 4 whose $SDN_4 = 3$ and $SN_4 = 3$. In the broadcast report, the SDN_4-th (i.e., third) SN equals to 4. Therefore, the client knows that Data Item 4's value is invalid.

Performance Evaluation and Summary

We have analytically compared the query costs of the proposed schemes with an optimal strategy OPT, which assumes perfect location information is available on mobile clients, and a baseline strategy NSI, which drops the entire cache contents when handoff is performed (Xu et al., 2003). A series of simulation experiments was also conducted to validate the analysis. Both analytical and experimental results show that the proposed location-dependent invalidation schemes substantially outperform the NSI method in most cases (Xu et al., 2003). In particular, the ISI scheme is very close to the optimal strategy for a small to medium number of scope distributions. While BVC cannot scale up to system size and ISI does not perform well in a system with a large number of scope distributions and a low query rate, the GBVC method showed the best scalability. Moreover, we observed that when the data update rate is much higher than the client movement rate, sophisticated location-dependent cache invalidation is not necessary.

Location-Dependent Cache Invalidation: A Geometric Model

In the previous section, we addressed the issue of location-dependent cache invalidation for a cell-based symbolic location model. In this section, we discuss the same issue under a geometric location model, where a location is specified as coordinates. We assume mobile clients can identify their locations using systems such as the Global Positioning System (**GPS**). Recall the valid scope of an item value is defined as the region within which the item value is valid, and the set of valid scopes for all of the item values of a data item is called the scope distribution of the item. In a two-dimensional space, a valid scope v can be represented by a geometric polygon $p(e_1, ..., e_i, ..., e_n)$, where e_i is the endpoint of the polygon.

In the following, we first introduce two basic location-dependent invalidation schemes for geometric location models, namely, *polygonal endpoints* (PE) and *approximate circle* (AC), to represent valid scopes with different overheads and levels of precision. We then introduce a new performance criterion, called *caching efficiency*, and present a generic method, called *cache-efficiency-based* scheme (CEB), to balance the overhead and the precision of a representation scheme.

Location-Dependent Invalidation Strategies

Similar to that under a symbolic location model, the major issue of location-dependent cache invalidation in a geometric space is the organization of invalidation information for different data values. We assume that when a data value is delivered from the server

Figure 4. An example of data items with different distributions

<table>
<tr><td>CID</td><td>1</td><td>2</td><td>3</td><td>4</td><td>5</td><td>6</td><td>7</td><td>8</td><td>9</td><td>10</td><td>11</td><td>12</td></tr>
<tr><td>(SDN)Scope Distribution #1</td><td>1</td><td>2</td><td>3</td><td>4</td><td>5</td><td>6</td><td>7</td><td>8</td><td>9</td><td>10</td><td>11</td><td>12</td></tr>
<tr><td>(SDN)Scope Distribution #2</td><td colspan="3">1</td><td colspan="3">2</td><td colspan="3">3</td><td colspan="3">4</td></tr>
<tr><td>(SDN)Scope Distribution #3</td><td colspan="2">1</td><td colspan="3">2</td><td colspan="2">3</td><td colspan="3">4</td><td colspan="2">5</td></tr>
</table>

to the client, its complete valid scope is attached so that the client can check the data validity against its location of response time. However, different methods might be employed to represent a valid scope in the client cache. In the following, we first describe two basic schemes, **PE** and **AC**, followed by a generic method based on a proposed caching efficiency criterion.

Polygonal Endpoints Scheme

The PE scheme is a straightforward way to record the valid scope of a data value. It records all the endpoints of the polygon representing the valid scope. However, when the number of the endpoints is large, the endpoints will consume a large portion of the client's limited cache space, effectively reducing the amount of space for caching the data itself. This may worsen the overall performance.

Approximate Circle Scheme

The PE scheme contains complete knowledge of the valid scope of a data value. Its performance suffers when a polygon has a large number of endpoints. An alternative is to use an inscribed circle to approximate the polygon instead of recording the whole polygon (e.g., the shadowed area as illustrated in Figure 5[a]). In other words, a valid scope can be approximated by the center of the inscribed circle and the radius value. As can be seen, the overhead of this scheme can be minimized. For example, suppose eight bytes are used to record a point in a two-dimensional space; a polygon having seven points needs 56 bytes, while a circle needs 12 bytes only: eight for the center and four for the radius. However, the inscribed circle is only a conservative approximation of a valid scope. When the shape of the polygon is thin and long, the imprecision introduced by the **AC** method is significant. This will lead to a lower cache hit ratio since the cache will incorrectly treat valid data as invalid if the query location is outside the inscribed circle but within the polygon.

Cache-Efficiency-Based Method

As we saw in the previous subsections, both the PE and the AC schemes may perform poorly due to either a high overhead or imprecision of the invalidation information. In this subsection, we present a generic method for balancing the overhead and the precision of valid scopes.

We first introduce a new performance criterion: *caching efficiency*. Suppose that the valid scope of a data value is v, and v_i' is a subregion contained in v (Figure 5). Let s be the data size, $A(v_i')$ the area of any scope of v_i', and $O(v_i')$ the storage overhead needed to record the scope v_i'. The caching efficiency of the data value with respect to a scope v_i' is defined as follows:

$$E(v_i') = \frac{A(v_i')/A(v)}{(s+O(v_i')/s} = \frac{A(v_i')s}{A(v)(s+O(v_i'))} \,. \tag{6}$$

If we assume that the cache size is infinite and the probabilities of a client issuing queries at different locations are uniform, $A(v_i')/A(v)$ is a data value's cache hit ratio when the client issues the query within the valid scope v, and v_i' is the approximated scope information stored in the client cache. In contrast, $(s+O(v_i'))/s$ is the cost ratio for achieving such a hit ratio. The rationale behind this definition is as follows: When none of the invalidation information is cached, $E(v_i')$ is 0 since the cached data is completely useless; $E(v_i')$ increases with more invalidation information attached. However, if too much overhead is therefore introduced, $E(v_i')$ would decrease again. Thus, a generic method for balancing the overhead and the precision of invalidation information works as follows:

For a data item value with a valid scope of v, given a candidate valid scope set $V' = \{v_1', v_2', ..., v_k'\}$, $v_i' \subseteq v$, $1 \le i \le k$, we choose the scope v_i' that maximizes caching efficiency $E(v_i')$ as the valid scope to be attached to the data.

Figure 5 illustrates an example where the valid scope of the data value is $v = p(e_1, e_2, ..., e_7)$, v_1', v_2', **and** v_3' are three different subregions of v, $A(v_1') = A(v) = 0.788$, $A(v_2') = A(v) = 0.970$, and $A(v_3') = A(v) = 0.910$. Assuming that the data size s is 128 bytes, eight bytes are needed to represent an endpoint and four bytes for the radius of an inscribed circle, we have $O(v) = 56$, $O(v_1') = 12$, $O(v_2') = 48$, and $O(v_3') = 40$. Thus, $E(v) = 0.696$, $E(v_1') = 0.721$, $E(v_2') = 0.706$ **and** $E(v_3') = 0.694$. As a result, we choose v_1' as the valid scope to be attached to the data.

In the CEB method, a practical issue is the generation of the candidate valid scope set V'. There are various ways to do this. As a case study, in this book we discuss the following method that can be handled by a general program[4]. Basically, we consider contained circles and subpolygons as candidate valid scopes. For circles, due to the lack

Figure 5. An example of possible candidate valid scopes ($v = p(e_1, e_2, ..., e_7)$)

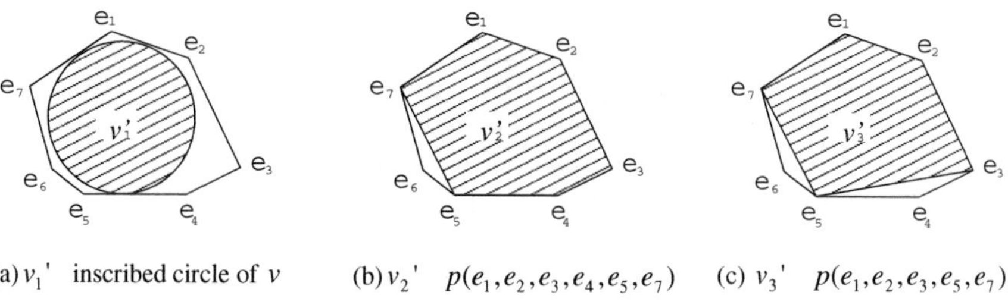

(a) v_1' inscribed circle of v (b) v_2' $p(e_1,e_2,e_3,e_4,e_5,e_7)$ (c) v_3' $p(e_1,e_2,e_3,e_5,e_7)$

of available geometric algorithms, we only consider the first-degree circle, that is, the inscribed circle of a polygon, which can be obtained using the medial axis approach (O'Rourke, 1994). For subpolygons, we generate a series of candidate polygons in a greedy manner. Suppose the current candidate polygon is v_i'. We consider all polygons resulting from the deletion of one endpoint from v_i', and choose as the next candidate v_{i+1}', the polygon that is bounded by v and has the maximal area. The pseudo algorithm is described in Zheng et al. (2002), where the generation of candidate valid scopes and the selection of the best valid scope are integrated.

Performance Evaluation and Summary

Simulation-based experiments have been conducted to evaluate the performance of the presented invalidation schemes (Xu, 2002; Zheng et al., 2002). The results showed that the CEB method, in a variety of system settings, obtains a better cache performance than PE and AC. This leads us to conclude that caching efficiency is an effective selection criterion for location-dependent invalidation information under a geometric location model. CEB is believed to achieve even better performance if better methods can be developed to generate candidate valid scopes.

Cache Replacement for Location-Dependent Data

We discussed cache replacement policies for general data access in a previous section, where access probability, among other factors, is considered the most important factor that affects cache performance. However, in location-dependent services, besides access probability, and so forth, there are two more unique factors, namely, data distance and valid scope area, which should be considered in cache replacement. In the following two paragraphs, we analyze respectively their impacts on cache performance.

- **Data Distance:** Data distance refers to the distance between the current location of a mobile client and the valid scope of a data value. In a location-dependent data service, the server responds to a query with the suitable value of the data item according to the client's current location. As such, when the valid scope of a data value is far away from the client's current location, this data will have a lower chance to become usable again since it will take some time before the client enters the valid scope area again; the data is useless before the user reaches the valid scope area. In this respect, we should favor ejecting the "farthest" data when replacement takes place. However, this reasoning is invalid in the following two cases. First, if the client continues to move away from a location, this location would have a smaller chance of being revisited even though the client's current location is very close to it. Thus, a directional data distance would make more sense. Second, with random movement patterns, the time it takes the client to traverse a distance is not always directly proportional to the distance. In summary, data distance, either directional or undirectional, may or may not affect cache performance, depending on the mobile client's movement and query patterns.

- **Valid Scope Area:** Valid scope area refers to the geometric area of the valid scope of a data value. For location-dependent data, valid scope areas can somehow reflect the access probabilities for different data values. That is, the larger the valid scope area of the data, the higher the probability that the client requests this data. This is because, generally, the client has a higher chance of being in large regions than small regions. Thus, we argue that a good cache replacement policy should also take this factor into consideration.

There exist some studies on location-dependent data caching. Data distance-based cache replacement policies, Manhattan distance and FAR, have been proposed (Dar et al., 1996; Ren & Dunham, 2000). In these two policies, the data that is farthest away from the client's current location is removed during replacement. Personè and Grassi (2003) proposed to cache and prefetch those zones that are adjacent to the current zone along the user's route. They claimed that when the user's route is highly predictable and stable, caching and prefetching significantly reduced the access latency and energy consumption. However, in all these studies, data distance was considered alone and not integrated with other factors such as access probability. Moreover, there are no existing studies that have considered the valid scope area in cache replacement policies.

PA and PAID Policies

Based on the analysis in the last section, a promising cache replacement policy should choose as its victim the data item with a low access probability, a small valid scope area, and a long distance if data distance is also an influential factor. Therefore, in the following, we present two cost-based cache replacement policies, probability area (PA) and probability area inverse distance (PAID), which integrate the factors that are supposed to affect cache performance. Our discussions shall be based on a geometric location model.

- **PA:** As the name suggests, for this policy, the cost of a data value is defined as the product of the access probability of the data item and the area of the attached valid scope. That is, the cost function for data value j of item i is as follows:

$$c_{i,j} = p_i \cdot A(v'_{i,j}) \,, \qquad (7)$$

  where p_i is the access probability of item i and $A(v'_{i,j})$ is the area of the attached valid scope $v'_{i,j}$ for data value j. The PA policy chooses the data with the least cost as its victim for cache replacement.

- **PAID:** Compared with PA, this scheme further integrates the data distance factor. For the PAID policy, the cost function for data value j of item i is defined as follows:

$$c_{i,j} = \frac{p_i \cdot A(v'_{i,j})}{D(v'_{i,j})} , \tag{8}$$

where p_i and $A(v'_{i,j})$ are defined in the same way as above, and $D(v'_{i,j})$ is the distance between the current location and the valid scope $v'_{i,j}$. Similar to PA, PAID ejects the data with the least cost during each replacement. Depending on different methods of calculating $D(v'_{i,j})$, we have two variations of PAID, that is, PAID-U and PAID-D. In PAID-U, the data distance is undirectional and is calculated regardless of the current direction of movement of the client. In PAID-D, the calculation of the data distance considers the client's current direction of movement: If the client is currently moving away from the valid scope, the distance is multiplied by a very large number δ (i.e., the longest distance in the system); otherwise, it is calculated normally as in PAID-U. This way, PAID-D favors keeping the data in the direction of movement of the mobile client.

Implementation Issues

In these two policies, there are three factors in total (i.e., access probability, valid scope area, and data distance) involved in computing the cost values for cached data. Among these three factors, the valid scope area of a data value can be simply obtained based on the attached valid scope. In order to estimate the access probability for each data item, the well-known exponential aging method is employed. Two parameters are maintained for each data item *i*: a running probability (p_i) and the time of the last access to the item (t_i^l). Initially, p_i is set to 0. When a new query is issued for data item *i*, p_i is updated using the following formula:

$$p_i^{\text{new}} = \alpha / (t^c - t_i^{\,l}) + (1-\alpha) p_i^{\text{old}}, \tag{9}$$

where t^c is the current system time and α is a constant factor to weigh the importance of the most recent access in the probability estimate.

Note that the access probability is maintained for each data item rather than for each data value. If the database size is small, the client can maintain the prob parameters (i.e., p_i and t_i^l for each item *i*) for all items in its local cache. However, if the database size is large, the prob information will occupy a significant amount of cache space. To alleviate this problem, we set an upper bound to the amount of cache used for storing the prob information (e.g., 5% of the total cache size), and use the LFU policy to manage the limited space reserved for the prob information.

For the PAID policy, we also need to compute the data distances between the current location and different valid scopes. Since a valid scope is normally a region (either a polygon or a circle in this chapter) rather than a single point, we introduce a *reference point* for each valid scope and take the distance between the current location and the reference point as the data distance. For a polygonal scope, the reference point is defined as the endpoint that is closest to the current location. For a circular scope, the reference point is defined as the point where the circumference and the line connecting the current location and the center of the circle meet.

Performance Evaluation and Summary

We have evaluated and compared the PA and PAID policies with the existing policies LRU, P_I, P_V, and FAR (Xu, 2002; Zheng et al., 2002). The results are summarized as follows. The PA and PAID policies demonstrated a substantial performance improvement over the existing policies. In particular, consideration of the valid scope area improves the performance significantly in all settings. However, the factor of data distance is sensitive to scope distributions, query patterns, and movement models.

Conclusion and Future Trend

Chapter Summary

This chapter discussed client-side caching techniques for a mobile computing environment. We considered the cache replacement issue in a realistic wireless data-dissemination environment, where there exist variable data sizes, data updates, and client disconnections. An optimal gain-based cache replacement policy, Min-SAUD, which incorporates various factors, namely, data item size, retrieval delay, access probability, update frequency, and cache validation delay, was introduced.

We also introduced the issue of location-dependent cache invalidation for LDISs. We first addressed this issue for a cell-based symbolic location model. Three schemes, namely, BVC, GBVC, and ISI, were presented to handle location-dependent invalidation. Next, we considered the location-dependent cache invalidation issue for a geometric location model. We introduced a new performance criterion, caching efficiency, and a generic method, CEB. The CEB method, based on the caching efficiency criterion, attempts to balance the overhead and the precision of the invalidation information when an approximation of a valid scope has to be decided.

Finally, we discussed the cache replacement issue for location-dependent data. We described two cache replacement policies, namely, PA and PAID, that consider the factors of valid scope area (for both methods) and data distance (for PAID only), and combine these factors with access probability.

In Table 1, we summarize the issues and factors addressed by each of the presented caching strategies. The Min-SAUD replacement policy considers various factors such as the access pattern, but it is confined to a single-cell environment and non-location-

Table 1. A summary of the presented caching strategies

	MIN-SAUD	*BVC,GBVC,ISI*	*CEB*	*PA,PAID*
Addressed Issue	Replacement	Invalidation	Invalidation	Replacement
Access Pattern	Addressed	Not Addressed	Not Addressed	Addressed
Update Pattern	Addressed	Not Addressed	Not Addressed	Not Addressed
Data Size	Addressed	Not Addressed	Addressed	Not Addressed
Location-Dependent Data	Not Addressed	Addressed	Addressed	Addressed
Location Model	N/A	Symbolic	Geometric	Geometric

dependent data. In contrast, the PA and PAID replacement policies focus on location-dependent data and incorporate the factor of access pattern with new factors unique to location-dependent data (i.e., data distance and valid scope area). Other factors such as the update pattern can be integrated into the PA and PAID policies in a way similar to the access pattern. In all the location-dependent invalidation policies so far, we have not considered the access pattern or the update pattern. Indeed, these two factors will affect cache performance. Obviously, it is not beneficial to attach invalidation information to data that are accessed infrequently and/or updated frequently.

Future Research Trend

- **Power-aware cache management:** Power constraint is one of the few notorious features in a mobile computing environment. Obviously, to cater for such a power-constrained environment, a promising cache management scheme should be power aware (Yin, Cao, Das, & Ashraf, 2002). However, this issue is not addressed adequately in the literature. We believe that power-aware cache management deserves further in-depth study.
- **Mobile LDISs:** Mobile LDISs have recently been gaining increasing attention from both industry and academia. While many traditional data management techniques are applicable to the implementation of an LDIS, they need to be reexamined and redevised in order to address the issues arising from the following three challenges: the constraints of mobile computing environments, the spatial property of location-dependent data, and the mobility of mobile users (Lee, Lee, Xu, & Zheng, 2002).
- **Wireless Internet:** Internet information services will eventually be mobilized and personalized so that users can have access to information without geographical and time limits. When mobile access meets Internet multimedia data, one of the critical issues is the data management of proxies bridging the Internet and wireless networks. Proxy cache management involving transcoding, interproxy cooperation, and request routing are interesting research topics to pursue.

Acknowledgments

The writing of this chapter is supported by a grant from Hong Kong Baptist University (Grant FRG/03-04/II-19). The authors would like to thank professor Dik Lun Lee at the Hong Kong University of Science and Technology, professor Wang-Chien Lee at Penn State University, and Dr. Qinglong Hu at IBM Silicon Valley Lab for valuable discussions on mobile cache management, and for their contributions in developing the algorithms presented in this chapter.

References

Acharya, S., & Muthukrishnan, S. (1998). Scheduling on-demand broadcasts: New metrics and algorithms. *Proceedings of the Fourth Annual ACM/IEEE MobiCom Conference*, (pp. 43-54).

Acharya, S., Alonso, R., Franklin, M., & Zdonik, S. (1995). Broadcast disks: Data management for asymmetric communications environments. *Proceedings of 1995 ACM SIGMOD Conference on Management of Data*, (pp. 199-210).

Acharya, S., Franklin, M., & Zdonik, S. (1996). Prefetching from a broadcast disk. *Proceedings of the 12th International Conference on Data Engineering*, (pp. 276-285).

Barbara, D. (1999). Mobile computing and databases: A survey. *IEEE Transactions on Knowledge and Data Engineering*, *11*(1), 108-117.

Barbara, D., & Imielinski, T. (1994). Sleepers and workaholics: Caching strategies for mobile environments. *Proceedings of 1994 ACM SIGMOD Conference on Management of Data*, (pp. 1-12).

Bukhres, O. A., & Jing, J. (1996). Performance analysis of adaptive caching algorithms in mobile environments. *Information Sciences*, *95*(1-2), 1-27.

Cao, G. (2000). A scalable low-latency cache invalidation strategy for mobile environments. *Proceedings of the Sixth Annual ACM/IEEE International Conference on Mobile Computing and Networking*, (pp. 200-209).

Dar, S., Franklin, M. J., Jonsson, B. T., Srivatava, D., & Tan, M. (1996). Semantic data caching and replacement. *Proceedings of the 22nd International Conference on Very Large Data Bases*, (pp. 330-341).

Imielinski, T., & Badrinath, B. R. (1994). Mobile wireless computing: Challenges in data management. *Communications of the ACM*, *10*(37), 18-28.

Khanna, S., & Liberatore, V. (2000). On broadcast disk paging. *SIAM Journal on Computing*, *29*(5), 1683-1702.

Lee, D. L., Lee, W.-C., Xu, J., & Zheng, B. (2002). Data management in location-dependent information services. *IEEE Pervasive Computing*, *1*(3), 65-72.

Lee, K. C. K, Leong, H. V., & Si, A. (1999). Semantic query caching in a mobile environment. *Mobile Computing and Communication Review*, *3*(2), 28-36.

O'Rourke, J. (1994). *Computational geometry in C*. Cambridge, UK: The Press of the University of Cambridge.

Personè, V., & Grassi, V. (2003). Performance analysis of caching and prefetching strategies for palmtop-based navigational tools. *IEEE Transactions on Intelligent Transportation Systems*, *4*(1), 23-34.

Ren, Q., & Dunham, M. H. (2000). Using Semantic caching to manage location dependent data in mobile computing. *Proceedings of the Sixth Annual ACM/IEEE International Conference on Mobile Computing and Networking*, 210-221.

Shim, J., Scheuermann, P., & Vingralek, R. (1999). Proxy cache design: Algorithms, implementation and performance. *IEEE Transactions on Knowledge and Data Engineering*, *11*(4), 549-562.

Song, H., & Cao, G. (2004). Cache-miss-initiated prefetch in mobile environments. *Proceedings of IEEE International Conference on Mobile Data Management (MDM),* (pp. 370-381).

Tan, K. L., & Cai, J. (1997). Broadcast-based group invalidation: An energy-efficient cache invalidation strategy. *Information Sciences*, *100*(1-4), 229-254.

Tan, K. L., Cai, J., & Ooi, B. C. (2001). An evaluation of cache invalidation strategies in wireless environments. *IEEE Transactions on Parallel and Distributed Systems (TPDS)*, *12*(8), 789-807.

Tassiulas, L., & Su, C. J. (1997). Optimal memory management strategies for a mobile user in a broadcast data delivery system. *IEEE Journal on Selected Areas in Communications*, *15*(7), 1226-1238.

Wu, K.-L., Yu, P. S., & Chen, M.-S. (1996). Energy-efficient caching for wireless mobile computing. *Proceedings of the 12th International Conference on Data Engineering,* (pp. 336-343).

Xu, J. (2002). *Client-side data caching in mobile computing environments*. PhD thesis, Hong Kong University of Science and Technology, Hong Kong, China.

Xu, J., Hu, Q., Lee, D. L., & Lee, W.-C. (2000). SAIU: An efficient cache replacement policy for wireless on-demand broadcasts. *Proceedings of the Ninth ACM International Conference on Information and Knowledge Management*, (pp. 46-53).

Xu, J., Hu, Q., Lee, W.-C., & Lee, D. L. (2001). An optimal cache replacement policy for wireless data dissemination under cache consistency. *Proceedings of the 30th ICPP Conference*, (pp. 267-274).

Xu, J., Hu, Q., Lee, W.-C., & Lee, D. L. (2004). Performance evaluation of an optimal cache replacement policy for wireless data dissemination. *IEEE Transactions on Knowledge and Data Engineering*, *16*(1), 125-139.

Xu, J., Liu, J., Li, B., and Jia, X. (2004, July/August). Caching and prefetching for web content distribution. *IEEE Computing in Science and Engineering (CiSE)* [Special issue], 6(4), 54-59.

Xu, J., Tang, X., & Lee, D. L. (2003). Performance analysis of location-dependent cache invalidation schemes for mobile environments. *IEEE Transactions on Knowledge and Data Engineering*, *15*(2), 474-488.

Yin, L., Cao, G., & Cai, Y. (2003). A generalized target-driven cache replacement policy for mobile environments. *Proceedings of the IEEE Symposium on Applications and the Internet (SAINT),* 14-21.

Yin, L., Cao, G., Das, C., & Ashraf, A. (2002). Power-aware prefetch in mobile environments. *Proceedings of IEEE International Conference on Distributed Computing Systems,* (pp. 571-578).

Zheng, B., Xu, J., & Lee, D. L. (2002). Cache invalidation and replacement policies for location-dependent data in mobile environments. *IEEE Transactions on Computers*, *51*(10), 1141-1153.

Endnotes

[1] In this case, TS simply drops the entire cache.

[2] The service time of a request is the time to serve the request when it is the only job in the system.

[3] It is assumed that the client accesses are numbered sequentially.

[4] In a real-life application, it is possible to use other ad hoc methods to generate candidates while using the proposed CEB method to guide the selection of the best valid scope.

Chapter III

Indexing and Clustering of Wireless Broadcast Data

Yon Dohn Chung, Dongguk University, Korea

Myoung Ho Kim, Korea Advanced Institute of Science and Technology, Korea

Abstract

This chapter describes some data management issues that are necessary for wireless data broadcasting. The major topics we include in this chapter are (a) broadcast data indexing and (b) broadcast data clustering. Mobile clients can access the wireless data in an energy-efficient way with the index on the broadcast channel, and the well-clustered broadcast data enables mobile clients to access the wireless data in a short latency.

Introduction

With the recent proliferation of mobile communication technology and portable computing devices, clients can move around without interruption of their information computing (Badrinath & Imielinski, 1994; Imielinski, Viswanathan, & Badrinath, 1994). We call this computing environment the mobile computing environment (in short, the mobile environment). Some major characteristics of mobile environments are as follows:

(1) The clients can move around.

(2) The bandwidth for wireless communication is physically limited.

(3) The usage of energy is restricted because clients use battery-powered portable computers.

(4) The wireless communication is less reliable than the wireline one.

(5) Mobile clients are frequently disconnected from the server in voluntary and involuntary ways.

(6) Portable devices are vulnerable to physical damage, for example, crash, theft, electromagnetic interference, and so on.

Owing to the characteristics mentioned above, the data-broadcasting approach is widely used for various applications in mobile computing environments (Barbara, 1999; Imielinski et al., 1994). In contrast to peer-to-peer communication, the broadcasting approach enables clients just to receive the data sent from the server without sending requests to the server. Some useful characteristics of data broadcasting in mobile computing environments are as follows:

- **Energy efficiency:** In wireless communications, the amount of energy consumption for sending data is much bigger than that of data receiving. For example, in the case of the *Hobbit* chip (Argade et al., 1993), the former is about one thousand times bigger than the latter. In this respect, the data-broadcasting approach is very energy-efficient because the clients receive data via broadcast channel without sending data (i.e., requests) to the server.
- **Bandwidth efficiency:** Different from wireline networks, the bandwidth of wireless networks is physically limited. Thus, we have to focus on the efficient utilization of communication bandwidth. In the data-broadcasting approach, many mobile clients share a single channel (i.e., broadcasting channel). Therefore, data broadcasting is said to be bandwidth efficient.
- **Scalability:** In peer-to-peer communication, we have to establish a channel (or more channels) for each client. However, the broadcasting approach uses only one single channel irrespective of the number of mobile clients. So, when using the broadcasting approach, we are free to add mobile clients to a server without additional channel allocation.

In this chapter, we consider data management issues that are necessary for wireless data broadcasting. The major topics we include in this chapter are (a) *broadcast data indexing* and (b) *broadcast data clustering.*

The rest of the chapter is organized as follows. The section entitled "Background" describes the background and some measures for wireless data broadcasting. In the section entitled "Broadcast Data Indexing," we introduce the indexing problem of wireless broadcast data and explain some indexing methods. In "Broadcast Data Clustering," we introduce the clustering issues on wireless broadcasting. We describe the clustering problem in some different views: clustering in uniform and nonuniform broadcasting environments, clustering multipoint queries, and clustering for partial-match queries. In the last section, we present a summary of the chapter and future research issues.

Background

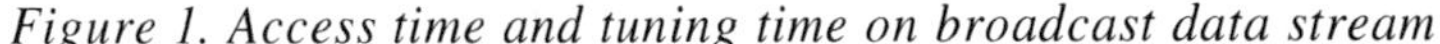

In data broadcasting, the server sends a data stream to many unspecific clients via a public channel, called broadcasting channel. The (mobile) clients tune to the broadcast channel and selectively receive data of their interests. Since the broadcasting channel is used by many clients, the server tries to construct the broadcast channel with data objects accessed by more clients. The broadcast data stream is usually repeated for some period (Imielinski et al., 1994). We call one cycle of broadcast data stream a *bcast*. The indexing and clustering methods that will be described in the following sections are based on the bcast unit.

In data broadcasting, the clients need the information for locating data objects, that is, temporal addresses for data objects on the broadcast data stream. Without this information, a mobile client has to read the entire broadcast data stream, most of which is not interesting to the client. Figure 1 shows a mobile client tuned to a broadcast channel and accessing two data objects: d1 and d3. The client tunes to the broadcast channel and reads the index information (denoted by 'a' in the figure). Then the client remains in a doze mode (i.e., energy-saving mode) until d1 arrives, reads d1 then sleeps until d3 arrives, then reads d3. In the figure, we call the duration (denoted by 'd' in the figure) from the start to the end of data access the *access time* (or *latency*), and the sum of periods (that is, a + b + c in the figure) for which the actual *tune-in* is required the *tuning time*. The

Figure 1. Access time and tuning time on broadcast data stream

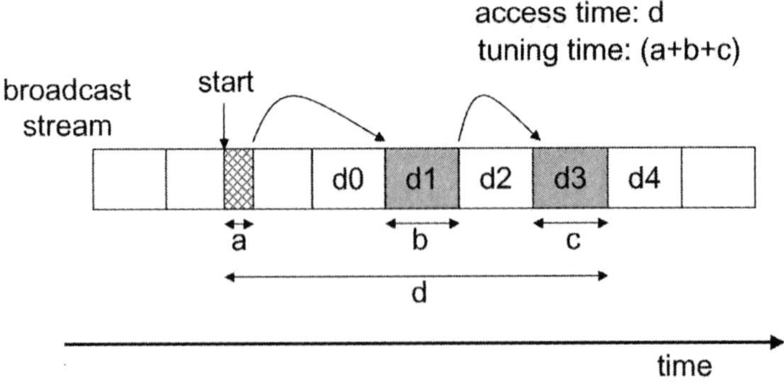

tuning time will determine the power consumed by the client to retrieve the required data. (Note that the Hobbit chip from ATT consumes 250mW in the full operation mode *active* mode and 50uW in the *doze* mode.)

Before explaining indexing and clustering of broadcast data, we describe some notations for wireless data broadcasting. The smallest logical unit of the broadcast is called a *bucket*. All buckets are of the same size, which is only for convenience and uniformity. The size of a bucket will be equal to some multiple of the *packet* size (the basic unit of message transfer in packet-switched networks). Both access time and tuning time will be measured in terms of the number of buckets.

Broadcast Data Indexing

In the broadcasting mode, mobile clients receive data from the server through the broadcasting channel. (In this chapter, we consider a single broadcasting channel since multiple channels are logically equivalent to a single high-capacity channel.) Because the clients do not send a request to the server, they are not informed of the delivery time (i.e., temporal address) of data on the broadcasting channel. The index on the broadcasting channel is the directory information of data broadcast by the server. Through the index information, mobile clients are able to determine the temporal addresses of data of their interests and probe the data in an energy-efficient way. Figure 2 shows a typical example of data probing on a broadcasting channel using index information compared

Figure 2. Broadcast data probing with and without index

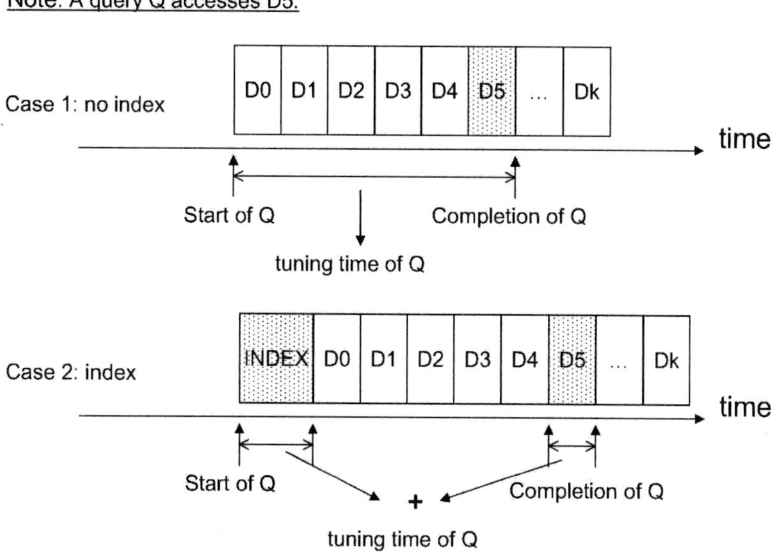

with the case of no index. In Case 2, after reading index information, the mobile client consumes little energy in the doze mode until the target data arrives.

There are some index methods for wireless broadcasting: index tree methods for single-attribute queries (Imielinski et al., 1994, 1997) and signature methods for multiple-attribute queries (Hu, Lee, & Lee, 2000; Lee & Lee, 1996). In this section, we mainly investigate tree-structured indexing methods. For signature-based methods, refer to Hu et al. (2000) and Lee and Lee (1996).

The index information will increase access time although it reduces the tuning. For benchmarks for comparison, we present two methods which are optimal in the one-dimensional space of access time and tuning time. We also illustrate them in Figure 3.

- **Optimal Access Time Method:** The best access time is obtained when no index is broadcast with data. The broadcast data stream is minimal in this way. However, the tuning time becomes the worst. In this method, both the average access time and the average tuning time are *Data*/2, where *Data* is the total size of broadcast data objects. Here, *Data* is equal to the size of a bcast.
- **Optimal Tuning Time Method:** The server broadcasts the index at the beginning of each bcast. The client that needs the data object with primary key *K* tunes into the broadcast channel and moves to the beginning of the next bcast (that is, waits in the doze mode until the next bcast broadcasts) to get the index. It then follows the index pointers to the data object with the primary key *K*. This method provides the worst access time because the client has to wait until the beginning of the next bcast even if the required data is just in front of it. However, its tuning time is minimal. The average access time is *Data* + *Index*, and the average tuning time *Index* + *1* + α, where *Index* is the size of index information (denoted by 'B' in Figure 3), α is the size of target data in buckets (denoted by 'C' in Figure 3), and 1 is the size

Figure 3. Two optimal methods

Note: A query Q accesses D5.

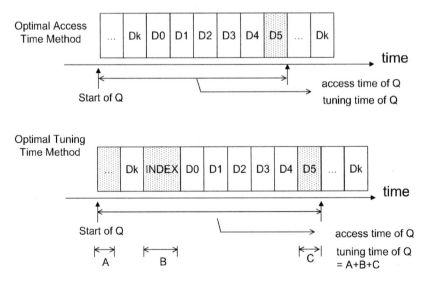

Optimal Access Time Method
... Dk D0 D1 D2 D3 D4 D5 ... Dk
time
access time of Q
Start of Q
tuning time of Q

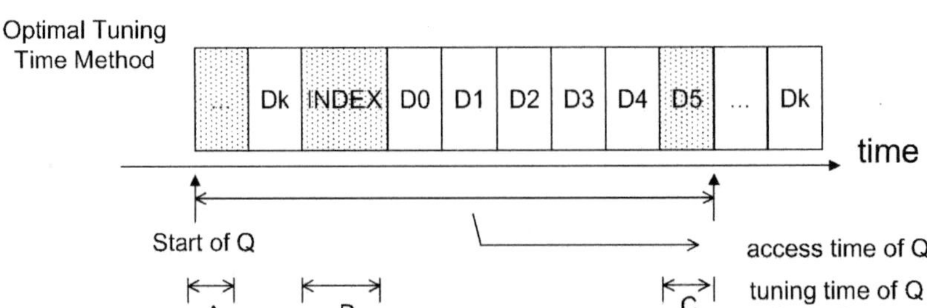

of one bucket where the initial probe (denoted by 'A' in Figure 3) is occurred. We assume every bucket contains the pointer (i.e., temporal address) to the next bcast.

If the tuning time is of no concern, then one can use the optimal access time method for achieving the least access time. If the access time is of no concern, the optimal tuning time method can be used for the least tuning time. However, in most cases, both the tuning time and the access time are of interest, so more sophisticated methods are needed. In the next section we introduce some indexing methods.

Tree-Structured Index Methods

The index on the broadcast data stream is the directory information of data on the air, that is, the temporal addresses of broadcast data objects. Therefore, main issues on the broadcast data indexing are (a) how to construct the index and (b) where to place the index onto the broadcast data stream. With respect to these two aspects, there have been some approaches that construct the index in a tree shape and distribute the nodes of the tree onto the broadcast data stream intermixed with data. Among tree-structured methods, we introduce the *(1, m)* indexing method and *distributed indexing* method in this section (Imielinski et al., 1994, 1997).

(1, m) Indexing

We use the terms *index buckets*, holding the index, and *data buckets*, holding the data. An *index segment* refers to the set of contiguous index buckets and a *data segment* refers to the set of contiguous data buckets.

Figure 4. (1, m) indexing

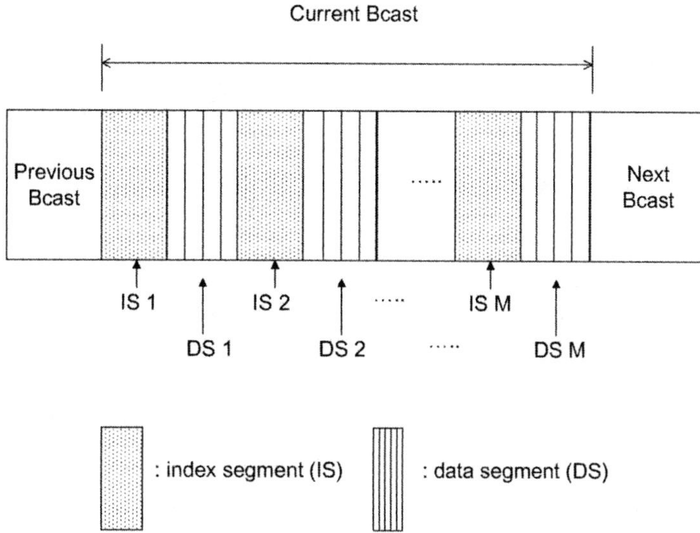

$(1, m)$ indexing is an index allocation method where the whole index is broadcast m times during one bcast. Figure 4 shows this indexing method. Since all buckets have a pointer to the beginning of the next bcast and the first bucket of each index segment contains the information on the data objects that were broadcast last, the clients that have missed the required data in the current bcast can tune into the next bcast. The data access protocol for data object with key K is as follows:

(1) Tune into the current bucket on the broadcast channel.

(2) Read the current bucket and move to the next nearest index segment.

(3) From the index segment, determine the time (i.e., address) when the target data object is broadcast. This process may be done through successive index probes by following index pointers. The client might go into doze mode between two successive index probes.

(4) Tune in again at the time of target data arrival and download the data object.

When assuming the probability distribution of the initial probe of the clients to be uniform within the bcast, the average access time and the average tuning time are analyzed as follows (Imielinski et al., 1997) (Here, n is the capacity of the bucket, i.e., the number of (search key, pointer) pairs an index bucket can contain.):

- Access Time = $0.5 \times ((M + 1) \times Index + (1/M + 1) \times Data)$
- Tuning Time = $2 + \lceil \log_n(Data) \rceil$

For finding the minimal access time, we differentiate the former formula (for the access time) with respect to m. Then, the optimum value of m^* can be computed as follows:

Figure 5. An index tree

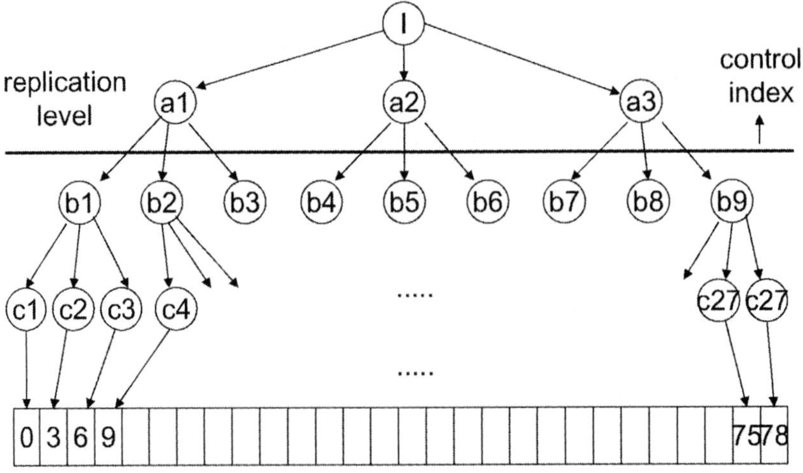

$$m^* = \sqrt{\frac{Data}{Index}}.$$

Therefore, we construct broadcast data stream by dividing the data set to be broadcast into m^* equal parts (i.e., m^* data segments), each of which being proceeded by the index.

Distributed Indexing

Distributed indexing is a technique in which the index is partially replicated. This method is based on the observation that there is no need to replicate the entire index between successive data segments. That is, it is sufficient to have only the portion of index that indexes the data segment which follows it (i.e., relevant index).

For describing the relevant index, we use a tree-structured index approach shown in Figure 5. The figure shows a set of data objects for broadcasting, which consists of 81 data buckets. Each rectangular box represents a collection of three data buckets. Owing to the space limitation, we depict the number of the first data bucket among three data buckets in the box (for example, the square box numbered 0 denotes it contains data buckets 0, 1, and 2). The index tree is shown above the data buckets. Each index bucket contains three index pointers, that is, three (key value: pointer) pairs. In the lowest level, we describe only one arrow for simplicity.

We consider three different index distribution (or replication) approaches. In all three approaches, the index is interleaved with data and the index segment describes only data in the data segment which immediately follows it. Three approaches are classified

Figure 6. Nonreplicated distribution

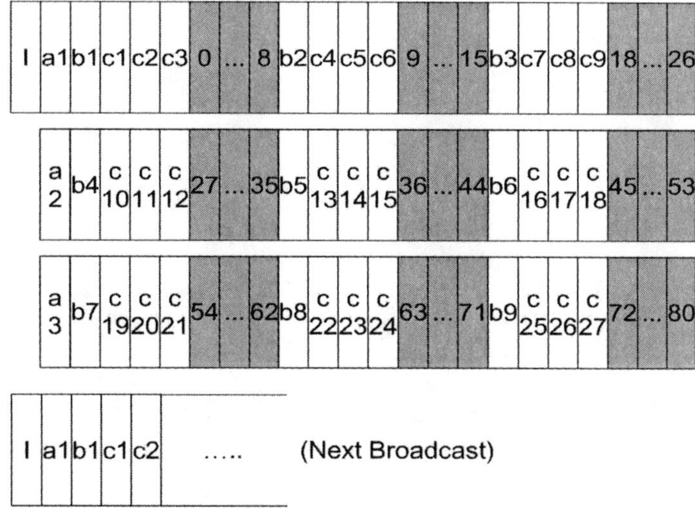

Figure 7. Entire path replication

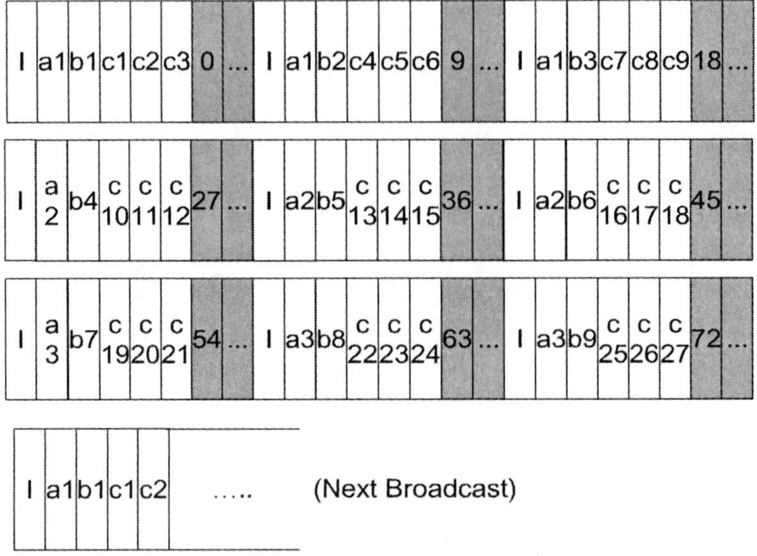

according to the degree of the replication of index information with the last one being called distributed indexing.

(1) **Nonreplicated distribution:** There is no replication, so all different index segments are disjoint. Each data segment is preceded by its relevant index, that is, the path information to the data segment in a depth-first traversal way. This is shown in Figure 6.

(2) **Entire path replication:** Before each data segment, the path from the root of the index tree to the lowest index bucket to the data segment is replicated. This approach is shown in Figure 7.

(3) **Partial path replication (distributed indexing):** The distributed indexing approach replicates some portion (specified by the *replication level*) of the path information. Figure 8 shows this approach. The index buckets below the replication level are placed without replication before each relevant data segment. Each index bucket above the replication level is replicated when it is the least common ancestor of the nonreplicated index buckets.

Figures 6, 7, and 8 show examples of the above three approaches based on Figure 5. In the figures, the index tree is composed of four levels. For the distributed indexing approach, we set the replication level as 2. We will show the steps for accessing data based on the above three approaches. We assume that a client requires Data Bucket 66 and makes the initial probe at Data Bucket 3.

- **Nonreplicated distribution:** The client reads Data Bucket 3 and obtains the address of the next bcast. In the next bcast, the client makes a sequence of index probes:

Figure 8. Partial path replication (distributed indexing)

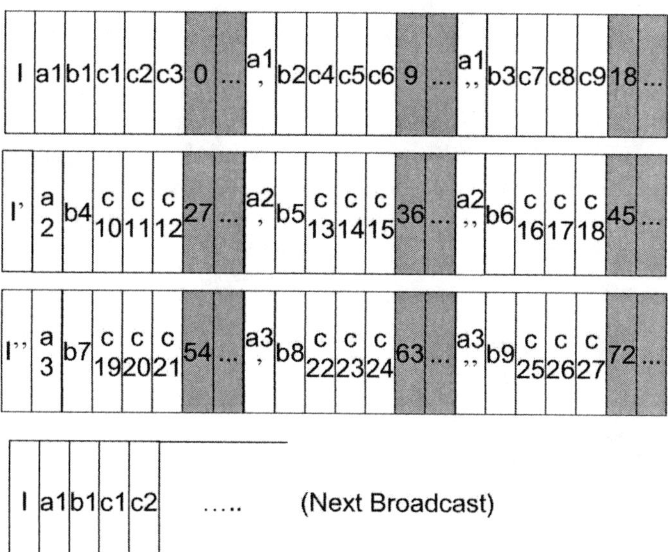

I, a3, b8, and c23. After reading c23 it obtains the address of Data Bucket 66, gets to the address, and downloads Data Bucket 66.

- **Entire path replication:** The client reads Data Bucket 3 and obtains the address of I which is broadcast before the second a1 (denoted by a1'). (In this approach I is replicated six times in one bcast.) Note that in this and the following approaches, every bucket contains the address of the next nearest replicated index bucket, and the second I is the nearest replicated index to Data Bucket 3. Then, the sequence of probes is the first a3, b8, c23, then Data Bucket 66.
- **Partial path replication (distributed indexing):** After reading Data Bucket 3, the client directly moves to the second a1 (a1') because a1' is the nearest replicated index of Data Bucket 3. The index information in a1' tells the client to move to the second I. Then the sequence of probes is as follows: (Data Bucket 3, the second a1), the second I, the first a3, b8, c23, and Data Bucket 66.

In the distributed indexing approach, there is some index information, called the *control index*, in replicated index buckets. (In the rest of this chapter, we will use the following two terms interchangeably: replicated index buckets and control index buckets.) The control index describes the scope of data the index bucket covers. For example, the control index of a1' is [*(8, the beginning of the next bcast), (26, the second I)*]. This means that:

if *(the key of target data is less than or equal to '8')*

the client moves to the next bcast;

else if *(the key of target data is greater than '26')*

the client moves to the second I;

else *find the corresponding index information in the current index;*

The distributed indexing approach gets the best of two extremes: nonreplication and entire path replication. In Imielinski et al. (1994, 1997), the authors analyzed the access time and tuning time performance of the distributed indexing approach. They proposed a formula for optimal replication level (Here, *r'* denotes the optimal replication level, *n* denotes the capacity of the bucket, and *k* denotes the number of levels in the index tree.):

$$r' = \left\lfloor \frac{1}{2} \times (\log_n(\frac{Data \times (n-1) + n^{k+1}}{n-1}) - 1) \right\rfloor + 1.$$

For a detailed description, refer to Imielinski et al. (1994, 1997).

An Efficient Index Replication

In the distributed indexing scheme, the control index buckets are replicated as many times as the fan-out of the index tree, where the replicated index buckets contain the same index information. For example, in Figure 8, a1, a1' and a1" are the three replicas that have the same index information for {b1, b2, b3}. By the way, b1 and b2 had passed over at the moment when the client reads index a1". Because a client can only access broadcast data that are or will be on the air, the addresses of b1 and b2 in a1" are not useful. In this section, we point out the problem of inefficient index replication in distributed indexing and describe some of its solutions.

Problem Description

We call the index replication scheme used in distributed indexing the *blind index replication scheme* (BIRS) since it replicates the same index information irrespective of delivery positions. We first measure the amount of index that is uselessly replicated in the BIRS. An index bucket contains *n* (the fan-out of the index tree) index slots. There are (*i* - 1) index slots that are uselessly replicated in the *i*th replicated index bucket. For example, a1' (the second replica of a1) has one useless index slot (i.e., b1) and a1" (the third replica of a1) has two useless slots (i.e., b1 and b2). Since one control index bucket generates $\sum_{i=1}^{n} (i-1)/n = \frac{(n-1)}{2}$ bucket(s) of bandwidth waste and the number of replicated control index buckets in level *i* is n^i, we can derive the following formula for the amount of bandwidth waste (in terms of the number of buckets) in the BIRS. Here, *r* denotes the replication level:

$$\frac{n-1}{2} \times \sum_{i=0}^{r-1} n^i = \frac{n^r - 1}{2}.$$

As shown in the formula, the bandwidth waste of BIRS significantly increases with a large fan-out and more levels of replication. The waste of index space on wireless channel means bandwidth inefficiency, which is not desirable in mobile computing systems.

The Criteria for Index Replication

The bandwidth inefficiency of the BIRS is caused by the temporal property of the index information. That is, an address in the index bucket indicates the time when the data (or index) to which the index points is delivered. So when replicating an index bucket, the address of the data or index that had already passed over becomes of no use. To overcome the bandwidth waste, Chung and Kim (2000) proposed a new index replication scheme that satisfies the following criteria:

- **Accessibility:** Based on a given time, an address is called *accessible* if the bucket that the address indicates has not passed over but comes in the future. In Figure 8, for example, the addresses of b2 and b3 in a1' are called accessible whereas that of b1 is not. Inaccessible addresses are useless for accessing the target data.
- **Energy efficiency:** An index pointer should provide good performance characteristics with respect to the tuning time. For example, insertion of the address of a1' into I' does not give any kind of energy efficiency in BIRS. This is because every

Figure 9. Neighborhood of control index bucket a1'

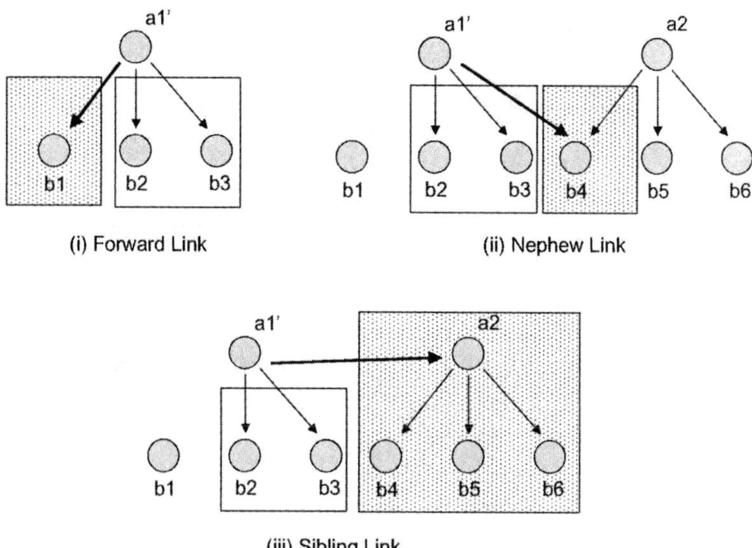

bucket (regardless of index or data bucket) basically has an offset to the next-nearest upper-level control index and I' is that of a1' (Imielinski et al., 1994).

- **Adjacency:** Let us consider the locality (Bellady, 1966), especially spatial locality, in choosing index buckets to link. Spatial locality means that storage references tend to be clustered so that once a location is referenced, it is highly likely that nearby locations will be referenced. This will be effective for processing range queries. Therefore, we think it is more desirable to link the index bucket whose coverage is *adjacent* to that of the control index to be replicated. Here, we say that the coverages of two index nodes are adjacent if the ranges of data in the subtrees rooted at the index nodes are continuous.

Efficient Index Replication Methods

Based on the three criteria mentioned above, three index replication strategies were proposed (Chung & Kim, 2000). The main differences among them are the scopes of links (i.e., index pointers). In Figures 5 and 8, for example, a1' has the addresses of {b1, b2, b3}, but b1 is not accessible. Therefore, an accessible, energy efficient, and adjacent address needs to be used instead of b1. All the possible index nodes that are adequate to be linked with respect to the given three criteria are: (a) b1 in the next bcast, (b) b4, and (c) a2. All the others do not satisfy at least one of the three criteria. These three nodes are also the only candidates whose covering areas are adjacent to that of a1'.

Figure 9 shows a control index, a1', and its neighbor index nodes. The coverage of a1' is Data 9 to Data 26, the coverage of b1 is Data 1 to Data 8, the coverage of b4 is Data 27 to Data 35, and the coverage of a2 is Data 27 to Data 53. Each shaded area represents a region that one index node covers and is adjacent to the coverage of a1'. One is on the left-hand side of a1' (Figure 9[i]) and the other two are on the right-hand side (Figure 9[ii] and Figure 9[iii]). The adjacent index in the left means that it had already passed over, so we have to link the same one in the next bcast (e.g., b1 in the next broadcast stream).

Forward Link Approach

In the *forward link* (FL) approach, forward addresses instead of inaccessible ones are used. The forward address is the address in the next bcast rather than that in the current bcast. In control index a1' the address of b1 in the next bcast replaces the address of b1 in the current bcast. Thus the index bucket a1' contains the addresses of ($b1_{next}$, b2, b3) instead of those of (b1, b2, b3). We use the subscript next, like I_{next}, to indicate that it is the bucket in the next bcast.

Let us look at the contents of index buckets when using the FL approach. The notation "Content (W) = (X, Y, Z)" means the index W contains the addresses of X, Y, and Z.

Content (I) = (a1, a2, a3),

Content (I') = ($a1_{next}$, a2, a3),

Content (I") = ($a1_{next}$, $a2_{next}$, a3),

Content (a1) = (b1, b2, b3),

Content (a1') = ($b1_{next}$, b2, b3),

Content (a1") = ($b1_{next}$, $b2_{next}$, b3),

...

However, this approach requires that bcasts should be static. In other words, two successive bcasts must be the same otherwise some forward addresses may be dangling.

Example 1. In Figure 5, suppose a client asks for Data 6 (which is under b1) when Data 15 is on the air at this time. Then each sequence of node accesses in the BIRS and the FL is as follows:

- BIRS: Data15 → a1" → I_{next} → $a1_{next}$ → $b1_{next}$ → $c3_{next}$ → $Data6_{next}$
- FL: Data15 → a1" → $b1_{next}$ → $c3_{next}$ → $Data6_{next}$

When a client asks for the data that is in the subtree rooted at b1, the client can remain in doze mode until b1 in the next bcast arrives by using the forward link in control index a1". In the BIRS, the client must traverse I_{next} and $a1_{next}$ additionally, which results in more energy consumption of the mobile unit.

The FL approach covers the left-hand-side region of a1' in Figure 9. The right-hand-side regions are covered by the approaches in the subsequent methods.

Nephew Link Approach

The *nephew link* (NL) approach adopts nephews as link pointers when replicating control index buckets. The nephew of a control index is the child of its next sibling, for

Figure 10. The nephew link approach

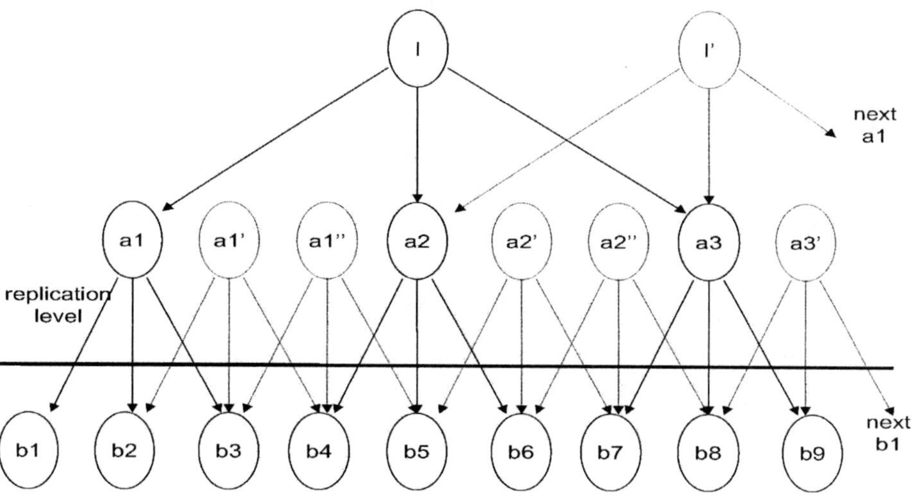

Figure 11. The sibling link approach

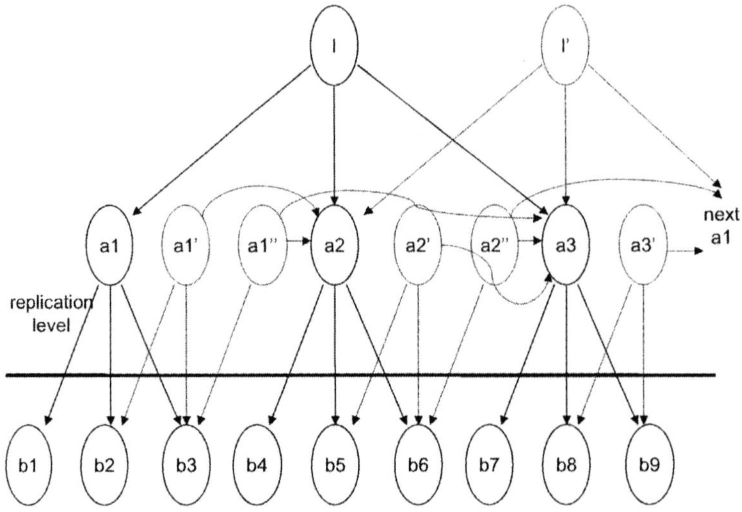

example, b4 is a nephew of a1. Figure 10 illustrates the NL approach applied to Figure 5. The contents of index buckets in the figure are as follows:

Content (I) = (a1, a2, a3),

Content (I') = (a2, a3, $a1_{next}$),

Content (I") = (a3, $a1_{next}$, $a2_{next}$),

Content (a1) = (b1, b2, b3),

Content (a1') = (b2, b3, b4),

Content (a1") = (b3, b4, b5),

...

Example 2. From the example in Figure 5, suppose a client asks for Data 27 that is in the subtree rooted at b4 at the time when Data 6 is on the air. Then each sequence of node accesses in the BIRS and the NL is as follows:

- BIRS: Data 6 → a1' → I' → a2 → b4 → c10 → Data 27
- NL: Data 6 → a1' → b4 → c10 → Data 27

In the NL approach the client can move directly to b4 after reading a1' because a1' has the pointer to b4 instead of b1. Therefore, two probes into index nodes I' and a2 are saved. Note that bcasts need not be static in the NL approach.

Sibling Link Approach

The *sibling link* (SL) approach uses the pointers to the sibling nodes instead of inaccessible ones. For instance, in Figure 9, the node a1' contains the pointer to a2 that is a sibling of a1. An example of the SL approach is depicted in Figure 11. Note that since the root node has no sibling, the second level nodes are linked in the root node. With this approach, the client can access a2 without traversing I' when it tunes in a1' and accesses the data that is in the subtree of a2. The contents of index buckets in the SL approach are as follows:

Content (I) = (a1, a2, a3),

Content (I') = (a2, a3, $a1_{next}$),

Content (I") = (a3, $a1_{next}$, $a2_{next}$),

Content (a1) = (b1, b2, b3),

Content (a1') = (b2, b3, a2),

Content (a1") = (b3, a2, a3),

...

When comparing the SL with the NL, the sibling index node has larger coverage than the nephew one but less benefit in the light of the tuning time. The following example shows the relationship between the NL and the SL.

Example 3. Consider the control index a1' in Figures 10 and 11.

- NL: a1' contains the pointers to {b2, b3, b4}. From Figure 10 we can see that the NL approach reduces two tuning steps compared with the BIRS (i.e., I' and a2) when a client accesses a data object under b4.
- SL: a1' contains {b2, b3, a2}. From Figure 11 we can see that the SL approach reduces one tuning step compared with the BIRS (i.e., I') when a client accesses the data under b4. (In fact this is true whenever a client accesses the data under a2.)

In the above example the coverage of b4 is three (i.e., the fan-out of the tree) times less than that of a2, although the former gives more tuning time reduction. More detailed comparisons and analyses are presented in the following subsection. Note that as in the NL approach, the SL approach does not require static bcasts.

Analysis

We assume that the initial probe positions and the target data objects are uniformly distributed in the bcast. Let r and n denote the replication level and the fan-out of the index tree, respectively. Suppose that an index tree is balanced and each node has the same number of children.

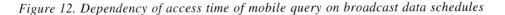

Figure 12. Dependency of access time of mobile query on broadcast data schedules

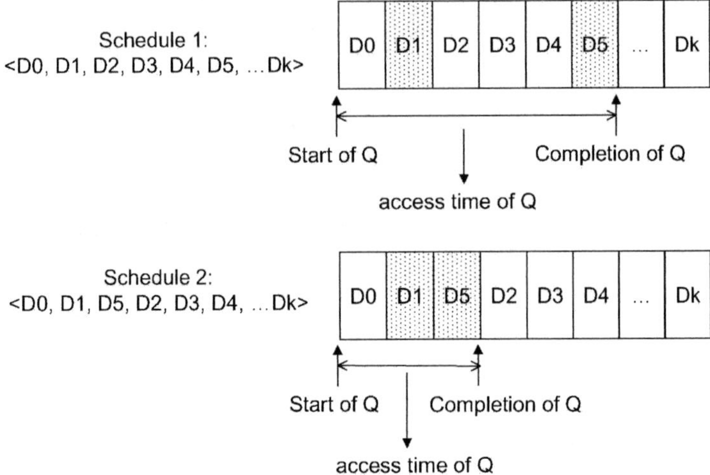

Then, the performance improvement (i.e., reduction of the number of index probes) is analyzed as follows (For detailed analysis, refer to Chung & Kim, 2000):

- FL: $\dfrac{(n-1)r(r+1)}{4n^r}$
- NL: $\dfrac{1}{n+1}$ (with large n^r)
- SL: $\dfrac{1}{2}$ (with large n)

Among the three proposed approaches, the tuning time reduction of the FL and NL decreases with large n or r while that of SL is independent of them.

Comparing the NL and the SL, we can observe that there is a trade-off between the coverage of a replicated index slot and its effect on tuning reduction. While the coverage of the upper-level index (b2) is bigger than the lower-level one (c4) as many times as the fan-out, the effect on tuning reduction of the latter is usually bigger than the former by one tuning step. Therefore, we can deduce that linking to upper-level index is more energy-efficient because the profit from the larger coverage compensates the loss from the less tuning benefit. Following this way, Chung and Kim (2000) have developed a new approach that links the most-upper-level indexes, that is, a1, a2, and a3, when replicating the control index buckets. The root index is the most-upper-level one, but it has no effect on tuning reduction because every control index has the address of I_{next}. That is, it does

Figure 13. An example of using the broadcast disks method

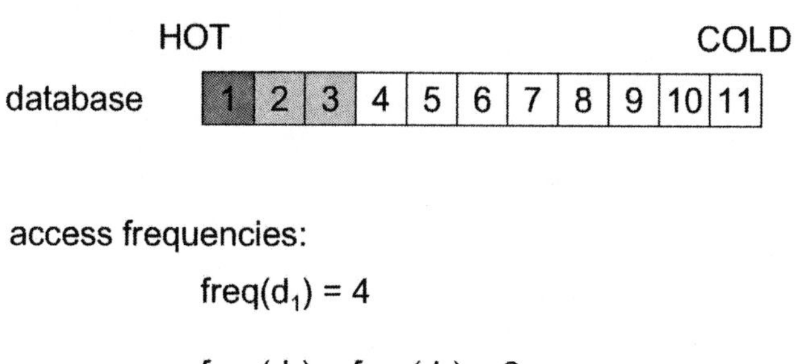

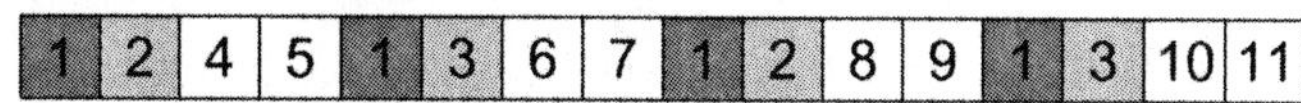

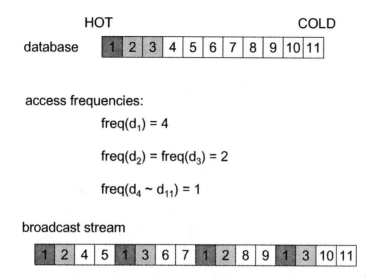

not satisfy the energy efficiency criterion. Although this approach will not be effective for sequential processing, it gives the best tuning reduction performance.

Broadcast Data Clustering

The scheduling of wireless broadcast data is one of the most important tasks for broadcast data organization. Here, the schedule of broadcast data represents the order of data broadcasting. The server clusters data objects and constructs a broadcast schedule according to the clients' data access patterns, for example, hot data versus cold data, correlation between data objects, and so on. Thus, the broadcast data scheduling is also called the clustering of broadcast data. The clustering of broadcast data has an effect on the access time performance of clients' queries, where the access time of a query is the period from the time of query start to the time of completion of data download. Figure 12 shows an example of query processing on a broadcast data stream. On two different schedules, we have different access times for the same mobile query.

In the past, there have been many clustering methods for wireless broadcasting. We classify the clustering methods according to the uniformity of broadcasting frequencies and the query types they support, that is:

- Clustering methods for uniform broadcasting versus nonuniform broadcasting.
- Clustering methods for single-point queries, multipoint queries, and so on.

In the following subsections, we formulate the clustering problem and investigate some clustering methods.

Broadcast Data Clustering for Nonuniform Broadcasting

In this section, we introduce a broadcast data clustering method, called *broadcast disks* (Acharya, Alonso, Franklin, & Zdonik, 1995), for nonuniform broadcasting environments. The basic concept of this method is to create the broadcast stream by assigning data objects to different *disks*, which are different in size and speed, and multiplexing the disks on the broadcast channel. Data objects stored on faster disks are broadcast more often than those on slower disks.

Figure 13 shows an example of a broadcast data stream using the broadcast disks method. Assume a list of data objects that has been partitioned into three disks, where data objects in Disk 1 are broadcast twice as frequently as those in Disk 2 and four times as frequently as those in Disk 3. Then, the broadcast frequencies of the data in Disks 1, 2, and 3 are 4, 2, and 1, respectively, and the broadcast stream is constructed as illustrated in the bottom of the figure. The data denoted by 1 is broadcast four times and the data denoted by 2 and 3 is broadcast two times, while those denoted by 4 to 11 are broadcast once.

Since we can access frequently broadcast data in a shorter access time, by broadcasting hot data more frequently than cold data we can optimally minimize the *total access time* (TAT) denoted as follows:

$$\text{TAT} = \sum_{\forall q_i} \text{AccessTime}(q_i) \times \text{Frequency}(q_i) \; .$$

Some caching strategies related to broadcast disks were also proposed. We will not cover them in this chapter. There is another nonuniform broadcast scheduling method (Su, Tassiulas, & Tsotras, 1998). It constructs the broadcast schedule by using the stochastic model. It considers the access frequencies of data objects and controls their delivery intervals.

Broadcast Data Clustering for Uniform Broadcasting

In this section, we introduce broadcast clustering methods for uniform broadcasting environments, that is, the broadcasting frequencies of data objects are the same. Note that in uniform broadcasting environments, the broadcast data scheduling is meaningless (i.e., a *zero-sum game*) for the single-point query because the average access time of any data object is half of the size of the broadcast stream. Thus, we consider the queries that access two or more data objects in uniform broadcasting environments.

In the following sections, we present broadcast data clustering methods for multipoint queries and those for partial-match queries. (The partial-match query is a special case of the multipoint query.) Both the multipoint query and the partial-match query access more than one data object in the broadcast data stream.

Clustering for Multipoint Queries

Figure 12 shows an example of multipoint query, where the mobile client accesses D1 and D5. As shown in the figure, the access time of the multipoint query depends on the broadcast schedule. In this subsection, we introduce two broadcast data-clustering methods for multipoint queries: the *gray code clustering method* (GCM; Chung & Kim, 2001) and the *affinity-based clustering method* (ACM; Chung, Bang, & Kim, 2002).

Notation

We first explain some relevant notation that will be used throughout the rest of the section. The data object is denoted by d_i, and $|d_i|$ is the size of d_i. The set of data in a bcast is denoted by D, and the size of one bcast is denoted by *BSize*. Thus, in this section, *BSize* is equal to $\Sigma_i |d_i|$, $\forall d_i \in D$ because uniform broadcasting is assumed. We use the symbol q_i for the query and Q for the set of queries. $D(q_i)$ denotes the set of data accessed by q_i, and $freq(q_i)$ denotes the frequency of q_i. The broadcast schedule denoted by δ is the broadcasting sequence of data with angle brackets at both ends.

Figure 14 shows a placement of the data set of query q_i, that is, $D(q_i)$. Here, $D(q_i) = \{d_{i1}, d_{i2}, \ldots, d_{ik}\}$ where k is the number of data objects that q_i accesses. The distance between two data objects d_{ij} and $d_{i(j+1)}$ is denoted by δ_{ij}. Note that since the data objects are on he air, the distance has the same semantics as the length of time.

Suppose $AT^{avg}(q_i, \sigma)$ is the average access time of q_i in σ. Then, the data clustering problem for wireless broadcasting is to find a broadcast schedule σ that minimizes the TAT, denoted by:

Figure 14. Placement of a query data set $D(q_i)$

$$D(q_i) = \{d_{i1}, d_{i2}, \ldots, d_{ik}\}$$

$$|D(q_i)| = k$$

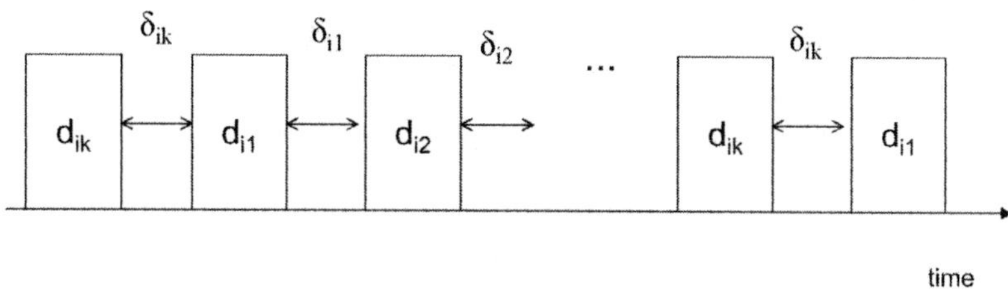

$$TAT(\sigma) = \sum_{\forall q_i \in Q} AT^{avg}(q_i, \sigma) \times freq(q_i).$$

The access time of each query varies depending on the start time of the query. In Chung and Kim (2001), the average access time of a query is analyzed as follows:

$$AT^{avg}(q_i, \sigma) = \sum_{q_i \in Q} \left(BSize - \frac{1}{2BSize} \sum_{j=1}^{k} (\delta_{ij})^2 \right).$$

Measure Definition

While the average access time $AT^{avg}(q_i, \sigma)$ is an intuitive measure for the performance of a query, it is too complex to manipulate how the performance of a broadcast schedule is analyzed. Thus, Chung and Kim (2001) defined a new measure, called the *query distance* (QD), which is the minimum distance within which all the relevant data objects for a given query can be accessed when the bcast is repeatedly broadcast.

Definition 1. Let $D(q_i)$ be $\{d_{i1}, d_{i2}, \ldots, d_{ik}\}$ in schedule σ. Then the QD of q_i in σ is defined as follows:

$$QD(q_i, \sigma) = BSize - MAX(\delta_{ij}), j = 1, 2, \ldots, k.$$

Figure 15. Graphical illustration of the QD

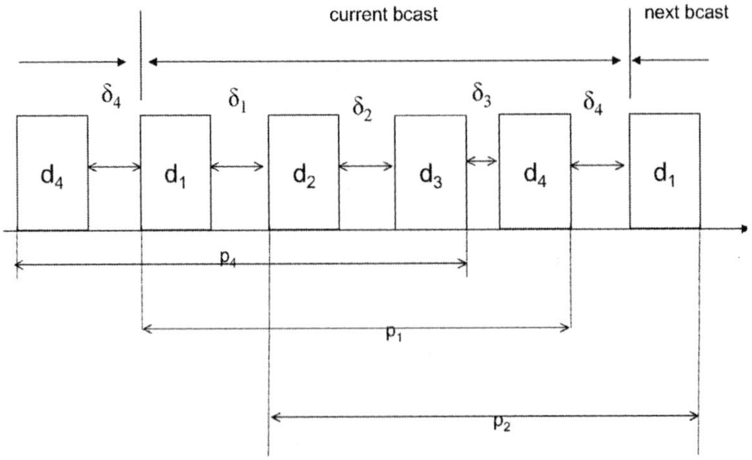

The meaning of QD is illustrated in Figure 15, where D(q) for a query q is $\{d_1, d_2, d_3, d_4\}$ in a broadcast schedule σ. (Note that in the figure there may exist other data objects between two data objects of D(q).) Then, the QD of q in the given schedule σ is the minimum among p_1, p_2, p_3, and p_4, where p_i is the access time when query q starts at the beginning of d_i, that is, $p_1 = BSize\text{-}\delta_4$, $p_2 = BSize\text{-}\delta_1$, $p_3 = BSize\text{-}\delta_2$, and $p_4 = BSize\text{-}\delta_3$.

Lemma 1. Given a query q_i and two schedules σ_1 and σ_2, if $QD(q_i, \sigma_1) > QD(q_i, \sigma_2)$, then $AT^{avg}(q_i, \sigma_1) > AT^{avg}(q_i, \sigma_2)$.

Proof: See Chung and Kim (2001).

Let the total QD in schedule σ, denoted by TQD(σ), be:

$$\sum_{q_i \in Q} \left(QD(q_i, \sigma) \times freq(q_i) \right).$$

Then, the problem of wireless data clustering can be redefined by using Lemma 1 as follows.

Definition 2. Given a set of data objects D and a set of queries Q, the wireless data clustering problem is to find a broadcast schedule σ_i such that $TQD(\sigma_i)$ is the minimum among all possible σ_i, $i = 1, \ldots$.

Theorem 1. The wireless data clustering problem in Definition 2 is *NP-complete.*

Proof: The proof easily follows transformation from the optimal linear arrangement problem (Garey & Johnson, 1976).

We describe two basic properties in the wireless broadcast schedule.

Property 1 (Interchangeability). Let there be two schedules, σ_1 and σ_2, such that:

$$\sigma_1 = < d_1, d_2, \ldots, d_{i-1}, d_i, d_{i+1}, \ldots, d_{j-1}, d_j, d_{j+1}, \ldots, d_{N-1}, d_N >$$
$$\sigma_2 = < d_1, d_2, \ldots, d_{i-1}, d_j, d_{i+1}, \ldots, d_{j-1}, d_i, d_{j+1}, \ldots, d_{N-1}, d_N >$$

Then, for any query q_i such that both d_i and d_j are in $D(q_i)$, $QD(q_i, \sigma_1) = QD(q_i, \sigma_2)$.

Definition 3. Given two schedules, σ_1 and σ_2, if $QD(q_i, \sigma_1)$ is equal to $QD(q_i, \sigma_2)$ for every query q_i, then we say that σ_1 is *distance equivalent* to σ_2, and is denoted by the equality, that is, $\sigma_1 = \sigma_2$.

Property 2 (Symmetry). If a schedule σ_2 is the mirror image of σ_1, that is:

$$\sigma_1 = < d_1, d_2, \ldots, d_{i-1}, d_i, d_{i+1}, \ldots, d_{N-1}, d_N >$$
$$\sigma_2 = < d_N, d_{N-1}, \ldots, d_{i+1}, d_i, d_{i-1}, \ldots, d_2, d_1 >,$$

then σ_1 is distance equivalent to σ_2, that is, $\sigma_1 = \sigma_2$.

The interchangeability property is a direct consequence from the definition of the QD. The property says that any two data objects in $D(q_i)$ are mutually interchangeable with respect to the QD of q_i. In the symmetry property, $QD(q_i, \sigma_1)$ is equal to $QD(q_i, \sigma_2)$ for all q_i because any δ in σ_1 is equal to that in σ_2.

The Gray Code Clustering Method

This method utilizes the gray coding scheme for clustering wireless data. The gray coding scheme is one of the schemes for linear mapping of multidimensional space. In the binary reflected gray code, numbers are coded into binary bit strings such that successive numbers differ in exactly one bit position. It is observed that difference in only one bit position has a relationship with locality (Faloutsos, 1986; Jagadish, 1990). The terms and notation below are from Faloutsos (1986).

Figure 16 shows an illustration of 3-bit binary reflected gray codes with corresponding binary codes. From now on, we use the term *gray code* for '*binary reflected gray code* throughout the chapter. The *gray value* of a binary string, denoted by $(.)_G$, is the *order* (or position) of the binary string in the gray code. For instance, the gray value of (110), denoted by $(110)_G$, is 4. It is the same as $(4)_{10}$ and $(100)_2$.

The conversion formulas of a gray code word to its order (i.e., gray value) described below are from Faloutsos (1986). Suppose the n-bit Gray code word is $(g_n, g_{n-1}, \ldots, g_1)$ and its order in binary is $(b_n, b_{n-1}, \ldots, b_1, b_0)$. Then:

Figure 16. Illustration of the 3-bit binary reflected gray code

gray value	gray code	binary code
0	000	000
1	001	001
2	011	010
3	010	011
4	110	100
5	111	101
6	101	110
7	100	111

$$b_n = 0,$$

$$b_n = \sum_{m=j+1}^{n} g_m \bmod 2, (0 \le j < n).$$

In this method, each data object is associated with a bit vector of dimension M, where M is the number of queries in Q. In the bit vector, the ith bit is set to 1 if the data object is accessed by query q_i. Otherwise, the bit is set to 0.

Definition 4. Let there be M number of queries in Q such that for all q_i, $i = 1, \ldots, (M - 1)$ in Q, $freq(q_i) \ge freq(q_{i+1})$. Let $\vec{d_i}$ denote the bit vector for data object d_i. Then, $\vec{d_i}$ consists of $(u_1, u_2, \ldots, u_M)$ such that:

$$u_k = \begin{cases} 1 & \text{if } d_i \in D(q_k) \\ 0 & \text{otherwise} \end{cases}.$$

Note that u_1 corresponds to the most frequently referenced query, u_2 corresponds to the same or next frequently referenced query, and u_M corresponds to the least one.

Example 4. Suppose that the set of queries is $\{q_1, q_2, q_3\}$ and the set of data objects in the bcast is $\{d_1, d_2, d_3, d_4, d_5, d_6, d_7, d_8\}$. Suppose also that the reference frequency and the query data set $D(q_i)$ of each query are:

$$freq(q_1) = 3, D(q_1) = \{d_1, d_2, d_4, d_5\}$$
$$freq(q_2) = 2, D(q_2) = \{d_4, d_5, d_6, d_7, d_8\}$$
$$freq(q_3) = 1, D(q_3) = \{d_2, d_3, d_5, d_6\}.$$

Then, the bit-vector representations of the data objects are as follows:

$$\vec{d_1} = (1,0,0),\ \vec{d_2} = (1,0,1),\ \vec{d_3} = (0,0,1),\ \vec{d_4} = (1,1,0)$$

$$\vec{d_5} = (1,1,1),\ \vec{d_6} = (0,1,1),\ \vec{d_7} = (0,1,0),\ \vec{d_8} = (0,1,0).$$

Here, the leftmost vector element is for q_1, the second element is for q_2, and the third element is for q_3.

The clustering method consists of two steps: (a) generating a bit vector for each data object and (b) sorting the bit vectors based on their gray values. Thus, the complexity

Figure 17. Data ordering based on gray values

gray value	bit vector	data objects
1	001	$\overrightarrow{d_3}$
2	011	$\overrightarrow{d_6}$
3	010	$\overrightarrow{d_7}$, $\overrightarrow{d_8}$
4	110	$\overrightarrow{d_4}$
5	111	$\overrightarrow{d_5}$
6	101	$\overrightarrow{d_2}$
7	100	$\overrightarrow{d_1}$

of the method is *MAX(O(NM), O(NlogN))*, where *N* is the number of data objects and *M* is the number of queries. The sequence of data objects corresponding to the sorted sequence of bit vectors is our broadcast schedule. For convenience, the bit vector is denoted by the binary code word, for example, 101 for (1, 0, 1).

Figure 17 shows the sorted result of the data objects in Example 4. Thus, a broadcast schedule (i.e., the broadcasting sequence of wireless data) based on our proposed gray code clustering method is as follows:

$$\sigma_{Gray} = < d_3, d_6, d_7, d_8, d_4, d_5, d_2, d_1 >.$$

The data objects having the same bit-vector representation, such as d_7 and d_8 in the figure, can be mutually interchangeable in the schedule (by Property 1). The reverse ordering of the above schedule is also distance equivalent to the original ordering (by Property 2), which means that the sorted sequence of the bit vectors in a nonincreasing order has the same effect with respect to our proposed method.

In the paper Chung and Kim (2001), the performance of GCM is analyzed and experimented. According to its result, GCM gains 10% to 30% of access time improvement over nonclustered broadcast schedules.

The Affinity-Based Clustering Method

We introduce another broadcast data clustering method for multipoint queries. This method defines and uses the affinity concept, thus we call the method the ACM in this chapter. Inherently, this method also uses the QD measure. After describing the definition of affinity concepts and some of their properties, we explain the clustering method.

The Data Affinity

The data affinity of two data records represents the degree that they are referenced together in a query, which is defined as follows:

Definition 5. The data affinity between two data records d_i and d_j is as follows (Here, $N_{data}(q_i)$ is the number of data objects that query q_i accesses.):

$$aff(d_i, d_j) = \begin{cases} \sum_{\forall q_k \in Q} \frac{has(q_k, d_i, d_j) \times freq(q_k)}{{}_{N_{data}(q_k)}C_2}, & \text{if } i \neq j \\ 0, & \text{otherwise} \end{cases}$$

$$has(q_k, d_i, d_j) = \begin{cases} 1, & \text{if } d_i \in q_k \wedge d_j \in q_k \\ 0, & \text{otherwise} \end{cases}$$

Property 3. The data affinity has the following properties:

- If there is no query that accesses both data records d_i and d_j, then $aff(d_i, d_j) = 0$.
- The affinity of two data records is symmetric, that is, $aff(d_i, d_j) = aff(d_j, d_i)$.
- The sum of affinity values of all data pairs equals to the sum of frequency values of queries that access more than one data record. That is:

$$\sum_{i=1}^{N-1} \sum_{j=i+1}^{N} aff(d_i, d_j) = \sum_{q_k \in Q \wedge N_{data}(q_k) > 1} freq(q_k).$$

Actually, the affinity of two data records is the sum of frequencies of queries that access both of them. Thus, the affinity of a pair of data that is coreferenced in many queries will be high. In the definition, we divide the sum by the number of all possible pairs of data that can be generated from the data records a query accesses. It is because the more data records are accessed by a query, each pair of data records has less impact on the clustering performance. In the method, data pairs with high affinity are preferentially clustered.

Definition 6. Let N be the number of data records. The affinity matrix (AM) is an N x N matrix whose elements are defined as $AM[i][j] = aff(d_i, d_j)$.

The Segment Affinity

The segment is a sequence of data records and is denoted by $S_i = (d_1, d_2, \ldots, d_k)$. The ACM generates a broadcast schedule by merging segments. Initially, one data record comprises a segment, and the segments are merged using the following rules:

Definition 7. The segment affinity of two segments, S_i and S_j, is:

$$SegAff(S_i, S_j) = \begin{cases} \sum_{d_k \in S_i} \sum_{d_l \in S_j} aff(d_k, d_l) \times \frac{MaxDist(d_k, d_l) - dist(d_k, d_l)}{MaxDist(d_k, d_l)}, & \text{if} \quad i \neq j \\ 0, \quad \text{otherwise} \end{cases}.$$

The segment affinity is the sum of data affinity of the data in two segments, where each data affinity is multiplied by a distance factor; *dist(*d_i, d_j*)* is the minimal distance between d_i and d_j on the currently generated broadcasting sequence and *MaxDist(*d_i, d_j*)* is the maximal one.

Definition 8. Let d be the interval between d_k and d_1. Then, *dist*(d_k, d_1) and *MaxDist*(d_k, d_1) is defined as:

$$dist(d_k, d_l) = \begin{cases} |d_k| + \delta + |d_l|, & \text{if} \ \delta < \frac{BSize - |d_k| - |d_l|}{2} \\ BSize - \delta, & \text{otherwise} \end{cases}$$

$$MaxDist\ (d_k, d_l) = \frac{BSize + |d_k| + |d_l|}{2}.$$

Note that in the formulas above, we consider the distance on the broadcast schedule, not on a segment. Also, since we have assumed broadcasts are repeated, the interval of two data records that are located at the head and tail of a broadcast stream is zero, that is, they are located closely together on the broadcast schedule.

The value of $\frac{MaxDist(d_k, d_1) - dist(d_k, d_1)}{MaxDist(d_k, d_1)}$ (in Definition 7) approaches to 1 when two data records (d_k and d_1) are closely located in the schedule, and approaches to 0 when they are located apart from each other.

In contrast to the data affinity, the segment affinity is asymmetric, for the intervals between data records vary according to the ordering of segments. Thus, when we merge segments for a broadcast schedule, we have to consider not only which segments to merge but also how to place them.

Definition 9. Let *N* be the number of segments. Then, the *segment affinity matrix* (SAM) is an *N* x *N* matrix whose elements are defined as follows:

$$SAM[i][j] = SegAff(S_i, S_j).$$

The Inverse Segment

The inverse schedule is the sequence of data that are placed in the reverse order of the original one. For example, the two schedules below are mutually inverse schedules. Since the access time of a query is determined by intervals of data records on a broadcast schedule and the intervals on a schedule are equal to those on the inverse schedule, the access time of a query on a broadcast schedule is equal to that on its inverse schedule.

$$\sigma_1 = < d_1, d_2, ..., d_{i-1}, d_i, d_{i+1}, ..., d_{n-1}, d_n >$$
$$\sigma_2 = < d_n, d_{n-1}, ..., d_{i+1}, d_i, d_{i-1}, ..., d_2, d_1 >$$

For considering this property in the process of clustering, we define the inverse segment, which is the reverse sequence of data records of a segment:

Definition 10. Let a segment S_i be $S_i = (d_1, d_2, ..., d_n)$. Then, the inverse segment of S_i, denoted by S_i^{-1}, is $S_i^{-1} = (d_n, d_{n-1}, ..., d_1)$.

Property 4. $SegAff(S_i, S_j) = SegAff(S_j^{-1}, S_i^{-1})$. The segment affinity of two segments S_i and S_j, merged in this order, is equal to that of their inverse segments merged in the reverse order. When considering all placements of two segments (S_i and S_j), the following eight cases are possible.

$$SegAff(S_i, S_j),\ SegAff(S_j, S_i)$$
$$SegAff(S_i^{-1}, S_j),\ SegAff(S_j^{-1}, S_i)$$
$$SegAff(S_i, S_j^{-1}), SegAff(S_j, S_i^{-1})$$
$$SegAff(S_i^{-1}, S_j^{-1}), SegAff(S_j^{-1}, S_i^{-1})$$

Figure 18. The algorithm of ACM

***Algorithm** ACM*
***INPUT:** a set of data records D; a set of queries Q*
***OUTPUT:** a broadcast schedule*
METHOD:
1) Make AM from D and Q;
2) Initialize each segment S_i to have a data record d_i;
3) Make SAM and ISAM;
*4) **DO***
5) Find Segment S_i and S_j, and their merging order where the segment affinity is maximized;
6) Merge S_i and S_j based on the order;
7) Recompute SAM and ISAM;
*8) **UNTIL** (All element values of SAM are 0);*
9) Merge the remaining segments;

According to Property 4, their segment affinities are:

$$SegAff(S_i, S_j) = SegAff(S_j^{-1}, S_i^{-1}),\ SegAff(S_i^{-1}, S_j) = SegAff(S_j^{-1}, S_i)$$
$$SegAff(S_i, S_j^{-1}) = SegAff(S_j, S_i^{-1}),\ SegAff(S_i^{-1}, S_j^{-1}) = SegAff(S_j, S_i).$$

Thus, we can be informed about the segment affinities of all cases from the four cases. Also, we can see $SegAff(S_i, S_j)$ and $SegAff(S_j, S_i)$ from the SAM, and hence we need to keep the segment affinity values of $SegAff(S_i^{-1}, S_j)$ and $SegAff(S_i, S_j^{-1})$ only. For these values, we use the *inverse segment affinity matrix* (ISAM), which is defined as follows.

Definition 11. Let N be the number of segments. The ISAM is an N x N matrix, whose elements are defined as follows:

$$ISAM[i][j] = \begin{cases} SegAff(S_i^{-1}, S_j), & \text{if} \quad i > j \\ SegAff(S_i, S_j^{-1}), & \text{otherwise} \end{cases}.$$

The Algorithm

After each segment is initialized with a single data record, the clustering method makes a broadcast schedule by merging segments. In merging segments, we select two segments and determine the merge order such that their segment affinity is maximized. Figure 18 is the algorithm of the ACM.

Let us explain the algorithm with an example. Suppose that there are seven data records and eight queries. The frequency values and data sets of queries are as follows:

$$D = \{d_1, d_2, d_3, d_4, d_5, d_6, d_7\}$$

$$freq(q_1) = 5,\ q_1 = \{d_2, d_3\},\ freq(q_2) = 2,\ q_2 = \{d_1, d_2\}$$

$$freq(q_3) = 4,\ q_3 = \{d_1, d_3\},\ freq(q_4) = 1,\ q_4 = \{d_1, d_2, d_3\}$$

$$freq(q_5) = 3,\ q_5 = \{d_4, d_6\},\ freq(q_6) = 3,\ q_6 = \{d_5, d_7\}$$

$$freq(q_7) = 2,\ q_7 = \{d_4, d_5, d_6\},\ freq(q_8) = 1,\ q_8 = \{d_4, d_5, d_6, d_7\}.$$

The AM is made from the sets of data records and queries (D and Q). For instance, since d_1 and d_2 is accessed by queries q_2 and q_4, the data affinity of d_1 and d_2 is:

$$AM[1][2] = aff(d_1, d_2) = \frac{freq(q_2)}{{}_2C_2} + \frac{freq(q_4)}{{}_3C_2} = 2.33.$$

In this way of computing affinity values, we can make the AM as Figure 19. Now, we make segments and the SAM. Initially, a segment is made up of a single data record, thus there are k segments if the number of data records is k. At each step of merging, we recompute the SAM and ISAM since the number of segments decreases.

Figure 20 shows the merging steps of this example. In the figure we use a structure for describing each segment. The structure contains the following information: the sequence of data records of this segment, the link to a segment whose segment affinity with this segment is maximal, its segment affinity value, and the merging order of the two segments. The order is denoted by one of flags NN, NI, IN, and II. NN means that two segments are merged as normal (not inverse) segment and normal segment. NI means that this segment is arranged as a normal segment and the segment to be merged is arranged as an inverse order. Likewise, IN means the inverse-normal order and II means the inverse-inverse order. Now, we show the steps of merging segments according to the algorithm.

In Step 1, all segments, SAM, and ISAM are initialized. Each segment contains the link to a segment that has to be merged first (i.e., the max segment affinity segment). Here, we can see the segment affinity of S_2 and S_3 is the maximum among all segment affinity values. Thus, we merge them in Step 2. After merging, the SAM and ISAM are recomputed. In this figure, those are described in the structures of segments. Note that some segment affinity values are changed after two segments are merged.

In Step 2, the structure of S_2 (which is the merge result of S_2 and S_3 in Step 1) indicates that the segment S_1 is the max-affinity segment with itself and the merging order is NI. Thus, we merge S_2 and S_1 in the NN order: merge (d_2, d_3) and (d_1) into (d_2, d_3, d_1). Here, the merge order of S1 is NI. With respect to S_1, the order NI means that we have to merge (d_1) and (d_3, d_2) into (d_1, d_3, d_2). Since (d_1, d_3, d_2) is symmetric with (d_2, d_3, d_1), we can merge the segments either way.

Figure 19. The AM of the example

	d_1	d_2	d_3	d_4	d_5	d_6	d_7
d_1	0	2.33	4.33	0	0	0	0
d_2	2.33	0	5.33	0	0	0	0
d_3	4.33	5.33	0	0	0	0	0
d_4	0	0	0	0	0.16	3.16	0.16
d_5	0	0	0	0.16	0	0.83	3.83
d_6	0	0	0	3.16	0.83	0	0.83
d_7	0	0	0	0.16	3.83	0.83	0

Figure 20. The segment merging process

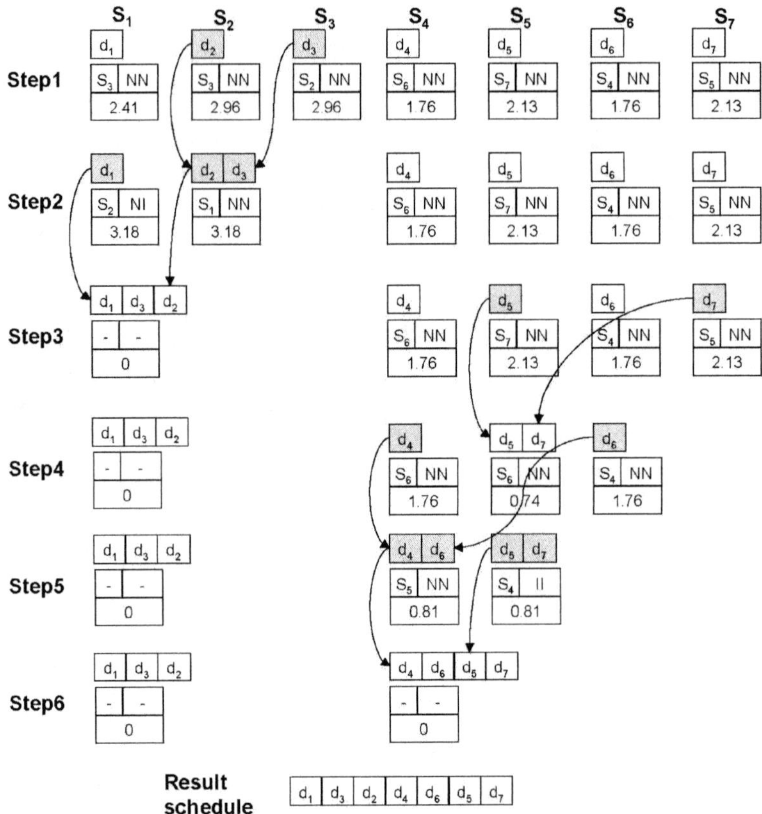

In Steps 3, 4, and 5, we merge the segments with the max-affinity values. Finally in Step 6, all affinity values are 0. Then, we merge the remaining segments in any orders as follows (These schedules are all equivalent–they have the same access time performance.):

$$< d_1, d_3, d_2, d_4, d_6, d_5, d_7 >,$$

$$< d_2, d_3, d_1, d_4, d_6, d_5, d_7 >,$$

$$< d_4, d_6, d_5, d_7, d_1, d_3, d_2 >,$$

and so on.

The complexity of the algorithm is determined by the number of merging steps. In the worst case, we have to merge n - 1 times, and in each time, we have to recompute the SAM and ISAM whose sizes are n^2. Thus, we can see that the complexity is $O(n^3)$. Since the clustering is processed once in the period of data broadcasting, this complexity is a reasonable one.

According to the authors' experiments, this method outperforms the GCM, especially for large amounts of mobile clients. For details on the experiments, refer to Chung et al. (2002).

Clustering for Partial-Match Queries

The GCM and ACM introduced in the previous sections consider that a query qualification for required data objects is based on a single attribute, that is, the key attribute. In this section, we introduce a broadcast data clustering method (Lee, Chung, Lee, & Kim, 2001) for partial-match queries. A partial-match query retrieves data objects by specifying some nonkey attributes (Kim & Pramanik, 1988). In other words, the partial-match query uses a content-based retrieval. It is known as a query type that has been widely used in various applications. Since the method also uses gray codes as its clustering mechanism, we call it *gray code clustering method for partial-match queries* (GCM-P).

In GCM-P, data objects and queries are represented as data signatures and query signatures using multi-attribute hashing. We use a running example for explaining how data objects and queries are represented as signatures based on multi-attribute hashing.

Suppose there is a wireless information system that broadcasts stock price information in real time. Each data object has four attributes: A_1 (*company*), A_2 (*amount of sale*), A_3 (*amount of purchase*), and A_4 (*price* [$]). For each attribute, we assume the following simple hash functions.

$$h_{\text{company}}(x) = \begin{cases} 00 \text{ if } x \text{ is a financial company} \\ 01 \text{ if } x \text{ is a manufacturing company} \\ 10 \text{ if } x \text{ is a computer and communication company} \\ 11 \text{ otherwise} \end{cases}$$

Figure 21. An example of data and their signature representations

data	company	amount of sale	amount of purchase	price ($)	data signature	gray value	binary value
d_1	New York Bank	5000	35000	69	000011111	21	31
d_2	LA Broadcasting	15000	25000	49	110110101	294	437
d_3	San Jose Computer	25000	15000	39	101001100	392	332
d_4	Dallas Machines	35000	5000	29	011100011	190	227
d_5	Texas Electronics	500	45000	19	010011010	236	154
d_6	Miami Industry	8000	31000	4	010011000	239	152
d_7	Hawaii Comm.	24000	19000	32	101001100	392	332

$$h_{\text{amount of sale}}(x) = \begin{cases} 00 \text{ if } x < 10000 \\ 01 \text{ if } 10000 \leq x < 20000 \\ 10 \text{ if } 2000 \leq x < 30000 \\ 11 \text{ if } 30000 \leq x \end{cases}$$

$$h_{\text{amount of purchase}}(x) = \begin{cases} 00 \text{ if } x < 10000 \\ 01 \text{ if } 10000 \leq x < 20000 \\ 10 \text{ if } 2000 \leq x < 30000 \\ 11 \text{ if } 30000 \leq x \end{cases}$$

$$h_{\text{price}}(x) = \begin{cases} 000 \text{ if } x < 5 \\ 001 \text{ if } 5 \leq x < 10 \\ 010 \text{ if } 10 \leq x < 20 \\ 011 \text{ if } 20 \leq x < 30 \\ 100 \text{ if } 30 \leq x < 40 \\ 101 \text{ if } 40 \leq x < 50 \\ 110 \text{ if } 50 \leq x < 60 \\ 111 \text{ if } 60 \leq x \end{cases}$$

A data signature is generated by concatenating the hashed l_i-bit vector of each attribute i. Therefore, the signature becomes l-bit binary bit vector, where $l = \sum_{i=1}^{k} l_i$ (k is the number of attributes). In the example, we assume that the attributes are arranged in this order: $<A_1, A_2, A_3, A_4>$, and l_1, l_2, l_3 = 2 and l_4=3. Based on the above hashing scheme, we can represent a data object *{company = 'New York Bank', amount of sale = '13000', amount of purchase = '2000', price = '22'}* as *010100011*. Similarly, we can represent a partial-match query *{amount of sale ≥ '30000', '10' ≤ price < '20' }* as ***11**010*, where * denotes a do-not-care condition.

The Clustering Algorithm

The GCM-P consists of two steps: (a) representing data objects as data signatures based on the given hash functions and (b) sorting data signatures based on their gray values. Suppose that there are seven data objects d_1 through d_7 (in Figure 21), and attributes and hash functions are the same as described before. Then, the clustering method works as follows.

In the first step of the proposed method, we represent the data object as data signature, which is described in the *data signature* column of the table. In the second step, we sort the data based on their gray values. As a result, we get the following schedules; σ_{Gray} is the result of the GCM-P and σ_{binary} is the result of the *binary coding method* (BCM). The BCM is the method which sorts the data signatures based on binary values.

$$\sigma_{Gray} = < d_1, d_4, d_5, d_6, d_2, d_3, d_7 >$$

$$\sigma_{binary} = < d_1, d_6, d_5, d_4, d_3, d_7, d_2 >$$

In the sorting step, the data objects of the same gray value or binary value are mutually interchangeable based on Property 1. (In this example d_3 and d_7 have the same gray value.) Also, because of Property 2, it does not matter whether the set of data is sorted in an increasing order or decreasing one. In the chapter we sort the data in a nondecreasing order.

The authors (Lee et al., 2001) analyzed the QD of a partial-match query based on σ_{Gray} and σ_{binary}. They also analyzed the AQD_{gain} (average QD gain) over the binary coding method as follows (Here, n denotes the number of bits used for signature representation.):

$$AQD_{gain}(n) = \frac{2(6^{n-1} - 1)}{5(3^n - 2^n)}.$$

Conclusion

Summary

In this chapter, we have introduced some data management issues in wireless data broadcasting. We especially focus on indexing and clustering-related research issues.

The indexing of wireless broadcast data deals with the construction and placement of index information. The index on the broadcast channel means the directory information which is the temporal addresses of broadcast data objects. With the index, mobile clients can access broadcast data objects in an energy-efficient manner. In the chapter we introduced some indexing methods: the (1, *m*) indexing method and the distributed indexing method. We pointed out a problem (i.e., inefficient index replication) in the previous method and proposed new index replication strategies.

In data broadcasting, the broadcasting order of data has significant impacts on clients' access time performance. We classified the broadcast data clustering methods based on query types and broadcasting frequencies. For nonuniform broadcasting environments, we introduced the broadcast disks method. The method determines the broadcasting frequency of each data object according to its access frequency. For uniform broadcasting environments, we introduced some clustering methods: GCM and ACM for multipoint queries and GCM-P for partial-match queries.

Future Research Issues

In addition to indexing methods, there are some approaches using signatures for improving tuning time performance (Hu et al., 2000; Lee & Lee, 1996). The signature is aggregate information containing the directory of broadcast data objects. Also, there have been some approaches to incorporate signatures into the index structure. On constructing index and signatures for the broadcast data stream, there are many research topics, some of which are

- indexing for multichannel broadcasting environments (Leong & Si, 1995),
- indexing of multi-attribute data objects (Hu et al., 2000),
- indexing for multipoint queries,
- indexing for partial-match queries, and
- combination of signature and index structures.

For clustering of broadcast data, we explained some methods according to various categories. However, they do not include the following research issues:

- clustering for multipoint queries in nonuniform environments,
- clustering for partial-match queries in nonuniform environments,
- clustering for multichannel environments,
- cache-aware clustering, and
- clustering with index.

References

Acharya, S., Alonso, R., Franklin, M., & Zdonik, S. (1995). Broadcast disks: Data management for asymmetric communication environments. *Proceedings of 1995 ACM SIGMOD Conference on Management of Data*, (pp. 199-210).

Argade, P. V., Ayneloglu, S., Berenbaum, A.D., DePaolis Jr., M.V., Franzop, R.T., Freeman, R.D., Inglis, D.A., Komoriya, G., Lee, H., Little, T.R., MacDonald, G.A., Mclellan, H.R., Morgan, E.C., Pham, H.Q., Ronkin, G.D., Scavvuzzo, R.J., & Woch, T.J. (1993). Hobbit: A higher-performance low-power microprocessor. *Proceedings of COMPCON*, (pp. 88-95).

Barbara, D. (1999). Mobile computing and databases: A survey. *IEEE Transactions on Knowledge and Data Engineering, 11*(1), 108-117.

Barbara, D. & Imielinski, T. (1994). Sleepers and workaholics caching strategies in mobile environments. *Proceedings of ACM SIGMOD*, (pp. 1-12).

Bellady, L. A. (1966). A study of replacement algorithms for virtual storage computers. *IBM Systems Journal, 5*(2), 78-101

Chung, Y. D., & Kim, M. H. (2000). An index replication scheme for wireless data broadcasting. *The Journal of Systems and Software, 51*, 191-199.

Chung, Y. D., & Kim, M. H. (2001). A wireless data clustering method for multipoint queries. *Decision Support Systems, 30*, 469-482.

Chung, Y. D., Bang, S. H., & Kim, M. H. (2002). An efficient broadcast data clustering method for multipoint queries in wireless information systems. *The Journal of Systems and Software, 64*, 173-181.

Faloutsos, C. (1986). Multiattribute hashing using gray codes. *Proceedings of ACM SIGMOD*, (pp. 227-238).

Garey, M. R., & Johnson, D. S. (1976) *Computers and intractability: A guide to the theory of NP-completeness*. San Francisco: Freeman Publishing.

Hu, Q., Lee, W. -C., & Lee, D. L. (2000). Power conservative multi-attribute queries on data broadcast. *Proceedings of ICDE, 157-166.*

Imielinski, T., Viswanathan, S., & Badrinath. B. R. (1994). Energy efficient indexing on air. *Proceedings of ACM SIGMOD*, (pp. 25-36).

Imielinski, T., Viswanathan, S., & Badrinath. B. R. (1997). Data on air: Organization and access. *IEEE Transactions on Knowledge and Data Engineering, 9*(3), 353-372.

Jagadish, H. V. (1990). Linear clustering of objects with multiple attributes. *Proceedings of ACM SIGMOD*, (pp. 332-342).

Kim, M. H., & Pramanik, S. (1988). Optimal file distribution for partial match retrieval. *Proceedings of ACM SIGMOD*, (pp. 173-182).

Lee, J. Y., Chung, Y. D., Lee, Y. J., & Kim, M. H. (2001). Gray code clustering of wireless data for partial match queries. *Journal of Systems Architecture, 47*, 445-458.

Lee, W. -C., & Lee, D. L. (1996). Using Signature techniques for information filtering in wireless and mobile environments. *Distributed and Parallel Databases Journal, 4*(3), 205-227.

Leong, H. V., & Si, A. (1995). Data broadcasting strategies over multiple unreliable wireless channels. *Proceedings of CIKM*, (pp. 96-104).

Su, C., Tassiulas, L., & Tsotras, V. J. (1998). Broadcast scheduling for information dissemination. *Wireless Networks, 5*(2), 137-147.

Chapter IV

Data Broadcasting in a Mobile Environment

A.R. Hurson, The Pennsylvania State University, USA

Y. Jiao, The Pennsylvania State University, USA

Abstract

The advances in mobile devices and wireless communication techniques have enabled anywhere, anytime data access. Data being accessed can be categorized into three classes: private data, shared data, and public data. Private and shared data are usually accessed through on-demand-based approaches, while public data can be most effectively disseminated using broadcasting. In the mobile computing environment, the characteristics of mobile devices and limitations of wireless communication technology pose challenges on broadcasting strategy as well as data-retrieval method designs. Major research issues include indexing scheme, broadcasting over single and parallel channels, data distribution and replication strategy, conflict resolution, and data retrieval method. In this chapter, we investigate solutions proposed for these issues. High performance and low power consumption are the two main objectives of the proposed schemes. Comprehensive simulation results are used to demonstrate the effectiveness of each solution and compare different approaches.

Introduction

The increasing development and spread of wireless networks and the need for information sharing has created a considerable demand for cooperation among existing, distributed, heterogeneous, and autonomous information sources. The growing diversity in the range of information that is accessible to a user and rapidly expanding technology have changed the traditional notion of timely and reliable access to global information in a distributed system. Remote access to data refers to both mobile nodes and fixed nodes accessing data within a platform characterized by the following:

- low bandwidth,
- frequent disconnection,
- high error rates,
- limited processing resources, and
- limited power sources.

Regardless of the hardware device, connection medium, and type of data accessed, users require timely and reliable access to various types of data that are classified as follows:

- Private data, that is, personal daily schedules, phone numbers, and so forth. The reader of this type of data is the sole owner or user of the data.
- Public data, that is, news, weather information, traffic information, flight information, and so forth. This type of data is maintained by one source and shared by many—a user mainly queries the information source(s).
- Shared data, that is, traditional, replicated, or fragmented databases. Users usually send transactions as well as queries to the information source(s).

  Access requests to these data sources can be on-demand-based or broadcast-based.

On-Demand-Based Requests

In this case users normally obtain information through a dialogue (two-way communication) with the database server—the request is pushed to the system, data sources are accessed, operations are performed, partial results are collected and integrated, and the final result is communicated back to the user. This access scenario requires a solution that addresses the following issues.

- **Security and access control.** Methods that guarantee authorized access to the resources.
- **Isolation.** Means that support operations off-line if an intentional or unintentional disconnection has occurred.

- **Semantic heterogeneity.** Methods that can handle differences in data representation, format, structure, and meaning among information sources and hence establish interoperability.
- **Local autonomy.** Methods that allow different information sources to join and depart the global information-sharing environment at will.
- **Query processing and query optimization.** Methods that can efficiently partition global queries into subqueries and perform optimization techniques.
- **Transaction processing and concurrency control.** Methods that allow simultaneous execution of independent transactions and interleave interrelated transactions in the face of both global and local conflicts.
- **Data integration.** Methods that fuse partial results to draw a global result.
- **Browsing.** Methods that allow the user to search and view the available information without any information processing overhead.
- **Distribution transparency.** Methods to hide the network topology and the placement of the data while maximizing the performance for the overall system.
- **Location transparency.** Methods that allow heterogeneous remote access (HRA) to data sources. Higher degrees of mobility argue for higher degrees of heterogeneous data access.
- **Limited resources.** Methods that accommodate computing devices with limited capabilities.

The literature is abounded with solutions to these issues (Badrinath, 1996; Bright, Hurson, & Pakzad, 1992, 1994; Joseph, Tauber, & Kaashoek, 1997; Satyanarayanan, 1996). Moreover, there are existing mobile applications that address the limited bandwidth issues involved in mobility (Demers, Pertersen, Spreitzer, Terry, Theier, & Welch, 1994; Fox, Gribble, Brewer, & Amir, 1996; Honeyman, Huston, Rees, & Bachmann, 1992; Joseph et al., 1997; Kaashoek, Pinckney, & Tauber, 1995; Lai, Zaslavsky, Martin, & Yeo, 1995; Le, Burghardt, Seshan, & Rabaey, 1995; Satyanarayanan, 1994, 1996).

Broadcast-Based Requests

Public information applications can be characterized by (a) massive numbers of users and (b) the similarity and simplicity in the requests solicited by the users. The reduced bandwidth attributed to the wireless environment places limitations on the rate of the requests. Broadcasting (one-way communication) has been suggested as a possible solution to this limitation. In broadcasting, information is provided to all users of the air channels. Mobile users are capable of searching the air channels and pulling the desired data. The main advantage of broadcasting is that it scales up as the number of users increases and, thus, eliminates the need to multiplex the bandwidth among users accessing the air channel. Furthermore, broadcasting can be considered as an additional storage available over the air for mobile clients. Within the scope of broadcasting one needs to address three issues:

- effective data organization on the broadcast channel,
- efficient data retrieval from the broadcast channel, and
- data selection.

The goal is to achieve high performance (response time) while minimizing energy consumption. Note that the response time is a major source of power consumption at the mobile unit (Imielinski & Badrinath, 1994; Imielinski & Korth, 1996; Imielinski, Viswanathan, & Badrinath, 1997; Weiser, 1993). As a result, the reduction in response time translates into reducing the amount of time a mobile unit spends accessing the channel(s) and thus has its main influence on conserving energy at the mobile unit.

Chapter Organization

In this chapter, we first introduce the necessary background material. Technological limitations are outlined and their effects on the global information-sharing environment are discussed. Issues such as tree-based indexing, signature-based indexing, data replication, broadcasting over single and parallel channels, data distribution, conflict, and data access are enumerated and analyzed next. Then we present solutions to these issues with respect to the network latency, access latency, and power management. Finally, we conclude the chapter and point out some future research directions.

Mobile Computing

The mobile computing environment is composed of a number of network servers enhanced with wireless transceivers—mobile support stations (MSSs) and a varying number of mobile hosts (MHs) free to move at will (Figure 1).

The role of the MSS is to provide a link between the wireless network and the wired network. The link between an MSS and the wired network could be either wireless (shown as a dashed line) or wire based. The area covered by the individual transceiver is referred to as a cell. To satisfy a request, an MH accesses the MSS responsible for the cell where the MH is currently located. It is the duty of the MSS to resolve the request and deliver the result back to the client. Once an MH moves across the boundaries of two cells, a handoff process takes place between the MSSs of the corresponding cells. The MH is normally small, lightweight, and portable. It is designed to be compact with limited resources relying on temporary power supplies (such as batteries) as its main power source.

Characteristics of the Mobile Environment

Wireless communication is accomplished via modulating radio waves or pulsing infrared light. Table 1 summarizes a variety of mobile network architectures. Mainly, three

Figure 1. Architecture of the mobile-computing environment

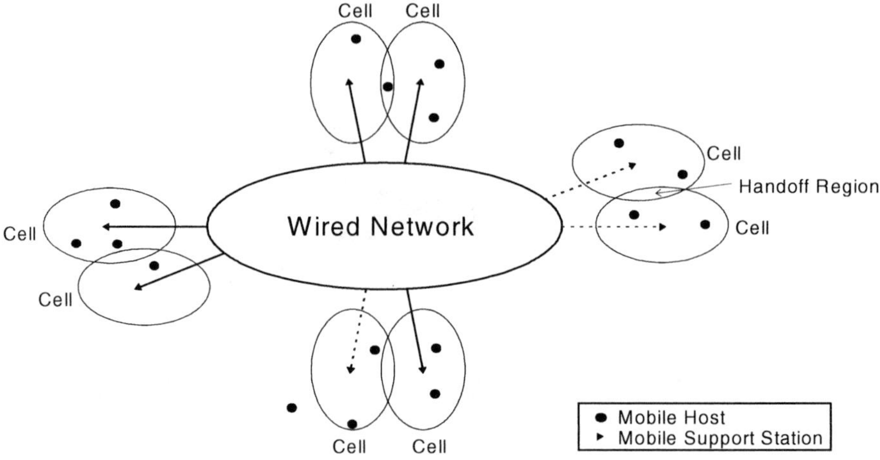

characteristics distinguish the mobile computing environment from traditional wired computing platforms, namely, wireless medium, mobility, and portability.

Wireless Medium

The common ground among all wireless systems is the fact that communication is done via the air (and not via cables). This fact changes a major underlying assumption behind the conventional distributed algorithms. The physical layer of the connection is no longer the reliable coaxial or optic cable. Communication over the air is identified by frequent disconnections, low data-rate, high cost, and lack of security (Alonso & Ganguly, 1992; Alonso & Korth, 1993; Chlamtac & Lin, 1997; Imielinski & Badrinath, 1994; Imielinski & Korth, 1996; Imielinski et al., 1997; Weiser, 1993).

Mobility

Mobility introduces new challenges beyond the scope of the traditional environment. Mobile devices can be used at multiple locations and in transition between these locations. Mobility results in several issues including disconnections due to handoff processes, motion management, location-dependent information, heterogeneous and fragmented networks, security, and privacy.

Portability

There are many variations of portable computer systems with different physical capabilities. However, they share many common characteristics such as limited memory, processing power, and power source. The ideal goal would be to develop a device that is compact, durable, lightweight, and that consumes a minimum amount of power. Table 2 highlights some limitations of the mobile environment.

Broadcasting

The cost of communication is normally asymmetric: Sending information requires 2 to 10 times more energy than receiving the information (Imielinski, Viswanathan, & Badrinath, 1994). In the case of accessing public information, instead of the two-way, on-demand, traditional communication pattern, popular public information can be generated and disseminated over the air channel. The MH requiring the information can tune to the broadcast and access the desired information from the air channel.

In general, data can be broadcast either on one or several channels. Broadcasting has been used extensively in multiple disciplines, that is, management of communication systems (Comer, 1991) and distributed database environments (Bowen, 1992). In this chapter, the term *broadcast* is referred to as the set of all broadcast data elements (the stream of data across all channels). A broadcast is performed in a cyclic manner. The MH can only read from the broadcast, whereas the database server is the only entity that can write to the broadcast.

In the data-broadcasting application domain, power consumption and network latency are proven constraints that limit "timely and reliable" access to information. The necessity of minimizing power consumption and network latency lies in the limitation of current technology. The hardware of the mobile units have been designed to mitigate this

Table 1. Mobile network architectures

Architecture	Description
Cellular Networks	• Provides voice and data services to users with handheld phones • Continuous coverage is restricted to metropolitan regions • Movement over a wide area may need user to inform the network of the new location • Low bandwidth for data-intensive applications • Could be based on either analog technology or digital technology
Wireless LANs	• A traditional LAN extended with a wireless interface • Serves small, low-powered, portable terminals capable of wireless access • Connected to a more extensive backbone network, such as a LAN or WAN
Wide Area Wireless Networks	• Special mobile radio networks provided by private service providers (RAM, ARDIS) • Provides nationwide wireless coverage for low-bandwidth data services, including e-mail or access to applications running on a fixed host
Paging Networks	• Receive-only network • No coverage problems • Low bandwidth • Unreliable

Satellite Networks	• Unlike the static, grounded MSSs, satellites are not fixed
	• Normally classified based on their altitudes (from earth) into three classes:
	  • Low Earth Orbit Satellites (LEOS)
	  • Medium Earth Orbit Satellites (MEOS)
	  • Geostationary Satellites (GEOS)

Table 2. Limitations of the mobile environment

Limitations	Concerns/Side Effects
Frequent Disconnections	• Handoff blank out in cellular networks • Long down time of the mobile unit due to limited battery power • Voluntary disconnection by the user • Disconnection due to hostile events (e.g., theft, destruction) • Roaming off outside the geographical coverage area of the window service
Limited Communication Bandwidth	• Quality of service (QoS) and performance guarantees • Throughput and response time and their variances • Efficient battery use during long communication delays
Heterogeneous and Fragmented Wireless Network Infrastructure	• Rapid and large fluctuations in network QoS • Mobility transparent applications perform poorly without mobility middleware or proxy • Poor end-to-end performance of different transport protocols across network of different parameters and transmission characteristics

limitation by operating in various operational modes such as active, doze, sleep, nap, and so forth to conserve energy. A mobile unit can be in active mode (maximum power consumption) while it is searching or accessing data; otherwise, it can be in doze mode (reduced power consumption) when the unit is not performing any computation. Along with the architectural and hardware enhancements, efficient power management and energy-aware algorithms can be devised to manage power resources more effectively. In addition, appropriate retrieval protocols can be developed to remedy network latency and hence to allow faster access to the information sources. In general, two issues need to be considered.

- The MH should not waste its energy in continuously monitoring the broadcast to search for information. As a result, the information on the broadcast should be organized based on a disciplined order. Techniques should be developed to (a) instruct the MH of the availability of the data element on the broadcast and (b) if the data element is available, instruct the MH of the location of the data element on the broadcast.
- An attempt should be made to minimize the response time. As will be seen later, this is achieved by shortening the broadcast length and/or reducing the number of passes over the air channel(s).

Data Organization on the Air Channel

Unlike the conventional wired environment, where a disk is assumed to be the underlying storage, data in the mobile environment are stored on air channel(s). A disk and an air channel have major structural and functional differences. The disk has a three-dimensional structure (disks can have a four-dimensional structure if multiple disks are used — redundant arrays of independent disks [RAID]). An air channel, on the other hand, is a one-dimensional structure. The disk has a random-access feature and the air channel is sequential in nature. Finally, the current raw data rate of a disk is generally much higher than that of the air channel.

Zdonik, Alonso, Franklin, and Acharya (1994) and Acharya, Alonso, Franklin, and Zdonik (1995) investigated the mapping of disk pages onto a broadcast channel and the effects of that mapping on the management of cache at the MH. In order to place disk pages onto the data channel, the notion of multiple disks with different sizes spinning at multiple speeds was used. Pages available on faster spinning disks get mapped more frequently than those available on slower disks. In cache management, a nonconventional replacement strategy was suggested. Such a policy assumed that the page to be replaced might not be the least-recently used page in the cache. This is justifiable since the set of pages that are most frequently in demand are also the most frequently broadcast. This work was also extended to study the effect of prefetching from the air channel into the cache of the MH. These efforts assumed the same granularity for the data items on air channel and disk pages: if a data item is to be broadcast more frequently (replicated), the entire page has to be replicated. In addition, due to the plain structural nature of the page-based environment, the research looked at the pages as abstract entities and was not meant to consider the contents of the pages (data and its semantics) as a means to order the pages. In object-oriented systems, semantics among objects greatly influence the method in which objects are retrieved and, thus, have their direct impact on the ordering of these objects or pages. In addition, the replication should be performed at the data item granularity level.

An index is a mechanism that speeds up associative searching. An index can be formally defined as a function that takes a key value and provides an address referring to the location of the associated data. Its main advantage lies in the fact that it eliminates the need for an exhaustive search through the pages of data on the storage medium. Similarly, within the scope of broadcasting, an index points to the location or possible availability of a data item on the broadcast, hence, allowing the mobile unit to predict the arrival time of the data item requested. The prediction of the arrival time enables the mobile unit to switch its operational mode into an energy-saving mode. As a result, an indexing mechanism facilitates data retrieval from the air channel(s), minimizing response time

Table 3. Advantages and disadvantages of indexing schemes

Advantages	Disadvantages
Provides auxiliary information that allows mobile users to predict arrival time of objects	Longer broadcast
Enables utilization of different operational modes (active, nap, doze, etc.)	Longer response time
Reduces power consumption (less tune-in time)	Computational overhead due to complexity in retrieval, allocation, and maintenance of the indexes

while reducing power consumption. Table 3 summarizes the advantages and disadvantages of indexing schemes.

The literature has addressed several indexing techniques for a single broadcast channel as well as parallel broadcast channels with special attention to signature-based indexing and tree-based indexing (Boonsiriwattanakul, Hurson, Vijaykrishnan, & Chehadeh, 1999; Chehadeh, Hurson, & Miller, 2000; Chehadeh, Hurson, & Tavangarian, 2001; Hu & Lee, 2000, 2001; Imielinski et al., 1997; Juran, Hurson, & Vijaykrishnan, 2004; Lee, 1996).

Signature-Based Indexing

A signature is an abstraction of the information stored in a record or a file. The basic idea behind signatures on a broadcast channel is to add a control part to the contents of an information frame (Hu & Lee, 2000, 2001; Lee, 1996). This is done by applying a hash function to the contents of the information frame, generating a bit vector, and then superimposing it on the data frame. As a result, a signature partially reflects the data content of a frame. Different allocations of signatures on a broadcast channel have been studied; among them, three policies, namely, *single signature, integrated signature*, and *multilevel signature*, are studied in Hu and Lee (2000) and Lee (1996).

During the retrieval, a query is resolved by generating a signature based on the user's request. The query signature is then compared against the signatures of the data frames in the broadcast. A successful match indicates a possible hit. Consequently, the content of the corresponding information frame is checked against the query to verify that it corresponds to the user's demands. If the data of the frame corresponds to the user's request, the data is recovered; otherwise, the corresponding information frame is ignored. In general, this scheme reduces the access time and the tune-in time when pulling information from the air channel.

Tree-Based Indexing

Two kinds of frames are broadcast on the air channel: data frames and index frames. The index frame contains auxiliary information representing one or several data attributes pointing to the location of data collection (i.e., information frames) sharing the same common attribute value(s). This information is usually organized as a tree in which the lowest level of the tree points to the location of the information frames on the broadcast channel.

A broadcast channel is a sequential medium and, hence, to reduce the mobile unit's active and tune-in time, and consequently to reduce the power consumption, the index frames are usually replicated and interleaved with the data frames. Two index replication schemes (namely, *distributed indexing* and *(1, m) indexing*) have been studied in Imielinski et al. (1997). In distributed indexing, the index is partitioned and interleaved in the broadcast cycle (Hu & Lee, 2000, 2001; Lee, 1996). Each part of the index in the broadcast is followed by its corresponding data frame(s). In (1, *m*) indexing, the entire index is interleaved *m* times during the broadcast cycle (Imielinski et al., 1997; Lee 1996) — the whole index is broadcast before every 1/*m* fraction of the cycle.

Previous work has shown that the tree-based indexing schemes are more suitable for applications where information is accessed from the broadcast channel randomly, and the signature-based indexing schemes are more suitable in retrieving sequentially structured data elements (Hu & Lee, 2000, 2001). In addition, tree-based indexing schemes have shown superiority over the signature-based indexing schemes when the user request is directed towards interrelated objects clustered on the broadcast channel(s). Furthermore, tree-based indexing schemes relative to signature-based indexing schemes are more suitable in reducing the overall power consumption. This is due to the fact that a tree-based indexing provides global information regarding the physical location of the data frames on the broadcast channel. On the other hand, signature-based indexing schemes are more effective in retrieving data frames based on multiple attributes (Hu & Lee, 2000). Table 4 compares and contrasts the signature- and tree-based indexing.

Data Organization on a Single Channel

An appropriate data placement algorithm should attempt to detect data locality and cluster related data close to one another. An object-clustering algorithm takes advantage of semantic links among objects and attempts to map a complex object into a linear sequence of objects along these semantic links. It has been shown that such clustering can improve the response time by an order of magnitude (Banerjee, Kim, Kim, & Garza, 1988; Chang & Katz, 1989; Chehadeh, Hurson, Miller, Pakzad, & Jamoussi, 1993; Cheng & Hurson, 1991a). In the conventional computing environment, where data items are stored on disk(s), the clustering algorithms are intended to place semantically connected objects physically along the sectors of the disk(s) close to one another (Cheng & Hurson, 1991a). The employment of broadcasting in the mobile computing environment motivates the need to study the proper data organization along the sequential air channel. Figure 2 depicts a weighted directed acyclic graph (DAG) and the resulting clustering sequences achieved when different clustering techniques are applied.

Table 4. Signature-based versus tree-based indexing

Feature	Signature-Based Indexing	Tree-Based Indexing
Less power consumption		✓
Longer length of broadcast	✓	✓
Computational overhead	✓	✓
Longer response time	✓	✓
Shorter tune-in time		✓
Random data access		✓
Sequentially structured data	✓	
Clustered data retrieval		✓
Multi-attribute retrieval	✓	

In order to reduce the response time, the organization of data items on an air channel has to meet the following three criteria.

- **Linear ordering.** The one-dimensional sequential access structure of the air channel requires that the object ordering be linear. In a DAG representation of a complex object, an edge between two nodes could signify an access pattern among the two nodes. The *linearity* property is defined as follows: If an edge exists between two objects, o_1 and o_2, and in the direction $o_1 \rightarrow o_2$, then o_1 should be placed prior to o_2.
- **Minimum linear distance between related objects.** In a query, multiple objects might be retrieved following their connection patterns. Intuitively, reducing the distance among these objects along the broadcast reduces the response time and power consumption.
- **More availability for popular objects.** In a database, not all objects are accessed with the same frequency. Generally, requests for data follow the 20/80 rule — a popular, small set of the data (20%) is accessed the majority of the time (80%). Considering the sequential access pattern of the broadcast channel, providing more availability for popular objects can be achieved by simply replicating such objects.

Figure 2. Graph and various clustering methods

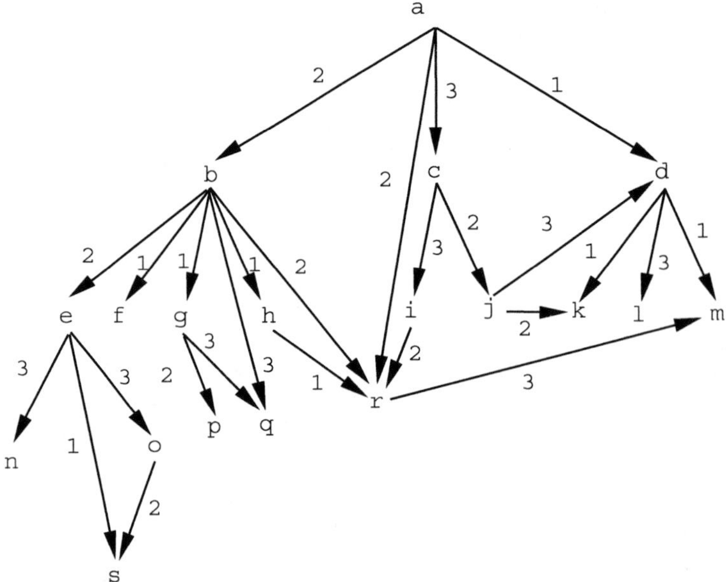

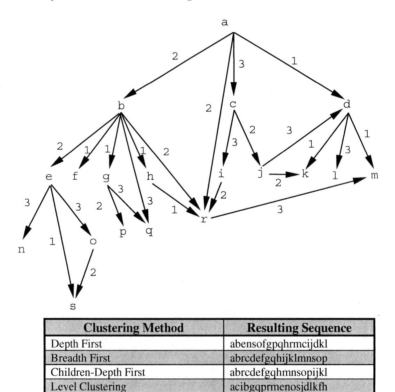

Clustering Method	Resulting Sequence
Depth First	abensofgpqhrmcijdkl
Breadth First	abrcdefgqhijklmnsop
Children-Depth First	abrcdefgqhmnsopijkl
Level Clustering	acibgqprmenosjdlkfh

Figure 3 depicts a directed graph and multiple linear sequences that satisfy the linear ordering property. The middle columns represent the cost of delays between every two objects connected via an edge. For the sake of simplicity and without loss of generality, a data unit is used as a unit of measurement. Furthermore, it is assumed that all data items are of equal size. The cost associated with an edge between a pair of data items is calculated by counting the number of data items that separate these two in the linear sequence. For example, in the abfgchdeij sequence, data items a and d are separated by the sequence bfgch and thus have a cost of 6. The rightmost column represents the total cost associated with each individual linear sequence. An optimal sequence is the linear sequence with the minimum total sum. In a query where multiple related objects are retrieved, a reduced average linear distance translates into smaller average response time. In this example, the best linear sequence achieves a total sum of 26.

Figure 3. Graph, linear sequences, and costs

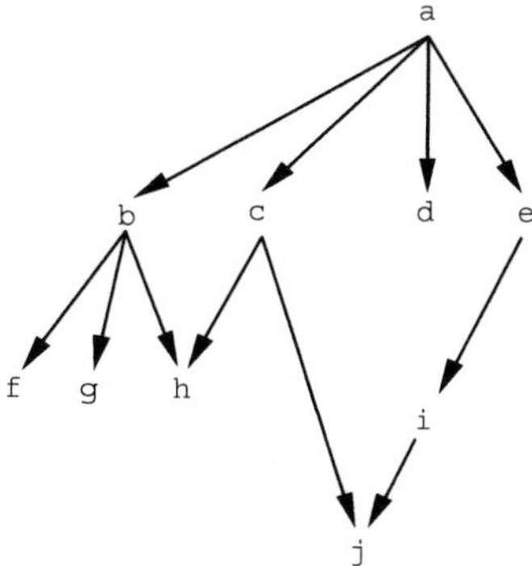

	Linear Sequence	Individual Costs											Total Cost
		ab	ac	ad	ae	bf	bg	bh	ch	cj	ei	ij	
1	abfgchdeij	1	4	6	7	1	2	4	1	5	1	1	33
2	abfgcheijd	1	4	9	6	1	2	4	1	4	1	1	34
3	abcdefghij	1	2	3	4	4	5	6	5	7	4	1	42
4	abgfeichjd	1	6	9	4	2	1	6	1	2	1	3	36
5	acdeijbhgf	6	1	2	3	3	2	1	6	4	1	1	30
6	adeicjbhgf	6	4	1	2	3	2	1	3	1	1	2	26
7	adecbihgfj	4	3	1	2	4	3	2	3	6	3	4	35
8	adecbhgfij	4	3	1	2	3	2	1	2	6	6	1	31
9	adecijbhgf	6	3	1	2	3	2	1	4	2	2	1	27
10	adbfgcheij	2	5	1	7	1	2	4	1	4	1	1	29
11	adceijbhgf	6	2	1	3	3	2	1	5	3	1	1	28
12	aeidcjbhgf	6	4	3	1	3	2	1	3	1	1	3	28
13	aedcbihgfj	4	3	2	1	4	3	2	3	6	4	4	36
14	aedcijbhgf	6	3	2	1	3	2	1	4	2	3	1	28

Data Organization on Parallel Channels

The broadcast length is a factor that affects the average response time in retrieving data items from the air channel — reducing the broadcast length could also reduce the response time. The broadcast length can be reduced if data items are broadcast along parallel air channels.

Formally, we attempt to assign the objects from a weighted DAG onto multiple channels, while (a) preserving dependency implied by the edges, (b) minimizing the overall broadcast time (load balancing), and (c) clustering related objects close to one another (improving the response time). As one could conclude, there are trade-offs between the second and third requirements: Achieving load balancing does not necessarily reduce the response time in accessing a series of data items.

Assuming that all channels have the same data rate, one can draw many analogies between this problem and static task scheduling in a homogeneous multiprocessor environment — tasks are represented as a directed graph $D \equiv (N, A)$, with nodes (N) and directed edges (A) representing processes and dependence among the processes, respectively. Compared to our environment, channels can be perceived as processors (PEs), objects as tasks, and the size of a data item as the processing cost of a task. There is, however, a major distinction between the two environments. In the multiprocessor environment, information is normally communicated among the PEs, while in the multi-channel environment there is no data communication among channels.

The minimum makespan problem, in static scheduling within a multiprocessor environment, attempts to find the minimum time in which n dependent tasks can be completed on m PEs. An optimal solution to such a problem is proven to be NP hard. Techniques such as graph reduction, max-flow min-cut, domain decomposition, and priority list scheduling have been used in search of suboptimal solutions. Similar techniques can be developed to assign interrelated objects closely over parallel channels.

Distribution of data items over the broadcast parallel air channels brings the issue of access conflicts between requested data items that are distributed among different channels. The access conflict is due to two factors:

- the receiver at the mobile host can only tune into one channel at any given time, and
- the time delay to switch from one channel to another.

Access conflicts require the receiver to wait until the next broadcast cycle(s) to retrieve the requested information. Naturally, multiple passes over the broadcast channels will have a significant adverse impact on the response time and power consumption.

Conflicts in Parallel Air Channels

Definition 1. A K-data item request is an application request intended to retrieve K data items from a broadcast.

It is assumed that each channel has the same number of pages (frames) of equal length and, without loss of generality, each data item is residing on only a single page. A single broadcast can be modeled as an $N \times M$ grid, where N is the number of pages per broadcast and M is the number of channels. In this grid, K data items ($0 \leq K \leq MN$) are randomly distributed throughout the MN positions of the grid. Based on the common page size and the network speed, the time required to switch from one channel to another is equivalent to the time it takes for one page to pass in the broadcast. Thus, it is impossible for the mobile unit to retrieve both the ith page on Channel A and $(i + 1)$th page on Channel B (where A ≠ B). Figure 4 is a grid model that illustrates this issue.

Definition 2. Two data items are defined to be in conflict if it is impossible to retrieve both on the same broadcast.

In response to a user request, the access latency is then directly dependent on the number of passes over the broadcast channels. One method of calculating the number of required passes over the broadcast channels is to analyze the conflicts between data items. For any particular data item, all data items in the same or succeeding page (column) and on a different row (channel) will be in conflict. Thus, for any specific page (data object) in the grid, there are $(2M - 2)$ conflicting pages (data items) in the broadcast (The last column has only $M - 1$ conflict positions, but it is assumed that N is sufficiently large to make this difference insignificant.) These $(2M - 2)$ positions are known as the conflict region.

For any particular data item, it is possible to determine the probability of exactly i conflicts occurring, or $P(i)$. Because the number of conflicts for any particular data item is bounded by $(M - 1)$, the weighted average of these probabilities can be determined by summing a finite series. This weighted average is the number of broadcasts (passes) required to retrieve all K data items if all conflicts between data items are independent.

Figure 4. Sample broadcast with $M = 4$, $N = 6$, and $K = 8$

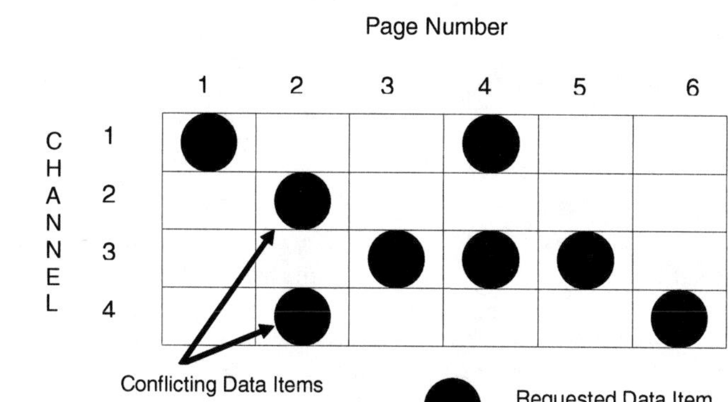

$$B = \sum_{i=0}^{M-1} (i + 1) * P(i) \quad (1)$$

Access Patterns

In order to reduce the impact of conflicts on the access time and power consumption, retrieval procedures should be enhanced by a scheduling protocol that determines data retrieval sequence during each broadcast cycle. The scheduling protocol we proposed is based on the following three prioritized heuristics:

1) Eliminate the number of conflicts
2) Retrieve the maximum number of data items
3) Minimize the number of channel switches

The scheme determines the order of retrieval utilizing a forest - an *access forest.* An access forest is a collection of trees (*access trees*), where each access tree represents a collection of access patterns during a broadcast cycle. Naturally, the structure of the access forest, that is, the number of trees and the number of children that any parent can have, is a function of the number of broadcast channels.

Definition 3. An access tree is composed of two elements: nodes and arcs.

- **Node.** A node represents a requested data item. The nodes are labeled to indicate its conflict status: mnemonically, C_1 represents when the data item is in conflict with another data item(s) in the broadcast and C_0 indicates the lack of conflict.

Each access tree in the access forest has a different node as a root-the root is the first accessible requested data item on a broadcast cycle. This simply implies that an access forest can have at most n trees where n is the number of broadcast channels.

- **Arcs.** The arcs of the trees are weighted arcs. A weight denotes whether or not channel switching is required in order to retrieve the next scheduled data item in the access pattern. A branch in a tree represents a possible access pattern of data items during a broadcast cycle with no conflicts. Starting from the root, the total number of branches in the tree represents all possible access patterns during a broadcast cycle.

This scheme allows one to generate all possible nonconflicting, weighted access patterns from all channels. The generated access patterns are ranked based on their weights-a weight is set based on the number of channel switches-and then the one(s) that allows the maximum number of data retrievals with minimum number of channel switches is selected. It should be noted that the time needed to build and traverse the access forest is a critical factor that must be taken into account to justify the validity of this approach. The following working example provides a detailed guide to illustrate the generation of the access patterns for each broadcast cycle.

1) **Search.** Based on the user's query, this step determines the offset and the channel number(s) of the requested objects on the broadcast channels. Figure 5 depicts a request for eight data items from a parallel broadcast channel of four channels.

2) **Generation of the access forest.** For each broadcast channel, search for the requested data item with the smallest offset (these objects represent the roots of an access tree). For the example, the data items with the smallest offsets are O_1, O_3, O_6 and O_8. Note that the number of access trees is upper bounded by the number of broadcast channels.

3) **Root assignment.** For each channel with at least one data item requested, generate a tree with root node as determined in Step 2. The roots are temporarily tagged as C_0.

4) **Child assignment.** Once the roots are determined, it is necessary to select the child or children of each rooted access tree: For each root, and relative to its position on the air channel, the algorithm determines the closest nonconflicting data items on each channel. With respect to a data item $O_{i,x}$ at location X on air channel i ($1 \le i \le n$), the closest nonconflicting data item is either the data item $O_{i,x+1}$ or the data item $O_{j,x+2}$, $j \ne i$. If the child is in the same broadcast channel as the root, the arc is weighted as 0; otherwise it is weighted as 1. Each added node is temporarily tagged as C_0. Figure 6 shows a snapshot of the example after this step.

5) **Root label update.** Once the whole set of requested data items is analyzed and the access forest is generated, the conflict labels of the nodes of each tree are updated. This process starts with the root of each tree. If a root is in conflict with any other

Figure 5. A parallel broadcast of four channels with eight requested data items

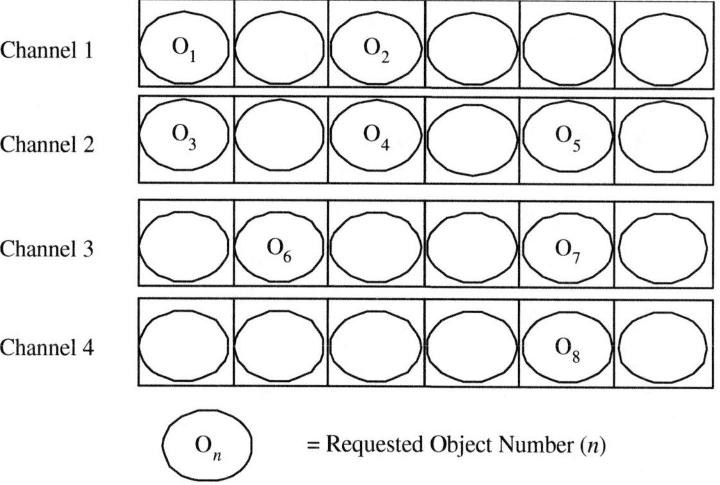

root(s), a label of C_1 is assigned to all the roots involved in the conflict, otherwise the preset value of C_0 is maintained.

6) **Node label update.** Step 5 will be applied to the nodes in the same level of each access tree in the access forest. As in Step 5, a value of C_1 is assigned to the nodes in conflict. Figure 7 shows the example with the updated labels.

7) **Sequence selection.** The generation of the access forest then allows the selection of the suitable access patterns in an attempt to reduce the network latency and power consumption. A suitable access pattern is equivalent to the selection of a tree branch that:
   - has the most conflicts with other branches,
   - allows more data items to be pulled off the air channels, and
   - requires the least number of channel switches.

The O_3, O_4, and O_5 sequence represents a suitable access pattern for our running example during the first broadcast cycle. Step 7 will be repeated to generate access patterns for different broadcast cycles. The algorithm terminates when all the requested data items are covered in different access patterns. The data item sequence O_1, O_2, and O_7 and data

Figure 6. Children of each root

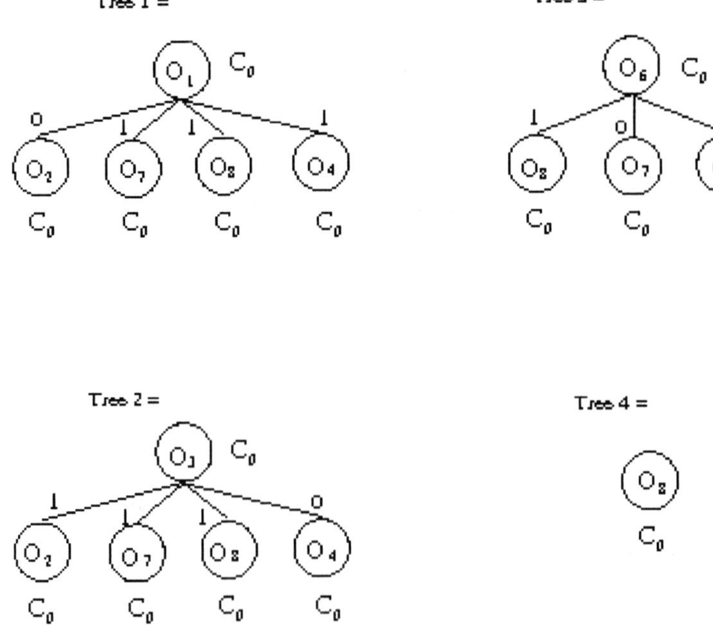

Figure 7. Final state of the access forest

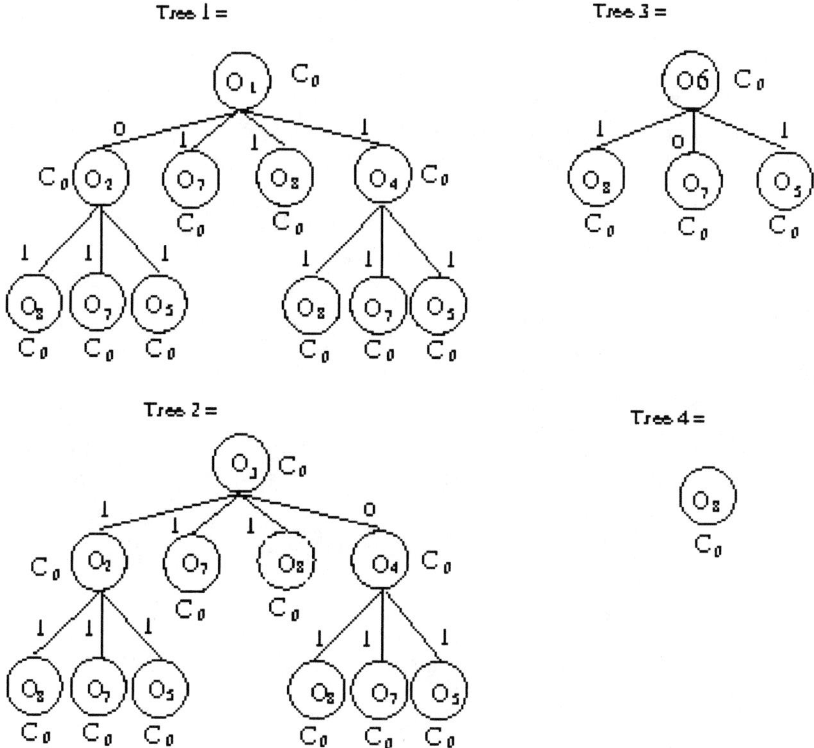

item sequence O_6 and O_8 represent the last two patterns for retrieving all of the data items requested in the example.

Data Organization on a Single Channel

As noted in the literature, the object-oriented paradigm is a suitable methodology for modeling public data that are by their very nature in multimedia format (Atkinson, Bancilhon, DeWitt, Dittrich, Maier, & Zdonik, 1989; Fong, Kent, Moore, & Thompson, 1991; Hurson, Pakzad, & Cheng, 1993; Kim, 1990). In addition, object-oriented methodology provides a systematic mechanism to model a complex object in terms of its simpler components.

In this section, without loss of generality, we model information units as objects. Object clustering has proven to be an effective means of data allocation that can reduce response times (Banerjee et al., 1988; Chang & Katz 1989; Chehadeh et al., 1993; Cheng & Hurson, 1991b; Lim, Hurson, Miller, & Chehadeh, 1997). In our research, we investigated two heuristic allocation strategies. The first strategy assumes a strict linearity requirement and deals with nonweighted DAGs. The second approach relaxes such restriction in favor of clustering strongly related objects closer to one another and consequently deals with weighted DAGs.

Strict Linearity: ApproximateLinearOrder Algorithm

Definition 4. An independent node is a node that has either one or no parent. A graph containing only independent nodes makes up a forest.

Heuristic Rules

1) Order the children of a node based on their number of descendants in ascending order.
2) Once a node is selected, all of its descendants should be visited and placed on the sequence in a depth-first manner, without any interruptions from breadth siblings.
3) If a node has a nonindependent child, with all of its parents already visited, the nonindependent child should be inserted in the linear sequence before any independent child.

The ApproximateLinearOrder algorithm implements these heuristics and summarizes the sequence of operations required to obtain a linear sequence. The algorithm assumes a greedy strategy and starts by selecting a node with an in-degree of zero and out-degree of at least one.

ApproximateLinearOrder Algorithm

traverse DAG using DFS traversal and as each node is traversed
append the traversed node *N* to the sequence
remove *N* from {nodes to be traversed}
if {nonindependent children of *N* having all their parents in the sequence} $\neq \varnothing$
  Set $\leftarrow$ {nonindependent children of *N* having all their parents in the sequence}
else
  if {independent children of *N*} $\neq \varnothing$
    Set $\leftarrow$ {independent children of *N*}
NextNode $\leftarrow$ node $\in$ *Set* | node has least # of descendants among the nodes in *Set*

Applying this algorithm to the graph of Figure 3 generates either the 5th or 11th sequence — dependent on whether c or d was chosen first as the child with the least number of independent children. As one can observe, neither of these sequences is the optimal sequence. However, they are reasonably better than other sequences and can practically be obtained in polynomial time. It should be noted that nodes not connected to any other nodes — nodes with in-degree and out-degree of zero — are considered harmful and thus are not handled by the algorithm. Having them in the middle of the sequence introduces delays between objects along the sequence. Therefore, we exclude them from the set of nodes to be traversed and handle them by appending them to the end of the sequence. In addition, when multiple DAGs are to be mapped along the air channel, the mapping should be done with no interleaving between the nodes of the DAGs.

Varying Levels of Connectivity: PartiallyLinearOrder Algorithm

In a complex object, objects are connected through semantic links with different degrees of connectivity. The different access frequency of objects in an object-oriented database reveals that some patterns are more frequently traversed than others (Fong et al., 1991). This observation resulted in the so-called PartiallyLinearOrder algorithm that assumes a weighted DAG as its input and produces a linear sequence. It combines the nodes (single_node) of the graph into multi_nodes in descending order of their connectivity (semantic links). The insertion of single_nodes within a multi_node respects the linear order at the granularity level of the single_nodes. The multi_nodes are merged (with multi_nodes or single_nodes) at the multi_node granularity, without interfering with internal ordering sequences of a multi_node. Figure 8 shows the application of the PartiallyLinearOrder algorithm.

PartiallyLinearOrder Algorithm

```
for every weight w_s in descending order
  for every two nodes N_1 & N_2 connected by w_s
    merge N_i & N_j into one multi_node
  for every multi_node MN
    w_m = w_s - 1
    for every weight w_m in descending order
      while ∃ adjacent_node AN connected to MN
        if ∃ an edge in both directions between MN & AN
        compute WeightedLinearDistance_MN_AN & WeightedLinearDistance_AN_MN
        merge MN & AN into one multi_node, based on the appropriate direction
```

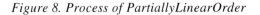

Figure 8. Process of PartiallyLinearOrder

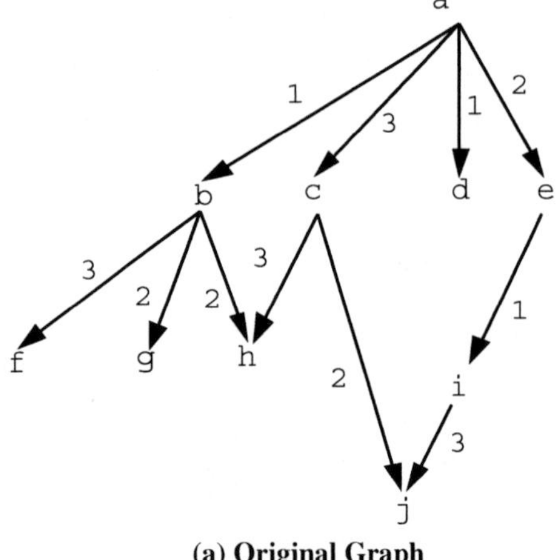

(a) Original Graph

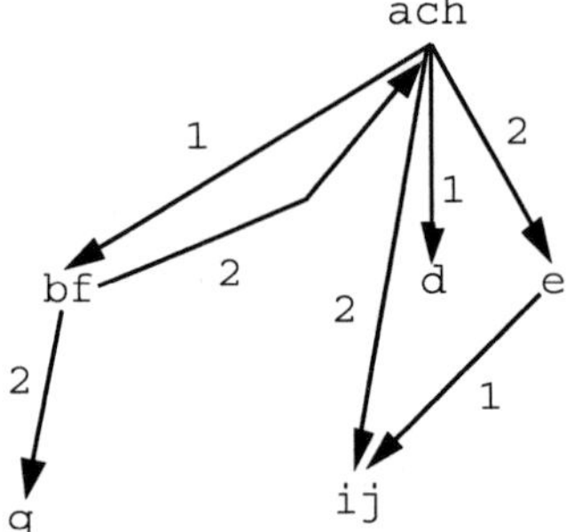

(b) First and Second Iterations

bfgacheijd

(c) Third Iteration

Performance Evaluation

Parameters

A simulator was developed to study the behavior of the proposed mapping algorithms based on a set of rich statistical parameters. Our test bed was an object-oriented financial database. The OO7 benchmark was chosen to generate the access pattern graphs. We used the NASDAQ exchange (NASDAQ, 2002) as our base model, where data is in both textual and multimedia (graphics — i.e., graphs and tables) formats. Table 5 shows a brief description of the input and output parameters. The simulator is designed to measure the average access delay for the various input parameters. Table 6 provides a listing of the input parameters along with their default values and possible ranges. The default values are set as the value of the parameter when other parameters are varied during the course of the simulation. The ranges are used when the parameter itself is varied.

Results

The simulator operates in two stages.

- Structuring the access-pattern object graph, based on certain statistical parameters, and mapping it along the air channel using various mapping algorithms. To get a wide spectrum of possible graphs, parameters such as (a) the percentage of nonfree nodes, (b) the depths of the trees within the graph, and (c) the amount of sharing that exists between trees through nonfree nodes that were varied. Varying these statistical parameters, we generated 500 access graphs that were used as part of our test bed. In addition, we simulated three mapping algorithms: a nonlinear, children-depth-first clustering algorithm (Banerjee et al., 1988), and the PartiallyLinearOrder and ApproximateLinearOrder algorithms.
- Generating queries and accessing the requested objects from the air channel. During each run, each query on average accesses 20 objects either through their semantic links or randomly (following the [C:R] value of the next-node ratio). The simulator measured the average access delay. Each point in the curves (Figure 9) is the average result of running the simulator 100,000 times. Finally, we assumed a broadcast data rate of 1Mbit/sec and showed the results in terms of seconds.

Impact of Number of Objects. ApproximateLinearOrder and PartiallyLinearOrder schemes performed better than the conventional children-depth first by taking the linearity issue into consideration. As expected in all three cases, the average access delay increased as the total number of objects increased. The mapping of additional nodes on the broadcast introduced extra delays between the retrievals of two consecutive objects. Taking a closer look at this effect, we observed that this extra delay is mainly due to an increase in the distance for objects that are retrieved randomly (not based on their

Table 5. Description of parameters

Parameter	Description
Input Parameters	
Number of Nodes	Number of objects within the graph (excluding replication)
Object Size	Sizes of objects (small/medium/large)
Object-Size Distribution	Distribution of the sizes of objects within the database
Next-Node Ratio	Connectivity to next node (random or connection)
Out-Degree Distribution	Distribution of the type of nodes based on their out-degrees
Level Distribution	Semantic connectivity of two objects (weak/normal/strong)
Percentage of Popular Objects	Percentage of objects requested more often than others
Replication Frequency	The number of times a popular object is to be replicated
Output Parameter	
Average Access Delay	In a single query, the average delay between accessing two objects

Table 6. Input parameter values

Parameter	Default Value	Ranges
Number of Nodes	5,000	400-8,000
Object Size (in Bytes) • Small • Medium • Large	 $2 \leq o < 20$ $20 \leq o < 7K$ $7K \leq o < 50K$	 2-20 20-7K 7K-50K
Object-Size Distribution [S:M:L]	1:1:1	0-6:0-6:0-6
Next-Node Ratio [C:R]	8:2	0-10:10-0
Out-Degree Distribution [0:1:2:3]	3:3:2:1	1-6:1-6:1-6:1-6
Level Distribution [W:N:S]	1:1:1	1-4:1-4:1-4
Percentage of Popular Objects	20%	10-50%
Replication Frequency	2	1-10

semantic links) since the goal of both algorithms is to cluster semantically related objects close to one another. The ApproximateLinearOrder algorithm outperformed the PartiallyLinearOrder algorithm since the latter attempts to cluster strongly connected objects closer to one another than loosely connected ones and, hence, compromises the linearity property for the loosely connected objects. This compromise overshadows the benefit and is amplified as the number of objects increased. To get a better insight on how our proposed schemes compare with the optimal case, two graphs with 10 nodes were constructed and the optimal sequences exhaustively generated. Using the same set of input values, the average access delay for both proposed schemes were simulated and compared against the average access delay for the optimal sequence. The results of ApproximateLinearOrder and PartiallyLinearOrder were 79% and 76%, respectively, of the access delay of the optimal case.

Size Distribution. In this experiment, we observed that the smallest average access delay took place when the air channel contained smaller data items. However, as the population of data items shifted toward the larger ones, the average access delay increased.

Next-Node Ratio. During the course of a query, objects are either accessed along the semantic links or in a random fashion. At one extreme, when all objects were accessed along the semantic links, the average access delay was minimal. The delay, however, increased for randomly accessed objects. Finally, where all the accesses are on a random basis, clustering (and linearity) does not improve the performance, and all mapping algorithms perform equally.

Out-Degree Distribution. This parameter indicates the number of children of a node within the graph — an out-degree of 0 indicates a sink node. Figure 9 shows the effect of varying the out-degree distribution within the graph structure. The point [9:0:0:0] indicates that all the nodes within the graph have an out-degree of 0, with no semantic link among the objects. This is similar to stating that any access to any object within the graph is done on a random basis. In general, the average access delay is reduced as more connectivity is injected in the access graph. It is interesting to note that it would be more

Figure 9. Average access delay versus connectivity

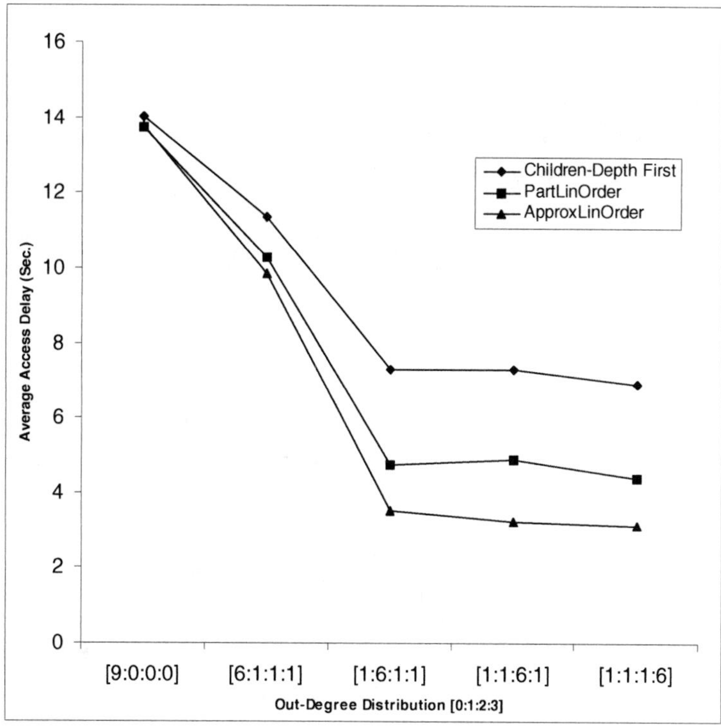

desirable to deal with more, but simpler, objects than with few complex objects on the air channel.

In separate simulation runs, the simulator was also used to measure the effect of varying the percentage of popular objects and the replication frequency. These two parameters have the same effect on the total number of objects on the air channel, however, from the access pattern perspective, the semantic of the accesses are different. In both cases, the average access delay increased as either parameter increased. We also observed and measured the average access delay for different degrees of connectivity among objects. The average access delay for objects connected through strong connections is about 4.3 seconds, whereas it is 7.3 and 7.6 seconds for normally and weakly connected objects, respectively. As would be expected, these results show that the improvement is considerable for the objects connected by a strong connection, but for a normal connection, the performance was close to that of the weak-connection case since the algorithm performs its best optimization for strongly connected objects.

Section Conclusion

In this section, two heuristically based mapping algorithms were discussed, simulated, and analyzed. Performing the mapping in polynomial time was one of the major issues of concern while satisfying linearity, locality, and replication of popular objects. The ApproximateLinearOrder algorithm is a greedy-based approximation algorithm that guarantees the linearity property and provides a solution in polynomial time. The PartiallyLinearOrder algorithm guarantees the linearity property for the strongest related objects and relaxes the linearity requirement for objects connected through looser links. Finally, it was shown that the proposed algorithms offer higher performance than the traditional children-depth-first algorithm.

Data Organization on Parallel Channels

Reducing the broadcast length is one way to satisfy timely access to the information. This could be achieved by broadcasting data items along parallel air channels. This problem can be stated formally as follows: Assign the data items from a weighted DAG onto multiple channels while (a) preserving dependency implied by the edges, (b) minimizing the overall broadcast time (load balancing), and (c) clustering related data items close to one another (improving the response time). Realizing the similarities between these objectives and the task-scheduling problem in a multiprocessor environment, we proposed two heuristic–based, static scheduling algorithms, namely the largest object first (LOF) algorithm and the clustering critical-path (CCP) algorithm.

The Largest Object First Algorithm

This algorithm relies on a simple and localized heuristic by giving priority to larger data items. The algorithm follows the following procedure: For each collection of data items, recursively, a "proper" node with in-degree of 0 is chosen and assigned to a "proper" channel; a "proper" channel is the one with the smallest overall size and a "proper" node is the largest node with in-degree of 0. The assigned node along with all of its out-edges are eliminated from the object DAG. This results in a set of nodes with in-degree of 0. These nodes are added to the list of free nodes and then are selected based on their sizes. This process is repeated until all the nodes of the DAG are assigned.

Definition 5. A free node is a node that either has an in-degree of 0 (no parent) or has all of its parents allocated on a channel. A free node is a candidate node available for allocation.

Assuming that there are *n* nodes in the graph, the algorithm requires the traversal of all the nodes and thus requires *n* steps. At each step, the algorithm searches for the largest available node whose parents have been fully allocated. This would require at most $O(n^2)$. Therefore, the overall running time of the algorithm is $O(n^3)$. The LOF algorithm respects the dependency among the nodes, if any, and achieves a better load balancing by choosing the largest object first. This algorithm, however, does not allocate objects based on the degree of connectivity and/or the total size of the descendent objects that could play a significant role in balancing the loads on the channels. In addition, this algorithm does not necessarily cluster related object on the parallel air channels.

LOF Algorithm

1) **repeat (2-4)** until all nodes are assigned
2) assign a free node with the largest weight whose parents are fully allocated to the least-loaded channel
3) remove all out-edges of the assigned node from the DAG
4) insert resulting free nodes into the list of free nodes

The Clustering Critical-Path Algorithm

A critical path is defined as the longest sequence of dependent objects that are accessed serially. A critical path is determined based on the weights assigned to each node. A weight is defined based on several parameters such as the size of the data item, the maximum weight of the descendents, the total weight, and the number of descendents.

Definition 6. A critical node is a node that has a child with an in-degree greater than 1.

Load Balancing

Critical Node effect. Allocate a critical node with the highest number of children with in-degrees greater than 1 first.

Number of children with in-degrees of 1. Allocate nodes with the highest number of children with in-degrees of 1 first. This could free up more nodes to be allocated in parallel channels.

Clustering Related Objects

The weight of a node should be made a function of the weights of the incoming and outgoing edges. The weight of each node is calculated based on Equation 2. It should be noted that:

- There is a trade-off between load balancing and clustering related objects: The allocation strategy for the purpose of load balancing could upset the clustering of related objects and vice versa. Therefore, we propose a factor to balance the two requirements. This factor takes a constant value $\in$ [0,1] and can be assigned to favor either requirement over the other.

- The size of a data item is a multiple of a constant value.
- The weight of an edge is a multiple of a constant value.

$$W = MWC + F\left[S + NCID1 + \sum_{i=1}^{NCIDM} SPC_i - NCIDM(S)\right] + (1-F)\left[(NMIW)MIW + \frac{1}{(NMOW)MOW}\right] \quad (2)$$

where

W	weight of a node
MWC	maximum weight among the node's children
F	factor of optimizing for load balancing versus clustering related objects
S	size of a node (object)
NCID1	number of children with in-degrees of 1
NCIDM	number of children with in-degrees greater than 1
SPC	size of all parent objects
MIW	maximum weight of incoming edges
NMIW	number of maximum-weighted incoming edges
MOW	maximum weight of outgoing edges
NMOW	number of maximum-weighted outgoing edges

The algorithm required to assign the weight of every node in the graph with time complexity of $O(n^2)$ (n is the number of nodes in the DAG) is as follows.

AssignWeights(DAG) Algorithm

1) for every node *i* (Starting at the leaf nodes and traversing the DAG in a breadth-first manner)
2) Calculate SPC_i
3) Calculate W_i

The CCP algorithm takes a DAG as its input and calls the AssignWeights Algorithm. The running time of the CCP algorithm is equal to the running time of AssignWeights plus the running time of the *repeat* loop. The loop has to be repeated *n* times and Line 4 can be done in $O(n)$. Therefore, the overall running time of the CCP algorithm is $O(n^2)$.

CCP(DAG) Algorithm

```
AssignWeights(DAG)
repeat until all the nodes have been processed
        Select the free node N with the largest weight
        if all parents of N are fully allocated on the channels
                place it on the currently least-loaded channel
        else
                Fill up the least-loaded channel(s) with nulls up to the end of the last
                allocated parent of N then place N on it.
```

Performance Evaluation

To evaluate the performance of the proposed algorithms, our simulator was extended to measure the average response time per data item retrieval. To measure the effectiveness of the algorithms across a more unbiased test bed, the degree of connectivity among the data items in the DAG was randomly varied, and 100 different DAGs were generated. In every DAG, the out-degrees of the nodes were determined within the range of 0 and 3. To limit the experimentation running time, a decision was made to limit the number of nodes of each DAG to 60. The weights connecting the nodes, similar to the experiment reported in previously, were categorized as strong, normal, and weak, and were uniformly distributed along the edges of a DAG.

The simulation is accomplished in two steps: In the first step, every DAG is mapped onto the air channels using the LOF and CCP algorithms. In the second step, the simulator simulates the process of accessing the air channels in order to retrieve the data items requested in a query. Among the requested data items, 80% were selected based on their semantic relationship within the DAG and 20% were selected randomly. Finally, the average response time was calculated for 100,000 runs.

1) **Number of Air Channels.** As anticipated, increasing the number of channels resulted in a better response time for both the LOF and CCP. However, this improvement tapered off as the number of channels increased above a certain threshold value, since, additional parallelism provided by the number of channels did not match the number of free nodes available to be allocated, simultaneously. In addition, as expected, the CCP method outperformed the LOF method—the CCP heuristics attempt to smooth the distribution of the objects among the air channels while clustering the related objects.

2) **Out-Degree Distribution.** In general, the CCP method outperformed the LOF method. When the out-degree distribution is biased to include nodes with larger out-degrees (i.e., making the DAG denser), the LOF performance degrades at a much faster rate than the CCP method. This is due to the fact that such bias introduces more critical nodes and a larger number of children per node. The CCP method is implicitly capable of handling such cases.

Figure 10. Load balancing versus clustering

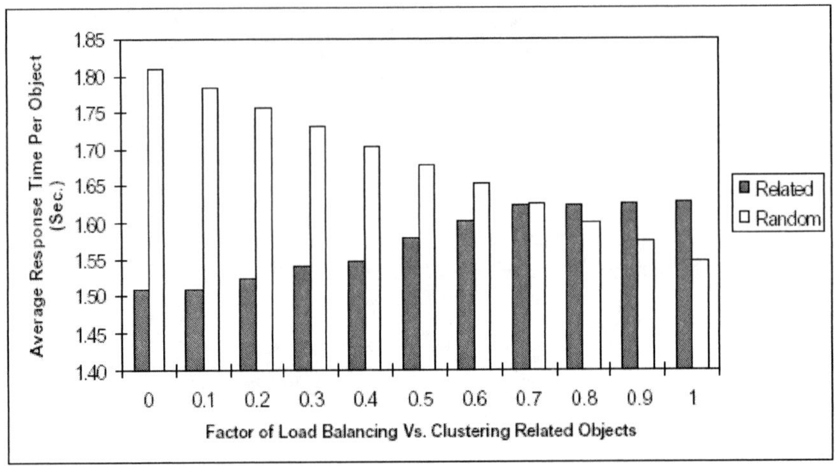

3) **Factor of Load Balancing versus Clustering Related Objects.** To get a better insight on the operations of the CCP method, we analyzed its behavior by varying the load balancing and degree of clustering (*F*; Equation 2). In this experiment, 80% of the data items requested by each query were related through certain semantic links and the rest were selected randomly. As can be seen (Figure 10), biasing in favor of clustering degrades the average response time for randomly selected data items. Optimization based on clustering increases the overall length of the broadcast, thus, contributing to larger response time for randomly accessed objects. For semantically related data items, however, decreasing *F* influenced the broadcast to favor the allocation of related data items closer to one another, thus improving the average response time. Such rate of improvement, however, declined as *F* reached a certain threshold value (0.2 in this case). At this point the behavior of the system reaches a steady state (the objects cannot be brought closer to one another). In different simulation runs, the ratio of randomly selected and semantically related data items varied in the ranges between 30/70% and 70/30% and the same behavior was observed. This figure can be productive in tuning the performance of the CCP method. Assuming a feedback channel is to be used to collect the statistics of the users' access pattern, *F* can be adjusted adaptively to match the access pattern. As an example, if the frequency of accessing data items based on their connection is equivalent to that of accessing data items randomly, then a factor value of 0.7 would generate the best overall response time.

Section Conclusion

This section concentrated on the proper mapping of data items on multiple parallel air channels. The goal was to find the most appropriate allocation scheme that would (a) preserve the connectivity among the data items, (b) provide the minimum overall broadcast time (load balancing), and (c) cluster related data items close to one another (improving the response time). Applying the LOF heuristic showed an improvement in load balancing. However, it proved short in solving the third aforementioned requirement. The CCP algorithm was presented to compensate this shortcoming. Relying on the critical path paradigm, the algorithm assumed several heuristics and showed better performance.

Energy-Efficient Indexing

In this section, we investigate and analyze the usage of indexing and indexed-based retrieval techniques for data items along the single and parallel broadcast channel(s) from an energy-efficient point of view. In general, index-based channel access protocols involve the following steps.

1) **Initial probe.** The client tunes into the broadcast channel to determine when the next index is broadcast.
2) **Search.** The client accesses the index and determines the offset for the requested data items.
3) **Retrieve.** The client tunes into the channel and pulls all the required data items.

In the initial probe, the mobile unit must be in active, operational mode. As soon as the mobile unit retrieves the offset of the next index, its operational mode could change to doze mode. To perform the *Search* step, the mobile unit must be in active mode, and when the unit gets the offset of the required data items, it could switch to doze mode. Finally, when the requested data items are being broadcast (*Retrieve* step), the mobile unit changes its operational mode to active mode and tunes into the channel to download the requested data. When the data is retrieved, the unit changes to doze mode again.

Figure 11. Inner-node structure of single-class and hierarchical schemes

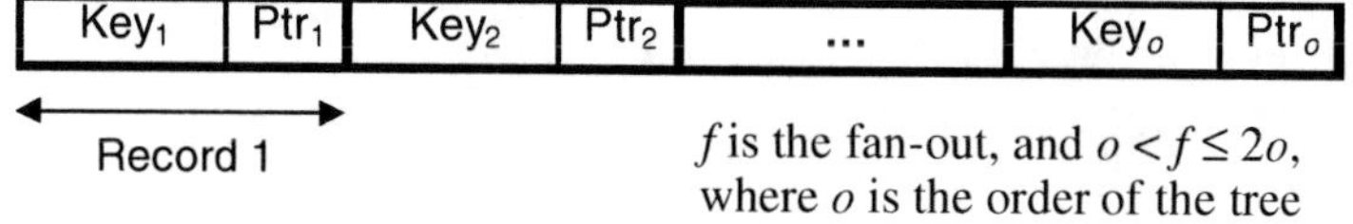

Figure 12. Leaf-node structure of single-class and hierarchical schemes

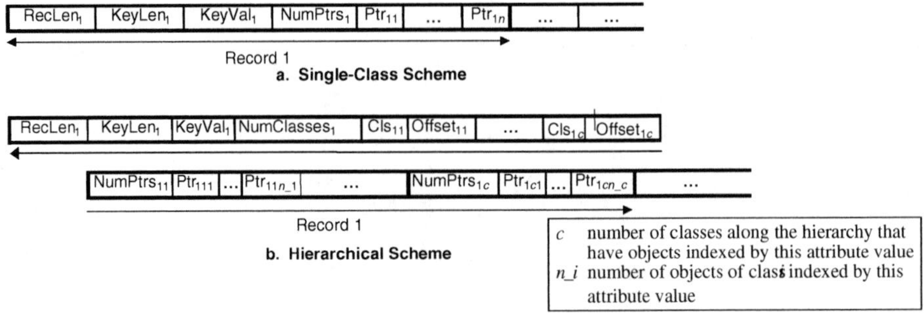

Object-Oriented Indexing

Object-oriented indexing is normally implemented as a multilevel tree. We can classify the possible implementation techniques into two general schemes: single-class indexing and hierarchical indexing. In the single-class scheme, multiple multilevel trees are constructed, each representing one class. In this case, the leaf nodes of each tree point to data items belonging only to the class indexed by that tree. A query requesting all objects with a certain ID has to navigate all these trees. On the other hand, the hierarchical scheme constructs one multilevel tree representing an index for all classes. The same query has to only navigate the common tree.

Data Indexing on a Single Air Channel

We assume an *air-channel page* as the storage granule on the air channel. Due to the sequential nature of the air channel, the allocation of the nodes of a multilevel tree has to follow the navigational path used to traverse the tree, starting at the root. Therefore, an ordering scheme is used to sequentially map the nodes on the air channel. Similarly, data items are allocated onto air channel pages following their index.

Storage Requirement

The overall storage requirement is the sum of the storage required by the inner and leaf nodes. For both schemes, the structure of the inner node is the same (Figure 11). An inner node is a collection of records, where each record is composed of a [*Key*, *pointer*] pair. Assume the order of the tree is o and the fan-out of every node is f ($o \leq f \leq 2o$, except for the root where $2 \leq f \leq 2o$). The leaf node structures of both schemes are shown in Figure 12. As can be seen, the main difference between the two schemes is that the hierarchical scheme requires a list of classes that have data items indexed by the index.

For the sake of simplicity, and without loss of generality, we assume that there are no overflow pages, furthermore assuming the following notations.

P	size of air-channel page
K	average number of distinct keys for an attribute
S	average size of a leaf-node index record in a single-class index
H	average size of a leaf-node index record in a hierarchical index
L	number of leaf-node pages
IN	number of inner-node pages for either scheme

$$L_{Single-Class} = \lceil K / \lfloor P / S \rfloor \rceil \quad (3)$$

$$L_{Hierarchical} = \lceil K / \lfloor P / H \rfloor \rceil \quad (4)$$

$$IN = 1 + \lfloor L / f \rfloor + \lfloor \lfloor L / f \rfloor / f \rfloor + \ldots \quad (5)$$

It should be noted that in the case of a single-class scheme, Equations 3 and 5 should be calculated for all the classes.

Timing Analysis

To perform the timing analysis, one has to consider the domain of a query. The cardinality of the domain of a query is the number of classes to be accessed by the query along the hierarchy. Our timing analysis evaluates the *response* and *active* time as the performance metrics. The response time is defined as the time elapsed between the first user access to the air channel and when the required information is retrieved. The active time is defined as the time during which the mobile unit has to be active accessing the channel. In the timing analysis, we use the number of pages as our unit of measurement. Finally, to support our protocols, we assume that every air-channel page contains control information indicating the location of the first page of the next index. This can simply be implemented as an offset (2 or 4 bytes).

a) **Hierarchical Method.** In this scheme, whether the domain of the query covers one class or all classes along the hierarchy, the same index structure has to be traversed. The protocol is shown below.

Hierarchical Protocol

1) Probe onto channel and get offset to the next index *active*
2) Reach the index *doze*
3) Retrieve the required index pages *active*

4) Reach the required data pages *doze*

5) Retrieve required data pages *active*

- **Response Time.** Assume I_H and D denote the size of the index and data, respectively. On average, it takes half the broadcast (the size of the broadcast is I + D) to locate the index from the initial probe. Once the index is reached, it has to be completely traversed before data pages appear on the broadcast. On average, it takes half the size of the data to locate and retrieve the required data items. Thus, the response time is proportional to:

$$\frac{I_H + D}{2} + I_H + \frac{D}{2} = \frac{3I_H}{2} + D = Broadcast + \frac{I_H}{2} \quad (6)$$

- **Active Time.** The mobile unit's modules have to be active to retrieve a page. Once the index is reached, a number of inner-node pages have to be accessed in order to get and retrieve a leaf-node page. The number of pages to be retrieved at the index is equal to the height of the index tree ($\log_f(D)$). Finally, the amount of the data pages to be read is equal to the number of data items to be retrieved that reside on distinct pages (NODP). Therefore, the active time is:

$$1 + \log_f(D) + NODP \quad (7)$$

b) **Single-Class Method.** In this scheme, we assume that the first page of every index contains information indicating the location of each index class. This structure can be implemented by including a vector of pairs [class_id, offset]. Assuming that the size of the offset and the class_id is 4 bytes each, the size of this structure would be 8c, where c is the number of class indexes on the broadcast.

Single-Class Protocol

1) Probe onto channel and get offset to the next index *active*

2) Reach the index *doze*

3) Retrieve offsets to the indexes of required classes *active*

4) for every required class

5) Reach the index *doze*

6) Retrieve the required index pages *active*

7) Reach the required data *doze*

8) Retrieve required data pages *active*

- **Response Time.** The size of a single index and its associated data are labeled as I_i and D_i, respectively. Since the total number of objects to be indexed is the same in single-class and hierarchical indexes, the sum of all D_i for all classes is equal to D. Assume a query references a set of classes where x and y stand for the first and last classes to be accessed. The average distance to be covered to get to x is half the distance covering the indexes and data between the beginning of y and the beginning of x. Once the index x is located, then all the indexes and data of all the classes between x and y (including those of x) have to be traversed. Once y is reached, its index and half of its data (on average) have to be traversed. Thus, the response time is proportional to:

$$\frac{\sum_{i=y}^{x-1}(I_i + D_i)}{2} + \sum_{i=x}^{y}(I_i + D_i) - \frac{D_y}{2} \tag{8}$$

Equation 8 provides a general means for calculating the average response time. However, the results are dependent on the location of the probe and the distance between x and y. It has been shown that the response time is lower bounded by half the size of the broadcast and upper bounded by slightly above the size of the broadcast. Further discussion on this issue is beyond the scope of this chapter and the interested reader is referred to Chehadeh, Hurson, and Kavehrad (1999).

- **Active Time.** Similar to the hierarchical case, the active time is dependent on the number of index pages and data pages to be retrieved. Therefore, the active time is the sum of the height of the trees for all the indexes of classes to be retrieved plus the number of the corresponding data pages. This is shown in the Equation 9. The 2 in the front accounts for the initial probe plus the additional page containing the index of classes (Line 3 in the protocol).

$$2 + \sum_{i=x}^{y}\left[\log_f(D_i) + NODP_i\right] \tag{9}$$

Performance Evaluation

Our simulator was extended to study both the response time and energy consumption with respect to the two allocation schemes. The overall structure of the schema graph determines the navigational paths among the classes within the graph. The relationships of the navigational paths within the graph influence the number and structure of indexes to be used.

- **Inheritance Relationship.** Within an inheritance hierarchy, classes at the lower level of the hierarchy inherit attributes of the classes at the upper level. Therefore,

Table 7. Input parameters

Parameter	Value (Default/Range)
Number of data items on Broadcast	5,120
Average Number of Classes Along Hierarchy	8
Percentage Distribution of Number of data items in Inheritance Hierarchy	25,25,25,25%
Percentage Distribution of Number of data items in Aggregation Hierarchy	40,30,20,10%
Distribution of data Size [S,M,L,VL]	16,512,3K,6K Bytes
Distribution of the data Sizes in Inheritance Hierarchy	VL,L,M,S
Distribution of the data Sizes in Aggregation Hierarchy	S,M,L,VL
Percentage of Classes to be Retrieved (Default/Range)	70% / [10-100%]
Average Number of data items to Retrieve Per Class	2
Fan-out in Index Tree	5
Average Number of data items with Distinct Key Attribute per Class	60% of data items per Class
Size of Air-Channel Page	512 Bytes
Broadcast Data Rate	1 M bits/sec
Power Consumption Active Mode	130 mW
Power Consumption Doze Mode	6.6 mW

Table 8. Number of index and data pages

	Aggregation/ Hierarchical	Aggregation/ Single [Eight Classes]	Inheritance/ Hierarchical	Inheritance/ Single [Eight Classes]
Index Pages	2,343	67,63,49,39,36,32,18,16	2,343	40,40,40,40,40,40,40,40
Data Pages	13,562	73,75,652,769,2504,28140,3206, 3502	34,015	12517,11520,5253,4252,637,440 ,25,20

data items belonging to the lower-level classes tend to be larger than those within the upper levels. The distribution of the number of data items is application dependent. In our analysis, and without loss of generality, we assumed the data items to be equally distributed among the classes of the hierarchy.

- **Aggregation Relationship.** In an aggregation hierarchy, data items belonging to lower classes are considered "part of" data items and those at the higher ends are the "collection" of such parts. Therefore, data items belonging to higher classes are generally larger than those belonging to the lower ones. In addition, the cardinality of a class at the upper end is smaller than a class at the lower end.

As a result, the organization of classes within the schema graph has its influence on the distribution of both the number and size of data items among the classes of the database. We assumed an average of eight classes for each hierarchy and categorize the sizes of data items as small, medium, large, and very large. Furthermore, 60% of the data items have distinct keys and the value of any attribute is uniformly distributed among the data items containing such attribute. Table 7 shows a list of all the input parameters assumed for this case.

Table 9. Response time degradation factor relative to the no-index scheme

Aggregation/ Hierarchical	Aggregation/ Single	Inheritance/ Hierarchical	Inheritance/ Single
1.17	1.05	1.1	1.02

The information along the broadcast channel is organized in four different fashions: the hierarchical and single-class methods for the inheritance and aggregation relationships. Table 8 shows the data and index page sizes for these organizations. Note that it is the number of data items (not data pages) that controls the number of index blocks. Within each indexing scheme, for each query, the simulator simulates the process of probing the air channel, getting the required index pages, and retrieving the required data pages. In each query, on average, two data items from each class are retrieved. The simulation measures the response time and amount of energy consumed.

a) **Response Time.** Placing an index along the air channel contributes to extra storage overhead and thus longer response time. Hence, the best response time is achieved when no index is placed, and the entire broadcast is searched. Table 9 shows the degradation factor in the average response time due to the inclusion of an index in the broadcast. The factor is proportional to the ratio of the size of the index blocks to that of the entire broadcast.

   Figure 13 shows the response time for all four different broadcast organizations. From the figure, one could conclude that for both the inheritance and aggregation cases, the response time of the hierarchical organization remained almost constant (with a slight increase, as the number of classes to retrieve increases). This is due to the fact that regardless of the number of classes and the location of the initial probe, all accesses have to be directed to the beginning of the index (at the beginning of the broadcast). The slight increase is attributed to the increase in the total number of objects to be retrieved — assuming that the objects to be retrieved

Figure 13. Response time versus number of retrieved classes

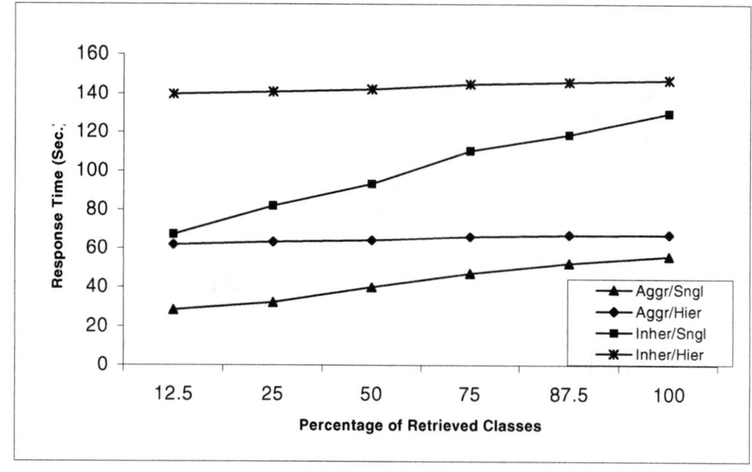

Figure 14. Detailed energy consumption

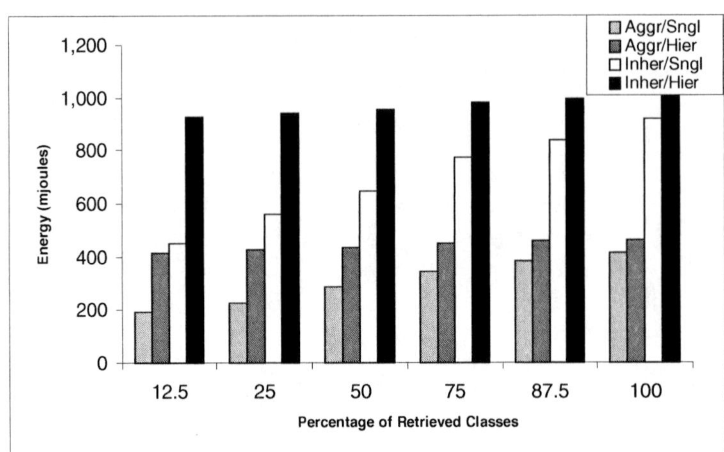

are distributed uniformly along the broadcast. It should be noted, however, that such an increase is only minor since the response time is mainly influenced by the initial procedure.

Two observations can be made: (a) the single-class method offers a better response time than the hierarchical case and (b) the response time for the single-class method increases as the number of retrieved classes increase. The first observation is due to the fact that in the single-class method, accesses do not have to be directed to the beginning of the broadcast. The second observation is due to the fact that an increase in the number of classes to be retrieved directly increases the number of index and data pages to be accessed.

Indexing based on the aggregation relationship offers lower response time than indexing based on the inheritance relationship since the distribution of the number of objects in the inheritance relationship is more concentrated on the larger objects. Having larger objects results in a longer broadcast, and, hence, it takes longer to retrieve the objects.

b) **Energy.** For each query, the amount of energy consumed is the sum of the energy consumed while the unit is in both active and doze modes. In the case where no index is provided, the mobile unit is in active mode during the entire probe. However, in the case where an index is provided, the active time is proportional to the number of index and data pages to be retrieved. As expected, the active time increases as the number of retrieved classes increases. The hierarchical method searches only one large index tree, whereas the single-class method searches through multiple smaller index trees. The number of pages to retrieve per index tree is proportional to the height of the tree. For a query spanning a single class, the single-class method produces a better active time than the hierarchical method. As the number of classes to be retrieved increases, the hierarchical tree is still traversed only once. However, more single-class trees have to be traversed, and, hence, results in an increase in the active time.

In both the single-class and the hierarchical methods, the aggregation case requires lower active time than the inheritance case since the inheritance case has larger objects, thus requiring the retrieval of more pages. For the sake of practicality, we utilized the power consumption data of the Hitachi SH7032 processor: 130 mW when active and 6.6 mW when in doze. Since power is the amount of energy consumed per unit of time, the total energy can be calculated directly using Equation 10:

$$Energy = (ResponseTime - ActiveTime)DozeModePower + (ActiveTime)ActiveModePower \tag{10}$$

Figure 14 details the energy consumed during the entire query operation in mjoules. The power consumption of the mobile unit is much higher while the unit is in active mode, with a ratio of 19.7. However, our experiments showed that, in general, the duration of the active-mode operations was much smaller than doze-mode operations. As a result, the energy consumed during doze time was the dominating factor. As can be seen from Figure 14, the single-class method is superior to the hierarchical method. This is very similar to the results obtained for the response time, and similarly, the power consumption of the single-class method is lower than that of the hierarchical method.

Data Indexing on Parallel Air Channels

Allocation of Object-Oriented Indexing on Parallel Air Channels

Figure 15 shows the allocation of the single-class and hierarchical-based schemes on the two parallel air channels. For the single-class indexing scheme (Figure 15a), the index and data of each class are distributed and placed along the channels. The hierarchical indexing scheme (Figure 15b) places the index on one channel, and divides and distributes the data among the channels. The most popular data items can be put in the

Figure 15. Allocation of single-class and hierarchical indexes on two parallel air channels

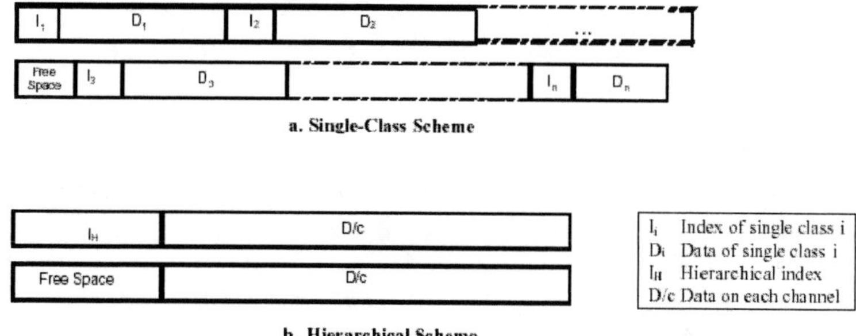

free space. Note that in both cases, similar to the single air channel, it is possible to interleave and distribute the index pages and associated data pages using a variety of methods.

Storage Requirement

In the case of broadcasting along parallel air channels, the storage requirement is the same as that for a single air channel.

Timing Analysis

In the case of parallel air channels, one has to account for switching between channels when analyzing access time and power consumption. During the switching time, the pages that are being broadcast on different channels cannot be accessed by the mobile unit. In addition, the mobile unit at each moment of time can tune into one channel—*overlapped page range*. By considering the average page size (512 bytes), communication bandwidth (1Mbit/sec), and switching time (the range of microseconds), we assume that the overlapped page range equals two pages. Finally, we assumed that the power consumption for switching between two channels is 10% of the power consumed in active mode. Equation 11 calculates the power consumption.

a) **Hierarchical Method.** The following protocol shows the sequence of operations.

Hierarchical Protocol

1)	Probe onto channel and retrieve offset to the next index	*active*
2)	Do {Reach the next index	*doze*
3)	Retrieve the required index pages	*active*
4)	Do {Reach the next possible required data page	*doze*
5)	Retrieve the next possible required data page	*active*
6)	}while every possible required data page is retrieved from the current broadcast	
7)	} while there are unaccessed data items because of overlapped page range	

$$EnergyConsumption = (ResponseTime - ActiveTime) * DozeModePower + (ActiveTime) * ActiveModePower + TheNumberOfSwitching * 10\% * ActiveModePower \quad (11)$$

- **Response Time.** I_H and D are used to denote the size of the hierarchy index and data, respectively. For the c-channel environment, the average size of data on each channel is D/c. To locate the index from the initial probe, it takes half the broadcast of one channel (the size of the broadcast is $I_H + D/c$). Once the index is reached, it has to be completely traversed before data pages appear on the broadcast and, on average, it takes half the size of the data to locate and retrieve the required data items. Thus, the response time from the initial probe to the first complete broadcast is proportional to

$$(I_H + D/c)/2 + I_H + (D/c)/2 = 3I_H/2 + D/c \quad (12)$$

Because of the overlapped page range, the mobile units may not be able to get all of the required data during one complete broadcast (e.g., because of conflicts). Therefore, it has to scan the next broadcast. Let P be the probability of the data that are in the same overlapped page range. The distance from the last location to the next index is also half the size of the data of one channel. Once the index is reached, the same process will occur. Thus, the response time from the last location of the previous broadcast until the mobile unit can acquire all of the required data is proportional to

$$P * (D/2c + I_H + D/2c) = P * (I_H + D/c) \quad (13)$$

As a result, on average, the response time is proportional to

$$(1.5 + P)I_H + (1 + P)D/c \quad (14)$$

- **Active Time.** The mobile unit has to be active during the first probe (to retrieve a page). Once the index is reached, a number of nonleaf node pages have to be accessed in order to get and retrieve a leaf-node page. The number of pages to be retrieved at the index is equal to the height of the index tree ($\log_f(D)$). The amount of the data pages to be read is equal to the *NODP*. Again, because of the overlapped page range among parallel air channels, the probability of accessing the index of the next broadcast has to be included. Therefore, the active time is proportional to

$$1 + \log_f(D) + NODP + P * \log_f(D) \quad (15)$$

b) **Single-Class Indexing Scheme**

Single-Class Protocol

1) Probe onto channel and retrieve offset to the next index *active*
2) Do {Reach the next index *doze*
3) Retrieve offsets to the indexes of required classes *active*
4) Reach the next possible index *doze*
5) Retrieve the next possible required index page *active*

6) Do {Reach the next possible index or data page *doze*

7) Retrieve the next possible index or data page *active*

8) } while not (all indexes and data of required classes are scanned)

9) } while there are some data pages which are not retrieved because of overlapped page range

- **Response Time.** As before, the size of a single index and its associated data are labeled as I_i and D_i, respectively. The response time is simply driven by dividing Equation 8 by the number of the air channels (Equation 16):

$$\frac{\sum_{i=y}^{x-1}(I_i+D_i)}{2c}+\frac{\sum_{i=x}^{y}(I_i+D_i)}{c}-\frac{Dy}{2c} \qquad (16)$$

Let P be the probability of the data that are in the same overlapped page range. Thus, the response time for getting the remaining required data items on the second broadcast probe is proportional to

$$P*(\frac{\sum_{i=y}^{x-1}(I_i+D_i)}{2c}+\frac{\sum_{i=x}^{y}(I_i+D_i)}{c}-\frac{Dy}{2c}) \qquad (17)$$

As a result, the response time is proportional to

$$(1+P)*(\frac{\sum_{i=y}^{x-1}(I_i+D_i)}{2c}+\frac{\sum_{i=x}^{y}(I_i+D_i)}{c}-\frac{D_y}{2c}) \qquad (18)$$

- **Active Time.** Similar to the hierarchical case, the active time is the sum of the height of the trees for all the indexes of the classes to be retrieved plus the number of the corresponding data pages. This is shown in Equation 19. The *2* at the beginning of the equation accounts for the initial probe plus the additional page containing the index of classes. Because of the overlapped page range among parallel air channels, the probability of accessing the index of the next broadcast has to be included. Therefore, the active time is proportional to

$$2+\sum_{i=x}^{y}\left[\log_f(D_i)+NODP_i\right]+\frac{P}{2}\sum_{i=x}^{y}\log_f(D_i) \qquad (19)$$

Performance Evaluation

Once again, our simulator was extended to study the response time and energy consumption of the single-class and hierarchical indexing schemes in parallel air channels based on the input parameters presented in Table 7.

a) **Response Time.** In the case of no indexing, the response time was constant and independent of the number of channels. This is due to the fact that without any indexing mechanism in place, the mobile unit has to scan every data page in sequence until all required data pages are acquired. Moreover, when indexing schemes are in force, the response time lessens as the number of channels increases.

For the inheritance and aggregation cases, the response time decreases as the number of channels increases. This is due to the fact that, as the number of channels increases, the length of the broadcast becomes shorter. However, the higher the number of channels, the higher the probability of conflicts in accessing data residing on different channels in the overlapped page range. As a result, doubling the number of channels will not decrease the response time by half.

For both the inheritance and aggregation indexing schemes, the single-class method offers a better response time than the hierarchical method. The single-class method accesses do not have to be started at the beginning of the broadcast. For the hierarchical method, on the other hand, any access has to be started from the beginning of the broadcast, which makes the response time of the hierarchical method longer. Indexing based on the aggregation relationship offers a lower response time than that of the inheritance relationship because the distribution of data items in the inheritance relationship is more concentrated on the larger data items.

b) **Energy.** The active time is proportional to the number of index and data pages to be retrieved. For broadcast data without an index, the active time is the same as the response time. In addition, for all four indexing schemes, the active time remains almost constant and independent of the number of air channels. This is because the active time is proportional to the number of index and data pages to be retrieved.

In general, the hierarchical method requires less active time than the single-class method. The hierarchical method searches only one large index tree, whereas the single-class method searches through multiple smaller index trees, and the number of pages to be retrieved per index tree is proportional to the height of the tree. In both the single-class and the hierarchical methods, the indexing based on an aggregation relationship requires lower active time than the inheritance method. This is simply due to the fact that the inheritance relationship resulted in larger data items, thus requiring the retrieval of more pages.

Figure 16. Detailed energy consumption

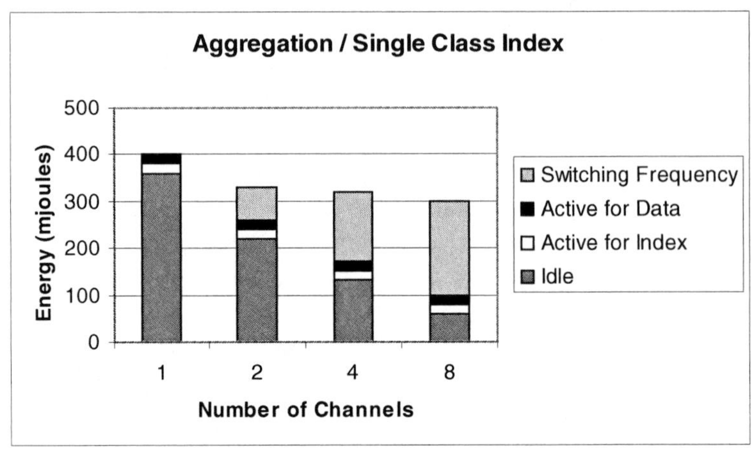

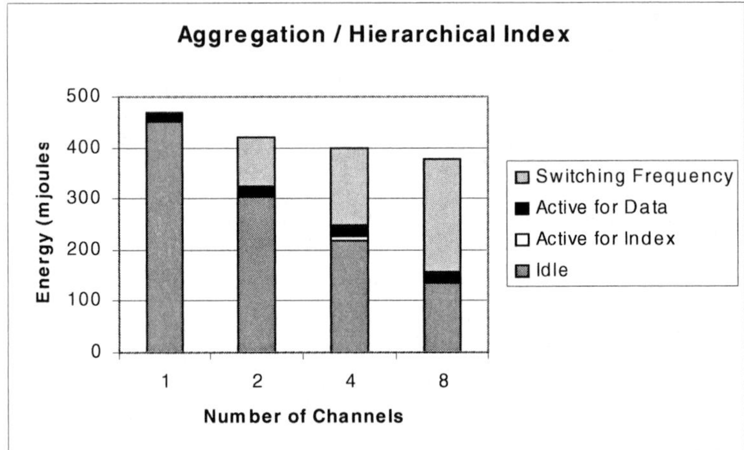

In a separate simulation run we observed the total energy consumption. It was concluded that the total energy consumption of broadcasting without any indexing schemes is much higher than that of broadcasting supported by indexing, and the energy consumption of the single-class method is lower than that of the hierarchical method. This is very similar to the results obtained for the response time. When indexing was supported, energy consumption, on average, decreased about 15 to 17 times in the case of the aggregation relationship and the inheritance relationship, respectively.

Figure 16 shows the detail of energy consumption for the aggregation relationship. As the number of channels increases, the energy consumption during idle time decreases. The energy consumption for retrieving indexes increases because the probability of the data being in the same overlapped page range increases. The higher this probability, the more the mobile unit has to get the index from the next broadcast. Finally, the energy consumption for switching between two different channels increases because the required data are distributed among the channels. The larger the number of channels, the more distributed is the data among the channels, and, consequently, the more frequent switching between channels.

Section Conclusion

This section investigated an energy-efficient solution by the means of applying indexing schemes to object-oriented data broadcast over single and parallel air channels. Two methods, namely, the hierarchical and single-class methods, were explored. Timing analysis and simulation were conducted to compare and contrast the performance of different indexing schemes against each other. It was shown that including an index degrades the response time moderately, however, such degradation is greatly offset by the improvement in energy consumption. For a single air channel, broadcasting with supported indexing schemes increased the response time when compared with broadcasting without indexing support. However, the response time is reduced by broadcasting data with an index along parallel air channels. Moreover, the response time decreased as the number of air channels is increased. Relative to nonindexed broadcasting, the mobile unit's energy consumption decreased rather sharply when indexing is supported. For a set of queries retrieving data items along the air channel(s), the single-class indexing method resulted in a faster response time and lower energy consumption than the hierarchical method.

Conflicts and Generation of Access Patterns

One of the problems associated with broadcasting information on parallel air channels is the possibility for conflicts between accessing data items on different channels. Because the mobile unit can tune into only one channel at a time, some data items may have to be retrieved on subsequent broadcasts. In addition, during the channel switch time, the mobile unit is unable to retrieve any data from the broadcast. Conflicts will directly influence the access latency and, hence, the overall execution time. This section is intended to provide a mathematical foundation to calculate the expected number of passes required to retrieve a set of data items requested by an application from parallel air channels by formulating this problem as an *asymmetric traveling salesman problem* (TSP). In addition, in an attempt to reduce the access time and power consumption, we propose heuristic policies that can reduce the number of passes over parallel air channels. Analysis of the effectiveness of such policies is also the subject of this section.

Enumerating Conflicts

Equation 1 showed the number of broadcasts (passes) required to retrieve *K* data items from *M* parallel channels if conflicts between data items are independent. To calculate $P(i)$, it is necessary to count the number of ways the data items can be distributed while having exactly *i* conflicts, then divide it by the total number of ways the *K* data items can be distributed over the parallel channels. In order to enumerate possible conflicting cases, we classify the conflicts as single or double conflicts as defined below.

Definition 7. A single conflict is defined as a data item in the conflict region that does not have another data item in the conflict region in the same row. A double conflict is a data item that is in the conflict region and does have another data item in the conflict region in the same row.

The number of data items that cause a double conflict, *d*, can range from 0 (all single conflicts) up to the number of conflicts, *i*, or the number of remaining data items, $(K - i - 1)$. When counting combinations, each possible value of *d* must be considered separately. The number of possible combinations for each value of *d* is summed to determine the total number of combinations for the specified value of *i*. When counting the number of ways to have *i* conflicts and *d* double conflicts, four factors must be considered.

- Whether each of the $(i - d)$ data items representing a single conflict is in the left or right column in the conflict region. Because each data item has two possible positions, the number of variations due to this factor is $2^{(i-d)}$.
- Which of the $(M - 1)$ rows in the conflict region are occupied by the $(i - d)$ single conflicts. The number of variations due to this factor is $\binom{M-1}{i-d}$.
- Which of the $(M - 1) - (i - d)$ remaining rows in the conflict region are occupied by the *d* double conflicts; $(i - d)$ is subtracted because a double conflict cannot occupy the same row as a single conflict. The number of variations due to this factor is $\binom{(M-1)-(i-d)}{d}$.
- Which of the $(MN - 2M + 1)$ positions not in the conflict region are occupied by the $(K - i - d - 1)$ remaining data items. The number of variations due to this factor is $\binom{MN-2M+1}{K-i-d-1}$.

Note that these sources of variation are independent from each other and, hence:

$$P(i) = \frac{\sum_{d=0}^{d \le MIN(i, K-i-1)} 2^{(i-d)} \binom{M-1}{i-d} \binom{(M-1)-(i-d)}{d} \binom{MN-2M+1}{K-i-d-1}}{\binom{MN-1}{K-1}} \quad (20)$$

If the conflicts produced by one data item are independent from the conflicts produced by all other data items, then Equation 20 will give the number of passes required to retrieve all K requested data items. However, if the conflicts produced by one data item are not independent of the conflicts produced by other data items, additional conflicts will occur which are not accounted for in our analysis. Equation 20 will thus underestimate the number of broadcasts required to retrieve all K data items.

Retrieving Data from Parallel Broadcast Air Channels in the Presence of Conflicts

The problem of determining the proper order to retrieve the requested data items from the parallel channels can be modeled as a TSP. Making the transformation from a broadcast to the TSP requires the creation of a complete directed graph G with K nodes, where each node represents a requested object. The weight w of each edge (i, j) indicates the number of broadcasts that must pass in order to retrieve data item j immediately after retrieving data item i. Since any particular data item can be retrieved in either the current broadcast or the next broadcast, the weight of each edge will be either 0 or 1. A weight of 0 indicates that the data item j is after data item i in the broadcast with no conflict. A weight of 1 indicates that data item j is either before or in conflict with data item i.

Simulation Model

The simulation models a mobile unit retrieving data items from a broadcast. A broadcast is represented as an N x M two-dimensional array, where N represents the number of data items in each channel of a broadcast and M represents the number of parallel channels. For each value of K, where K represents the number of requested data items ($1 \leq K \leq NM$), the simulation randomly generates 1,000 patterns representing the uniform distribution of K data items among the broadcast channels. The K data items from each randomly generated pattern are retrieved using various retrieval algorithms. The number of passes is recorded and compared. To prevent the randomness of the broadcasts from affecting the comparison of the algorithms, the same broadcast is used for each algorithm in a particular trial and the mean value is reported for each value of K. Finally, several algorithms for ordering the retrieval from the broadcast, both TSP related and non-TSP related, were analyzed.

Data Retrieval Algorithms

Both exact and approximate TSP solution finders and two heuristic based methods were used to retrieve the data items from the broadcast.

a) **TSP Methods.** An exact TSP solution algorithm was used to provide a basis for comparison with the other algorithms. These algorithms are simply too slow and too resource intensive. While a better implementation of the algorithm may somewhat reduce the cost, it cannot change the fact that finding the exact solution will require exponential time for some inputs. Knowing the exact solution to a given TSP does, however, allow us to evaluate the quality of a heuristic approach. A TSP heuristic based on the assignment problem relaxation requires far less CPU time and memory than the optimal tour finders, so it is suitable for use on a mobile unit. A publicly available TSP solving package named TspSolve (Hurwitz& Craig, 1996) was used for all TSP algorithm implementations.

b) **Next Data Item Access.** The strategy used by this heuristic is simply to always retrieve the next available data item in a broadcast. This can be considered as a greedy approach. It is also similar to the nearest neighbor approach to solving TSP problems.

c) **Row Scan.** A simple *row scan* heuristic was also used. This algorithm simply reads all the data items from one channel in each pass. If a channel does not have any requested data in it, it is skipped. This algorithm will always require as many passes as there are channels with requested data items in them. The benefit of this algorithm is that it does not require any time to decide on an ordering. It can thus begin retrieving data items from a broadcast immediately. This is especially important when a large percentage of the data items in a broadcast are requested.

Results

As expected, the TSP methods provide much better results than both the two heuristic-based algorithms. Our simulations showed that the TSP heuristic performed almost exactly as well as the optimal TSP algorithm. This is a very interesting observation because it means that one can use a fast heuristic to schedule retrievals of data items from the broadcast without any performance degradation.

In Figure 17, the TSP methods show that the number of broadcasts required to retrieve all *K* requested data items from a broadcast is much greater than the number of broadcasts

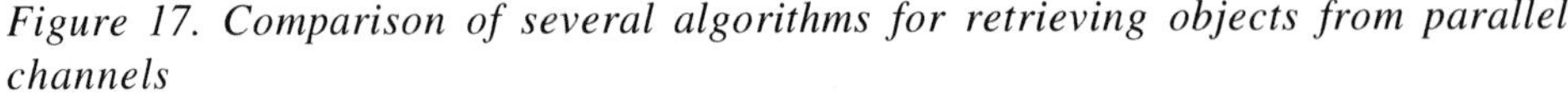

Figure 17. Comparison of several algorithms for retrieving objects from parallel channels

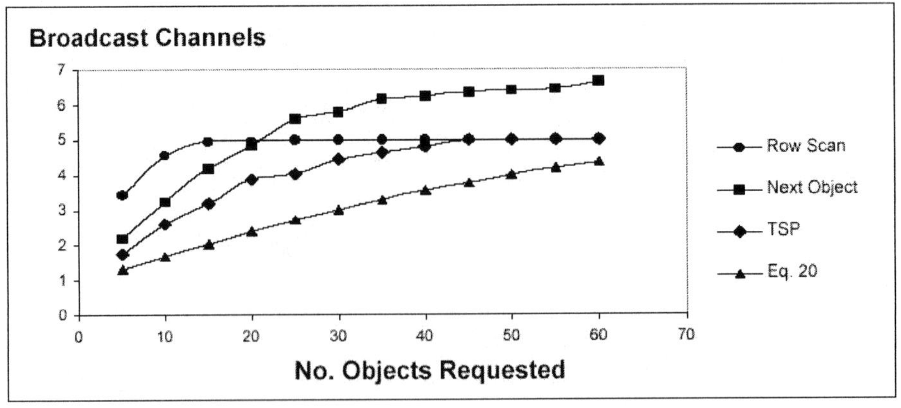

predicted by Equation 20 — Equation 20 was based on the assumption that the conflicts among the requested data items are independent. Figure 17 used five parallel channels and 20 pages per channel. It is also interesting to note that the straightforward row scan nearly matches the performance of the TSP-based algorithms when more than about 45% of the total number of data items is requested. In this case, there are so many conflicts that it is virtually impossible to avoid having to make as many passes as there are parallel channels. When this occurs, it is better to do the straightforward row scan than to spend time and resources running a TSP heuristic.

Optimal Number of Broadcast Channels

More channels mean that a given amount of information can be made available in a shorter period of time at the expense of more conflicts. The simulation results showed that it is always advantageous to use more broadcast channels. While there will be more conflicts between data items, this does not quite counteract the shorter broadcast length of the many-channel broadcasts. This was especially evident when only a few data items in a broadcast were being accessed.

Ordered Access List

The scope of the general access protocol for indexed parallel-channel configuration in the presence of conflicts was extended in order to use heuristics that can generate the ordered access list of requested data items that reduces

- the number of passes over the air channels and
- the number of channel switches.

During the *Search* step, the index is accessed to determine the offset and the channel of the requested data items. Then, a sequence of access patterns is generated. Finally, the *Retrieval* step is performed following the generated access patterns.

Extended Retrieval Protocol

1) Probe the channel and retrieve the offset to the next index
2) Access the next index
3) Do {Search the index for the requested object
4) Calculate the offset of the object
5) Get the channel on which the object will be broadcast
6) } while there is an unprocessed requested object
7) Generate access patterns for the requested objects (using retrieval scheme)
8) Do {Wait for the next broadcast cycle

9) Do {Reach the first object as indicated by the access pattern
10) Retrieve the object
11) } while there is an unretrieved object in the access pattern
12) } while there is an unprocessed access pattern

Performance Evaluation

We extended the simulator to emulate the process of accessing data from a hierarchical indexing scheme in parallel air channels. Moreover, the simulator also analyzes the effect of conflicts on the average access time and power consumption.

Our retrieval scheme, based on the user request, generates a retrieval forest representing all possible retrieval sequences. However, as expected, the generated retrieval forest grows exponentially with the number of requested data items. The key observation needed to reduce the size of the tree is to recognize that each requested data item has a unique list of children, and the number of children for a particular data item is limited to the number of channels. The simulator takes advantage of these observations to reduce the size of the retrieval tree and the calculation time without sacrificing accuracy.

The generation of the user requests was performed randomly, representing a distribution of K data items in the broadcast. In various simulations runs, the value of K was varied from one to $N \times M$—in a typical user query of public data, K is much less than $N \times M$. Finally, to take into account future technological advances, parameters such as transmission rate and power consumption in different modes of operation were fed to the simulator as variable entities.

The simulator calculates the average active time, the average idle time, the average query response time, the average number of broadcast passes, the number of channel switches, and the energy consumption of the retrieval process. As a final note, the size of the index was 13.52% of the size of the broadcast (not including the index) and the number of channels varied from 1 to 16 (2 to 17 when an independent channel was used for transmitting the index).

Simulation Model

For each simulation run, a set of input parameters, including the number of parallel air channels, the broadcast transmission rate, and the power consumption in different

Table 10. Improvement of proposed algorithm versus row scan (10 data items requested)

# of Channels	# of Passes	Response Time	Energy
2	48.0%	28.0%	2.7%
4	68.0%	43.6%	3.1%
8	72.3%	46.5%	3.3%
16	71.8%	40.8%	3.4%

operational modes, was passed to the simulator. The simulator was run 1,000 times and the average of the designated performance metrics was calculated. The results of the simulations where an indexing scheme was employed were compared against a broadcast without any indexing mechanism. Two indexing scenarios were simulated.

- **Case 1.** The index was transmitted with the data in the first channel (index with data broadcast).
- **Case 2.** The index was transmitted over a dedicated channel in a cyclic manner.

Results

A comparison between the extended retrieval protocol against the row scan algorithm was performed. The index transmission was performed in a cyclic manner on an independent channel, and the number of requested data items was varied between 5 and 50 out of 5,464 securities within the NASDAQ exchange database. The simulation results showed that, regardless of the number of parallel air channels, the proposed algorithm reduces both the number of passes and the response time compared to the row scan algorithm. Moreover, the energy consumption was also reduced, but only when the number of data items retrieved was approximately 15 or less (Table 10).

Relative to the row scan algorithm, one should also consider the expected overhead of the proposed algorithm. The simulation results showed that in the worst case, the overhead of the proposed algorithm was slightly less than the time required to transmit one data page.

a) **Response Time.** Figures 18 to 20 show the response times in terms of the number of data items requested and the number of broadcast channels. Three cases were examined.

   - **Case 1. Data and index are intermixed on broadcast channel(s).** Figure 18 shows the response times for different numbers of broadcast channels when retrieving the full range of existing data items from the broadcast. It can be

Figure 18. Response time (Case 1)

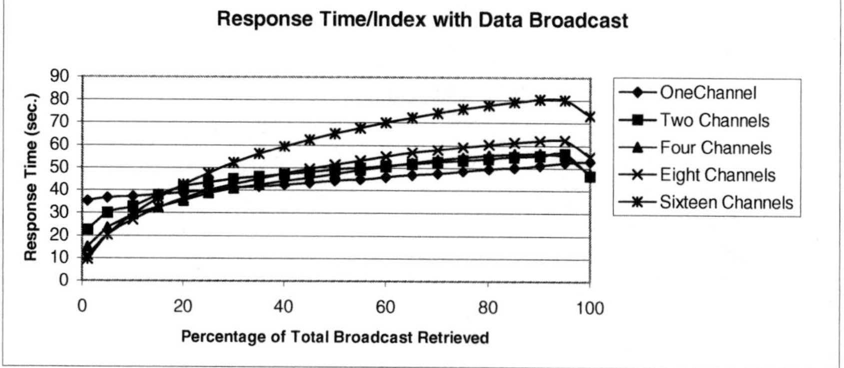

Figure 19. Response time (Case 2)

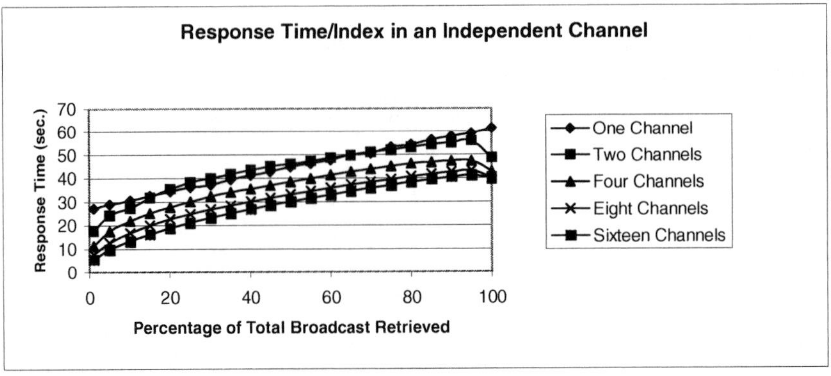

concluded that when a few data items are requested, the response time decreases as the number of channels increases. After a certain threshold point, the response time increases as the number of channels increases. This is due to an increase in the number of conflicts and hence an increase in the number of passes over the broadcast channels to retrieve the requested data.

- **Case 2. Index is broadcast over a dedicated channel in cyclic fashion.** Similar to Case 1, Figure 19 depicts the simulation results when retrieving the full range of existing data from the broadcast. Again, as expected, the response time decreases as the number of channels increases. Comparing the results for one-channel and two-channel configuration, we can conclude that in some instances the two-channel configuration is not as effective — there is the possibility of conflicts, many of which unavoidably cause an increase in the number of passes and hence longer response time.

Figure 20. Response time (Case 3)

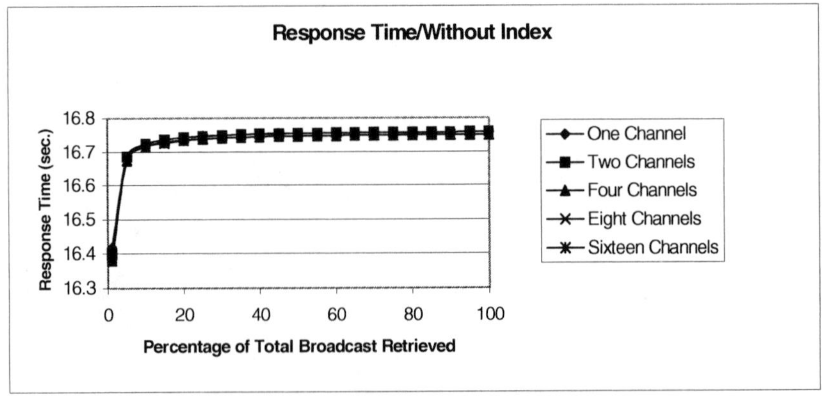

- **Case 3. No indexing is employed.** From Figure 20 one can conclude that the response time remains relatively constant regardless of the number of channels used. In this organization, the user must scan the same amount of data regardless of the user query and number of parallel channels.

In general, employment of an indexing scheme reduces the response time when retrieving a relatively small number of data items. As the percentage of data items requested increases, the number of conflicts increases as well. The proposed retrieval protocol tries primarily to reduce the conflicts in each pass of the broadcast; however, when the number of potential conflicts increases considerably, some conflicts become unavoidable, causing an increase in the number of passes and hence an increase in the response time. When the percentage of requested data approaches 100%, the response time reduces. This proves the validity of the proposed scheduling algorithm since it generates the same retrieval sequence as the row scan method.

b) **Switching Frequency.** Again, three cases were examined.

- **Case 1 & Case 2. Employment of indexing schemes.** Figure 21 shows the switching frequency for Case 1 and Case 2—the switching pattern is not affected by the indexing policy employed. From this figure one can conclude that the switching frequency increases as the number of channels and number of data items retrieved increase. This can be explained by an increase in the number of conflicts; as the proposed method tries to reduce the number of conflicts, the switching frequency will increase. Also, as stated previously, an increase in the number of channels increases the number of conflicts as well. One can notice that when the percentage of data items requested exceeds 50%, the switching frequency begins to decrease. This is due to the fact that the proposed method does not attempt to switch channels as often to avoid the conflicts as the number of conflicts increases substantially.
- **Case 3. No indexing is employed.** When no indexing technique is utilized, the row scan method is employed, producing a constant switching frequency

Figure 21. Switching frequency (Case 1 and Case 2)

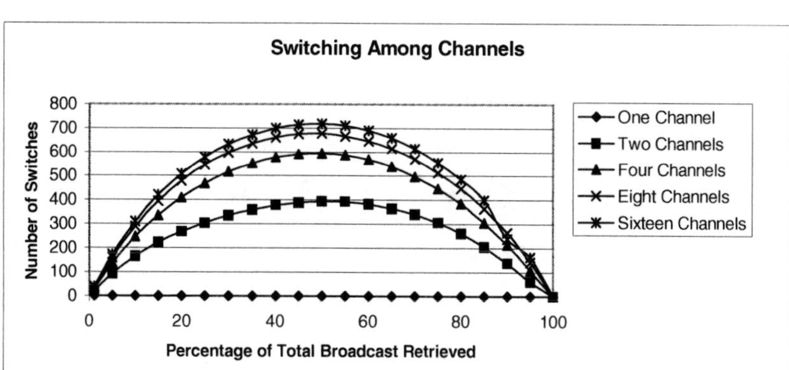

Figure 22. Energy consumption (Case 1)

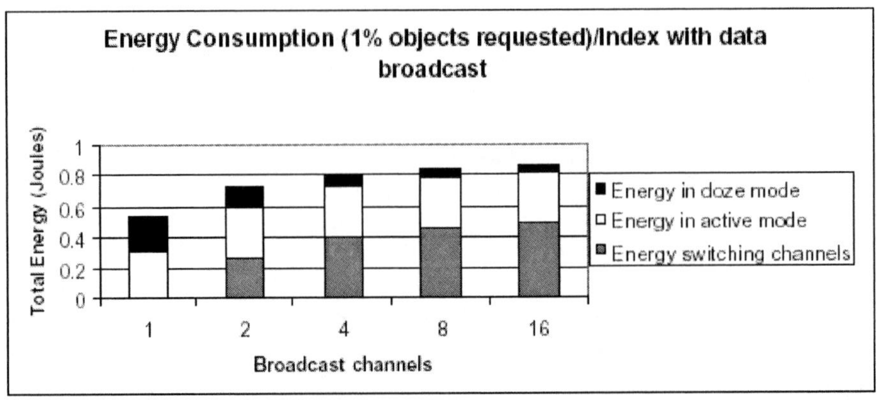

Figure 23. Energy consumption (Case 2)

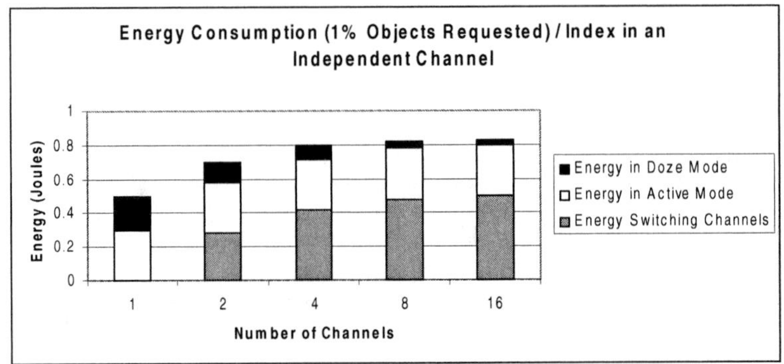

Figure 24. Energy consumption (Case 3)

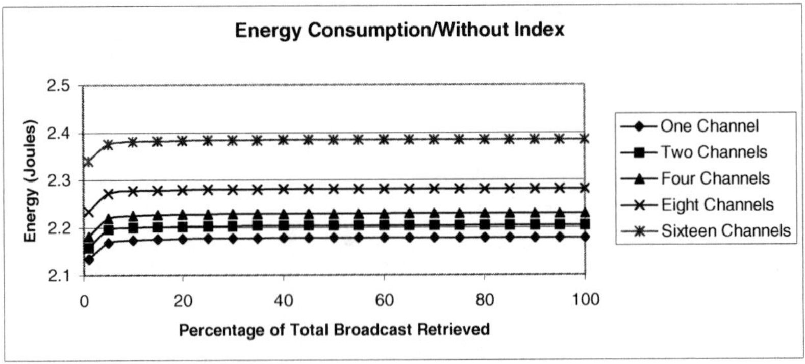

Figure 25. Number of passes over the parallel channels

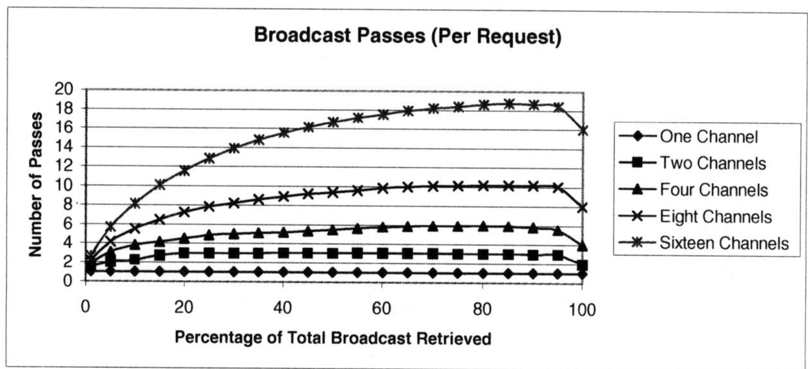

independent of the number of data items requested. The switching frequency is, at the most, equal to the number of total channels employed in the simulation.

c) **Energy Consumption.** Figures 22 and 23 depict detailed energy consumption when 1% of data items on the broadcast are requested. It can be observed that the energy consumption is almost the same; however, Case 1 consumes more energy than Case 2 in doze mode.

- **Case 1 & Case 2. Employment of indexing schemes.** In general, due to the increase in the number of channel-switching frequency, the energy consumption increases as the number of channels increases. In addition, we noted that the energy consumption increases when up to 50% of the broadcast data items are requested, then it decreases as the number of requested data items increases. This is directly related to the channel-switching frequency. In Case 1, in many instances, the mobile unit must wait in doze mode while the index is retransmitted.
- **Case 3. No indexing is employed.** When no indexing technique is used (Figure 24), the energy consumption varies only minimally due to the nature of the row scan algorithm employed.

From these figures we can observe that both Case 1 and Case 2 consume less power than Case 3 when a small percentage of data items is retrieved (around 1%). When the percentage of data items requested increases, the number of conflicts, the switching frequency, and, consequently, the energy consumption increase.

d) **Number of Passes.** As a note, the number of passes is independent of the index allocation scheme. Therefore, the number of passes for Cases 1 and 2 is the same.

- **Case 1 & Case 2. Employment of indexing schemes.** The increase in the number of passes is directly related to the increase in the number of channels and increase in the number of data items requested (Figure 25). An increase in the number of channels implies an increase in the number of conflicts, and, hence, the higher possibility of unavoidable conflicts, resulting in an increase in the number of passes. It can be noticed that when the number of data items requested is large, the number of passes exceeds the number of channels available. This is due to the priority order of the heuristics used in the proposed retrieval algorithm. In general, it is improbable that a query for public data requests a lot of data items from the broadcast channels. Our experience showed that for a query requesting up to 50 data items, the proposed method reduces the number of passes compared to Case 3.

- **Case 3. No indexing is employed.** In contrast, when no indexing technique is employed, the number of passes required is a function of the number of air channels. In the worst case we need N passes where N is the number of broadcast channels.

Section Conclusion

Conflicts directly influence the access latency and, hence, the overall execution time. This section provided a mathematical foundation to calculate the expected number of passes required to retrieve a set of data items requested by an application from parallel air channels. In addition, in an attempt to reduce the access time and power consumption, heuristics were used to develop access policies that reduce the number of passes over the parallel air channels. Analysis of the effectiveness of such policies was also the subject of this section.

Conclusion and Future Research Directions

This chapter aims to address the applicability and effectiveness of data broadcasting from two viewpoints: energy and response time. Within the scope of data broadcasting, we discussed different data allocation schemes, indexing approaches, and data retrieval methods for both single and parallel air channels. Comparisons of different algorithms were demonstrated through simulation results.

The scope of this research can be extended in many directions. For instance, we assumed that the resolution of queries happens on an individual basis at the mobile unit. It may be possible to reduce computation by utilizing a buffer and bundling several queries together, processing them as a whole. Our proposed scheduling scheme was based on three prioritized heuristics. It is interesting to investigate a new set of heuristics that can reduce the switching frequency while retrieving a large percentage of data items from the broadcast.

Acknowledgment

This work would have not been possible without the sincere effort of many students who participated in the development of conceptual issues as well as simulation results. We would like to thank them. In addition, this work in part has been supported by the Office of Naval Research and the National Science Foundation under the contracts N00014-02-1-0282 and IIS-0324835, respectively.

References

Acharya, S., Alonso, R., Franklin, M., & Zdonik, S. (1995). Broadcast disks: Data management for asymmetric communication environments. *Proceedings of ACM SIGMOD International Conference on the Management of Data*, (pp. 199-210).

Alonso, R., & Ganguly, S. (1992). Energy efficient query optimization. *Technical Report MITL-TR-33-92*, Princeton, NJ: Matsushita Information Technology Laboratory.

Alonso, R., & Korth, H. F. (1993). Database system issues in nomadic computing. *Proceedings of ACM SIGMOD Conference on Management of Data*, (pp. 388-392).

Atkinson, M., Bancilhon, F., DeWitt, D., Dittrich, K., Maier, D., & Zdonik, S. (1989). The object-oriented database system manifesto. *Proceedings of Conference on Deductive and Object-Oriented Databases*, (pp. 40-57).

Badrinath, B. R. (1996). Designing distributed algorithms for mobile computing networks. *Computer Communications*, *19*(4), 309-320.

Banerjee, J., Kim, W., Kim, S.-J., & Garza, J. F. (1988). Clustering a DAG for CAD databases. *IEEE Transactions on Software Engineering, 14*(11), 1684-1699.

Boonsiriwattanakul, S., Hurson, A. R., Vijaykrishnan, N., & Chehadeh, C. (1999). Energy-efficient indexing on parallel air channels in a mobile database access system. *Proceedings of the Third World Multiconference on Systemics, Cybernetics, and Informatics, and Fifth International Conference on Information Systems Analysis and Synthesis, IV,* (pp. 30-38).

Bowen, T. F. (1992). The DATACYCLE architecture. *Communication of ACM, 35*(12), 71-81.

Bright, M. W., Hurson, A. R., & Pakzad, S. (1992). A taxonomy and current issues in multidatabase systems. *IEEE Computer*, *25*(3), 50-60.

Bright, M. W., Hurson, A. R., & Pakzad, S. (1994). Automated resolution of semantic heterogeneity in multidatabases. *ACM Transactions on Database Systems*, *19*(2), 212-253.

Chang, E. E., & Katz, R. H. (1989). Exploiting inheritance and structure semantics for effective clustering and buffering in an object-oriented DBMS. *Proceedings of ACM SIGMOD Conference on Management of Data*, (pp. 348-357).

Chehadeh, Y. C., Hurson, A. R., & Tavangarian, D. (2001). Object organization on single and parallel broadcast channel. *Proceedings of High Performance Computing*, (pp. 163-169).

Chehadeh, Y. C., Hurson, A. R., & Kavehrad, M. (1999). Object organization on a single broadcast channel in the mobile computing environment [Special issue]. *Multimedia Tools and Applications Journal, 9*, 69-94.

Chehadeh, Y. C., Hurson, A. R., & Miller L. L. (2000). Energy-efficient indexing on a broadcast channel in a mobile database access system. *Proceedings of IEEE Conference on Information Technology*, (pp. 368-374).

Chehadeh, Y. C., Hurson, A. R., Miller, L. L., Pakzad, S., & Jamoussi, B. N. (1993). Application of parallel disks for efficient handling of object-oriented databases. *Proceedings of the Fifth IEEE Symposium on Parallel and Distributed Processing*, (pp. 184-191).

Cheng, J.-B. R., & Hurson, A. R. (1991a). Effective clustering of complex objects in object-oriented databases. *Proceedings of ACM SIGMOD Conference on Management of Data*, (pp. 22-27).

Cheng, J.-B. R., & Hurson, A. R. (1991b). On the Performance issues of object-based buffering. *Proceedings of International Conference on Parallel and Distributed Information Systems*, (pp. 30-37).

Chlamtac, I., & Lin, Y.-B. (1997). Mobile computing: When mobility meets computation. *IEEE Transactions on Computers, 46*(3), 257-259.

Comer, D. C. (1991). *Internetworking with TCP/IP Volume I: Principles, Protocols, and Architecture* (2nd ed.). Englewood Cliffs, NJ: Prentice Hall.

Demers, A., Pertersen, K., Spreitzer, M., Terry, D., Theier, M., & Welch, B. (1994). The bayou architecture: Support for data sharing among mobile users. *Proceedings of IEEE Workshop on Mobile Computing Systems and Applications*, (pp. 2-7).

Fong, E., Kent, W., Moore, K., & Thompson, C. (1991). *X3/SPARC/DBSSG/OODBTG Final Report*. Available from NIST.

Fox, A., Gribble, S. D., Brewer, E. A., & Amir, E. (1996). Adapting to network and client variability via on-demand dynamic distillation. *Proceedings of ASPLOS-VII*, Boston, Massachusetts, (pp. 160-170).

Honeyman, P., Huston, L., Rees, J., & Bachmann, D. (1992). The LITTLE WORK project. *Proceedings of the Third IEEE Workshop on Workstation Operating Systems*, (pp. 11-14).

Hu, Q.L., & Lee, D. L. (2000). Power conservative multi-attribute queries on data broadcast. *Proceedings of IEEE International Conference on Data Engineering (ICDE 2000)*, (pp. 157-166).

Hu, Q. L., & Lee, D. L. (2001). A hybrid index technique for power efficient data broadcast. *Distributed and Parallel Databases Journal, 9*(2), 151-177.

Hurson, A. R., Chehadeh, Y. C., & Hannan, J. (2000). Object organization on parallel broadcast channels in a global information sharing environment. *Proceedings of IEEE Conference on Performance, Computing, and Communications*, (pp. 347-353).

Hurson, A. R., Pakzad, S., & Cheng, J.-B. R. (1993). Object-oriented database management systems. *IEEE Computer, 26*(2), 48-60.

Hurwitz, C. & Craig, R. J. (1996). *Software Package Tsp_Solve 1.3.6.* Available from http://www.cs.sunysb.edu/~algorithm/implement/tsp/implement.shtml.

Imielinski, T., & Badrinath, B. R. (1994). Mobile wireless computing: Challenges in data management. *Communications of the ACM, 37*(10), 18-28.

Imielinski, T., & Korth, H. F. (1996). Introduction to mobile computing. In T. Imielinski and H. F. Korth (Eds.), *Mobile computing* (pp. 1-43). Boston: Kluwer Academic.

Imielinski, T., Viswanathan, S., & Badrinath, B. R. (1994). Energy efficient indexing on air. *Proceedings of ACM SIGMOD Conference on Management of Data*, (pp. 25-36).

Imielinski, T., Viswanathan, S., & Badrinath, B. R. (1997). Data on air: Organization and access. *IEEE Transactions on Computer, 9*(3), 353-372.

Joseph, A. D., Tauber, J. A., & Kaashoek, M. F. (1997). Mobile computing with the rover toolkit [Special issue]. *IEEE Transactions on Computers, 46*(3), 337-352.

Juran, J., Hurson, A. R., & Vijaykrishnan, N. (2004). Data organization and retrieval on parallel air channels: Performance and energy issues. *ACM Journal of WINET, 10*(2), 183-195.

Kaashoek, M. F., Pinckney, T., & Tauber, J. A. (1994). Dynamic documents: Mobile wireless access to the WWW. *IEEE Workshop on Mobile Computing Systems and Applications*, 179-184.

Kim, W. (1990). *Introduction to object-oriented databases.* Cambridge, MA: MIT Press.

Lai, S. J., Zaslavsky, A. Z., Martin, G. P., & Yeo, L. H. (1995). Cost efficient adaptive protocol with buffering for advanced mobile database applications. *Proceedings of the Fourth International Conference on Database Systems for Advanced Applications.*

Lee, D. L. (1996). Using signatures techniques for information filtering in wireless and mobile environments [Special issue]. *Distributed and Parallel Databases, 4*(3), 205-227.

Lee, M. T., Burghardt, F., Seshan, S., & Rabaey, J. (1995). InfoNet: The networking infrastructure of InfoPad. *Proceedings of Compcon*, (pp. 779-784).

Lim, J.B., & Hurson, A. R. (2002). Transaction processing in mobile, heterogeneous database systems. *IEEE Transactions on Knowledge and Data Engineering, 14*(6), 1330-1346.

Lim, J. B., Hurson, A. R., Miller, L. L., & Chehadeh, Y. C. (1997). A dynamic clustering scheme for distributed object-oriented databases. *Mathematical Modeling and Scientific Computing,* 8, 126-135.

Munoz-Avila, A., & Hurson, A. R. (2003a). Energy-aware retrieval from indexed broadcast parallel channels. *Proceedings of Advanced Simulation Technology Conference (High Performance Computing)*, (pp. 3-8).

Munoz-Avila, A., & Hurson, A. R. (2003b). Energy-efficient objects retrieval on indexed broadcast parallel channels. *Proceedings of International Conference on Information Resource Management*, (pp. 190-194).

NASDAQ World Wide Web Home Page. (2002). Retrieved May 11, 2004, from http://www.nasdaq.com

Satyanarayanan, M. (1996). Fundamental challenges in mobile computing. *Proceedings* of *15th ACM Symposium on Principles of Distributed Computing*, (pp. 1-7).

Satyanarayanan, M., Noble, B., Kumar, P., & Price, M. (1994). Application-aware adaptation for mobile computing. *Proceedings of the Sixth ACM SIGOPS European Workshop,* (pp. 1-4).

Weiser, M. (1993). Some computer science issues in ubiquitous computing. *Communications of the ACM, 36*(7), 75-84.

Zdonik, S., Alonso, R., Franklin, M., & Acharya, S. (1994). Are disks in the air just pie in the sky? *Proceedings of Workshop on Mobile Computing Systems and Applications*, (pp. 1-8).

Chapter V

Broadcast Data Placement over Multiple Wireless Channels

Dimitrios Katsaros, Aristotle University of Thessaloniki, Greece

Yannis Manolopoulos, Aristotle University of Thessaloniki, Greece

Abstract

The advances in computer and communication technologies made possible an ubiquitous computing environment were clients equipped with portable devices can send and receive data anytime and from anyplace. Due to the asymmetry in communication and the scarceness of wireless resources, data broadcast is widely employed as an effective means in delivering data to the mobile clients. For reasons like heterogeneous communication capabilities and variable quality of service offerings, we may need to divide a single wireless channel into multiple physical or logical channels. Thus, we need efficient algorithms for placing the broadcast data into these multiple channels so as to reduce the client access time. The present chapter discusses algorithms for placing broadcast data to multiple wireless channels, which cannot be coalesced into a lesser number of high-bandwidth channels, assuming that there are no dependencies among the transmitted data. We give an algorithm for obtaining the optimal placement to the channels and explain its limitation since it is computationally very demanding and thus unfeasible. Then, we present heuristic schemes for obtaining suboptimal solutions to the problem of reporting on their implementation cost and their relative performance.

Introduction

The technological achievements in the field of computer communications and the ever-decreasing sizes of wireless devices enabled the proliferation of wireless data applications. Mobile clients equipped with laptops, palmtops, personal digital assistants (PDAs), and other portable devices are able to access a variety of information stored in the databases of servers located in fixed wireline networks. The mobile clients roam inside the coverage area of the wireless network requesting various data; types of data that may be of interest to the clients include stock quotes, weather information, traffic conditions, and airline schedules, to name a few.

There are two basic delivery methods in wireless applications: the unicast (or point-to-point) and broadcast methods. In the former, each client establishes a connection with the server and poses a request; in response, the server sends the requested data to the client using the established connection. This delivery method implements the classic client-server paradigm of communication encountered in traditional wireline networks. The broadcast delivery method (Wong, 1988) differs because all clients monitor the same channel (*broadcast channel* or *downlink channel*) in order to acquire the information transmitted by the server. The contents of the transmission are determined by the server, based either on the estimation of client preferences (pure broadcast systems; Acharya, Alonso, Franklin, & Zdonik, 1995) or on the client requests acquired through *uplink channels* (on-demand broadcast systems; Aksoy & Franklin, 1999).

Although, many current systems are based on unicast delivery, the broadcast method is increasingly appealing. Unicast delivery is a waste of resources because each datum must be transmitted for each client that requests it. Thus, the network and server load increases with every client. In contrast, broadcasting is advantageous because of its excellent scalability, that is, a single broadcast satisfies all pending requests for it. Moreover, wireless environments are characterized by the asymmetry in communication, that is, the broadcast channel capacity is much greater than the uplink channel capacity. Therefore, broadcasting is able to exploit the high bandwidth of the downlink channel. Concrete examples of broadcast delivery include the cache-satellite distribution systems (Armon & Levy, 2004), where satellites broadcast popular data, for example, Web pages, to clients (e.g., humans or proxy servers), and the mobile Infostations (Iacono & Rose, 2000).

Multiple-Channel Broadcast Environments

The most common assumption about broadcasting is that there exists a single physical channel for the broadcasted data. There are, though, many scenarios where a server has access to multiple low-bandwidth physical channels which cannot be combined to form a single high-bandwidth channel (Prabhakara, Hua, & Oh, 2000; Yee & Navathe, 2003). Possible reasons for the existence of multiple broadcast channels include application scalability, fault tolerance, reconfiguration of adjoining cells, heterogeneous client-communication capabilities, and so forth. In the following paragraphs, we give example scenarios of the aforementioned reasons.

- **Application scalability.** Consider an application running on the server that needs to be scaled in order to support a larger number of clients. In this case, it may need to acquire additional physical channels. If these channels are in noncontiguous frequencies, they may have to be treated as separate channels when broadcasting data.
- **Fault tolerance.** Suppose that a transmitting station can have more than one server with a transmission capability, and let A, B and C be servers which broadcast data in three noncontiguous frequency ranges all in the same cell. If servers B and C crash, then frequencies assigned to them should be allocated to server A.
- **Reconfiguration of adjoining cells.** Suppose that there are two adjacent cells whose servers transmit in different frequency ranges and, at some point in time, we decide to "merge" the two cells and use one of the two servers to serve the newly generated cell. Then, the frequency range of the other server should be migrated and added to the residual server. In this case also, the latter server gets multiple physical channels.
- **Heterogeneous clients.** The mobile clients may have heterogeneous communication capabilities, precluding the existence of a single high-speed transmission channel.

There are several concerns related to the exploitation of a multichannel broadcast system. The first consideration is related to the capability of the server to concurrently transmit in all channels; the second is related to the capability of the client to simultaneously listen to multiple channels and also perform instantaneous "hopping" among channels. Since the interest of this chapter is to focus on the data placement problem, we will make simplifying assumptions, considering that the server is able to concurrently transmit to all channels, and the clients are able to listen simultaneously to all channels and perform instantaneous hopping from channel to channel. Moreover, we do not assume any kind of dependencies between broadcasted data. Techniques for broadcast scheduling when there exist dependencies among the data can be found in other chapters of this book.

The two major issues in broadcast dissemination are how and what the server transmits and how the client retrieves. Of particular interest are the solutions that enable the mobile clients to get the disseminated data efficiently, that is, with short access latency and with minimum power expenditure. The former is quantified by the *query access time*, which is equal to the time elapsed between when the client starts seeking for an item until it gets it. The latter is quantified by the *tuning time*, which is the time the client spends actively listening to the broadcast channel. The access time is directly related to the size of the broadcast. On the other hand, providing information to the clients for *selective auto tuning*, that is, indexes (Imielinski, Viswanathan, & Badrinath, 1997), reduces the tuning time. However, including such information increases the overall size of the broadcast, which in turn increases the access time. The trade-off between these two performance measures is obvious.

The chapter's focus is the data placement issues, so we assume that the clients have complete knowledge of the broadcast schedule. In other words, they know a priori the arrival time and channel of all broadcasted data items. Hence, we are only interested in minimizing the client access time and we do not assume the existence of index packets in the broadcast. Though, in order to make the broadcast schedule "predictable," we are interested in placement schemes, which guarantee that

- the broadcast is cyclic, that is, the schedule has a beginning and an end, and
- the interarrival time between successive transmissions of an item is constant for all the broadcast cycles.

Under the aforementioned assumptions, the problem addressed by this chapter can be captured by the following question:

Given a number of identical broadcast channels and knowledge of the data item probabilities, how can we decide the contents of the multiple channels in order to reduce the average access time?

The aim of the present chapter is to provide an answer to the above question and it is organized as follows. "Background" describes in detail the assumed architectural model, whereas "Problem Formulation" defines the problem in mathematical terms. Optimal and heuristic algorithms for the problem of the data placement over multiple wireless channels are provided in "An Optimal Solution for Data Placement" and "Heuristic Approaches for Data Placement," respectively. Relevant work on multiple broadcast wireless channels is discussed in the next section, and the chapter concludes with a discussion about future trends and with a summary of its contributions.

Background

We consider a generic architecture of a mobile computing environment depicted in Figure 1 (Agrawal & Zeng, 2003; Dunham & Helal, 1995). Our system serves a geographical area, called the *coverage area*, where mobile clients or users (MUs) can move. The coverage

Figure 1. Generic architecture of a wireless system

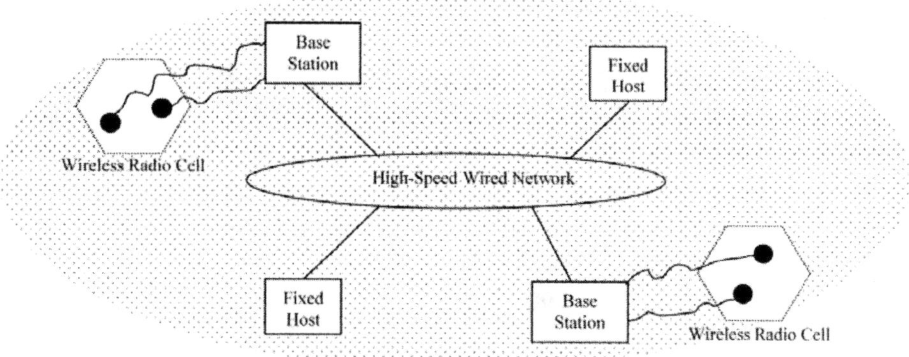

area served by the wireless system is partitioned into a number of nonoverlapping regions, called *cells*. At the heart of the system lies a fixed backbone (wireline) network. A number of *fixed hosts* are connected to this network. Fixed hosts are general-purpose computers storing databases which are of interest to the mobile clients. Fixed hosts are not equipped to manage mobile units, although they can be configured to do so. Each cell is served by one *base station* (called simply *server*, in the sequel), which is connected to the fixed network and is equipped with wireless transmission and receiving capability. The server is responsible for converting the network signaling traffic and data traffic to the radio interface for communication with the mobile unit, and also for transmitting paging messages to the clients. Each server is assigned a number of identical broadcast channels and is able to transmit to all these channels concurrently.

In addition, we assume that every fixed host maintains a full replica of the database, and part of this database is broadcast by the server. The data are equal-sized items and the broadcast of each one of them takes one time unit, that is, a *tick*. The server has a priori knowledge of the client access probabilities and, based on these, it decides the contents of the broadcast, which is usually comprised of the most popular data items. Although, such a priori knowledge seems like a strong assumption, there are several methods for determining the data access probabilities. The works Sakata and Yu (2003) and Yu, Sakata, and Tan (2000) describe statistical methods for the estimation of these probabilities, whereas the works Nicopolitidis, Papadimitriou, and Pomportsis (2002) and Stathatos, Roussopoulos, and Baras (1997) present more practical approaches. The broadcast is a sequence of data blocks (containing data) and there are no index data.

Mobile clients are battery-powered, portable computers which use radio channels to communicate with the server to gain access to the fixed or wireless network. The clients have complete a priori knowledge of the broadcast schedule and they are able to listen simultaneously to all channels and perform instantaneous hopping from channel to channel. They continuously monitor the broadcast channels in order to acquire the data of interest. We also assume that the size of a cell is such that the average time between when a client starts seeking for an item until it gets it from the broadcast is much smaller than the time required by the client to traverse the cell. Therefore, a user will seldom submit a query and exit a cell before receiving the response. The clients request one item per query.

Problem Formulation

Due to the nice characteristics of the generic paradigm of broadcast disks described in Acharya et al. (1995), that is, *fixed interarrival times* and *fixed-length broadcast cycle*, we adopt that model to implement in our broadcast channels. According to this model, the server partitions the database into groups and each group is broadcast into one channel. The contents of each group are cyclically broadcast in a round-robin fashion. Thus, we are left with the question of what to put into each group of items. In the next paragraphs, we will formulate this question into the problem of *Broadcast Program Generation on Multiple Channels.*

We consider a database D of equal-sized data items for which the access probabilities are given, and a set C of k broadcast channels (C_1, C_2, ..., C_k). Without loss of generality, we assume that the channels with small indexes will finally accommodate less items. We need to partition the items into these k wireless channels. Let our database be comprised of n items, that is, D = d_1, d_2, ..., d_n. The items are ordered from the most popular to the least popular and have access probabilities $P = p_1, p_2, \ldots, p_n$, respectively, where:

$$\sum_{i=1}^{n} p_i = 1. \tag{1}$$

This ordering means that for any two coordinates (p_i, p_j) of the vector of probabilities P, where $1 \leq i < j \leq n$, it holds that $p_i \leq p_j$. Since the items have equal size, the broadcast bandwidth needed to allocate each data item is the same and is denoted as a time slot. The interval between two consecutive broadcasts of the same item will be denoted as s_i, and its reciprocal as b_i. The latter estimates the probability that the item d_i will be selected for broadcast at each time slot. The average access time for item d_i will be denoted as a_i, whereas the average access time for all items at D will be:

$$a_{\text{total}} = \sum_{i=1}^{n} p_i * a_i . \tag{2}$$

Wong (1988) proved that, for the case of a single channel, when the data items are equisized, the average access time can be minimized if each data item is equally spaced and for any two items d_i and d_j, the following equation holds:

$$\frac{\sqrt{p_i}}{\sqrt{p_j}} = \frac{b_i}{b_j}. \qquad (3)$$

We can easily show that in the case of k broadcast channels, the following holds: $\sum_{i=1}^{n} b_i = k$. Therefore, the analytical, minimum average access time for the database D broadcast on k channels is given by the following formula:

$$a_{\text{minimum}} = \frac{1}{2k}\left(\sum_{i=1}^{n}\sqrt{p_i}\right)^2, \qquad (4)$$

assuming that the access frequencies p_i sum up to 1. Otherwise, the analytical, minimum average access time is given by the following formula:

$$a_{\text{minimum}} = \frac{1}{2k}\left(\sum_{i=1}^{n}\sqrt{\frac{p_i}{\sum_{j=1}^{n} p_j}}\right)^2. \qquad (5)$$

This lower bound is difficult to achieve due to the difficulty in approximating Equation 3. Therefore, the average access time for each channel and, subsequently, the total average access time are different than that presented in Equation 5. Indeed, under the assumption that all interarrival times for an item d_i are equal, it follows that the average access time for d_i is:

$$a_i = \frac{s_i}{2}, \qquad (6)$$

and the average access delay for all items of channel C_{ij}, assuming that these are d_i, d_{i+1}, ..., d_j, will be:

$$a_{ij} = \frac{1}{2}(j-i+1)\sum_{m=i}^{j} p_m \, , \ j \geq i \, . \tag{7}$$

Therefore, the total average access time will be:

$$a_{\text{true\_total}} = \frac{1}{2}\sum_{m=1}^{k} N_m * P_m \, , \tag{8}$$

where N_m and P_m are the number of items and the popularity of the *m*th channel, respectively.

As mentioned above, generating a broadcast program for *k* channels can be viewed as a partition problem and can be characterized by an assignment G:[1... *n*]→[1... *k*] of items to the channels. Since more popular pages should be transmitted more frequently and, thus, should be accommodated into channels with a smaller number of items (i.e., channels with a small index), it is obvious that there can exist no pair d_i and d_j, with $i < j$, in the above ordering such that $G(d_i) > G(d_j)$. This means that either the two items will be accommodated into the same channel or the more popular of the two items will be accommodated into a smaller channel. Thus, creating a broadcast program for *k* channels is equivalent to determining a partition of the interval [1... *n*].

Definition 1 (Broadcast Program Generation on Multiple Channels problem). Suppose that we have a database of *n* equisized items and a set of *k* broadcast wireless channels. Assume also that we are aware of the access frequency of each item. Then, the problem of *Broadcast Program Generation on Multiple Channels* is to find a partition of the *n* items into *k* groups and subsequently assign each group to a channel, such that the average access time for all items, that is, Equation 8, is minimized.

Optimal Solution for Data Placement

There exists an optimal solution to the problem described by Definition 1. This solution has been formulated into two different-in-nature algorithms. The first was presented in Peng and Chen (2003) and is based on the *A* optimization method*, whereas the second was presented in Yee, Omiecinski, and Navathe (2001) and is based on a dynamic programming approach. Denoting with $a_{\text{optimal\_minimal}}(i, j)$, the optimal solution (i.e., the minimal average access delay) for allocating items *i* to *n* on *j* channels, it is obvious that, trivially, $a(i, 1) = a_{ij}$ (Equation 7). Then, the recurrence formula for the determination of the minimal average access delay is given by the following equation:

$$a_{\text{optimal\_minimal}}(i,k) = \min_{l \in \{i,i+1,\ldots,n\}} \{a_{il} + a_{\text{optimal\_minimal}}(l,k-1)\}. \quad (9)$$

We can prove that the following propositions hold.

Proposition 1. The time complexity of the dynamic programming algorithm is $O(k^*n^2)$.

Proposition 2. The space complexity (storage of intermediate, partial solutions) of the dynamic programming algorithm is $O(k^*n)$.

Obviously, the time complexity of the optimal dynamic programming solution is too high. In applications where the access frequencies change quite frequently or new items are to be broadcast, this approach turns out to be inapplicable. Therefore, various heuristics have been proposed in order to generate the partitioning very fast, producing average access time very close to the optimum. In the next section, we survey these approaches.

Heuristic Approaches for Data Placement

The procedure of determining a partition can be "top down," "bottom up," or "one scan." In the top-down (Hsu, Lee, & Chen, 2001; Peng & Chen, 2003; Yee, Navathe, Omiecinski, & Jermaine, 2002) approach, we start from a large partition, possibly including all the items, and gradually split it into smaller pieces. In the bottom-up (Hwang, Cho, & Hwang, 2001; Katsaros & Manolopoulos, 2004), we start with many small partitions which gradually grow, whereas the one-scan (Vaidya & Sohail, 1999) approach makes a single scan over the vector $P = p_1, p_2, \ldots, p_n$ of the access probabilities, assigning items to channels based on some criterion. The top-down and bottom-up make multiple passes over P (or parts of it) and make splitting or concatenating decisions based on the computation of Equation 8 over P (or portions of it).

The Bucketing Scheme

The *bucketing scheme* (Vaidya & Sohail, 1999) is the simplest approach for the assignment of items to channels. It makes a single scan over the vector P of access probabilities of the data and assigns them to channels based on the following criterion. Let A_{min} (A_{max}) denote the minimum (maximum) value of $\sqrt{p_i}$, $1 \le i \le n$. Let $\delta = A_{max} - A_{min}$. If for the item

d_i it holds that $\sqrt{p_i} = A_{min}$, then d_i is assigned to channel C_k. Any other item d_i is assigned to channel C_{k-j} $(1 \leq j \leq k)$ if $(j - 1)*\delta/k < (\sqrt{p_i} - A_{min}) \leq (j*\delta/k)$. This partitioning criterion is suitable for the case when the access probabilities are uniformly distributed over the "probability interval" and performs poorly for skewed access patterns. On the other hand, this approach has the lowest complexity and never tries any "candidate" partitions in order to select the most appropriate.

Algorithm Bucketing (int n, int k, float vector P)
//n: number of items, k: number of channels, P: access probabilities

BEGIN

$A_{min} = \text{minimum}(\sqrt{p_i}),\ 1 \leq i \leq n;$

$A_{max} = \text{maximum}(\sqrt{p_i}),\ 1 \leq i \leq n;$

$\delta = A_{max} - A_{min};$

for(i = 1; $i \leq n$; $i = i + 1$){

if($\sqrt{p_i} == A_{min}$) then

item d_i is assigned to channel C_k;

else

if($(j - 1)*\delta/k < (\sqrt{p_i} - A_{min}) \leq (j*\delta/k)$)

item d_i is assigned to channel C_{k-j}, where $1 \leq j \leq k$;

}

END

Therefore, we can easily deduce the following proposition.

Proposition 3. The complexity of the bucketing scheme is $O(n)$.

The Growing Segments Scheme

The *growing segments* (Hwang et al., 2001) scheme starts with an initial "minimal" allocation assigning one item to each channel, which acts as the initial "seed" partition. Then, it enlarges each segment by including a number of items equal to the user-defined parameter *increment* and computes which of these enlargements gives the greatest reduction in average delay. Next, it selects the corresponding partition as the new seed partition and continues until the partition covers the whole P. The parameter *increment* is very important and has the following trade-off associated with it: the greater the value of *increment*, the lower complexity the algorithm has and the lower quality the produced broadcast program has. Assuming that a set of points S = (r_0,r_1, ..., r_k), satisfying the inequalities $0 = r_0 < r_1 < ... < r_k = n$, denotes a partition of the interval (0, n], we have the following pseudocode for the growing segments algorithm.

Algorithm Growing Segments (int n, int k, float vector P)

//n: number of items, k: number of channels, P: access probabilities

```
BEGIN
// getNextIncrement(): returns an appropriate value for the next increment
// minDelay(); returns the minimum expected delay for a partition, i.e., Equation 8
increment = getNextIncrement();
r_0 = 1;
for(i = 1; i < = k; ++ i)      // initial setup of (r_0, r_1, ..., r_k)
      r_i = r_i-1 + increment;
while(r_k < n){
      increment = getNextIncrement();
      for(i = 1; i < = k; ++ i){
          p_0 = r_0;
          for(x = 1; x < = k; ++ x)
               if (x ≥ i) p_x = r_x + increment;
               else p_x = r_x;
          S_i = (p_0, p_1, ... , p_k);
      }
Find a partition (r_0, r_1, ..., r_k) among S_is such that
      minDelay(r_0, r_1, ..., r_k) = min{ minDelay(S_x) } for x = 1, 2, ..., k.
}
return (r_0, r_1, ... , r_k);
END
```

Since the computation of Equation 8 takes $\Theta(n)$ time, and the algorithm makes $\Theta((n - k)/$increment$)$ steps, computing at each step $\Theta(k)$ candidate partitions, we can easily deduce the following proposition.

Proposition 4. The complexity of the growing segments method is $O(n^2*(k/\text{increment}))$.

The Variant-Fan-Out Tree Scheme

The *variant fan-out with the constraint K* (VF^K) (Peng & Chen, 2003) scheme adopts a top-down approach. It starts with an initial allocation where all the items have been assigned to the first channel. Then repetitively, it determines which channel incurs the largest cost so far and partitions its contents into two groups. The partitioning is done by the routine *Partition*, which tries all possible partitions that respect the property that no channel can have more items than its next channel. The first group remains in the current channel and the newly created group is allocated to the next channel, shifting all the other channels downwards. This procedure repeats until all available channels are allocated. The cost of this algorithm is $O(k)$ times the cost of the *Partition* procedure. The cost of this procedure depends on the number of items of the channel that is to be partitioned, which in turn depends on the distribution of the access probabilities. Let the reduction in access delay, which will occur if we split a channel C_{ij} containing the items d_i, d_{i+1}, ..., d_j into two channels C_{ip} and C_{pj} containing the items d_i, d_{i+1}, ..., d_p and items d_{p+1}, d_{p+2}, ..., d_j, be denoted by $\delta(p)$, then $\delta(p) = a_{ij} - (a_{ip} + a_{pj})$. Then, the pseudocode for VF^K is shown on the next page.

```
Algorithm Variant Fan-out with the Constraint K (int n, int k, float vector P)
//n: number of items, k: number of channels, P: access probabilities

BEGIN
Create table AT with k rows;
AT(1).B = 1;        // AT(1).B records the beginning of channel
AT(1).E = n;        // AT(1).E records the end of channel
AT(1).LC = a_1n;    // AT(1).LC records the average access delay of channel
for(each row i in table AT AND i ≥ 2){
    AT(i).B = 0;  AT(i).E = 0;  AT(i).LC = 0;
}
pivot = 1;
repeat{
    Choose row i from table AT such that AT(i).LC is maximal among all unmarked rows;
    if (i == 1 or i == pivot){
        j = Partition(p_AT(i).B, p_AT(i).B+1, ..., p_AT(i).E);
        {Update table AT accordingly and unmark all rows; pivot ++;}
    }
    else{
        j = Partition(p_AT(i).B, p_AT(i).B+1, ..., p_AT(i).E);
```

```
        if (AT(i – 1).E –AT(i –1).B) < (j –AT (i).B)
            {Update table AT accordingly and unmark all rows; pivot ++;}
        else{
            Mark row i;
            Merge (p_AT(i).B, p_AT(i).B+1, ..., p_j ) with (p_j+1, p_j+2, ..., p_AT(i).E);
        }
    }
}until (pivot == k)
END
```

Procedure Partition $(p_i, p_{i+1}, \ldots, p_j)$.

BEGIN

Determine p^* such that $'(p^*) = \max_{\forall p \in \{i, i+(j-i+1)/2-1\}} \{\delta(p)\}$;

Assign items $p_{p^*+1}, p_{p^*+2}, \ldots, p_j$ into a new channel;

Return p^*;

END

Therefore, we can easily deduce the following proposition.

Proposition 5. The complexity of the VF^K method is $k^*(O(k^*\log(k)) + O(n))$.

The Greedy Scheme

The *greedy scheme* (Yee et al., 2002) adopts the top-down approach and it is very similar to the VF^K scheme. It performs several iterations. At each iteration, it chooses to partition the contents of the channel whose split will bring the largest reduction in access time. The partitioning point is determined by calling the routine *Partition* (see above). Thus, at each iteration, the greedy scheme computes (if not already computed) and stores the optimal split points for all channels that have not been split so far. Hence, it differs from VF^K in two aspects. First, it differs in the partitioning criterion (recall that VF^K splits the channel which incurs the largest access time). Second, after each split, it will compute and store the optimal split points of every channel.

Algorithm Greedy (int n, int k, float vector P)

//n: number of items, k: number of channels, P: access probabilities

```
BEGIN
numPartitions = 1;
while (numPartitions < k){
    for each partition r with data items i through j {
    // Find the best point to split in partition r
        for(s = i;  s < = j; s = s + 1)    // Initialize the best split point for this partition
                                           //as the ûrst data item. If we ûnd a better one
                                           //subsequently, update the best split point.
            if ( (s == i) OR (localChange > C_ij^s) )
                localS = s;
                localChange = C_ij^s;      // Initialize the best solution as the one for the ûrst
                                           // partition. If we ûnd a better one subsequently,
                                           // update the best solution.
    if ( (r == 1) OR (globalChange > localChange) )
        globalChange = localChange;
        globalS = localS;
        bestpart = r;
    }
    split partition bestpart at point globalP;
    numPartitions = numPartitions + 1;
}
END
```

We can easily verify that the following proposition holds.

Proposition 6. The complexity of the greedy method is $O((n + k)*\log(k))$.

The Data-Based Scheme

The *data-based scheme* (DB; Hsu et al., 2001) is similar to VF^K, but avoids taking the local optimal decision of the *Partition* routine of VF^K, which splits a channel into two. DB has several phases. At each phase, it decides the contents of a particular channel starting from the smallest channel, which will accommodate the more frequently accessed data. First, it determines which is the maximum allowable number of items that can be accommodated into the considered channel. This number can be computed with the help of Lemmas 1 and 2 (see also Hsu, Lee, & Chen, 2001).

Lemma 1. If we denote the number of items allocated to channel C_i as $|C_i|$, then it holds that $|C_1| \leq |C_2| \leq \ldots \leq |C_k|$.

Lemma 2. For the optimal allocation, it holds that $|C_{i-1}| \leq |C_i| \leq \dfrac{n - \sum_{j=1}^{i-1} |C_j|}{k - i + 1}$.

Then, it computes the average access delay for all the allowable allocations of items into the considered channel and selects the allocation with the minimum cost. The computation of the average access delay takes also into account the delay that will be incurred due to the items that will be allocated to the rest of the channels. This is the difference from VF^K. The above procedure continues until the allocation of all the items into the channels is completed.

Algorithm Data Based (int *n*, int *k*, float vector *P*)

//*n*: number of items, *k*: number of channels, *P*: access probabilities

```
BEGIN
for(i = 1; i ≤ k - 2; i = i + 1){
    Calculate the range of the number of data items in channel i;
    Determine the number of data items for allocating in channel i
    such that the expected average access time is minimal;
}
Allocate the remainder data items into the last two channels so that
the average access time of all items is minimal;
END
```

Assuming that z_i is equal to the number of items allocated to channels C_1 to C_{i-1}, and y_i is equal to the range of items (as determined by Lemma 2), we can easily deduce the following proposition.

Proposition 7. The complexity of the data-based method is $\sum_{i=1}^{k} (y_i * (n - z_i)$.

The CascadedWebcasting Scheme

The *CascadedWebcasting scheme* (Casc; Katsaros & Manolopoulos, 2004) starts from a very basic intuition about the partitioning, claiming that there are three "classes" of items:

- a practically constant number of items with *high probability*,
- items belonging to a few *large groups*, and
- leftover items, which contribute *negligibly* to the total delay.

Using this intuition about the partitioning, Casc is a bottom-up scheme that uses "predetermined" initial seed partitions, which subsequently are greedily concatenated until the number of partitions becomes equal to the number of available broadcast channels. Initially, the seed partitions $P_1^0, P_2^0, \ldots, P_v^0$ are generated, with sizes equal to 2^0, 2^1, 2^2, ... Then, $\nu - k$ merging steps ($\lceil \nu = \log_2(n+1) \rceil$) are performed. At the *i*th merging step ($1 \le i \le \nu - k$), the algorithm tries $\nu - i - 1$ concatenations and selects the one which incurs the least expected cost. Due to the way the partitions are concatenated and the size of the initial seed partitions, it is obvious that at each step of the algorithm, the size of a partition is always smaller than the size of its successive partition, thus respecting Lemma 1. ™f the number n of items is not equal to $2^\nu - 1$, but it is $2^\nu - 1 < n\ (= 2^\nu - 1 + \beta\ ^2)$ $< 2^{\nu+1} - 1$ for some $\beta > 0$, then we treat the first $2^\nu - 1$ items with the procedure mentioned above and simply append the last ² items to the last channel. The pseudocode for Casc is presented below.

Algorithm CascadedWebcasting (int n, int k, float vector P)

//n: number of items, k: number of channels, P: access probabilities

BEGIN

$v = \lceil \log_2(n+1) \rceil$;

create v seed partitions, $P_1^0, P_2^0, \ldots, P_v^0$ with $size(P_i^0) = 2^{i-1}$;

$P_1 = P - (P_1^0 \cup P_2^0 \cup, \ldots, \cup P_v^0)$;

$P^0 = \{P_1^0, P_2^0, \ldots, P_v^0\}$; //the set of seed partitions

for($i = 1; i \le v - k; i = i + 1$){ //perform $v - k$ merging steps

for($j = 1; j \le v - i - 1); j = j + 1$)

$C^j = \{C_1^i(=P_1^{i-1}), \ldots, C_j^i(=merge(P_j^{i-1}, P_{j+1}^{i-1})), \ldots, C_{v-i-1}^i(=P_{v-i}^{i-1})\}$;

$P^i = minDelay(C^j)$; //The partition incurring the minimum delay

}

$P^{final} = \{P_1^{½-k}, \ldots, P_k^{½-k} \cup P_1\}$;

return P^{final};

END

It is easy to prove the following proposition (see Katsaros & Manolopoulos, 2004).

Proposition 8. The complexity of Casc is dominated by $O(n)$.

A comprehensive performance evaluation of the aforementioned algorithms has been conducted in Katsaros and Manolopoulos (2004). There, it was recognized that the greedy scheme, VF^K, and DB perform very close to optimum with respect to the reduction of the average access delay, but they incur significant execution cost. The fastest running algorithms are bucketing and Casc, with the latter producing partitions not far from the optimum with respect to the average access delay.

Relevant Work on Data Broadcasting over Multiple Wireless Channels

There are quite a lot of research efforts in finding efficient methods for broadcasting dependent data on multiple (or single) wireless channels. We mention the most important of them. In Chehadeh, Hurson, and Kavehrad (1999), Chung and Kim (2001), Lee and Lo (2003), Lee, Lo, and Chen (2002), and Liberatore (2004), the issue of scheduling dependent data on a single broadcast channel is investigated, whereas in Huang, Chen, and Peng (2003) and Juran, Hurson, Vijaykrishnan, and Kim (2004), the issue of dependent data scheduling over multiple channels is investigated.

Power conservation is a key issue for mobile computers. Air-indexing techniques for the broadcast data can be employed in order to help clients reduce the time they remain tuned into the broadcast channel(s). Various indexing techniques have been proposed and some of them have been described in chapters of this book.

The *index tree* (Imielinski et al., 1997) is the application of the traditional disk-based B-tree indexing technique in the context of wireless channels. Also, the well-known *signature tree* indexing method has been proposed for the case of broadcast data indexing (Lee & Lee, 1996). An amalgamation of the aforementioned techniques, the *hybrid index tree*, is proposed in Hu, Lee, and Lee (2001). The aforementioned techniques proposed balanced indexing structures. To achieve better performance for skewed queries, various techniques investigated the approach of creating unbalanced indexes. For instance, the index proposed in Shivakumar & Venkatasubramanian (1996) is a generalization of the *binary alphabetic tree* requiring that the number of available wireless channels is equal to the number of tree levels, whereas a generalization of the classic *binary Huffman tree* for the case of broadcast channels was proposed in Chen, Wu, and Yu (2003).

While many studies, like ours, consider data scheduling and indexing separately, some works addressed the issue of allocation of both data and index over multiple channels. In Prabhakara et al. (2000), various broadcast and client access methods are proposed, whereas the works appearing in Lo and Chen (2000) and Hsu, Lee, and Chen (2002) examined the issue of allocating index and data at the same time, so that the average access time is minimized. Finally, an interesting indexing structure for multiple broadcast channels is presented in Lee and Jung (2003).

Conclusion

Although the concept of broadacast delivery is not new, recently, the dissemination of data items by broadcast channels has attracted considerable attention due to the excellent scalability it offers. There are many scenarios where a server has access to multiple low-bandwidth physical channels which cannot be combined to form a single high-bandwidth channel. Possible reasons for the existence of multiple broadcast channels include application scalability, fault tolerance, reconfiguration of adjoining cells, and heterogeneous client-communication capabilities. Thus, it becomes a necessity to devise appropriate methods for data allocation over multiple wireless channels.

The present chapter explored the issue of generating broadcast programs for multiple wireless channels when the data access probabilities and the number of wireless channels are given, assuming that there are no dependencies among the data to be broadcasted. Initially, we gave the mathematical formulation for the problem and presented an optimal solution for it based on dynamic programming. Then, we presented six heuristic approaches which present different trade-offs between optimality in terms of average access delay and execution time.

Acknowledgments

This work has been funded through the bilateral program of scientific cooperation between Greece and Turkey (Γ.Γ.Ε.Τ.) and from TUBITAK Grant No. 102E021.

References

Acharya, S., Alonso, R., Franklin, M. J., & Zdonik, S. B. (1995). Broadcast disks: Data management for asymmetric communications environments. *Proceedings of the ACM International Conference on Management of Data (SIGMOD)*, (pp. 199-210).

Agrawal, D. P., & Zeng, Q.-A. (2003). *Introduction to wireless and mobile systems.* Brooks/Cole (Thomson Learning Inc.).

Aksoy, D., & Franklin, M. J. (1999). RxW: A scheduling approach for large-scale on-demand data broadcast. *IEEE/ACM Transactions on Networking, 7*(6), 846-860.

Armon, A., & Levy, H. (2004). Cache satellite distribution systems: Modeling, analysis, and efficient operation. *IEEE Journal on Selected Areas in Communications, 22*(2), 218-228.

Chehadeh, Y. C., Hurson, A. R., & Kavehrad, M. (1999). Object organization on a single broadcast channel in the mobile computing environment. *Multimedia Tools and Applications*, *9*(1), 69-94.

Chen, M.-S., Wu, K.-L., & Yu, P. S. (2003). Optimizing index allocation for sequential data broadcasting in wireless mobile computing. *IEEE Transactions on Knowledge and Data Engineering*, *15*(1), 161-173.

Chung, Y. D., & Kim, M.-H.(2001). Effective data placement for wireless broadcast. *Distributed and Parallel Databases*, *9*(2), 133-150.

Datta, A., Vandermeer, D. E., Celik, A., & Kumar, V. (1999). Broadcast protocols to support efficient retrieval from databases by mobile users. *ACM Transactions on Database Systems, 24*(1), 1-79.

Dunham, M. H., & Helal, A. (1995). Mobile computing and databases: Anything new? *ACM SIGMOD Record, 24*(4), 5-9.

Hsu, C.-H., Lee, G., & Chen, A. L. P. (2001). A near optimal algorithm for generating broadcast programs on multiple channels. *Proceedings of the ACM International Conference on Information and Knowledge Management (CIKM)*, (pp. 303-309).

Hsu, C.-H., Lee, G., & Chen, A. L. P. (2002). Index and data allocation on multiple broadcast channels considering data access frequencies. *Proceedings of the International Conference on Mobile Data Management (MDM)*, (pp. 87-93).

Hu, Q. L., Lee, W.-C., & Lee, D. L. (2001). A hybrid index technique for power efficient data broadcast. *Distributed and Parallel Databases*, *9*(2), 151-177.

Huang, J.-L., Chen, M.-S., & Peng, W.-C. (2003). Broadcasting dependent data for ordered queries without replication in a multi-channel mobile environment. *Proceedings of the IEEE Conference on Data Engineering (ICDE)*, (pp. 692-694).

Hwang, J.-H., Cho, S., & Hwang, C.-S. (2001). Optimized scheduling on broadcast disks. *Proceedings of the International Conference on Mobile Data Management (MDM)* (pp. 91-104).

Iacono, A. L., & Rose, C. (2000). Infostations: New perspectives on wireless data networks. In S. Tekinay (Ed.), *Next generation wireless networks* (Vol. 598). NJ: Kluwer Academic Publishers.

Imielinski, T., Viswanathan, S., & Badrinath, B. R. (1997). Data on air: Organization and access. *IEEE Transactions on Knowledge and Data Engineering, 9*(3), 353-372.

Juran, J., Hurson, A. R., Vijaykrishnan, N., & Kim, S. (2004). Data organization and retrieval on parallel air channels: Performance and energy issues. *ACM/Kluwer Wireless Networks*, *10*(2), 183-195.

Katsaros, D., & Manolopoulos, Y. (2004). Broadcast program generation for webcasting. *Data and Knowledge Engineering*, *49*(1), 1-21.

Lee, B., & Jung, S. (2003). An efficient tree-structured index allocation method over multiple broadcast channels in mobile environments. *Proceedings of the Database and Experts Systems Applications Workshop (DEXA)* (pp. 433-443).

Lee, G., & Lo, S.-C. (2003). Broadcast data allocation for efficient access of multiple data items in mobile environments. *ACM/Kluwer Mobile Networks and Applications, 8*(4), 365-375.

Lee, G., Lo, S.-C., & Chen, A. L. P. (2002). Data allocation on wireless broadcast channels for efficient query processing. *IEEE Transactions on Computers, 51*(10), 1237-1252.

Lee, W.-C., & Lee, D. L. (1996). Using signature techniques for information filtering in wireless and mobile environments. *Distributed and Parallel Databases, 4*(3), 205-227.

Liberatore, V. (2004). Circular arrangements and cyclic broadcast scheduling. *Journal of Algorithms, 51*(2), 185-215.

Lo, S.-C., & Chen, A. L. P. (2000). Optimal index and data allocation in multiple broadcast channels. *Proceedings of the IEEE Conference on Data Engineering (ICDE)*, (pp. 293-302).

Nicopolitidis, P., Papadimitriou, G. I., & Pomportsis, A. S. (2002). Using learning automata for adaptive push-based data broadcasting in asymmetric wireless environments. *IEEE Transactions on Vehicular Technology, 51*(6), 1652-1660.

Peng, W.-C., & Chen, M.-S. (2003). Efficient channel allocation tree generation for data broadcasting in a mobile computing environment. *ACM/Kluwer Wireless Networks, 9*(2), 117-129.

Prabhakara, K., Hua, K. A., & Oh, J. (2000). Multi-level multi-channel air cache designs for broadcasting in a mobile environment. *Proceedings of the IEEE Conference on Data Engineering (ICDE)*, (pp. 167-176).

Sakata, T., & Yu, J. X. (2003). Statistical estimation of access frequencies: Problems, solutions and consistencies. *ACM/Kluwer Wireless Networks, 9*(6), 647-657.

Shivakumar, N., & Venkatasubramanian, S. (1996). Efficient indexing for broadcast based wireless systems. *ACM/Baltzer Mobile Networks and Applications, 1*(4), 433-446.

Stathatos, K., Roussopoulos, N., & Baras, J. S. (1997). Adaptive data broadcast in hybrid networks. *Proceedings of the 23rd International Conference on Very Large Data Bases (VLDB)*, (pp. 326-335).

Vaidya, N., & Sohail, H. (1999). Scheduling data broadcast in asymmetric communication environments. *ACM/Baltzer Wireless Networks, 5*(3), 171-182.

Wong, J. W. (1988). Broadcast delivery. *Proceedings of the IEEE, 76*(12), 1566-1577.

Yee, W. G., & Navathe, S. B. (2003). Efficient data access to multi-channel broadcast programs. *Proceedings of the ACM International Conference on Information and Knowledge Management (CIKM)*, (pp. 153-160).

Yee, W. G., Navathe, S. B., Omiecinski, E., & Jermaine, C. (2002). Bridging the gap between response time and energy-efficiency in broadcast schedule design. *Proceedings of the International Conference on Extending Database Technology (EDBT)* (pp. 572-589).

Yee, W. G., Omiecinski, E., & Navathe, S. B. (2001). *Efficient data allocation for broadcast disk arrays* (Tech. Rep. No. GIT-CC-02-20). Georgia Institute of Technology, Atlanta.

Yu, J. X., Sakata, T., & Tan, K.-L. (2000). Statistical estimation of access frequencies in data broadcasting environments. *ACM/Baltzer Wireless Networks, 6*(2), 89-98.

Section II

Location Management

Chapter VI

Modeling and Management of Location and Mobility

Wenye Wang, North Carolina State University, USA

Abstract

Location modeling represents inclusive mobile objects and their relationship in space, dealing with how to describe a mobile object's location. The goal of mobility modeling, on the other hand, is to predict or statistically estimate the movement of mobile objects. With the increasing demand for multimedia applications, location-aware services, and system capacity, many recognize that modeling and management of location and mobility is becoming critical to locating mobile objects in wireless information networks. Mobility modeling and location management strongly influence the design and performance of wireless networks in many aspects, such as routing, network planning, handoff, call admission control, and so forth. In this chapter, we present a comprehensive survey of mobility and location models, and schemes used for location-mobility management in cellular and ad hoc networks, which are discussed along with necessary, but understandable, formulation, analysis, and discussions.

Introduction

One of the most salient features of wireless communications is that users can deploy a variety of wireless devices to communicate with others regardless of their location. While mobility support provides flexibility and convenience, it introduces many challenging issues in network design, planning, and performance evaluation. With the increasing demand for multimedia applications, location-aware services, and system capacity, many recognize that modeling and management of location and mobility is becoming critical to locating mobile objects in wireless information networks. Mobility modeling and location management strongly influence the choice and performance of mobility and resource management algorithms, such as routing, handoff, and call admission control, in a variety of wireless networks. For these reasons, it is important to understand mobility modeling and location management mechanisms, and the manner in which these mechanisms depend on the characteristics of the network and mobile environments. This chapter is concerned with issues in, and methods for, mobility modeling, location management, and applications in wireless wide-area networks (WWAN), wireless local-area networks (WLAN), and ad hoc networks.

The movement pattern of users plays an important role in system design, network management, and performance analysis of mobile and wireless networks. Therefore, the objective of mobility modeling is to estimate the current and future locations of a mobile user upon the arrival of a connection request, which involves many parameters such as moving speed, call duration time, distance between the last known position and destination, and geographical conditions. Management of location, however, deals with the problem of how to register or update the new location of a mobile user with the system, and how to locate a mobile terminal given the information in system databases. Location modeling represents inclusive mobile objects and their relationship in space. In other words, location modeling deals with how to describe a mobile object's location, which is, in turn, related to mobility modeling since the goal of mobility modeling is to predict or statistically estimate the location of mobile objects.

The location of a mobile object can be modeled or described by different methods depending on the network infrastructure. In cellular networks, a base station serves as an access point in delivering radio services. Since each base station covers one cell in cellular networks, the location of a mobile object is limited to one cell throughout a wireless system. That means, as long as we know in which cell a mobile stays, its location is determined in terms of a cell. Inside a cell, determining the exact positions of mobile nodes rather than finding the residing cell is considered a *geolocation* problem. This is similar to the localization problem in WLANs and ad hoc networks in which the location of mobiles cannot be represented by the cell in which a mobile stays. In wireless ad hoc networks, mobile nodes communicate with each other directly rather than through base stations as in cellular networks. Since each mobile node has a very limited transmission range, communications between any two nodes, which are not within the other's coverage, can only be accomplished through intermediate nodes with routing functions. Therefore, routing is very important in ad hoc networks because the communication of mobile objects relies on routing paths, where mobility models must be considered in evaluating new protocols with respect to average end-to-end delay, transmission efficiency, and so forth.

In what follows, we will mostly focus on mobility models and management of locations for different types of wireless networks. Numerous mobility models for a variety of applications will be introduced, which will let readers gain an in-depth understanding of this subject. Current and proposed location management algorithms for cellular networks are presented to demonstrate how mobility is supported in WWANs. In addition, localization algorithms are described for WLANs. These algorithms and mechanisms are discussed along with necessary, but understandable, formulation, analysis, and related discussions.

Mobility Modeling

There exist a variety of mobility models that can find applications in different kinds of wireless networks. Since mobility models are designed to mimic the movement of mobile users in real life, many parameters need to be considered. Most of the existing mobility models describe the behaviors of mobile terminals (MTs) without considering previous records. In this context, we will focus on this type of mobility models, the so-called synthetic models, and we will introduce several developments that take traces or profiles into account, and also combinational models.

Overview

Mobility models can be categorized into different groups based on the following criteria (Bettstetter, 2001; Camp, Boleng, & Davies, 2002).

- **Dimension:** The movement of MTs can be described in one-, two-, and three-dimension. For example, one-dimensional models may be appropriate to represent the behaviors of vehicles on highways and two-dimensional models are very useful in depicting terrestrial movements.
- **Scale of mobility:** Different levels of details may be used to describe a movement, including micromobility and macromobility. Micromobility models describe how MTs behave inside an autonomous network or inside a cell. In this case, there is no boundary crossing between two networks and user records can be found in a local database. Conversely, macromobility models are concerned with how MTs move from one network to other networks. The intersystem-information query is necessary for supporting mobility beyond an MT's home network.
- **Randomness:** Since the movement of an MT is random, depending on many unpredictable factors, it is helpful to describe the movement by varying randomness in different parameters. These parameters include direction, speed, and residence time in a certain area like a cell.

- **Geographical constraints:** Mobility models can be very specific for particular scenarios. In cellular networks, mobility models are divided into three types: indoor, outdoor pedestrian, and vehicular. For an indoor environment, a three-dimensional model can be used to describe the movement in a high-rise building, where the vertical movements in elevators are considered and a grid model can be used to describe movement on the same floor.
- **Destination oriented:** Some mobility models are developed to track or describe the route or the path of a movement. For example, if it is known that a mobile user will go to a particular location, it is very likely to estimate the position and the time of his or her traveling.
- **Change of parameters:** In order to capture the movement of a user in different conditions, kinetic mobility models are designed to derive the next movement of a mobile user based on his or her current actions. For example, if a driver steps on the brake in a car, we can expect a decrease in speed.

It is worth mentioning that there are many mobility models, which are derived from the models mentioned above. Also, there are some hybrid models which combine two or more attributes. More than ten mobility models will be introduced in this chapter by describing their motivations, representations, and applications.

Fluid-Flow Models

A mobility model for an individual mobile object requires the description of time-varying movement. The original mobility model is derived from Brownian motion with a drift model (Camp et al., 2002). In the one-dimensional Brownian model, the probability at location x at time t of a mobile starting at location x_0 at time t_0, $p(x|x_0, t)$, can be described as:

$$P_x(x\,|\,x_0,t) = \frac{1}{\sqrt{2\pi D(t-t_0)}}\exp\left\{-\frac{[x-x_0-v(t-t_0)]^2}{2D(t-t_0)}\right\},\quad t \ge t_0, \qquad (1)$$

where D is the diffusion constant (length^2/time), a parameter which represents the acceleration degree of the movement; v is the drift velocity (length/time) which represents the average moving speed of a mobile terminal. High D and v indicate a very active movement, whereas low D and v indicate very little change in the location with regards to changes in time.

In this mobility model, moving direction is not considered, which can be of any direction available. Depending on the selection or constraint of velocity, moving direction, and acceleration speed, many other mobility models are developed for different application scenarios. In cellular networks, especially for WWANs in which each cell covers an area with a radius of more than 10 km, mobile users usually move with a high speed that does not change frequently, such as that of vehicles on the highway. The objective of a mobility model is to estimate the number of MTs crossing over the border of a cell. The simple fluid-flow model describes the movement of a mobile user with the following assumptions (Mark & Zhuang, 2003; Thomas, Gilbert, & Mazziotto, 1988):

a) the mobile objects are uniformly distributed in the cell,

b) the movements of mobile users are not correlated and the directions of these movements are uniformly distributed on $[0, 2\pi]$, and

c) the mobile objects are uniformly distributed on the surface of the cell.

Let V be the mean velocity of MTs and r be the user density in the cell. Therefore, if the area of a part of the cell is A, the number of MTs on this part of the cell is rA. We denote L as the perimeter of the cell, which can be written as $L = 6R$ for hexagonal architecture. Then the number of MTs crossing the cell border per unit of time is:

$$\frac{dN_c}{dt} = \frac{V\rho L}{\pi}. \tag{2}$$

If we assume that r is constant all over the cell of area A, we can compute the crossover rate for a given MT:

$$\frac{dM_c}{dt} = \frac{1}{\rho A} \bullet \frac{V\rho L}{\pi} = \frac{VL}{\pi A}. \tag{3}$$

If we consider that all cells are grouped in clusters so that the crossing rate of a region with a cluster of cells per unit of time can be obtained with the area of the cell cluster approximated by a hexagon of radius R_{cluster}, then:

$$\frac{dM_R}{dt} = \frac{V\rho L_{\text{cluster}}}{\pi} \approx \frac{6\rho V\sqrt{N}R_{\text{cluster}}}{\pi}, \tag{4}$$

where N is the number of cells per cluster. This result can be regarded as the total handoff rate if handoff is required for MTs crossing the boundary of clusters.

Random-Walk Model and Derivatives

The random walk model was developed for such an environment in which there are many entities moving in random directions with unpredictable speed. In this model, a mobile object travels from its current location to a new location by randomly choosing a direction and speed in which to travel. The speed is usually used from a range of $[V_{min}, V_{max}]$ and the direction is chosen between $[0, 2\pi]$. In a random-walk mobility model, the time is divided into mini time slots. Therefore, this model is a discrete mobility model. The movement or current locations of a mobile object can be described either based on time, moving steps, or distances. In simulation, the movement is often observed in a simulation region. If a mobile reaches the boundary of the simulated region, it will be "absorbed" or "bounced" off.

The random-walk model is basically a memoryless pattern because it does not consider previous locations and speed values (Camp et al., 2002; Wang & Akyildiz, 2000). The only factor that influences future locations and speed is the current status. There are many derivatives of random-walk models which have been widely used in the design of cellular networks, especially for microcell or picocell systems in which moving users can freely move from their current locations to neighboring cells.

One- and Two-Dimensional Random-Walk Models

Let us assume that the user moves according to a symmetric, one-dimensional random walk, the space is divided into cells of the same length, and each cell has two neighbors. Figure 1a shows the cell partition in a one-dimensional coverage area. The numbers represent the distance of each cell from Cell 0, in which the mobile user is residing.

The two-dimensional space is divided into hexagonal cells of the same size (Akyildiz, Ho, & Lin, 1996). Each cell has six neighbors. Figure 1b shows the cell partition in a two-dimensional coverage area. The size of each cell is determined by service providers based on the number of subscribers, the number of channels in each cell, and the geographical condition of the coverage area. Note that if the size of cells is small, the probability of moving out of a cell will be high and vice versa. As shown in Figure 1, each cell is surrounded by neighboring cells. There are two neighboring cells for a current cell and six cells for a current cell in the one- and two-dimensional models, respectively. In this discrete-time random-walk model, it is assumed that at discrete time t, a user moves to one of the neighboring cells with probability q or stays at the current cell with probability $1 - q$. If the user decides to move to another cell, there is equal probability for each one of the neighboring cells to be selected as the destination. This probability is 1:2 in the one-dimensional model and 1:6 in the two-dimensional model.

Compared to the fluid-flow model introduced in the previous section, the random-walk model is more appropriate for picocell or microcell systems because most of the mobile users are likely to be pedestrians, changing moving directions frequently. The fluid-flow model is more suitable for vehicle traffic such that a continuous movement with infrequent speed and direction changes is expected. In wireless cellular networks, the ID of a cell is broadcast periodically so that the MT knows exactly its cell location at any time.

This model can also be represented by a Markov chain, in which a state i is defined as the distance between the current location of the MT and its center cell, which is the most recent location in which it interacts with the system. Transitions from a state to one of its two neighboring states represent movements of the terminal away from a cell. Let us

Figure 1. (a) One-dimensional model and (b) two-dimensional model

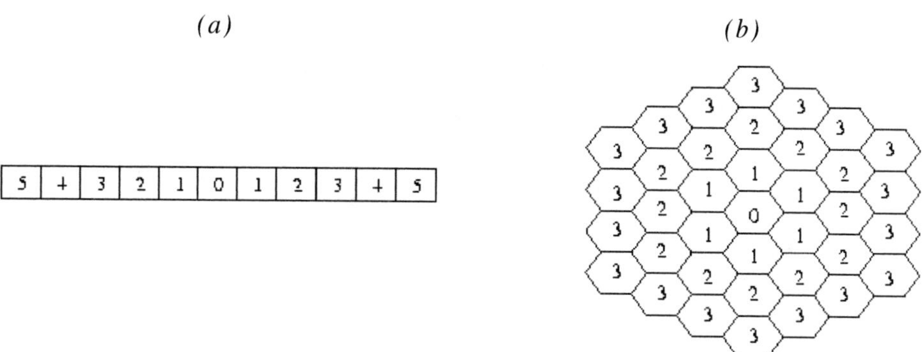

denote $a_{i,i+1}$ and $b_{i,i-1}$ as the probabilities at which the distance of the terminal from its center cell increases and decreases, respectively. Given that the maximum distance of an autonomous network is C, the Markov chain model is given in Figure 2.

The transition probabilities for the one-dimensional model are given as:

$$a_{i,i+1} = \begin{cases} q & \text{if } i = 0 \\ \frac{q}{2} & \text{if } 1 \le i \le C \end{cases} \text{ and } \quad b_{i,i-1} = \frac{q}{2}. \tag{7}$$

The transition probabilities of the two-dimensional model are:

$$a_{i,i+1} = \begin{cases} q & \text{if } i = 0 \\ q(\frac{1}{3} + \frac{1}{6i}) & \text{if } 1 \le i \le C \end{cases} \text{ and } \quad b_{i,i-1} = \frac{q}{2}(\frac{1}{3} - \frac{1}{6i}). \tag{8}$$

Figure 2. Markov chain model

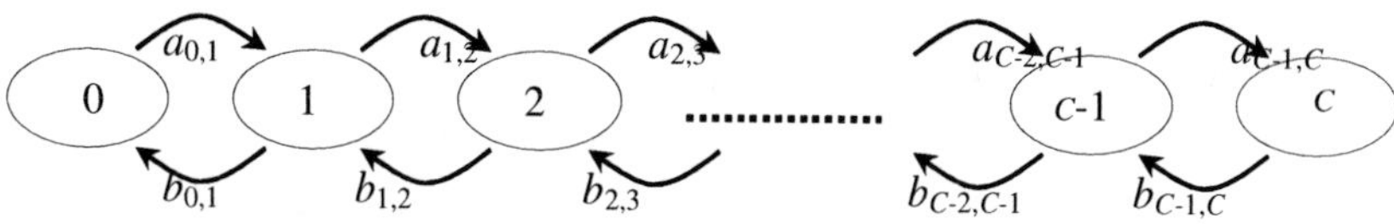

Figure 3. Two-dimensional Markov model

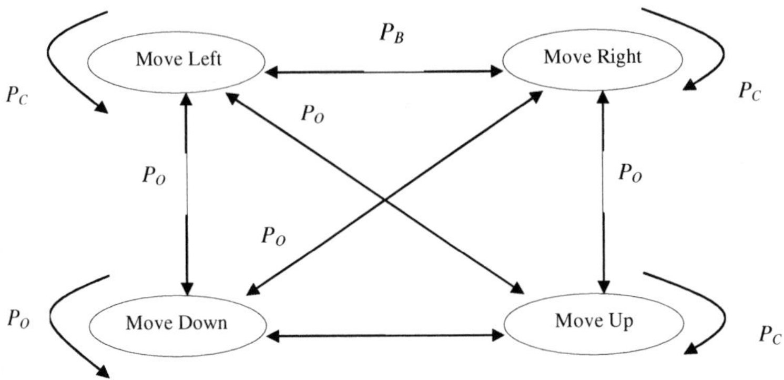

Two-Dimensional Markov Mobility Model

The boundary-crossing movement is modeled by a Markov state-transition diagram in which the direction of each movement is considered in four directions as shown in Figure 3. P_B is the probability that an MT will move back to its serving area, and P_C is the probability that the next move will be in the same direction as the previous move. The probability that the MT moves in any other direction is P_O, which is equal to $(1- P_B- P_C)/2$. These probabilities represent the locality in the users' movements. Higher probability P_B means a high degree of locality in the user's movement, which is regarded as lower mobility.

Combination Model

The synthetic models described in previous sections aim to realistically represent the behaviors of mobile objects without the use of traces. However, since traces are those mobility patterns that are observed in real-life environments, they can be used to describe moving behaviors in the similar environment. The combinational model characterizes different scales of mobility based on the synthetic model and empirical results (Brown & Mohan, 1997). It consists of micro and macro movements. At the micro level, it is assumed that the mobile objects move around in the vicinity of their current locations or frequently visited places such as home and office; this micro mobility is described by a random-walk model. Given an object at the origin at time $t = 0$, we have:

$$Prob\{d(t) < x\} = 1 - e^{-x^2/2\alpha t}, \tag{9}$$

where $d(t)$ is the distance traveled by the object during time period t and a is the random-walk speed parameter. If a mobile makes S steps every T units of time, $a = S^2/T$. Based on a set of empirical data obtained from an extensive study of trips and vehicles, it is found that the user occasionally makes large, macro moves. These moves are radically symmetric and exponentially distributed, that is:

$$Prob\{distance\ of\ a\ big\ move < x\} = 1 - e^{-\gamma x}, \tag{10}$$

where $1/\gamma$ is the average distance that a mobile moved. While parameters may be varying depending on the experiments, this model presents a framework that describes user behavior in different circumstances. The speed parameter in the random-walk model and average moved distance depend on empirical data collection.

Random-Waypoint Model and Derivatives

The random-waypoint model is a commonly used mobility model for simulations in wireless networks, especially for ad hoc networks (Bettstetter, Hartenstein, & Perez-Costa, 2003; Camp et al., 2002). This model is implemented in the network simulation tools ns-2 and GloMoSim and used in the performance evaluation of ad hoc networking protocols. A mobile node in a given simulation area randomly chooses a destination point, that is, the waypoint, and moves with constant speed on a straight line to this point. After it reaches the waypoint, the mobile node will pause for a certain period of time, which is the so-called "pause time," and change its speed or moving direction. Then it chooses a new destination and speed, moves at constant speed to this destination, and so on. Depending on how destination points, pause time, speed, and direction are chosen, the movement of mobile nodes can be obtained through simulations.

This stochastic process can be described by using a triplet M (P_i, V_i, T_i), where P_i is the current position of a mobile node, or a waypoint, V_i is the newly chosen speed at which a mobile node will move from this waypoint, and T_i is the pause time at this waypoint. Let us look at a one-dimensional environment which is a line segment $[0, a]$. Assume random waypoints are uniformly placed on this segment, that is, the probability density function (pdf) of a point's location, P_x, is:

$$f_{P_x}(x) = \begin{cases} 1/a & \text{for } 0 \le x \le a \\ 0 & \text{else} \end{cases}. \tag{11}$$

In wireless networks, the most interesting parameter is the traveling distance during a certain period of time. Therefore, we denote the distance between two consecutive waypoints as L_i, which is given by:

$$L_i = \|P_i - P_{i-1}\|. \tag{12}$$

The expected value of traveling distance $E[L]$, which is the time average of a moving node, is:

$$E[L] = \lim_{m\to\infty} \frac{1}{m}\sum_{i=1}^{m} L_i. \tag{13}$$

By using the definition of expected value, we have:

$$E[L] = \int_0^a l f_L(l) = \frac{1}{3}a, \tag{14}$$

where $f_L(l)$ is the derivative of a location probability $P(L £ l)$ given by:

$$f_L(l) = \frac{\partial}{\partial l} P(L \le l) = -\frac{2}{a^2}l + \frac{2}{a}. \tag{15}$$

For a two-dimensional environment, we can consider a rectangular area of size $a \times b$ and again derive the distribution of traveling distance L. The pdf of waypoints $\mathrm{P} = (\mathrm{P}_x, \mathrm{P}_y)$ with spatial uniform distribution is then given by:

$$f_{\mathrm{P}_x\mathrm{P}_y}(x, y) = \begin{cases} 1/(ab) & \text{for } 0 \le x \le a \text{ and } 0 \le y \le b \\ 0 & \text{else} \end{cases}. \tag{16}$$

The distance between two points $\mathrm{P}_1 = (\mathrm{P}_{x1}, \mathrm{P}_{y1})$ and $\mathrm{P}_2 = (\mathrm{P}_{x2}, \mathrm{P}_{y2})$ is:

$$L = \|\mathrm{P}_2 - \mathrm{P}_1\| = \sqrt{|\mathrm{P}_{x1} - \mathrm{P}_{x2}|^2 + |\mathrm{P}_{y1} - \mathrm{P}_{y2}|^2} = \sqrt{L_x^2 + L_y^2}. \tag{17}$$

Here the random variable $L_x = |\mathrm{P}_{x1} - \mathrm{P}_{x2}|$ represents the random distance between two uniformly distributed coordinates P_{x1} and P_{x2} on a one-dimensional line segment $[0, a]$. The same principle applies to the vertical movement. That means, we are assuming that both distances are independent of each other, thus the joint pdf can be obtained by taking the product of the pdf of two distance variables. After simplification, we can obtain:

$$f_L(l) = \frac{4l}{a^2b^2} \bullet f_0(l), \tag{18}$$

in which $f_0(l)$ is:

$$f_0(l) = \begin{cases} \frac{\pi}{2}al - al - bl + \frac{1}{2}l^2 & \text{for } 0 \le l \le b \\ ab \arcsin\frac{b}{l} + a\sqrt{l^2 - b^2} - \frac{1}{2}b^2 - al & \text{for } b < l < a \\ ab \arcsin\frac{b}{l} + a\sqrt{l^2 - b^2} - \frac{1}{2}b^2 - & \\ ab \arccos\frac{a}{l} + b\sqrt{l^2 - a^2} - \frac{1}{2}a^2 - \frac{1}{2}l^2 & \text{for } a \le l \le \sqrt{a^2 + b^2} \\ 0 & \text{otherwise.} \end{cases} \tag{19}$$

The expected value of L can be obtained by substituting Equation 18 into the definition as used in Equation 14.

$$E(L) = \frac{1}{15}\left[\frac{a^3}{b^2} + \frac{b^3}{a^2} + \sqrt{a^2 + b^2}\left(3 - \frac{a^2}{b^2} - \frac{b^2}{a^2}\right)\right] + \frac{1}{6}\left[\frac{b^2}{a} arcosh\frac{\sqrt{a^2 + b^2}}{b} + \frac{a^2}{b} arcosh\frac{\sqrt{a^2 + b^2}}{a}\right] \tag{20}$$

Note that for $b \to 0$, the two-dimensional rectangle becomes a one-dimensional segment, and Equation 20 will have the result of $E(L) = a/3$, which is the same as in Equation 14.

Smooth Random-Mobility Model

The smooth random-mobility model (Bettstetter, 2001) is proposed as a microscale mobility model that captures random movements in two-dimension scenarios. Two stochastic processes are used to describe the change over time: change of speed and change of direction. The idea of this model is to consider the correlation of speed or direction change, that is, new speed and direction are related to previous values so that there is no sudden speed change and sharp turnings. The speed is changed incrementally by the current acceleration of the mobile user, and the direction change is smooth through several steps toward the target direction. In this model, a mobile user moves with a constant speed v until a new target speed is decided by a random process, then accelerates or decelerates until the desired speed is achieved. The speed behavior of a mobile user at time t is represented by three parameters: current speed $v(t)$, current acceleration $a(t)$, and a target speed $v^*(t)$. In addition, there are three static speed parameters that characterize a certain group of users: maximum speed V_{max}, a set of preferred speeds $V_P = \{v_0, v_1, \ldots\}$, and maximum values for acceleration or deceleration A_{max}.

It is assumed that the frequency of speed-change events is a Poisson process. Therefore, a speed-change event occurs with a certain probability p_{v*} each time. During time period Δt, the arrival rate of events is $1 = p_{v*}/\Delta t$. The probability of a speed change at time t, $p(t)$, is:

$$p(t) = \frac{p_{v^*}}{\Delta t} \bullet e^{-p_{v^*} t / \Delta t}. \tag{21}$$

Let t_o be the time at which a speed change occurs and a new target speed $v^*(t_o)$ is chosen. The acceleration is determined based on a mobile's current speed. If $v^*(t) > v(t_o)$, then:

$$p(a(t)) = \begin{cases} \dfrac{1}{a_{max}} & \text{for } 0 < a \le a_{max} \\ 0 & \text{else} \end{cases}. \tag{22}$$

Otherwise, the acceleration is determined by:

$$p(a(t)) = \begin{cases} \dfrac{1}{a_{min}} & \text{for } a_{min} \le a < 0 \\ 0 & \text{else} \end{cases}. \tag{23}$$

Then the mobile object follows an increase or decrease of its speed at each step by $v(t) = v(t\text{-}\Delta t) + a(t) * \Delta t$ until the target velocity is achieved. Besides velocity control, the direction change is also considered in this model. Assume that each mobile object has an initial direction j (t = 0), which is chosen from a uniform distribution:

$$p(\varphi) = \frac{1}{2\pi} \quad 0 \le \varphi < 2\pi. \tag{24}$$

Once a mobile object is intended to change its direction, a new target direction φ^* is chosen from Equation 24. The difference between the new target direction and the old direction is $\Delta\varphi(t^*) = |\varphi^*(t^*) - \varphi(t^*)|$, which is uniformly distributed between -p and p. Thus, the direction behavior of a mobile object at time t is also described by three values: current direction $\varphi(t)$, direction change $\Delta\varphi(t)/\Delta t$, and target direction $\varphi(t^*)$. In order to increase the smoothness, $\Delta\varphi(t^*)$ can be further divided into many small time segments of curves, and incremental change can be even smaller at each step.

The random processes for speed change and direction change can be considered separately without showing the correlation between these two factors. For simplicity, we can consider a "stop-turn-and-go," which characterizes the fact that a stop by a mobile object is often followed by a direction change. That is, whenever a node comes to a stop, $v(t) = 0$, we choose a target direction with high probability p(j) instead of using a uniform distribution:

$$P(\Delta\varphi) = \begin{cases} P_{\varphi^*}/2 & \text{for } \Delta\varphi = \pm\pi/2. \\ 1 - P_{\varphi^*} & \text{for } \Delta\varphi = 0 \\ 0 & \text{else} \end{cases}, \tag{25}$$

where P_{j*} is the probability that the mobile will make a turn, which must be higher than that in the usual direction control where $v \neq 0$. This reflects the assumption that a turn occurs with a higher probability after a stop than that for a nonstop move.

Gaussian-Markov Model

Observing that fluid-flow models and random-mobility models have a memoryless nature, which is not suitable to describe a movement to a predetermined destination, a Gaussian-Markov model was proposed to capture the correlation of a mobile's velocity in time (Liang & Hass,1999). The Gaussian-Markov model has been used in the research of cellular networks as well as ad hoc networks. A mobile's velocity is assumed to be correlated in time and modeled by a Gauss-Markov process. In this model, the time is divided into time slots with Δt in each of them. Therefore, velocity and direction are presented in a discrete process. Let the velocity at time slot $k\Delta t$ be v_k. The correlation between two consecutive speeds is:

$$v_k = \alpha v_{k-1} + (1-\alpha)\mu + \sqrt{1-\alpha^2}\, u_{k-1}, \tag{26}$$

where $0 \leq a \leq 1$ with $a = exp(-b\Delta t)$, a parameter that affects the randomness of velocity; b determines the degree of memory in the mobility pattern, m is the asymptotic mean of v_k when k approaches infinity, and u_k is an independent, uncorrelated, and stationary Gaussian process, with zero mean and standard deviation same as v_k., that is, $s_u = s$. Given a known velocity of a mobile object, we can recursively expand Equation 26 to express v_k in terms of the initial velocity v_0:

$$v_k = \alpha^k v_0 + (1-\alpha^k)\mu + \sqrt{1-\alpha^2}\sum_{i=0}^{k-1}\alpha^{n-i-1}u_i . \quad (27)$$

Therefore, the position of an MT at time k is represented by coordinates (x_k, y_k):

$$\begin{aligned} x_k &= x_{k-1} + v_{k-1}\cos\theta_{k-1} \\ y_k &= y_{k-1} + v_{k-1}\sin\theta_{k-1}, \end{aligned} \quad (28)$$

where q_{k-1} is the direction of the mobile object at the $(k - 1)$th time interval. In order to describe finite movement states in simulations, a commonly used method is to consider an observation area within which a mobile object is monitored. If the mobile encounters the edge of the simulation area, there are two ways to handle the following movement: force a mobile to move back to the observation area or let the mobile exit, which is an absorb state in the model.

Geographic-Based Models

Although mobile objects can move in random directions at random speed, in reality, a mobile's movement is restricted to the geographic environments. For cellular wireless networks, mobile users are either pedestrians or vehicle drivers. Therefore, the movement of a mobile user follows streets or transportation facilities. Even in an ad hoc network for civilian users, the traveling behaviors of MTs are regulated by specific geographic conditions. For example, car-to-car communication on the highway is one of the promising applications of ad hoc networks; users move along the highway in two directions until they exit the highway. Therefore, geographic-based mobility models are designed to describe moving behavior under particular circumstances.

The simplest model that considers geographic condition is the city-section mobility model, or Manhattan model, under which the simulation area is divided into streets and buildings. Mobile objects will move along the streets without entering any buildings. The moving velocity and moving directions are thus restricted by the street conditions. An example of the Manhattan model is shown in Figure 4, in which a mobile starts from an initial position A. After a certain period of time, it reaches an intersection from which it is possible to move in three directions except in a U-turn. Note that this model is different

Figure 4. Manhattan mobility model

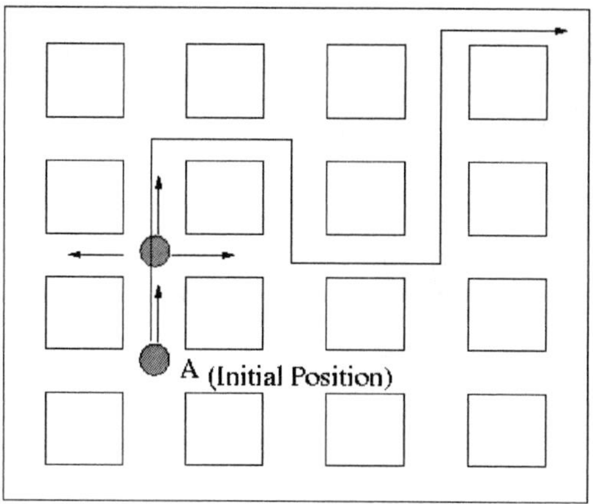

from the random-walk model in that there are four possible directions with equal probability in which the mobile terminal can move in the random-walk model compared to three possible directions in the Manhattan model. The mobiles must follow existing paths and have restricted moving speed. Many derivatives of this basic model to make it more realistic include considering the speed limit on each street, one-way streets, pause time due to traffic lights, rush-time traffic, and so on.

A mobility model for a three-dimensional indoor environment takes into account vertical motion (Kim, Kwon, & Sung, 2000). Consider a K-story building ($K \geq 3$) in which users move on square-shaped building floors, with horizontal motion speed V and vertical speed V'. Horizontal and vertical speeds are uniformly distributed with $[0, V_{max}]$ and $[0, V'_{max}]$, respectively. It is also assumed that users move straight until they change directions such as right, left, or back, and then continue to move straight again.

On each floor, there are two regions: the vertical region and the flat region. The former is the area in which a user is standing and where he or she will make a vertical move such as in elevators. The latter is the region in which users can only move horizontally because there is no exit to other floors. Direction changes according to a Poisson process, and the direction-selection ratio at the turning point in horizontal motion is distributed with certain probabilities to the left, back, and right. If the turning point is located in the vertical region and users arrive there, they move horizontally or vertically with a probability of a and b (a + b = 1), respectively. Let us denote the distances traveled along horizontal paths and in vertical directions with X and H, respectively. If turning points can be anywhere, then the mean elapsing time between two neighboring turning points will be $E[X/V]$ and $E[H/V']$, respectively. For handoff design and analysis, the number of cell crossings per time unit is important to us. Assume the area of a cell is A. There are no cell crossings when users encounter the outer wall during their movement. Therefore, the mean number of cell crossings per time unit in horizontal motions is $E[V]/d\,(1 - 1/A)$, where d is the one-side length of cells. The mean number of cell crossings per time unit, $E[M]$, is

$$E[M] = \frac{\alpha E[X]/d \bullet (1 - \frac{1}{A}) + \beta}{\alpha \frac{E[X]}{E[V]} + \beta \frac{H}{E[V']}} \quad (29)$$

Group-Mobility Models

In previous sections, individual mobile users' behaviors are considered independent of each other. In reality, however, it is very likely that the behavior of a mobile object is influenced by others. For example, a mobile user wants to take a routine path from home to work site. When hearing from the radio that there is an accident on the way to work, he or she may turn to another path. Another example is if a group of people are searching for a missing object, each individual needs to cooperate with others in the group rather than going to a predefined destination. In order to increase the possibility of success in finding the target, they will start in a distributed way, for example, each individual may take a different path. When one of them finds a valuable clue and notifies his or her teammates, the other people will change their original path and concentrate on the most likely path. Thus, it is necessary to model the moving behavior of mobile objects by taking into account the relationship among mobile objects, which is of particular interest for ad hoc networks because of the need for peer-to-peer communications.

The reference-point group model (RPGM) is one of the group models into which mobile nodes are organized (Hong, Gerla, Pei, & Chiang, 1999). In each group, there is a referencing point that defines the moving pattern of a whole group, including directions, moving speed, change of acceleration, and so on. Individual mobile nodes are randomly

Figure 5. Reference-point group-mobility model

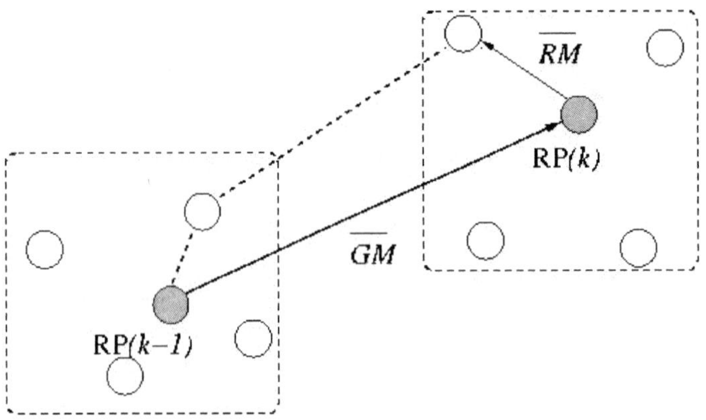

distributed around the reference point and they are allowed to move independently of other nodes in the group. However, the group motion is decided by the reference point. Let RP(k) denote the position of a reference point at time k in Figure 5. The movement of the reference model follows a group motion vector $\overline{GM}$. Each node has its own random motion vector $\overline{RM}$, which is uniformly distributed within a certain radius centered at the reference point, and its moving direction is also uniformly distributed as in the fluid-flow model. Therefore, the new position of a node at time k can be generated by combining the group motion $\overline{GM}$ and individual movement $\overline{RM}$ as shown in Figure 5. By defining anchor locations of reference points, moving speed, and directions, this model can be used to describe a group of mobile users in various applications.

Kinematics Mobility Models

The motion of a mobile object is very closely related to the environment in which it moves. On a highway, a mobile vehicle's movement can be predicted based on its direction and average speed or speed limit, for the vehicle's acceleration is dependent on its current speed and its driving-control action. If a vehicle's velocity in a given direction is high, the acceleration is less because of resistance, which can be represented by a linear parameter b. In the kinematics mobility model, vehicular motions are described by physical laws that affect the motion. Although in reality the motion of a vehicle is a continuous movement, we divide the time into time slots of T at which samples are measured and changes are made; also, the sampling frequency is $1/T$ and each time moment is k/T. At sample time k, the motion in one-dimension environments such as a highway is given by:

$$\ddot{x}(k) = -\alpha \dot{x}(k) + w(k) + u(k), \tag{30}$$

where $x(k)$ represents the position at time k; $\dot{x}(k)$ and $\ddot{x}(k)$ represent the speed, the first-order derivatives of $x(k)$, and acceleration, the second derivatives, respectively, and $w(k)$ is a white Gaussian process with $E[w(k)] = 0$. The last item, $u(k)$, is a deterministic function representing driving command, which is input by the driver to control the direction of a moving vehicle. This value directly influences the mean speed of a vehicle. If $u(k) = 0$, the vehicle will not move but stay at its current position $x(k) = 0$. A positive value of $u(k)$ means that a vehicle moves toward a positive direction, whereas a negative value means the opposite direction. The random acceleration is correlated in time and a is the reciprocal of the random acceleration-time constant.

In reality, the mobile position is a two-dimensional vector. A four-state space model can be used to describe the location state of the vehicle, which is given by McGuire, Plataniotis, and Venetsanopoulos (2003):

Table 1. Comparison of mobility models

	Indoor Environment	Outdoor Environment	Infrastucture (I)/ Ad hoc (A)	Group Activity (G) / Individual Activity (I)
Fluid Flow		x	I	G & I
Smooth Random		x	I	I
Random Walk	x	x	I & A	I
Random Waypoint	x	x	A	I
Gaussion-Markov		x	I & A	I
3-D Indoor	x		I	I
Manhattan		x	I	I
Group		x	A	G
Kinematics		x	I & A	I

$$X(k) = [p_x(k),\ v_x(k),\ p_y(k),\ v_y(k)]^T, \tag{31}$$

where $(p_x(k), p_y(k))$ are the location coordinates of the mobile at time k, and $(v_x(k), v_y(k))$ are the velocities of the mobile object in the x and y directions. Let us use $U(k) = [u_x(k), u_y(k)]^T$ and $W(k) = [w_x(k), w_y(k)]^T$ to denote the two-dimensional driving control and white Gaussian signals. The resulting discrete-time dynamic model is given by $X(k+1) = AX(k) + BU(k) + W(k)$, where A and B are given as follows:

$$A = \begin{bmatrix} 1 & \frac{1-\exp(-\alpha T)}{\alpha} & 0 & 0 \\ 0 & \exp(-\alpha T) & 1 & 0 \\ 0 & 0 & 0 & \frac{1-\exp(-\alpha T)}{\alpha} \\ 0 & 0 & 0 & \exp(-\alpha T) \end{bmatrix}$$

$$B = \begin{bmatrix} \frac{\exp(-\alpha T)-1+\alpha T}{\alpha^2} & 0 \\ \frac{1-\exp(-\alpha T)}{\alpha^0} & 0 \\ 0 & \frac{\exp(-\alpha T)-1+\alpha T}{\alpha^2} \\ 0 & \frac{1-\exp(-\alpha T)}{\alpha^0} \end{bmatrix}. \tag{32}$$

The dimension of the dynamic state vector of $X(k)$ can be extended to two states with position and velocity, and it can also be extended to three states with position, velocity, and acceleration. In cellular systems, the distance between an MT and a base station can be estimated by the receiving signal strength at the MT. That means, by taking the three largest measurements from the base station, the location of an MT can be determined.

Comparison of Mobility Models

The goal of a mobility model is to accurately describe the movement pattern of a mobile object, which is random in dynamic mobile environments. The mobility models discussed above can be applied in different scenarios. In Table 1, their applications are marked with "x".

Management of Locations in Wireless Networks

Overview

In cellular networks, each base station (BS) handles incoming and outgoing connection requests from mobile users residing within its coverage, which is called a *radio cell.* A cluster of cells can be grouped together, controlled by a mobile switching center (MSC). These MSCs are usually interconnected by wires for high-speed data transmission. The geographic coverage of a group of cells, which is also the area in the control of an MSC, is called a location area (LA) or a registration area (RA). In wireless networks, it is very important to maintain the latest location information of each MT in the system so that a connection can be established when an incoming or an outgoing service request is received. However, mobile users change their positions from time to time. The identity of mobile users and their billing information are stored in a centralized database, which is called the home location register (HLR). When an MT moves into a different region network, it must update or register with a local database, a visitor location register (VLR). In other words, the VLR in a visiting network keeps the most recent location information of a mobile user, and it will communicate with the HLR to renew location information.

Location management is a technique that updates the location of MTs during the course of their movement and determines the locations of MTs for call delivery. In particular, it includes two phases: location update and paging. During location update, or location registration, MTs send location-update requests to an MSC to establish or refresh their location information in VLRs. For paging and call delivery, the system needs to search for the called MTs for message delivery. Location update and paging involve the algorithms or strategies to send location-update messages and to find MTs based on known information in network databases. Location registration and call delivery are related to the procedures and signaling messages for implementation.

Once an MT is located in a cell, the system needs to find the exact position of the MT inside the cell, which is referred to as a geolocation issue. Recently this topic received more attention because of the requirements of the U.S. Federal Communications Commission (FCC) and increasing deployment of Wireless Fidelity (Wi-Fi) technologies. For WLANs and peer-to-peer networks, where there are no centralized databases to store user identity and location information, management of location means to position mobile nodes. Many localization methods are based on received radio signal strength, time of arrival, time difference of arrival (TDOA), or angle of arrival (AOA) by considering path loss attenuation with distance in radio propagation. The well-known Global Positioning System (GPS) estimates distance from the Radio Frequency (RF) signal time of flight using the TDOA, or the absolute amount of time by using three satellites for triangulation. Moreover, specific localization algorithms are designed for indoor and outdoor environments because radio transmission mechanisms are different, which is caused by the size of obstacles and incident objects, as well as walls, foliage coverage, and so forth.

In this section, we start with the location update and paging algorithms recommended in cellular systems specifications, followed by discussions on recent advances on this issue. Then, several localization algorithms for indoor and outdoor environments directed at improving the accuracy of localization are given.

Modeling and Description of Location

As a matter of fact, the modeling of locations of mobile users can be part of the mobility model, whereas traditionally, mobility models are focused on stochastic behaviors of mobile users. Meanwhile, location modeling is concerned with how to represent the current position of a mobile user and its relationship with others in wireless systems, which involves two issues. One is how to represent the current location with respect to network architecture, and the other is the semantics of description language. In this context, we consider location modeling as the first case, but not the semantics issue because that is not directly related to technical aspects of system design. In current wireless systems, the location of a mobile user can be specified at three levels. In other words, a mobile user's current position can be represented as follows:

- **Location Area:** In cellular networks, an MSC controls a group of BSs. When an MT moves from one MSC to another, the location registration process is triggered. The location of an MT is presented in terms of an LA, whose ID is stored in the HLR.
- **Cell ID:** In order to maintain an active connection for a mobile user, it is most important to know in which cell the mobile user is located. This information is especially useful in resource and mobility management. The network knows in which cell an MT resides by sending polling messages, which can be acquired without additional cost during call origination or termination, or through location services (LCS) management.
- **Position:** Although it is not necessary to know the exact position of a mobile user inside a cell, because radio connection can be maintained as long as the MT is covered by a BS, it is useful to know the location of an MT with respect to

geographical circumstances, such as a street number, the distance to a city sign, and so forth, for location-aware services as described previously.

Among these three levels of location information, the ultimate goal of location modeling is to find the current and next cells into which a mobile user will possibly move for service delivery. Therefore, location modeling is used for location management, which deals with how a mobile user informs the system about its current position and how the system tracks a mobile user given existing information. The prediction of future locations, which involves estimation of velocity, real-time monitoring, and specific environments, is beyond the scope of this chapter.

Basic Location-Update and Paging Algorithms in Cellular Networks

Location management is a two-stage process that provides the network with the capability to discover the current attachment point of a mobile user for call delivery. The first stage is *location update* or *registration*. In this stage, an MT periodically notifies the network of its new access point, allowing the network to authenticate an MT's identity and revise the user's location profile. The second stage of location management is *paging*, during which the network is queried for the user location profile so that the current position of a called MT can be found (Akyildiz, McNair, Ho, Uzunalioug1u, & Wang, 1999; Amotz & Ilan, 1995).

The often-used strategy is as follows: the service area of a network is divided into several LAs so that the MT informs the network when it enters a new LA by monitoring the LA Identification (LAI) through the public broadcast channels as shown in Figure 6. Usually, MTs send location-update requests when they move from one LA to another. Each LA may consist of a number of cells for the microcell systems, while it may be one single cell for the macrocell systems. The result of location update or registration is that the network

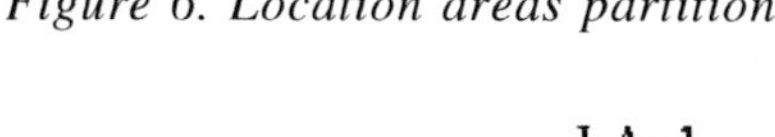

Figure 6. Location areas partition

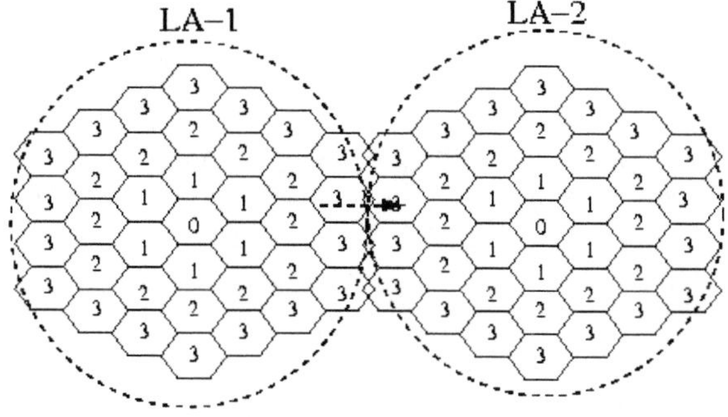

is able to keep track of MTs in a specific region that is constrained by the distance, time, or variance of movement.

As the LA can be determined through retrieving the MT's record in the HLR, the MT's current residing cell can be found by sending a polling message to all BSs encompassed in an LA. When an incoming call arrives, the MSC, which is associated with the LA of an MT's last registration, sends a paging message via the paging channel to the BSs with the called MT's ID. The serving BS of the MT then responds to the network. As a result, the network knows in which cell the MT is residing so that the incoming call can be delivered to the MT. These two stages are closely related in that one process affects the other. If an LA is very small, such as only one cell, an MT needs to update with the network whenever it moves from one cell to another, thus causing high signaling cost of location update. The network is always aware of the current position of a mobile user. Meanwhile, the paging cost is the minimum because it is not necessary to send any polling message to locate the MT. On the contrary, if an LA is extremely large and a mobile user does not update with the network for a long time, then when the MT moves to a new location, the network will lose the location of the MT because a mobile user may change its position from time to time. In this case, the paging cost would be very high whereas the location update cost would be minimized. Therefore, location update and paging are dependent on each other and both of them need to be considered in the design of location management.

Dynamic Location-Update and Delay-Constrained Paging Algorithms

The standard LA-based location update method does not allow adaptation to the mobility characteristics of the MTs. Therefore, in recent years, there have been many efforts in designing dynamic location-update and paging algorithms which take into account user mobility and optimize the total signaling cost of location update and paging. For instance, the MT can update its location or send a location registration request to the network based on the distance between its current location and the position from which the previous request was sent; this is referred to as the *distance-based* location update scheme.

Moreover, the MT can send a registration request upon a predefined timer, that is, the location information is updated whenever the timer is expired; this is called the *time-based* scheme. Another location update scheme is called the *movement-based* scheme, in which an MT performs a location update when the number of movements since the last location registration equals to a predefined threshold, which is referred to as the *movement threshold.* These schemes are aimed at designing LAs so that the total signaling overhead of location update and paging can be reduced. Considering that the moving patterns are different from one user to another, some dynamic location-management schemes, which are so-called *mobility-based* schemes, are designed based on the mobility scale.

Distance-Based Location Update

We consider the last cell in which an MT performed location update, originated a call, or received a call, as the *center cell*. In distance-based scheme, location update is performed when an MT's current distance away from its center cell exceeds a predefined threshold d as shown in Figure 7a. This location update scheme guarantees that the terminal is located in an area that is within a distance d from the center cell. Therefore, an LA is not static and it is independent of each individual user. The location update is implemented depending on the user's movement because the location update rate, that is, the number of location updates per unit of time, would be low if the terminal is a low-mobility user, which means it does not move around very often. If a terminal moves very fast, then the location update rate would be high because the distance threshold will be exceeded frequently. Therefore, this method is able to reduce unnecessary operations of location update for low-mobility users.

In order to determine if a terminal is located in a particular cell, the network needs to send a polling signal to the target cell and wait until a time-out occurs (Akyildiz et al., 1996). If a reply is received before time-out, the destination terminal is in the target cell. Otherwise, this polling reports failure, that is, the terminal is not in this cell. This polling process determines the round-trip delay from the time a polling message is sent to the time a response is received, and is regarded as a polling cycle. A maximum paging delay of m polling cycles means that the network must be able to locate the MT in m polling cycles. When an incoming call arrives, the network needs first to partition an LA into many subareas and poll each subarea one after another until the terminal is found. We denote subarea j by A_j, where j is the order in which this subarea will be polled. A subarea contains one or more rings and each ring is a set of cells that are away from the center cell by the same distance. Given a threshold distance of d and a maximum paging delay of m, the number of subareas, denoted by N, is:

$$N = \min(d + 1, m). \tag{33}$$

As a result, the LA can be partitioned according to the following steps.

1) Determine the number of rings in each subarea by using the following formula:

$$\eta = \left\lfloor \frac{d+1}{N} \right\rfloor.$$

2) Assign h rings to each subarea so that subarea A_j $(1 \leq j \leq N - 1)$ is assigned rings $R_{h(j-1)}$ to R_{hj-1}.

3) Assign the remaining rings to subarea A_N.

Thus, the rings that are closer to the center cell are polled first because there is a higher probability of finding a terminal in a cell that is closer to the center cell than in a cell that is further away. The total signaling overhead, $C(d, m)$, can be given by:

$$C(d,m) = C_u(d) + C_p(d,m), \quad (34)$$

where $C_u(d)$ and $C_p(d, m)$ are the average update cost and paging cost per terminal, respectively. Assume that the costs for performing a location update and polling a cell are U and V, respectively; then the equation above can be further written as:

$$C(d,m) = Prob(d)U + Prob(call) \bullet g(d)V, \quad (35)$$

where $Prob(d)$ is the probability of the occurrence of a location update given the update threshold is d. $Prob(call)$ is the probability of a call arrival during the time interval of location update and $g(d)$ is the number of cells that are within a distance of d from the center cell. The calculation of $Prob(d)$ and Prob($call$) are dependent on the mobility models and call arrival patterns (Lin & Park, 1997; Rose & Yates, 1995; Wang & Akyildiz, 2000).

Movement-Based Location Update

An MT performs a location update whenever it completes a predefined number of movements across cell boundaries. This number m is referred to as the movement threshold. The distance-based location update is different from the movement-based scheme in that sometimes, MTs move back and forth without going further as shown in Figure 7. A terminal may not initiate any location update operation because the distance

Figure 7. Distance-based and movement-based location update

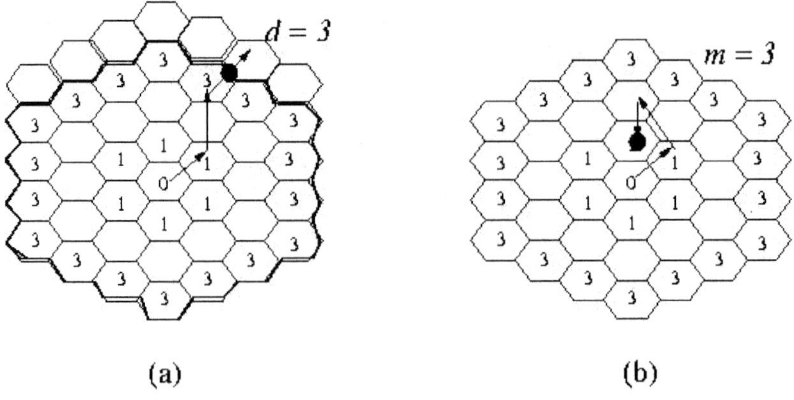

threshold has not been reached, even though the movement threshold may be exceeded. Also, it requires that MTs have information about the distance relationship among all cells. As an example, distance threshold $d = 3$ and movement threshold $m = 3$ are shown in Figure 7a and Figure 7b.

Wgen an incoming call arrives, the network pages the cells within a distance of d_m, which is traveled by a terminal due to its m cell-boundary crossings. The paging process is very similar to that in the distance-based scheme. The signaling cost for location update is:

$$C_u = U\sum_{i=1}^{\infty} i \sum_{j=im}^{(i+1)m-1} \alpha(j), \tag{36}$$

where a(j) is the probability that there are j boundary crossings between two call arrivals. Assume that the probability density of the cell residence time has the Laplace-Stieltjes transform $f_r^*(s)$ and a mean of $1/l_r$. The call arrival to each MT is a Poisson process with rate l_c. The call-to-mobility ratio is then defined as $q = l_c/l_r$. Based on these parameters, the expression of a(j) has been derived as:

$$\alpha(j) = \begin{cases} 1 - \dfrac{1}{\theta}[1 - f_r^*(\lambda_c)] & j = 0 \\ \dfrac{1}{\theta}[1 - f_r^*(\lambda_c)]^2 [f_r^*(\lambda_c)]^{j-1} & j > 0. \end{cases} \tag{37}$$

The average paging cost C_p is determined by:

$$C_P = V\sum_{k=0}^{N-1} \rho_k w_k, \tag{38}$$

where r_k and w_k represent the probability that the MT is residing in subarea A_k and the number of cells pooled before the terminal is successfully located, respectively. Therefore, the total signaling cost, $C_T = C_U + C_P$, can be obtained by substituting Equations 36 and 38.

Time-Based Location Update

In a time-based location update scheme, an MT performs location updates periodically at a constant time interval DT. If a location update occurred at location A at time 0 as shown in Figure 8, subsequent location updates will occur at locations B and C if the MT

Figure 8. Time-based location update and paging

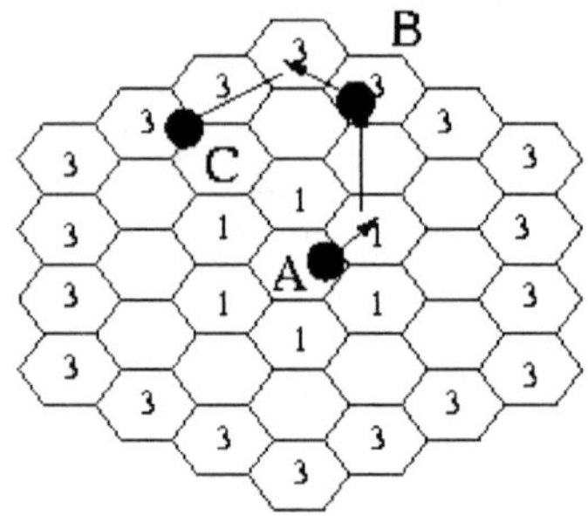

moves to these locations at times ΔT, $2\Delta T$, and $3\Delta T$, respectively. Time-based methods, as opposed to distance-based schemes, do not require location or distance information during the time between location updates. This feature is able to reduce location update cost for MTs that are not actively calling out and receiving calls; it is also helpful in reducing energy consumption of the MTs.

The total cost of the time-based scheme is closely related to the time between the last known location and the next update or paging event, which is called the *roaming interval*. Since paging costs are primarily generated by incoming calls, we assume that paging events form a Poisson process. The time interval $s(t)$ can be described by a probability density distribution (Lei & Rose, 1998; Tabbane, 1997). The cumulative distribution function is:

$$S_t(T) = 1 - \exp(-\lambda_c T), \tag{39}$$

given the last-known user location is $L(t = 0)$ at time $t = 0$ and the user is supposed to register at time $t = \Delta T$ if there is no incoming and outgoing call connection. Therefore, the probability of the roaming interval being terminated is $1 - S_t(\Delta T)$ due to the location update and $S_t(\Delta T)$ due to incoming calls. The total signaling cost is:

$$C_T = (1 - S_t(\Delta T)) \bullet U + V \int_0^{\Delta T} C_p(\tau)\lambda_c \exp(-\lambda_c \tau) d\tau, \tag{40}$$

where C_p(t) is the paging cost of successfully finding the target MT, which is a function of time during which an MT has traveled.

Mobility-Based Location Update

A dynamic LA assignment scheme can also be designed based on mobility patterns in which an MT carried by a pedestrian can register its location at the lowest LA level, and can quickly register at a higher LA level when that pedestrian enters a taxi, automobile, or other vehicle. LAs at higher levels of the system's registration hierarchy are larger than those at lower registration levels. Smaller LAs may consist of clusters of microcells while larger LAs may be formed by clusters of macrocells or spot beams. The location updating rate tends to remain fixed over a wide range of terminal mobility. This feature avoids excessive location-updating signaling traffic (Lin & Park, 1997). A similar strategy is to combine automatic updates by the users either when they make significant moves or when they go extended periods without network interaction. The user mobility consists of micro and macro movements. In general, at the micro level the user is modeled by a random walk, and the macro level model is based on empirical data. Therefore, it is able to avoid generating a rapid stream of updates when the user is traveling at high speed.

Intersystem Location Update and Paging

Diverse mobile services have stimulated an enormous number of people to deploy mobile devices such as cellular phones and portable laptops as their communications means for high-quality services. This demands developing effective techniques by which the locations of mobile users can be traced so as to deliver services with the minimum overhead in terms of resource consumption and processing latency. Efficient location management techniques for intersystem roaming, which consists of location update or registration and paging, thus have become one of the most challenging issues for next-generation wireless systems as the number of mobile users, different types of handsets, and various technologies increase. As shown in Figure 9, there are two systems, which may use different protocols such as personal communication service (PCS) systems and

Figure 9. Boundary location area and boundary location register

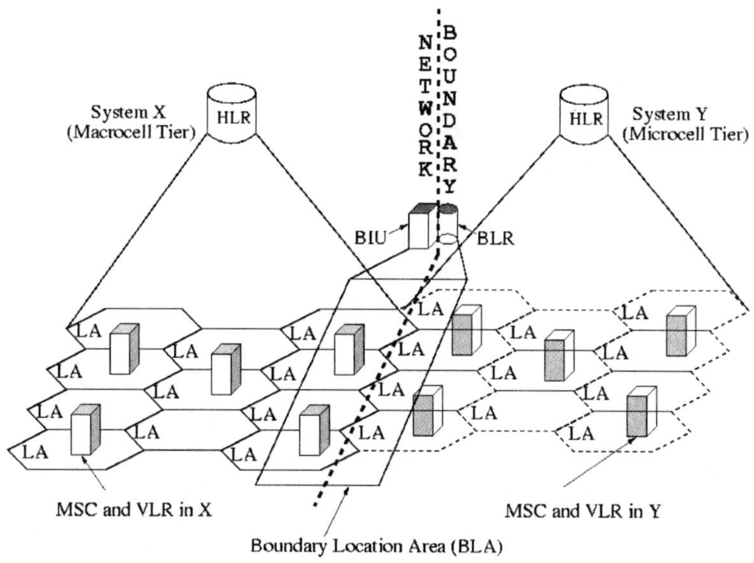

Wi-Fi in the microcell tier. Each hexagon represents an LA within a stand-alone system and each LA is composed of a cluster of microcells. The terminals are required to update their location information with the system whenever they enter a new LA; therefore, the system knows the residing LA of a terminal all the time. In the macrocell tier there may be different systems such as Global System for Mobile Communications (GSM) and Universal Mobile Telecommunications System (UMTS) in which one LA can be one macrocell.

Boundary Location Area (BLA) and Boundary Location Register (BLR) Concepts

The intersystem location update is concerned with updating the location information of an MT performing intersystem roaming. The intersystem paging is concerned with searching for the called terminal roaming between different service areas. The goal of intersystem location management is to reduce the signaling cost while maintaining quality of service (QoS) requirements. For example, reducing call loss is one of the key issues in maintaining the call connection; decreasing the paging delay is critical to reducing the call set-up time. For intersystem location update, we consider a boundary region called *boundary location area* (BLA) existing at the boundary between two systems in different tiers. As illustrated in Figure 9, systems X and Y are in the macrocell and microcell tiers, respectively. There is an HLR for each system and a user is permanently associated with an HLR in his or her subscribed system. The BLA is controlled by a boundary interworking unit (BIU), which is connected to MSCs and VLRs in both systems. The BIU is responsible for retrieving a user's service information and transforming message formats. Also, the BIU is assumed to handle some other issues such as the compatibility of air interfaces and the authentication of mobile users. The BIU is aware of roaming users' information such as service requirements and bandwidth consumption since all the roaming users are processed through the BIU. Furthermore, the BIU is connected to the LAs adjacent to system Y, and it sends the necessary information for intersystem roaming users to the cells in those LAs periodically. The configuration of a BIU depends on the two adjacent systems that this BIU is coordinating.

The BLA is considered as a dynamic region depending on each MT's profile, such as on speed and bandwidth requirement. When an MT enters the BLA, it sends a registration request to the BIU and the BIU forwards this request to the system toward which the MT is moving. According to the BLA concept, an MT is allowed to register and update its location before the MT receives or makes calls in the new system. In addition to the concept of BLA, we designate a *boundary location register* (BLR) to be embedded in the BIU. A BLR is a database cache that maintains the roaming information of MTs moving between different systems. The roaming information is captured when the MT requests a location registration in the BLA. The BLRs enable the intersystem paging to be implemented within the appropriate system that an MT is currently residing in, thus reducing the paging costs. Therefore, the BLR and the BIU are accessible to the two adjacent systems and are collocated to handle the intersystem roaming of MTs. On the contrary, the VLR and the MSC provide roaming information within a system and deal with the intrasystem roaming of MTs. There is only one BLR and one BIU between a pair of neighboring systems, but there may be many VLRs and MSCs within a stand-alone system.

BLA Location Update and BLR Paging

We define the BLA of an MT to be the region in which the MT sends a location registration request to the new system toward which that MT is moving. A new location update mechanism is used such that the MT will report its location when its distance from the boundary is less than *update distance* d_{xy}. This location update scheme guarantees that the MT updates its location information in an area that is within a distance threshold away from the boundary of two systems (tiers), X and Y. As a result, a BLA can be determined by an update policy based on the MT's update distance.

Delay-Constrained Paging Algorithms

In current PCS systems, paging is a fundamental operation for locating an MT. As the demand for wireless services grows rapidly, the signal traffic caused by paging increases accordingly, consuming limited radio resources. In a paging process, the system searches for the MT by sending poll messages to the cells close to the last reported location of the MT at the arrival of an incoming call. Delays and costs are two key factors in the paging issue (Rose & Yates, 1995; Wang, Akyildiz, Stuber, & Chung, 2001). Of the two factors, *paging delay* is very important in reducing the set-up relay of service delivery. *Paging cost*, which is measured in terms of cells to be polled before the called MT is found, is related to the efficiency of bandwidth utilization and should be minimized under the delay bound. Many paging schemes have been proposed to resolve this conflict. In cellular networks, paging is composed of two steps: find the serving MSC of the called MT and find the cell of the MT. With the current paging strategy, the paging messages are broadcast by each BS in the LA where the MT registered. Therefore, the signaling cost of paging is the maximum because every cell in the LA broadcasts the paging messages. This scheme is so-called classical *blanket polling*. At the same time, the delay of locating the MT is minimized because all the cells in an LA are searched at the same time.

Analytical Model

We assume that each LA consists of the same number of cells, N, in the system. The worst-case paging delay is considered as the delay bound, D, in terms of polling cycle. When D is equal to 1, the system should find the called MT in one polling cycle, requiring all cells within the LA to be polled simultaneously. The paging cost, C, which is the number of cells polled to find the called MT, is equal to N. Thus, the worst case occurs when the called MT is found in the last polling cycle, which means the paging delay would be at its maximum and equal to N polling cycles. However, the average paging cost may be minimized if the cells are searched in decreasing order of location probabilities.

Figure 10. Partition of a location area into paging areas under delay bound D

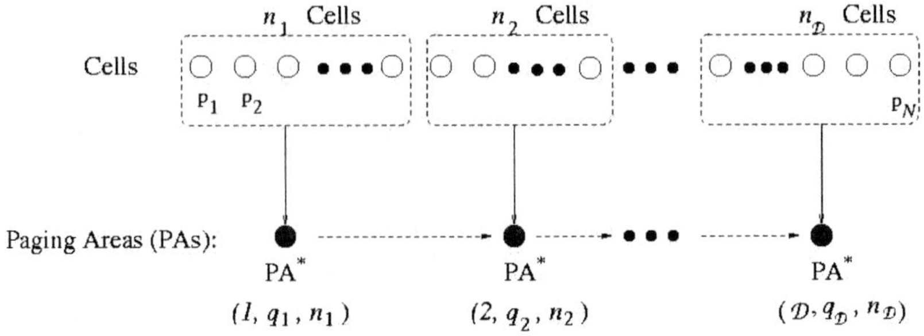

We consider the partition of *paging areas* (PAs) given that $1 < D < N$, which requires grouping cells within an LA into smaller PAs under the delay bound D. Suppose, at a given time, the initial state P is defined as $\boldsymbol{P} = [p_1, p_2, \ldots, p_j, \ldots, p_N]$, where p_j is the location probability of the jth cell to be searched in a decreasing order of probability. We use triplets $\text{PA}_p(i, q_i, n_i)$ to denote the PAs under the paging scheme P in which i is the sequence number of the PA; q_i is the location probability that the called MT can be found within the ith PA and n_i is the number of cells contained in this PA. In Figure 10, an LA is divided into D PAs because the delay bound is assumed to be D. Thus, the worst-case delay is guaranteed to be D polling cycles. The system searches the PAs one after another until the called MT is found. Accordingly, the location probability q_i of the ith PA is:

$$q_i = \sum_{j \in \text{PA}(i)} p_j \, . \tag{41}$$

If the called MT is found in the ith PA, the average paging cost under delay bound D, $E[C(D)]$, and the average delay, $E[T(D)]$, are computed as follows:

$$E[C(D)] = \sum_{i=1}^{D} q_i \sum_{k=1}^{u} n_k \quad \text{and} \quad E[T(D)] = \sum_{i=1}^{D} i q_i \, . \tag{42}$$

The location probability distributions of the MTs can be obtained in many ways. Basically, there are three methods being used to find the probabilities: *geographical-based computation*; *empirical-data-based estimation;* and *mathematical model-based prediction* (Wang et al., 2001).

Paging Area and Paging Algorithms

The paging starts at the last location where the MT contacted the system because the system is aware of the location of the MT whenever a call is connected to this MT. In a massive system, it is not realistic to send polling signals to all cells in the system simultaneously. The most common method is to partition the coverage area of the system into many subareas, which are called PAs, consisting of specific cells. A PA is designed specially for paging the cell of the called MT. The system searches these PAs one after another until the called MT is found.

A *selective paging scheme* is a multistep method, in which the paging may be more than one step (Akyildiz et al., 1996). In each step, the system selects a subset of the cells as the PA. The paging process terminates as soon as the MT is successfully found. The total number of PAs is defined as $l = min(h, d)$, where h is the maximum paging delay and d is the distance from the last location of the called MT. This paging scheme can be used together with a distance-based location update scheme which enables the MT to update its location when it is at a threshold distance d away from the last location. It is assumed that the system polls all the cells in a PA simultaneously. As a result, the called MT can be found if it is less than l polling cycles, which is the round-trip delay for sending a polling signal and receiving the response.

Another PA-based tracking scheme is named the *probability-criterion-based tracking scheme* (Lei & Rose, 1998). Here MTs are paged over personally constructed PAs. The size of PAs is an increasing function of the time since the last time the MT contacted the system. The shape of the PA depends on the particular probability distribution of the MT location. It is assumed that only a fixed set of cells with a high probability greater than a threshold probability denoted by P_C will be polled. If the MT is not present in these cells, then a paging failure is declared rather than continuing until the MT is found. This scheme is analyzed with the timer-based location update scheme with the timer value, which is the maximum time for an MT to wait before reporting its location to the system. Other paging schemes are proposed from the standpoint of designing the sequence of paging. The objective is to design an optimal searching sequence for polling the probable locations of the called MT to minimize the signaling cost of paging under the time constraint. Given the probability distribution of the location where the MT is located when it receives an incoming call, a *highest-probability-first* (HPF) scheme is designed, sequentially polling cells in a decreasing order of probabilities to minimize the mean number of cells being searched. It has been proved that the paging cost can be minimized by appropriately grouping cells in PAs (Rose & Yates, 1995).

Localization in Wireless Networks

With recent advances in wireless communications and low-power electronics, accurate localization has been a growing area of research with many potential applications, such as legal position location on highways, ad hoc networks, and even sensor networks. In the United States, the FCC has required wireless service providers to support a mobile phone callback feature and cell-site location mechanism since 1998 (Hellebrandt & Mathar, 1999; Rappaport, Reed, & Woerner, 1996). Location systems not only provide safety to cellular phone users, but provide new services and revenue sources for wireless carriers as well, such as location-aware services. In distributed wireless networks, accurate position of a sensor or a moving object can also improve routing efficiency by selective flooding or selective forwarding of data only in the direction of the destination.

Figure 11. Angle of arrival localization

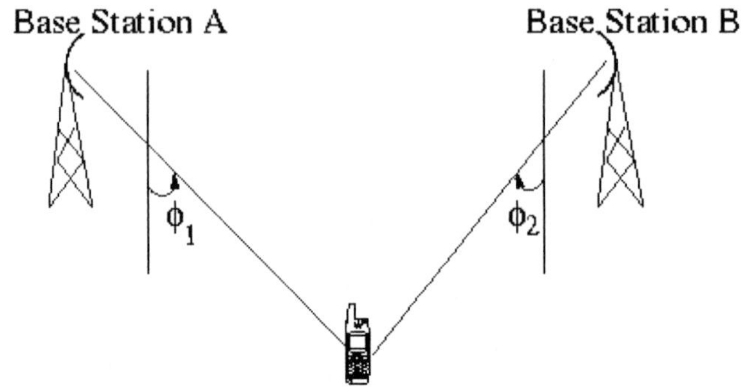

In 1993, GPS, which is based on the NAVSTAR satellite constellation, was deployed. NAVSTAR satellites, 24 in all, orbit 11,000 nautical miles above the earth. They are continuously monitored by three stations located worldwide. The satellites transmit signals that can be detected by anyone with a GPS receiver. Using the receiver, one can determine his or her location with great precision. The principle behind GPS is the measurement of distance between the receiver and the satellites. If three satellites are visible to the receiver, triangulation can be used to find the observer's location. The GPS accuracy is also affected by satellite geometry, which refers to where the satellites are located relative to each other. A typical civilian GPS receiver provides 60- to 225-feet accuracy, depending on the current status of selective availability, the number of satellites available, and the geometry of those satellites. However, a typical civilian GPS receiver's accuracy can be improved to 15 feet or better through a process known as Differential GPS (DGPS). DGPS works by using the cooperation of two receivers, one that is stationary and one that is traveling around making differential positioning measurements. The stationary receiver measures the timing errors and then provides the traveling receiver with the correct information by radio or other means. In most cases, the price of the receiver varies according to the level of accuracy desired.

As the deployment of cellular systems increases, geolocation in PCS systems, which relies on the existing infrastructure of cellular BSs, can offer position estimates of MTs as they transmit radio signals. A variety of locating techniques are based on the measurement of signal strength, AOA, and/or TDOA. Most of them are used in outdoor environments. For indoor environments, the performance of these algorithms is limited. RF-based and acoustic-based localization algorithms are designed for more accurate estimates of GPS-less mobile nodes or sensors, such as the RADAR system (Bahl & Padmanabhan, 2000), which is based on the idea that the received signal strength indicator (RSSI) is a function of mobile users' locations. In RADAR, a radio map of the building is created, which is a database of locations within the building and the corresponding signal strength of the beacons from various access points. A mobile object measures the signal strength of each reachable access point. Then it looks up the

Figure 12. Time of arrival and time difference of arrival

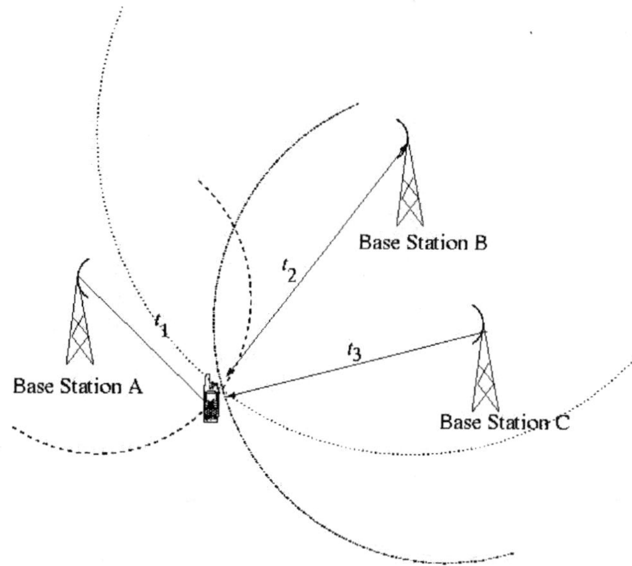

radio map for the closest match of the signal strength, thus determining the approximate location of its position.

AOA: This method is also called direction of arrival (DOA). It is usually performed at a BS by using multiple antennas in the direction of the arriving mobile signal. The location of a mobile object can be found by the intersection of two lines of bearing (LOBs), each formed by a radial from a BS to the MT as shown in Figure 11. AOA systems must be designed to account for multipath signals since they are sensitive to array calibration and require line of sight, which is impractical in urban environments. In reality, many pairs of LOBs may be used for making an accurate location estimation. For example, three BSs located at different points can be used to deduce the location at the intersection of two circles, which is known as triangulation, using analytic geometry.

Another method is called time of arrival (TOA), which uses absolute time of arrival at a certain BS. Since electromagnetic waves travel at a constant speed of light ($c = 3 * 10^8$ m/s), the distance from an MT to the receiving BS is directly proportional to the propagation time. If the traveling time of a signal from an MT to the fixed BS is t_1, then the receiver lines at the range of R_1, where $R_1 = ct_1$. The TOA measurements from two BSs will narrow a position to two points, and measurement from a third BS is required to resolve the precise position. There are two problems with the TOA method: synchronization and time stamp. TOA requires that transmitters and receivers have precisely synchronized clocks. Moreover, every transmitting signal must have a time stamp in order for the receiver to discern the distance the signal has traveled. Therefore, TDOA is proposed for practical systems.

TDOA: This uses the time that a signal travels as an indirect method of calculating distance. The idea is to determine the relative position of a mobile object by examining the difference in time at which the signal arrives at multiple receivers. With a minimum of three BSs receiving a signal from a handset, the difference in time at each receiver can be used to triangulate the position of the mobile object as shown in Figure 12. This method is called hyperbolic trilateration.

Unlike direct TOA measurements, TDOA measurements do not require precise synchronization at both receivers and transmitters. As long as the fixed receivers have an accurate clock, the location of an MT can be determined. To achieve accurate positioning, the BSs must be precisely synchronized in time, which is usually done by GPS. There are also many other techniques such as the combination of TDOA and AOA in cellular systems and location fingerprinting for indoor environments with WLANs. Location fingerprinting refers to techniques that match the fingerprint of some characteristics of the signal that is location dependant. The fingerprints of different locations are stored in a database and matched to measured fingerprints at the current location of the user under consideration.

Summary

In wireless networks, mobile users are allowed to move around and maintain their communications regardless of their locations, which means they may change their positions from time to time. A key design issue is the modeling of user mobility and location management so that a network can deliver incoming and outgoing calls and satisfy QoS. Numerous mobility models and location management algorithms have been devised.

The simplest mobility models are the fluid-flow model and the random-walk model. These are appropriate for cellular systems with different coverage areas when the mobile users are pedestrians or vehicles. When WLAN and ad hoc networks are introduced with applications for indoor environments and distributed architectures, traditional fluid and random-walk models are not good choices. Many other models such as random-waypoint models, group-mobility models, and many variants or derivatives of fluid-flow and random-walk models have been widely discussed, analyzed, and used in performance evaluations.

In a cellular wireless networks, location management techniques are used to store up-to-date location information and are queried for more accurate and efficient locating. These techniques, mainly location update and paging, have been the foundation of mobility management in cellular systems for intrasystem roaming and intersystem roaming. A class of schemes that adapt to user mobility in location management for reducing signaling overhead are referred to as dynamic location management to locate mobile users in a particular cell. Meanwhile, cellular geolocation and localization in WLANs aim to provide accurate positions within a small area such as a small cell or a building. Basic methods using radio signaling strengths at receivers are used in practical systems such as AOA and TDOA. Hybrid schemes that combine the idea of using location databases in cellular networks and signaling measurements are being developed and evaluated.

References

Akyildiz, I. F., Ho, H. S. M., & Lin, Y.-B. (1996). Movement-based location update and selective paging for PCS networks. *IEEE/ACM Transactions on Networking*, *4*(4), 629-638.

Akyildiz, I. F., McNair, J., Ho, J. S. M., Uzunaliouglu, H., & Wang, W. (1999). Mobility management in next-generation wireless systems. *Proceedings of the IEEE*, *87*(8), 1347-1384.

Amotz, B.-N., & Ilan, K. (1995). Mobile users: To update or not to update? *ACM/Baltzer Wireless Networks*, *1*(2), 175-186.

Bahl, P., & Padmanabhan, V. N. (2000). RADAR: An in-building RF-based user location and tracking system. *Proceedings of the IEEE Conference on Computer Communications (INFOCOM)*, 775-784.

Bettstetter, C. (2001). Mobility modeling in wireless networks: Categorization, smooth movement, and border effects. *Mobile Computing and Communications Review*, *5*(3), 55-67.

Bettstetter, C., Hartenstein, H., & Perez-Costa, X. (2003 Sept). Stochastic properties of the random waypoint mobility model. *ACM-Kluwer Wireless Network, Special Issue on Modeling and Analysis of Mobile Networks, 10*(5), (forthcoming).

Brown, T. X., & Mohan, S. (1997). Mobility management for personal communications systems. *IEEE Transactions on Vehicular Technology*, *46*(2), 269-278.

Camp, T., Boleng, J., & Davies, V. (2002). A survey of mobility models for ad hoc network research. *Wireless Communications & Mobile Computing (WCMC), Special Issue on Mobile Ad Hoc Networking: Research, Trends and Applications, 2*(5), 483-502.

Hellebrandt, M., & Mathar, R. (1999). Location tracking of mobiles in cellular radio networks. *IEEE Transactions on Vehicular Technology*, *48*(3), 1558-1562.

Hong, X., Gerla, M., Pei, G., & Chiang, C. (1999). A group mobility model for ad hoc wireless networks. *Proceedings of the ACM International Workshop on Modeling and Simulation of Wireless and Mobile Systems (MSWiM)*, (pp.53-60).

Kim, T. S., Kwon, J. K., & Sung, D. K. (2000). Mobility modeling and traffic analysis in three-dimensional high-rise building environments. *IEEE Transactions on Vehicular Technology*, *49*(5), 1633-1640.

Lei, Z., & Rose, C. (1998). Wireless subscriber mobility management using adaptive individual location areas for PCS systems. *Proceedings of the IEEE International Conference on Communications (ICC)*, (pp. 1390-1394).

Liang, B., & Hass, Z. (1999). Predictive distance-based mobility management for PCS networks. *Proceedings of the IEEE International Conference on Computer Communications (INFOCOM)*, (pp. 1377-1384).

Lin, Y., & Park, K. (1997). Reducing registration traffic for multi-tier personal communications services. *IEEE Transactions on Vehicular Technology, 46*(3), 597-602.

Mark, J. W., & Zhuang, W. (2003). *Wireless communications and networking*. Upper Saddle River, NJ: Prentice Hall.

McGuire, M., Plataniotis, K. N., & Venetsanopoulos, A. N. (2003). Environment and movement model for mobile terminal location tracking. *Wireless Personal Communications: An International Journal, 24*(4), 483-505.

Rappaport, T. S., Reed, J. H., & Woerner, B. D. (1996). Position location Using wireless communications on highways of the future. *IEEE Communications Magazine, 34*(10), 33-41.

Rose, C., & Yates, R. (1995). Minimizing the average cost of paging under delay constraint. *ACM/Baltzer Wireless Networks, 1*(2), 211-219.

Tabbane, S. (1997). Location management methods for third-generation mobile systems. *IEEE Communications Magazine, 35*(8), 72-84.

Thomas, R., Gilbert, H., & Mazziotto, G. (1988). Influence of the movement of the mobile station on the performance of the radio cellular network (Paper 9.4). *Proceedings of the Third Nordic Seminar on Digital Land Mobile Radio Communication,* Copenhagen, Denmark.

Wang, W., & Akyildiz, I. F. (2000). Inter-system location update and paging schemes for multitier wireless networks. *Proceedings of the ACM Mobile Computing and Networking Conference (MobiCom),* (pp. 99-109).

Wang, W., Akyildiz, I. F., Stuber, G., & Chung, B.-Y. (2001). Effective paging schemes with delay bounds as QoS constraints in wireless systems. *ACM/Kluwer Wireless Networks, 7*(10), 455-466.

Chapter VII

Mobility Management in Mobile Computing and Networking Environments

Samuel Pierre, Ecole Polytechnique de Montreal, Canada

Abstract

This chapter analyzes and proposes some mobility management models and schemes by taking into account their capability to reduce search and location update costs in wireless mobile networks. The first model proposed is called the built-in memory model*; it is based on the architecture of the IS-41 network and aims at reducing the home-location-register (HLR) access overhead. The performance of this model was investigated by comparing it with the IS-41 scheme for different call-to-mobility ratios (CMRs). Experimental results indicate that the proposed model is potentially beneficial for large classes of users and can yield substantial reductions in total user-location management costs, particularly for users who have a low CMR. These results also show that the cost reduction obtained on the location update is very significant while the extra costs paid to locate a mobile unit simply amount to the costs of crossing a single pointer between two location areas. The built-in memory model is also compared with the forwarding pointers' scheme. The results show that this model consistently* *outperforms the forwarding pointers' strategy. A second location management model to manage mobility in wireless communications systems is also proposed. The results show that significant cost savings can be obtained compared with the IS-41 standard location-management scheme depending on the value of the mobile units' CMR.*

Introduction

Mobile communication networks are made possible by the convergence of several different technologies, specifically computer networking protocols, wireless-mobile communication systems, distributed computing, and the Internet. With the rapidly increasing ubiquity of laptop computers, which are primarily used by mobile users to access Internet services such as e-mail and the World Wide Web (WWW), support of Internet services in a mobile environment has become a growing necessity. Mobile Internet providers (IPs) attempt to solve the key problem of developing a mechanism that allows Internet protocol (IP) nodes to change physical locations without changing IP addresses, thereby offering Internet users the so-called "nomadicity." Furthermore, advances in wireless networking technologies and portable information devices have led to a new paradigm of computing called *mobile computing*. According to this concept, users who carry portable devices have access to information services through a shared infrastructure regardless of their physical location or movements. Such a new environment introduces new technical challenges in the area of information access. Traditional techniques to access information are based on the assumptions that the host locations' distributed systems do not change during computation. In a mobile environment, these assumptions are rarely valid or appropriate.

Mobile computing is distinguished from classical, fixed-connection computing due to the following elements: (a) the mobility of nomadic users and the devices they use, and (b) the mobile resource constraints such as limited wireless bandwidth and limited battery life. The mobility of nomadic users implies that the users might connect from different access points through wireless links and might want to stay connected while on the move, despite possible intermittent disconnections. Wireless links are relatively unreliable and currently are two to three times slower than wired networks. Moreover, mobile hosts powered by batteries suffer from limited battery life constraints. These limitations and constraints provide many challenges to address before we consider mobile computing to be fully operational. This remains true despite the recent progress in wireless data communication networks and handheld device technologies.

In next-generation systems supporting mobile environments, mainly due to the huge number of mobile users in conjunction with the small cell size, the influence of mobility on the network performance is strengthened. More particularly, the accuracy of mobility models becomes essential to evaluate system design alternatives and network implementation costs. The device location is unknown a priori and call routing in general implies mobility management procedures. The problems which arise from subscriber mobility are solved in such a way that both a certain degree of mobility and a sufficient quality of the aspired services are achieved.

This chapter analyzes the problem of managing users' mobility in the context of mobile computing and networking environments. Mobility management implies two major components: *handover management* and *location management.*

Handover management is the way a network functions to keep mobile users connected as they move and change access points within the network. Generally, there are two types of handover: intracell handover and intercell handover. Intracell handover occurs when a user experiences degradation of signal strength within a cell. This leads to a choice of new channels with better signal strength at the same *base transceiver station* (BTS), also called *base station* (BS). Intercell handover occurs when a user moves from a cell to another. In this case, the user's connection information is transferred from the former BTS to the latter one. The following procedure occurs for both intracell and intercell handovers. First, the user initiates a handover procedure. Then, the network or the mobile unit (depending on the unit that controls the handover operation) provides necessary information and performs routing operations for the handover. Finally, all subsequent calls to the user are transferred from the former connection to the latter one.

Location management is the process used by a network to find the current attachment point of a mobile user for call delivery (Akyildiz & Wang, 2002). The first step of the procedure is the *location registration*. In this phase, the mobile user periodically notifies the network of its new access point. The notifications allow the network to authenticate users and update their location profiles. The second step is the *call delivery*. When a call destined to a user reaches the network, a search for the user's profile is usually conducted in a local database. Then, the call is forwarded to the user according to his profile.

This chapter aims to analyze different mobility management schemes and protocols in order to state their applicability to handle some key issues related to emerging mobile environments. The main concerns include the search for efficient and cost-effective location management schemes allowing to provide services and applications to users with an acceptable quality of service. The next section summarizes background and related work. Then we propose some new location management schemes, evaluate the performance of these schemes, and present some numerical examples. Finally, the chapter concludes and outlines future research directions.

Background and Related Work

Mobility is the primary advantage offered by *personal communication systems* (PCS). Location management is one of the most important issues of mobility management. Location management techniques essentially consist of partitioning the coverage area into many *location areas* (LAs) which are sets of cells. *Mobile Units* (MUs) within a cell communicate with a cell BS through wireless links. BSs, in turn, are connected to the wireline network through a *mobile switching center* (MSC) which serves a single LA. Each MSC is identified by a unique address. This address is stored in the memory of the MUs that are roaming in the MSC's LA and is broadcasted by the cell BSs within that particular LA. The MU compares the broadcasted address with the address stored in its memory. When these two addresses differ, the MU recognizes that it has moved to a new LA and sends a registration message to the MSC of the new LA. Then, the MSC forwards this message to the network database. In PCS, the wireline network uses the *Signaling System Number 7* (SS7) to carry user information and signaling messages between the MSCs and the location databases.

Main Standards and Basic Procedures

Two major standards are used for location management, namely IS-41 (Gallagher & Randall, 1997; TIA/EIA, 1996) and Global System for Mobile Communications [GSM] (Mouly & Paulet, 1992). This chapter only considers the IS-41 standard, which uses a two-level database architecture consisting of an HLR and some *visitor location registers* (VLRs). Each network comprises a single HLR and many VLRs. The HLR is a centralized database containing the profiles of its assigned subscribers. Most of the current PCS manufacturers implement a combined MSC and VLR with one VLR per MSC. A VLR stores the profiles of the MUs that are currently residing in its associated LA. Figure 1 illustrates the PCS architecture and signaling network.

Two main procedures are used in the IS-41 location management scheme: *location update* and *location search*. A location update occurs when an MU moves to a new LA; a location search occurs when a fixed or mobile host wants to communicate with an MU

Figure 1. PCS architecture and signaling network

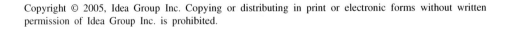

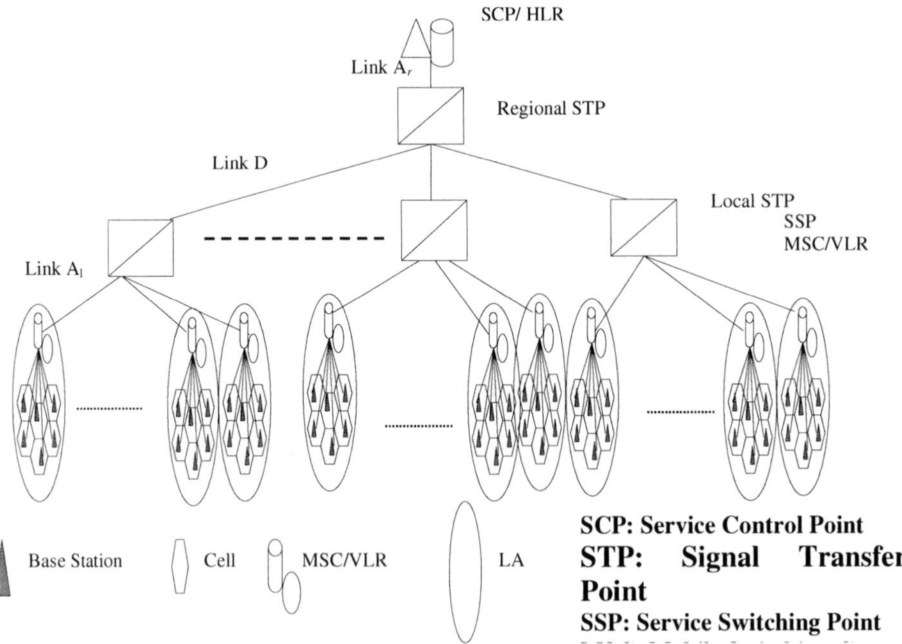

whose current LA is unknown. In IS-41, the HLR is queried for every location search or update, resulting in tremendous strain on the use of the network resources as the number of PCS subscribers increases.

We describe the location update and location search procedures as specified in revision C of the IS-41 standard (TIA/EIA, 1996), along with some additional models that were proposed to augment it (Cayirci & Akyildiz, 2002; Lin, Lee, & Chlamtac, 2002). Figure 2 illustrates the location update procedure of the IS-41 standard which can be described as follows. When an MU enters a new LA, it registers at its MSC or VLR. Then, the new MSC/VLR sends a *registration notification* message (REGNOT) to the network database HLR through an Single Transfer Point (STP) node. The STP node executes the global title translation (GTT) procedure using the MU's identification number in order to determine the HLR of the MU. Upon receiving the registration message, the HLR sends a *registration cancellation* message (REGCANC) to the MU's previous MSC/VLR. The previous MSC deletes the MU's profile in its associated VLR and sends a *cancellation acknowledgment* message (regcanc) to the HLR. The HLR acknowledges the location update by sending a regnot message that includes the MU's profile to the new MSC/VLR. When the new MSC/VLR receives the regnot message, it starts providing service to the MU.

The IS-41 location search procedure is presented in Figure 3. In this figure, a call is initiated to the mobile unit B (MU-B) through an originating MSC/VLR which sends a *location request* (*LOCREQ*) message to the HLR of the called MU through an STP node using the GTT technique. Upon receiving the request, the HLR sends a *routing request* (ROUTREQ) message to the MU-B's current MSC/VLR, which then allocates a temporary location directory number (TLDN) for the call and returns it to the HLR in a routreq

Figure 2. Location update procedure according to the IS-41 scheme

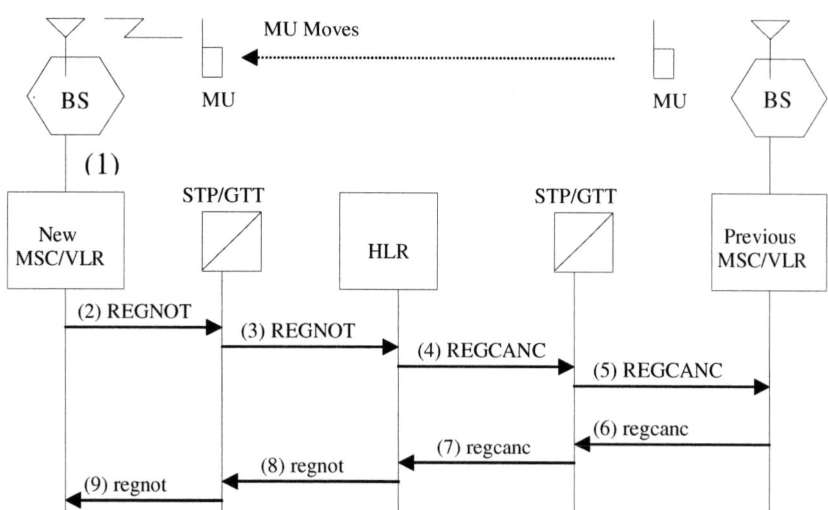

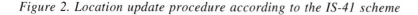

message. The HLR relays the TLDN to the originating MSC/VLR in a locreq message which routes the call to the TLDN. Finally, the current VLR/MSC launches a paging process to find the MU-B's current cell. At this point, the communication has been established.

In these two procedures, the network database HLR is queried every time the MU moves to a new LA or receives a call. Several strategies have been proposed to reduce the location update load and location search costs (Escalle, Giner, & Oltra, 2002; Mao, 2002; Suh, Choi, & Kim, 2000).

Advanced Mobility Management Schemes

A new signaling protocol was proposed by Wang & Akyildiz (2001) for intersystem roaming in next-generation wireless networks. According to this protocol, LAs that are on the boundary of two adjacent systems (*X* and *Y*) are called *peripheral location areas* (PLAs). Therefore, they are PLAs in systems *X* and *Y*. The intersystem location registration is controlled by a *boundary interworking unit* (BIU), which ensures the compatibility between the two systems and maintains a database of the roaming information of the mobile terminals (MTs) moving between the two networks. The underlying principle of this protocol is that the MT can request a location registration of intersystem roaming when it is in a PLA. As a result, it may finish signaling transformation and authentication before it arrives at the new system. Using this protocol, the HLR is not involved unless the MT goes from a PLA to a non-PLA. Moreover, Boundary Location Register (BLR) provides MTs with up-to-date location

Figure 3. Location search procedure according to the IS-41 scheme

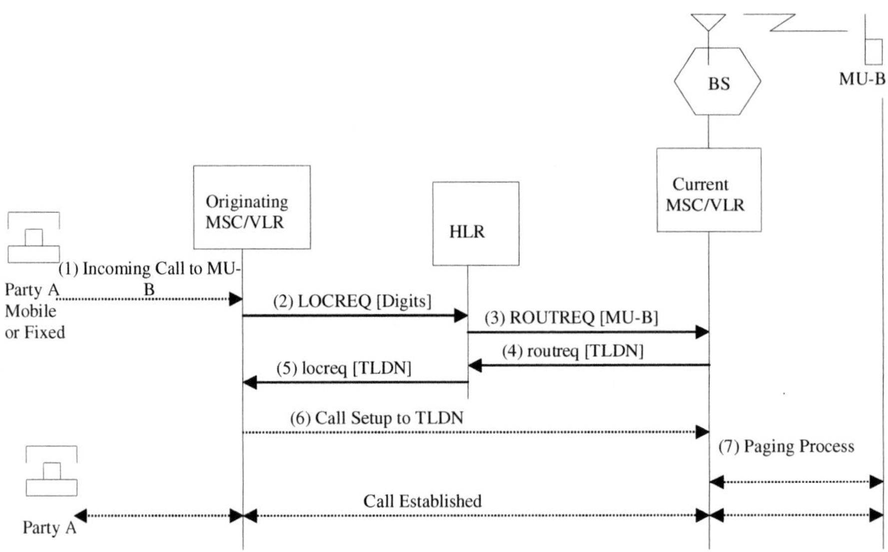

information; the incoming calls of the intersystem-roaming MTs are delivered to the serving MSC/VLR directly, rather than delivering it to the previous system. The numerical results show that the BLR protocol can reduce signaling costs, the latency of location registration, and call delivery, as well as call-loss rates for the MTs moving across different networks.

The models generally used to determine the optimal distance threshold are often based on certain simplified assumptions that do not give an accurate representation of a realistic cellular network (structured topology and configuration, geometric or exponential cell residence time distribution, symmetric random walk as a movement model, etc.). Wong and Leung (2001) overcome these drawbacks by focusing on the determination of the optimal update boundary for the distance-based location update algorithm, in a realistic environment, in order to minimize the expected total cost between call arrivals. The proposed model is applicable to arbitrary cell topologies; the call residence time can follow a general distribution and the movement history is taken into account. An implementation is described using an arbitrary cell topology. The location update is decided upon a simple table lookup.

Numerical results show that the proposed model gives a more accurate update boundary (distance threshold) in real wireless cellular environments compared with those derived from a hexagonal cell configuration with random-walk movement pattern. This is due to the fact that the network can maintain a better balance between the processing due to location update and the ratio bandwidth used for paging between call arrivals. The main drawback is the use of a Poisson distribution for the call arrival rate, although this may not be the case in real mobile environments.

The user mobility pattern (UMP) proposed by Cayirci and Akyildiz (2002) for location updates and paging where MTs keep track of their UMPs in a data structure called *user mobility pattern history* (UMPH). The UMP is a list of cells expected to be visited by the MT, and the UMPH records the mobility history of an MT. The model proposed by these authors differs from the other user profiles or history-data-based location update techniques in two main aspects. The first aspect is that according to this model, MTs are responsible to predict and register the UMPs, reducing the signaling traffic (for maintaining UMPH) and increasing the resolution of the data in UMPH. The second aspect is that an effective selective paging can be executed, based on call delivery times, by using the UMP nodes (pairs of cell identification and expected call entry time).

The performance of the UMP scheme is compared with the time-based and movement-based location update techniques, the blanket, the selective, and the velocity paging techniques. The UMP technique creates less location update traffic than the other techniques when reasonable time intervals and movement thresholds are used. It consistently outperforms other paging performance techniques.

Cayirci and Akyildiz (2003) proposed the *traffic-based static LAD* (TB-LAD) scheme where the mobile traffic between the cells is predicted according to the characteristics of the crossing loads. Then, by using these traffic expectations, the traffic-based cell-grouping technique groups cells into LAs such that the neighbor cell with higher intercell traffic is assigned to the same LAs. Since the number of cells in an LA is a fixed parameter in TB-LAD, paging traffic is undisrupted. However, a better design of an LA reduces the number of location updates. Therefore, the TB-LAD decreases the number of location updates without increasing the number of paged cells during a call delivery.

To investigate this concept, traffic data was collected in a metropolitan area. Then, the relation of inter-LA traffic with the cell size and the number of cells in an LA was analyzed, and the performance was compared with the one of a *proximity-based LAD* (PB-LAD). Experimental results show that the TB-LAD reduces inter-LA traffic from 27% to 36% on the average over PB-LAD. TB-LAD outperforms PB-LAD when the average cell size exceeds 2,000 m and the average number of cells within an LA is superior to nine. In the metropolitan city where the experiments were conducted, the optimal solution was obtained with an LA size larger than 13 cells and a cell size larger than 2,500 m.

A trade-off cost analysis was made for the movement-based location update and paging in wireless mobile networks without using simplifying assumptions. A general framework was proposed by Fang (2003) for the study of such problems, and analytical results were derived to obtain the crucial quantity and the average number of location updates during an interservice time (used in all cost analyses under general assumptions). The analytical approach and results developed in his paper can be very useful to design an optimal mobility management scheme for future wireless mobile networks. In fact, the study shows that the total cost of the location update and paging is a convex function of the movement threshold.

Location Management with Mobile IP

Mobility support in the IP protocol was developed by the Internet Engineering Task Force (IETF), leading to the Mobile IP protocol and all-IP networks (Chiussi, Khotimsky, & Krishnan, 2002). Currently, two versions of Mobile IP are available, Versions 4 (IPv4) and 6 (IPv6). An MN is a mobile node able to move from one subnet to another without requiring a change of IP address. The MN accesses the Internet via a home agent (HA) or a foreign agent (FA). The correspondent node (CN) is a node that connects with the

Figure 4. Mobile IP network architecture

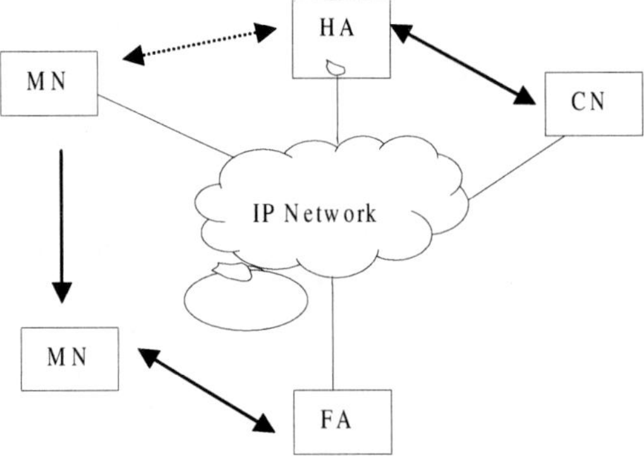

MN. The HA is a local router on the MN's home network, and the FA is a router on the visited network. Figure 4 illustrates a Mobile IP network architecture.

The following operations are introduced by the Mobile IP protocol:

1. *Discovery*: How an MN finds an agent (HA or FA).
2. *Registration*: How an MN registers with its HA.
3. *Routing and Tunneling*: How an MN receives datagrams when visiting a foreign network.

Location management operations include agent discovery, movement detection, forming care-of-address (CoA), and location (binding) update. Handover operations include routing and tunneling.

Discovery

To detect movement from one subnet to another, the MN uses two methods: advertisement lifetime and network prefix.

Advertisement Lifetime: The first detection method is based upon the lifetime field of the Internet Control Message Protocol (ICMP) router advertisement message. An MN records the lifetime indicated in any agent advertisement until that lifetime expires. If the MN has not maintained contact with its FA when the lifetime elapses, it must solicit a new agent.

Network Prefix: The second method uses the network prefix-detection movement. In some cases, an MN can determine whether or not a newly received agent advertisement was received on the same subnet by using its CoA. If the prefixes differ, the MN can assume that it has moved. This method is not available if the MN is currently using an FA's CoA. Once it has discovered a new FA and obtained a new CoA, the MN performs the location update procedure as follows:

1. The MN registers a new CoA with its HA by sending a binding update.
2. The MN notifies its CN of its current binding information.
3. If the binding update has an expiration date, the CN and the HA send a binding request to the MN to obtain the MN's current binding information.

Registration

When visiting any network away from home, each MN must have an HA. The MN registers with its HA in order to track the MN's current IP address. Two IP addresses are associated to each MN: one for location and one for identification purposes. The new IP address associated to an MN while it visits a foreign link is called its CoA. The association between the current CoA and the MN's home address is maintained by a mobility binding, so that packets destined to the MN may be routed using the current CoA, regardless of the MN's current point of attachment to the Internet. Each binding has a predetermined lifetime period, which is negotiated during the MN's registration, after which time the registration is deleted. The MN must reregister within this period in order to continue servicing this CoA. Depending on its method of attachment, the MN sends location registration messages (when moving between two subnets or once the lifetime elapses) directly to its HA, or through an FA which forwards the registration to the HA. In either case, the MN exchanges registration request and registration reply messages based on IPv4 as follows.

1. The MN registers with its HA using a registration request message (the request may be relayed to the HA by the current FA).
2. The HA creates or modifies a mobility binding for that MN with a new lifetime.
3. The appropriate agent (HA or FA) returns a registration reply message containing necessary information on the request status, including the lifetime granted by the HA.

Routing and Tunneling

The process of routing datagrams for an MN through its HA often results in the utilization of paths that are significantly longer than optimal. Route-optimization techniques for Mobile IP employ the use of tunnels to minimize inefficient path use. For example, when the HA tunnels a datagram to the CoA, the MN's home address is effectively shielded from intervening routers between its home network and its current location. Once the datagram reaches the agent, the original datagram is recovered and delivered to the MN. Currently, there are two protocols for routing optimization and tunnel establishment: route optimization in Mobile IP and the tunnel establishment protocol.

Route Optimization in Mobile IP: The route optimization aims to define extensions to basic Mobile IP protocols that allow better routing so that datagrams can travel from a CN to an MN without first going to the HA. These extensions provide a means for nodes to cache the binding of an MN, and then tunnel datagrams directly to the CoA indicated in that binding, bypassing the MN's HA.

Tunnel Establishment Protocol: In this protocol, Mobile IP is modified in order to perform among arbitrary nodes. Upon establishing a tunnel, the encapsulating agent (HA) transmits protocol data units (PDUs) to the tunnel endpoint (FA) according to a set of parameters. The process of creating or updating tunnel parameters is called tunnel establishment. Generally, the establishment of parameters includes a network address for the MN. In order to use tunnel establishment to transmit PDUs, the HA must determine the appropriate tunnel endpoint for the MN. This is done by consulting a table that is indexed by the MN's IP address. After receiving the packets, the FA "decapsulates" the PDUs and sends them to the MN.

Mobile IP provides simple scalable mobility solutions. However, it causes excessive signaling traffic and long delays. Xie and Akyildiz (2002) introduced a distributed and dynamic, regional location management mechanism for Mobile IP to distribute traffic and dynamically adjust the regional network boundaries more efficiently. (A distributed gateway foreign agent (GFA) system architecture is proposed where each FA can function as either an FA or a GFA). This distributed system may allocate signaling load more uniformly. An active scheme is adopted by the distributed system to dynamically optimize the regional network size of each MN according to its current traffic load and mobility.

The distributed and dynamic scheme proposed by Xie and Akyildiz (2002) can perform optimally for all users from time to time, and the system robustness is enhanced. Since the movement of MNs does not follow a Markov process, a novel, discrete, analytical model for cost analysis, and also a new iterative algorithm to find out the optimal number of FAs in a regional network, which consumes terminal network resources, was proposed. The proposed model is not plagued by constraints on the shape and the geographic location of Internet subnets.

Analytical results demonstrate that the signaling bandwidth is significantly reduced through the proposed distributed system architecture compared with the IETF Mobile IP regional registration scheme. It also demonstrates that the dynamic scheme has significant advantages under time-variant user parameters in cases where it is difficult to predetermine the optimal regional network size. The location management scheme requires that all FAs be capable of functioning as both FAs and GFAs. This increases the processing capability requirements of each mobile agent. There is additional processing load on the MTs, such as the estimation of the average packet arrival rate and subnet residence time.

Lee and Akyildiz (2003) proposed a cost-efficient scheme for route optimization to reduce the signaling costs caused by route optimization. Link-cost functions represent the network resources used by the routing path; signaling costs reflect the signaling and processing load incurred by route optimization. A Markovian decision model was used in order to find an optimal sequence for route optimization. In order to simplify the decision process, the model was restricted to intradomain handoff, which resulted in a decision rule. The optimal sequence is obtained by following the decision rule at each decision stage (minimizing the total cost). The performance of the optimal sequence is compared with the other sequences' action with route optimization (ARO) and action without route optimization (NRO). Simulation results show that the optimal sequence provides the lowest costs among the given sequences.

The recent advent of voice-over IP (VoIP) services and their fast growth is likely to play a key role in successful deployment of IP-based convergence of mobile and wireless networks. Kwon, Gerla, and Das (2002) have focused on mobility management issues regarding VoIP services in wireless access technologies. Different mobility management schemes are explored with a focus on Mobile IP and Session Initiation Protocol (SIP; Lin & Chen, 2003; Wu, Lin, & Lan, 2002); these two approaches are compared. The shadow registration concept is also presented; it aims at reducing distribution time in interdomain handoff for VoIP sessions in mobile environments. Considering the functionality of authentication, authorization, and accounting (AAA), the signaling message flow is illustrated for the two approaches in the presence and absence of shadow registration. Finally, an analytic comparison between the two approaches (Mobile IP and SIP) in terms of delay at initial registration and distribution in intradomain- and interdomain-handoff delay is presented.

Based on the previous analyses, numerical results show that the disruption for the Mobile IP handoff approach is smaller than the SIP approach in most situations. However, SIP shows shorter disruptions when the MN and the CN are nearby. Even though the smooth handoff is not taken into consideration in the disruption analyses, it is argued that it will play an important role in reducing disruption in interdomain handoff in the Mobile IP approach.

Misra, Das, Dutta, McAuley, and Das (2002) presented two enhancements to the Intradomain Mobility Management Protocol (IDMP) for IP-based hierarchical mobility management so that it can be used in a 4th Generation (4G) cellular environment. Thus, it would be highly relevant to develop IP-layer-fast handoff and paging solutions that would work across heterogeneous access technologies.

To minimize packet loss during intradomain handoffs, a time-bound localized multicasting approach is presented. By proactively informing its associated mobility agent (MA) of an imminent change, an MN enables the MA to multicast packets for a limited time span to a set of neighboring subnets. Subnet agents (SAs) buffer such multicast packets for a short while. Then, if the MN enters its subnet, the SA is able to immediately forward these packets to the mobile, thus eliminating packet-loss delays.

Next-generation mobile and wireless networks are already under preliminary deployment and they are likely to use the current existing infrastructure for economic reasons. Global roaming in current and next-generation networks is a major issue for the integration and the interoperability of different systems. Zahariadis, Vaxevanakis, Tsantilas, Zervos, and Nikolaou (2002) proposed a hierarchical cell architecture consisting of an infrastructure that is either installed or under development. Soft horizontal-mobility management mechanisms and vertical handover (for Wireless Local Area Networks [WLAN], 2nd Generation [2G], and satellite networks) are discussed. An enhanced roaming scenario for next-generation networks is initiated by the MT and supported by an all-mobile IP network.

Mao (2002) presented an *intralocation-area location update* (intra-LA-LU) strategy in order to reduce paging traffic in mobile networks while keeping the standard location area update unchanged within the LA. The intra-LA-LU is performed whenever the MT changes its location between the anchor cell (the MT residing cell) and the rest of the cell in the LA. For call delivery, either the anchor cell or the other cells of the LA are paged to locate the MT. The proposed analytical model considers a continuous-time Markov chain to describe the MT movement.

Compared to the conventional location-tracking scheme, one can think that the proposed strategy will add an extra cost by performing the inter-LALUs. However, numerical results indicate that the savings of paging costs is much more significant than the newly added location update costs. The proposed strategy is suitable for users roaming in the LAs associated with their homes or workplaces. If the location of a mobile subscriber is uniformly distributed within an LA, the proposed strategy should not be used in order to avoid the intra-LA-LU, which is not suitable in this case.

Personal number (PN) or follow-me service allows users to access telecommunication services from any terminal in any location within the service area. In the existing systems, users have to manually register a phone number every time they enter a new area, which is an unsuitable solution.

Lin et al. (2002) proposed an enterprise approach for *automatic follow-me service* (AFS). The significance of the proposed approach is that AFS can be integrated with existing follow-me databases to automate PN services offered by different Public Switched Telephone Network (PSTN) service providers. AFS automatically connects calls to a user at any location with appropriate communications terminals. The authors showed how to implement AFS with the VoIP and Bluetooth technologies. More specifically, the AFS utilizes VoIP to communicate with the follow-me database in the public network, and Bluetooth is used to implement radio-tracking mechanisms. Then, the impact of polling frequency on power consumption and call misrouting was presented. The analysis provided shows that, based on the AFS cost functions, the optimal polling frequency can be found efficiently. One of the future extensions consists of developing automatic polling-frequency-adjustment heuristics based on the proposed analytic model.

New Mobility Management Models and Schemes

This section introduces three new mobility management models: the *built-in memory model* (Safa, Pierre, & Conan, 2002), the *global location management scheme* (Safa et al.) and the Mobile IP network architecture (Diha & Pierre, 2003). Two of these models are based on the IS-41 standard and aim to improve upon this standard to some extent.

Built-In Memory Model

The built-in memory model is based on the IS-41 standard with additional small, built-in MU memory and a pointer table for each LA. In this model, we define an MU's anchor LA as the LA for which the MU's data location was updated at the network database HLR. The MU built-in memory stores the address of its anchor LA, that is, the MU built-in memory data and the MU location data at the HLR are the same. The pointer table comprises two columns: the MU identification number (MIN) and the MU's current LA. The LA pointer table stores the current LA addresses of the MUs which consider this particular LA as their anchor LA. When the MU moves to a new LA, the new LA queries the MU's anchor LA to update the pointer table, that is, to create a pointer between the MU's anchor LA and the new LA. Consequently, no location update operation is performed at the HLR level. When the MU is called, its HLR is queried to determine its anchor LA. If it no longer resides in that LA, the call is forwarded to the MU's current LA by passing over a single pointer.

Figure 5 illustrates the built-in memory model. We assume that an MU joins the network in the location area LA_1 and registers at the MSC/VLR of this LA. Then, the MSC/VLR of the LA_1 location area queries the HLR to update the MU's location data (Figure 5a). Once the HLR is updated, the MU also updates its built-in memory. In other words, the location area LA_1 becomes the MU's anchor LA. Figure 5b shows how the MU moves

Figure 5. Illustration of the built-in memory model

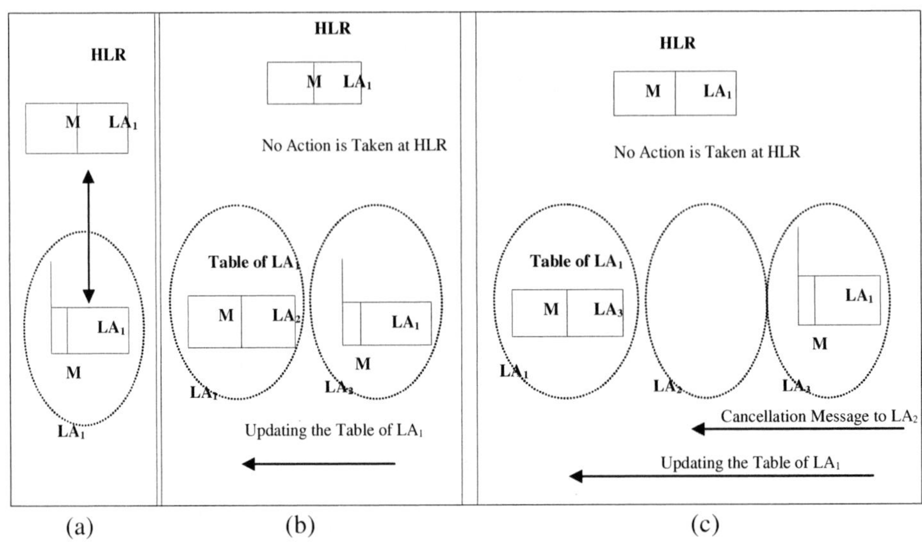

to a new location area LA_2. Then, instead of accessing the HLR, the MSC/VLR of the new location area LA_2 sends a query to the MU's anchor LA (LA_1) in order to establish a pointer between the MU's anchor LA (LA_1) and the current LA (LA_2). A pointer is established by updating any existing MU's location information in the pointer table of the MU's anchor LA, or adding this information to the table if there is none. If the MU leaves or returns to its anchor LA, the MSC/VLR of the MU's new LA sends a message to the MU's previous LA. However, when the MU's movement does not involve its anchor LA, the MSC/VLR of the MU's new LA sends two messages: a cancellation message to the MU's previous LA and a pointer updating message to the MU's anchor LA. This scenario is shown in Figure 5c. After the MU moves to the location area LA_3, the MSC/VLR of this LA sends a cancellation message to the MSC/VLR of the location area LA_2 and an updating message to the MU's anchor LA (LA_1) in order to establish a pointer from LA_1 to LA_3.

The location update procedure in the memory built in model is shown in Figure 6. It is assumed that an MU joins the network in the location area LA_i, registers at the MSC/VLR of LA_i, and updates its location data in the HLR and its built-in memory, thus location area LA_i is the anchor LA of the MU.

A New Global Location Management Scheme

A second approach, called the global location management scheme, adds two tables to the conventional signaling network architecture (Figure 1), which are respectively identified as *location data* and *pointer tables*. A location data table is stored on an Local

Figure 6. Example of the updating procedure

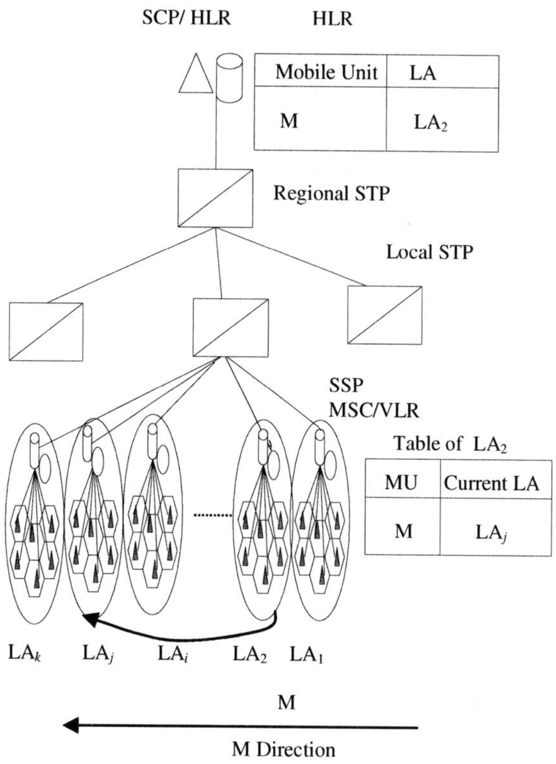

Signal Transfer Point (LSTP) node to serve all LAs connected to this LSTP. It contains the location information of some selected MUs, generally the ones which are frequently called from these LAs. This can significantly reduce the call-delivery procedure costs when the called MU has a profile in the location data table. Moreover, if this is not the case, there are no extra costs. In general, the use of a location data table serving many areas presents the following advantages:

- reduces network traffic by minimizing the number of updating queries sent to the data tables or to the HLR,
- saves table-installation costs,
- reduces data-storing redundancies and memory space wastes, and
- increases the frequency of locating an MU without accessing the network database. (For example, consider the case where an MU is often called from location area LA_1 but rarely from a neighboring location area LA_2. Since the same location data table may serve the two LAs, the called MU will be found locally even though it is called from LA_2.)

Figure 7. Architecture of the signaling network used in the global location management scheme

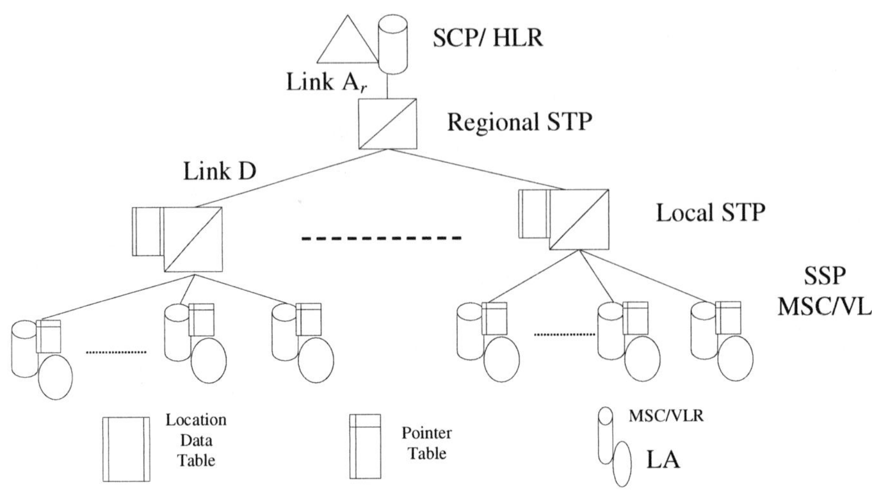

A pointer table is added to each LA. In order to explain the usefulness of the pointer table, we introduce the concept of the "anchor location area" of an MU as the LA in which the MU's location information is updated at the HLR. A pointer table of an anchor LA contains a pointer to the current LA of all the MUs having this LA as anchor LA. At this point, we must distinguish between two kinds of MU movements: an intra-STP move such that the new and the old LAs are connected to the same LSTP, and an inter-STP move such that the new and the old LAs belong to two different LSTPs.

We assume that each MU has a built-in memory that stores the address of its anchor LA and the addresses of the STP nodes which store the MU location information in their location data tables. This built-in memory is updated in either of two cases: The MU location information in the network database HLR is updated or the MU location information is added to a location data table. When the MU's movement is intra-LSTP, the pointer table of its anchor LA is queried to update the pointer to create a pointer between the anchor LA and the new LA. Hence, no location update operation is performed at the HLR. When the MU's movement is inter-LSTP, its new LA becomes its anchor LA and all information about the MU is then deleted from the previous anchor LA. The detailed location update and location search procedures of the proposed scheme will be presented further. They operate according to the signaling network architecture shown in Figure 7. It is implicitly assumed that when an MU's location information is added to a location data table, the MU is informed of this fact when it is called from any LA served by that location data table. Consequently, this operation does not require any additional costs.

When an MU moves to a new LA, a location update procedure is performed as illustrated in Figures 8 and 9, respectively, for intra-LSTP and inter-LSTP movements. For intra-LSTP movements, its anchor LA is updated instead of the HLR according to the following steps.

Figure 8. Updating procedure for an intra-LSTP move

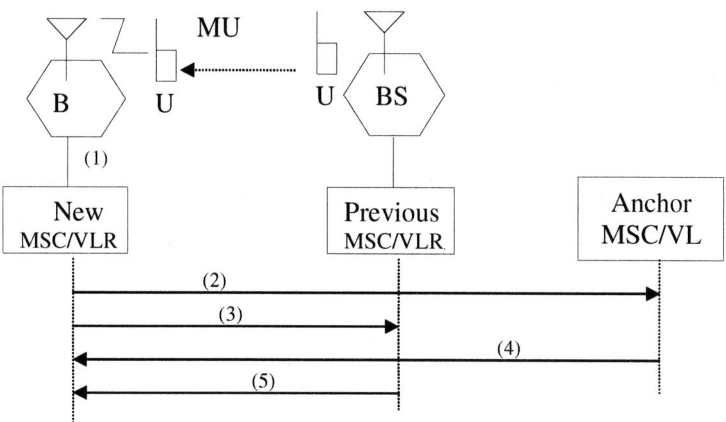

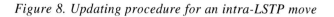

Figure 9. Updating procedure for an inter-LSTP move

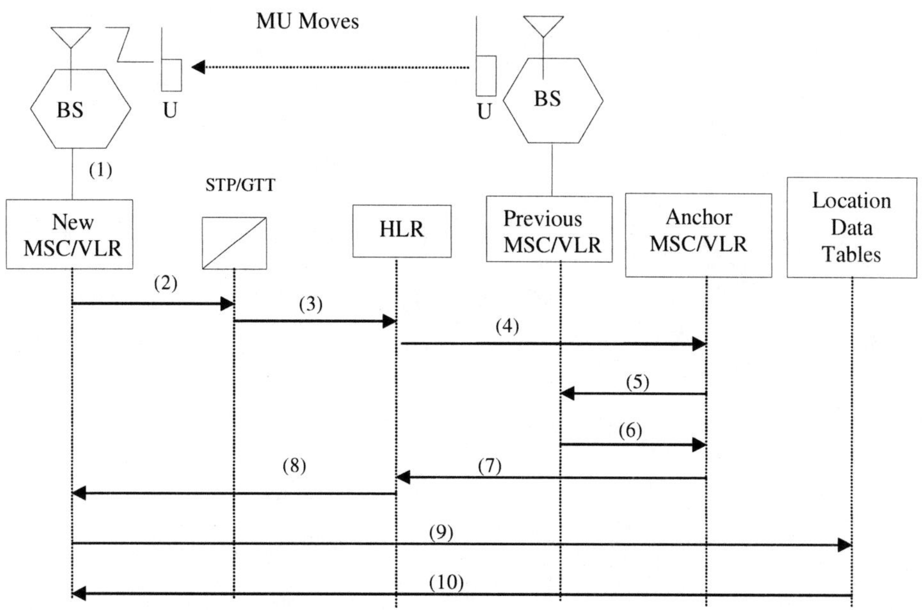

1. The MU moves to a new LA and sends a location update message to the MSC/VLR of this new LA.
2. The MSC of the new LA registers the MU with its associated VLR and sends a cancellation message to the previous LA.

Figure 10. Searching procedure (Scenario 1)

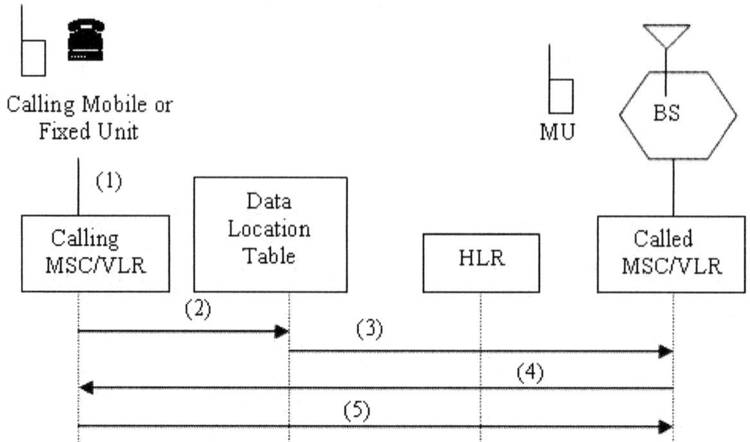

Figure 11. Location search procedure (Scenario 2)

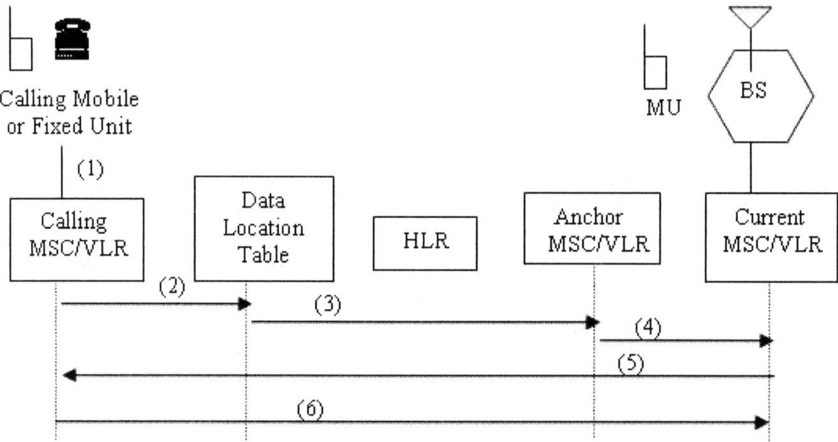

3. The new LA queries also the MU's anchor LA in order to create a pointer from the anchor LA to the new LA. In other words, no location update operation is performed at the HLR. The anchor LA is the LA which stored the MU's address in the built-in memory of the MU and in the HLR.
4. and 5. The new LA receives an acknowledgement from both the previous and the anchor LAs.

When the MU's movement is inter-LSTP, the location update procedure shown in Figure 9 is performed. The steps of this procedure are described as follows.

Figure 12. Location search procedure (Scenario 3)

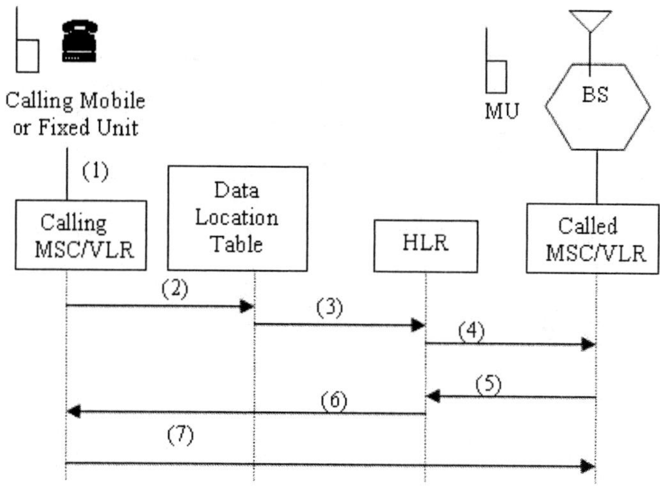

Figure 13. Location search procedure (Scenario 4)

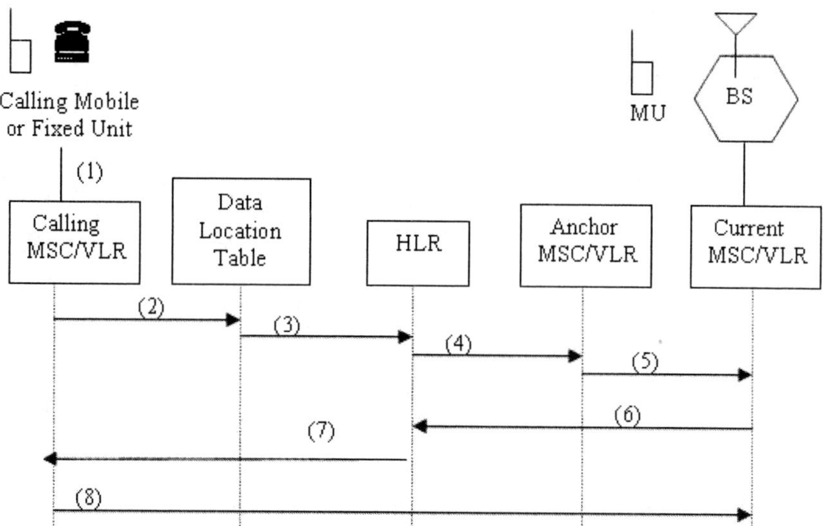

1. The MU moves to a new LA and sends a location update message to the MSC/VLR of this new LA.
2. The MSC of the new LA registers the MU in its associated VLR and sends a registration notification message to the HLR via the STP.
3. The STP uses the MU's identification number and executes the GTT procedure to determine the MU's HLR. The registration message is then forwarded to the HLR.

4. The HLR sends a registration cancellation message to the MU's anchor LA.
5. The MU's anchor LA sends a cancellation message to the previous (old) MSC/VLR.
6. The old MSC deletes the MU's profile in its associated VLR and sends a cancellation acknowledgment message to the MU's anchor LA.
7. The anchor LA sends an acknowledgment to the HLR and deletes the MU's profile in its pointer table.
8. The HLR sends a registration confirmation message to the new MSC/VLR and provides the profile of the MU in this message. The new MSC/VLR becomes the MU's anchor LA.
9. The new MSC/VLR sends an update message to location data tables whenever necessary. (The MU provides the MSC/VLR with the addresses of those location data tables as stored in its built-in memory.)
10. After updating the MU data, the data location table sends an acknowledgement message to the new LA.

The location search procedure involves determining the current serving LA of a called MU. Figures 10, 11, 12, and 13 show four distinct, possible scenarios which must be followed by this procedure according to the proposed location management scheme.

Scenario 1: The first scenario, shown in Figure 10, addresses the case where the called MU has a record in the location data table and it is roaming in its anchor LA (i.e., the LA address stored in the location data table). The steps of Figure 10 are described as follows.

1. A call is initiated to an MU, forwarded to the MSC of the calling unit.
2. The MSC sends a location request to its associated location data table which determines the anchor LA of the called MU.
3. The request is then forwarded to the anchor LA of the called MU.
4. The called MU's MSC assigns a TLDN to the call and sends it to the calling MSC.
5. The calling MSC sets up a connection to the called MSC using this TLDN.

Scenario 2: The second scenario, shown in Figure 11, is similar to the first one. However, we assume that the called MU has a record in the location data table of the calling MU, but is not roaming in its anchor LA. In this case, a pointer should be crossed, at the destination side, to reach the current LA of the called MU. Then, the MU's current LA assigns the call a TLDN and sends it to the calling LA, which establishes a connection to the called MSC using this TLDN.

Scenario 3: The third scenario, shown in Figure 12, is the IS-41 call delivery scenario. In this case, the called MU does not have any records in the location data table, and the HLR is queried in order to determine the current LA of the called MU TLDN.

Scenario 4: The fourth scenario, shown in Figure 13, illustrates the situation where the called MU does not have any record in the location data table and is not roaming in its anchor LA. In this case, a pointer should be traversed to reach the current LA of the called MU. Then, the MU's current MSC/VLR assigns a TLDN to the call and returns it to the HLR, which forwards it to the calling MSC/VLR before the connection is established.

A New Mobile IP Network Architecture

Figure 14 illustrates a new Mobile IP network architecture proposed by Diha and Pierre (2003). The architecture introduces the following main features.

- Multiple connections of MNs and CNs to an FA or HA with different arrival rates in the network.
- The different procedures associated with an MN (registration, discovery, tunneling, and routing) represent different tasks with a specific priority.
- Multiprocessor agent (HA or FA). In this chapter, HA is emphasized. Also the HA is redundant to allow failure recovery. A main processor dispatches the different tasks arriving to an agent. A set of faster processors is defined to handle high-priority tasks.
- A processor can breakdown with a probability p and restarts with a probability $1 - p$.

Based on this architecture, a set of new algorithms was defined to manage mobility with IP in mobile networks: registration algorithm, discovery algorithm, routing and tunneling algorithm, and task-scheduling and assignment algorithm.

Figure 14. Mobile IP network architecture

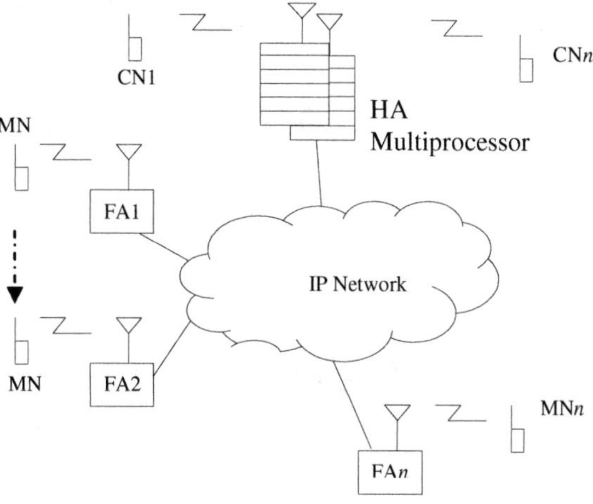

Registration Algorithm

The registration procedure is a task running on the HA with the highest priority. It can preempt any other mobility management task for a given user. For example, during a tunneling procedure, if a registration request is received for the same user, the tunneling process will be delayed until the registration is completed. The different stages of the algorithm are described as follows.

1. The MN sends a registration request to the HA (it may imply the FA).
2. The HA verifies if a task other than the registration is in process for the same user. If so, this task is preempted by the registration task.
3. The HA sends a response to the MN (it may imply the FA).
4. If the request is accepted, the procedure is completed; otherwise, the MN reattempts registration by sending a new request.

Discovery Algorithm

The discovery algorithm also introduces the notion of priority and it is based on lifetime expiration. The different steps are described as follows.

1. If the lifetime has expired and the MN notices the presence of an FA, it sends a registration request to the FA.
2. The FA verifies if higher priority tasks (e.g., registration procedure) are simultaneously executed for the same MN. If that is the case, the discovery process is delayed until the high-priority task execution is completed.
3. The FA returns a response to the MN.
4. If the registration succeeds, the MN sends new location information to its HA for location update.

Routing and Tunneling Algorithm

The new routing and tunneling algorithm also introduces the notion of priority. Thus, during a tunneling procedure, if a registration procedure is received for the same user, the location procedure will be suspended until registration is completed. The algorithm steps are the following.

1. The HA receives data for an MN.
2. The HA verifies if registration is requested for the same user. In that case, the HA suspends the tunneling process until registration is completed.
3. The HA verifies if the MN is in the local network. If so, the packets are delivered through a regular IP packets-delivery procedure; otherwise, the HA forwards the packets to the MN via its current FA using its CoA.

Task Scheduling and Assignment Algorithm

The scheduling part of the algorithm is based on the Earliest Deadline First (EDF) algorithm (Johnson & Perkins, 2001). The tasks are sorted according to deadlines and assigned to processors. If a task is critical, it is assigned to a faster processor. A task is assigned to a processor only if its current utilization rate is less than one. This ensures that a processor is not used at its full capacity while others are not used. The algorithm verifies that the targeted processor is not broken down before assigning it a task. Figure 15 shows the task scheduling and assignment algorithm.

Performance Analysis

This section analyzes empirical results associated with some mobility management models proposed in the previous section. Then, their performance is evaluated by comparing them with other schemes.

Figure 15. Task scheduling and assignment algorithm

```
Given n tasks TSK1, TSK2, ... TSKi, ... TSKn
        with the priorities p1, p2, ... pi, ... pn.
         S a set of faster processors.
         ps the threshold for a critical task
         (if pi > ps then TSKi is critical).
         u(i) the utilization rate of the task TSKi.
         U(j) a vector of utilization rate of current
         tasks of the processor Pj.

BEGIN
     Initialize i to 1.
     Initialize U(j) to 0.
     Initialize S to 1 (at least 1 faster processor).
     Sort the n tasks by descendant utilization rate
     order on the main processor P0.

     WHILE i ≤ n DO
          j = min{k|U(k) + u(i) ≤ 1}
          IF ((pi > ps) &&
          (∃ PR ∈ S | (PR is not broken down
          && ∑U(PR) < 1)) ) THEN
               Assign TSKi to processor PR
          ELSE IF Pj is not broken down THEN
               Assign TSKi to processor Pj
          i ← i + 1
     END

END
```

Built-In Memory Model

To evaluate the performance of the location update scheme in the built-in memory model, we use a timing diagram similar to the one used by Jain and Lin (1995). The steady-state case between two consecutive phone calls was considered. In this analysis, MUs are classified by their CMR. The CMR is defined as the average number of calls to an MU per time unit, divided by the average number of times the MU changes LAs per time unit (or mean arrival rate/ mean mobility rate). If we assume that the incoming calls to an MU has a mean arrival rate λ, and the time that the user resides in an LA has mean 1/u, then, the CMR may be expressed by the following:

$$\mathrm{CMR} = \frac{\lambda}{\mu}.$$

The objective is to determine the classes of MUs for which the memory-built-in model yields net reductions in signaling traffic and database loads. We define a(K) as the probability that the MU moves across K LAs between two phone calls. In order to evaluate a(K), the timing diagram shown in Figure 16 is used, where t_c denotes the interval between two consecutive phone calls to a mobile unit M. We suppose that the MU resides in a location area LA_0 when the first call arrives. After the first call, the MU visits another K LAs and remains in the location area LA_j for a period $T_j (0 \leq j \leq K)$. Let t_i $(0 \leq i \leq K)$ be the moment when the MU enters a new LA. T_i is then the interval between t_i and t_{i+1}. Let t_m denote the interval between the arrival of the first call and the time when the MU moves out of location area LA_0. Let T_i $(0 \leq i \leq K)$ be independent, identically distributed, random variables with a general distribution $F_M(T_i)$, the density function $f_M(T_i)$, and the mean $E[T_i]$. The Laplace transform of T_i is then:

$$f_M^*(s) = \int_{t=0}^{\infty} e^{-st} f_M(t)dt. \quad (1)$$

Let $d_m(t)$ be the density function of t_m. Based on the random observer property (Mitrani, 1987), we can show that:

$$d_m(t) = \frac{1}{E[T_i]} \int_{x=t}^{\infty} f_M(x)dx = \frac{1}{E[T_i]}[1 - F_M(t)]. \quad (2)$$

Furthermore, we assume that the MU's residence time in an LA is exponentially distributed with parameter m. Hence, the density function of the MU residence-time random variable T_i is $f_M(t) = \mu e^{-\mu t}$, and the expected residence time of an MU at an LA is $E[T_i] = \frac{1}{\mu}$. If the call arrivals to an MU are Poisson processes with mean arrival rate l, then, the interarrival time between two calls t_c is exponentially distributed with density function $f_c(t) = \lambda e^{-\lambda t}$ and $E[t_c] = \frac{1}{\lambda}$. Thus, the Laplace transform of the distribution of t_m is:

$$d_m^*(s) = \int_{t=0}^{\infty} e^{-st} d_m(t)dt = \int_{t=0}^{\infty} e^{-st} \mu[1 - F_M(t)]dt$$

$$= \frac{\mu}{s}[1 - f_M^*(s)]. \tag{3}$$

The probability a(*K*) that the MU moves across *K* LAs between two phone calls can be derived using Equations 1, 2, and 3:

$$\alpha(K) = \frac{\mu}{\lambda}[1 - f_M^*(\lambda)]^2 [f_M^*(\lambda)]^{k-1}. \tag{4}$$

Figure 16. Timing diagram

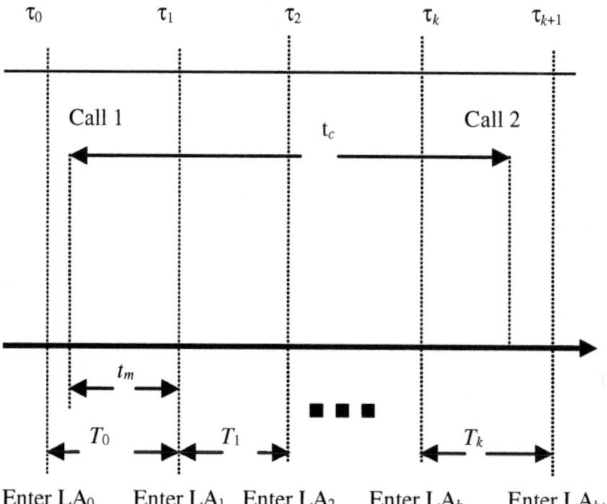

Let N denote the average number of location update operations performed among K moves. From Equation 4, N can be derived as follows:

$$N = \sum_{j=0}^{\infty} i\alpha(i) = \frac{\mu}{\lambda} = \frac{1}{CMR}. \quad (5)$$

Let M and L denote, respectively, the costs of IS-41 location update and location search procedures. Let m be the cost of the location update operation in the proposed built-in memory model. Let T denote the cost of traversing a link (pointer) between two LAs. We denote total costs between two consecutive calls for various operations used in this analysis as follows:

- U_{IS41}: Total cost of location update operations using the IS-41 scheme
- S_{IS41}: Total cost of location search operations using the IS-41 scheme
- $Total_{\text{IS41}}$: Total cost of location update and location search operations using the IS-41 model
- U_{BM}: Total cost of location update operations using the built-in memory strategy
- S_{BM}: Total cost of location search operations using the built-in memory model
- $Total_{\text{BM}}$: Total cost of location update and location search operations using the built-in memory model

The average number of location search operations executed between two consecutive phone calls is one. In the worst-case scenario, the cost of the location search procedure in the built-in memory model equals the cost of location search in the IS-41 scheme, plus the cost of traversing a pointer. Then, the total costs can be easily calculated with the following:

$$U_{\text{IS41}} = \frac{M}{CMR}, \quad (6)$$

$$Total_{\text{IS41}} = U_{\text{IS41}} + S_{\text{IS41}} = \frac{M}{CMR} + L, \quad (7)$$

$$U_{\text{BM}} = \frac{m}{CMR}, \text{ and} \quad (8)$$

$$Total_{\text{BM}} = U_{\text{BM}} + S_{\text{BM}} = \frac{m}{CMR} + L + T. \quad (9)$$

As a first approximation, we consider the values of the operation costs as follows. We observe that the location update procedure and the location search procedure in the IS-41 scheme involve the same number of messages between HLR and VLR databases. Therefore, we assume that $M = L$. Using the same reasoning, we observe that the number of messages in the built-in memory location-update procedure is about 4 times the number of messages required to cross links between two LAs served by the same LSTP, that is, we set $T = m/4$. Finally, we normalize $M = 1$. Then, from Equations 6, 7, 8, and 9, we obtain:

$$\frac{U_{BM}}{U_{IS41}} = \frac{m}{M} = m, \tag{10}$$

$$\frac{Total_{BM}}{Total_{IS41}} = \frac{4(m + CMR) + m * CMR}{4(M + CMR)}, \text{ and} \tag{11}$$

$$\frac{S_{BM}}{S_{IS41}} = \frac{L + \frac{m}{4}}{L} = 1 + \frac{m}{4}. \tag{12}$$

For simulation purposes, we assume approximate values for m. Figure 17 shows the total location update cost in the built-in memory model compared to the total location update costs in the IS-41 model when $m = 0.2$ and $m = 0.5$. We observe that for $m = 0.2$, the built-in memory model results in a cost reduction of 80% while the extra cost paid to locate an

Figure 17. Total location update cost: IS-41 versus built-in memory when (a) m = 0.2 and (b) m = 0.5

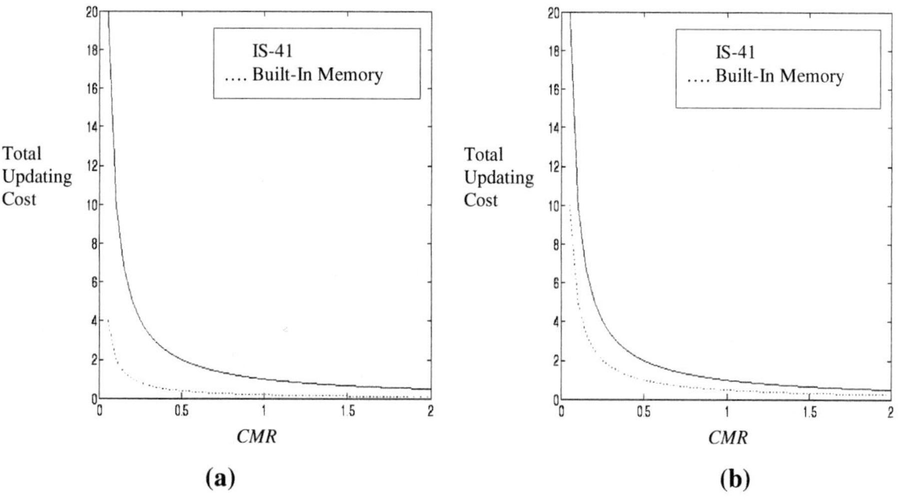

Figure 18. Reductions obtained with the built-in memory model

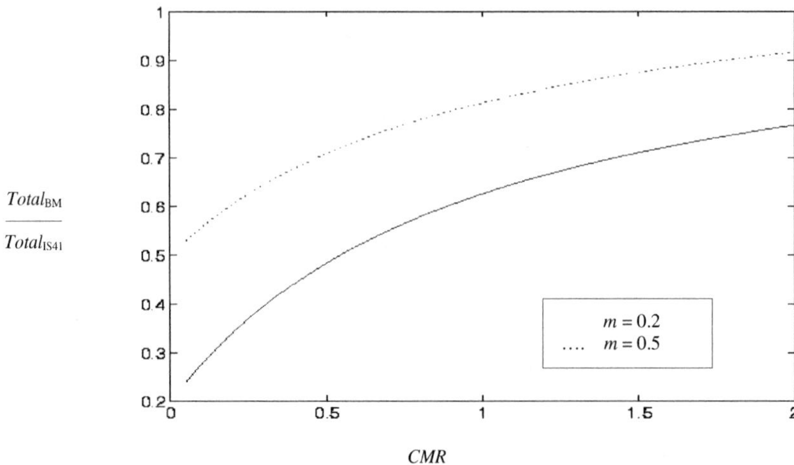

MU is, in the worst case, 5%. For $m = 0.5$, the total reduction of the location update cost is 50% and the extra cost required to locate the MU is 12.5%. The reduction level depends on the *CMR* of each MU. When *CMR* is low, significant cost savings are obtained with the location update in the built-in memory model. However, when *CMR* tends to infinity, the location update cost in both models tends to zero. This can be explained as follows. As *CMR* tends to infinity, the MU never moves out of an LA, so location updates are not performed.

Figure 18 shows that the reduction obtained in the total cost varies between 22% and 78% when $m = 0.2$, and between 8% and 52% when $m = 0.5$. To gain a better understanding of these results, we analyze the lower and upper bounds for the relative cost $Total_{BM}/Total_{IS41}$ given in Equation 11. We note that the lower and upper bounds of this performance measure occur as $CMR \to 0$ and $CMR \to \infty$, respectively. Then, we can write:

$$m \leq \frac{Total_{BM}}{Total_{IS41}} \leq 1 + \frac{m}{4}. \qquad (13)$$

Equation 13 can be explained as follows. When the *CMR* is low, the mobility rate is high compared to the call arrival rate, and the total costs of both schemes are dominated by the cost of the location update procedure. Since the built-in–memory, location-update procedure results in significant cost reduction, it outperforms the IS-41 scheme. Conversely, when the *CMR* is high, the call arrival rate is high compared to the mobility rate, and the cost of the location search procedure dominates. As the average number of incoming calls to an MU increases, the $Total_{BM}$ value approaches the $Total_{IS41}$ value.

When the $CMR \to \infty$, the MU never moves out of its LA. In this case, it is likely to update the HLR which will point to the MU's current LA. Consequently, the cost of the location search procedure will be equal in both schemes and the upper bound of $Total_{BM}/Total_{IS41}$ will be 1. Otherwise, the $Total_{BM}/Total_{IS41}$ upper bound will be $1 + m/4$ where $m/4$ represents the cost associated with moving between two LAs in the memory-built-in scheme.

The Global Location Management Scheme

This section investigates classes of users for which the global location management scheme yields a net reduction in signaling traffic and database loads. All users can be classified according to their CMR. For each target MU, the following quantities are defined.

l: Average number of calls to a target MU per time unit

m: Average number of times the user changes LA per time unit

$1/m$: Average LA residence time for a target MU

p: Probability that the called MU has a profile in the location data table

q: Probability that the new LA (VLR/MSC) is served by the same LSTP as the previous VLR (intra-LSTP movement)

r: Probability that the called MU is found in its anchor LA

The costs of the various operations used in the proposed architecture are denoted as follows.

U_{intra}: Location update operation costs when the MU's move is intra-LSTP

U_{inter}: Location update operation costs when the MU's move is inter-LSTP

U_{global}: Estimated cost of a location update operation

S_1: Cost of a location search operation using Scenario 1 (i.e., when the called MU has a record in the location data table and is found in its anchor LA)

S_2: Cost of a location search operation using Scenario 2 (when the called MU has a record in the location data table and is not found in its anchor LA)

S_3: Cost of location search operation using Scenario 3 (when the MU has no record in the location data table and is found in its anchor LA)

S_4: Cost of location search operation using Scenario 4 (when the MU has no profile in the location data table and is not found in its anchor LA)

S_{global}: Estimated cost of a location search operation

C_{global}: Total cost for location search and location update operations

The estimated cost for the location update procedure is given by:

$$U_{global} = qU_{intra} + (1 - q)U_{inter}. \quad (14)$$

The estimated cost for the location search procedure is given by:

$$S_{global} = p(rS_1 + (1 - r)S_2) + (1 - p)(rS_3 + (1 - r)S_4). \quad (15)$$

The total cost per time unit for location search and location update is given by:

$$C_{global} = mU_{global} + lS_{global}. \quad (16)$$

In order to compute the cost of the location update procedure based on the global scheme using the reference network architecture (Figure 7) and the two procedures presented in Figures 8 and 9, we define the following costs for crossing various network elements.

A_l: Cost of transmitting a message on A-link between Service Switching Post (SSP) and LSTP

D: Cost of transmitting a message on D-link between LSTP and Regional Signal Transfer Point (RSTP)

A_r: Cost of transmitting a message on A-link between RSTP and Service Control Point (SCP)

L: Cost of processing and routing a message by LSTP

R: Cost of processing and routing a message by RSTP

C_H: Cost of a database update or query at the HLR

C_v: Cost of a database update or query at the VLR

Based on the location update procedure shown in Figure 8, the location update cost for an intra-LSTP move is given by:

$$U_{intra} = 8A_l + 4L + 2C_v. \quad (17)$$

For an inter-LSTP move, the cost of a location update operation equals the cost of the location update according to the IS-41 standard, plus the cost of updating the location data tables that store the location data of the moving MU, and the cost of updating the MU's previous anchor LA. According to Figure 9, this cost is given by:

$$U_{\text{inter}} = 4(A_l + L + A_r + D + R) + 4A_l + 2L + 2C_V + C_H + \sum_{k\in E} U_k , \quad (18)$$

where E is the set of location data tables to be updated after an inter-LSTP move and U_k is the cost of updating a location data table:

$$U_k = 2(A_l + L + R + 2D).$$

The estimated cost of the location update procedure can be derived using Equations 14, 17, and 18 as:

$$U_{\text{global}} = q(8A_l + 4L + 2C_v) + (1 - q)(4(A_l + L + A_r + D + R) + 4A_l + 2L + 2C_V + C_H + \sum_{k\in E} U_k)$$

$$= 8qA_l + 4qL + 2qC_v + 8A_l + 6L + 4A_r + 4D + 4R + 2C_V + C_H + \sum_{k\in E} U_k$$

$$- 8qA_l - 6qL - 4qA_r - 4qD - 4qR - 2qC_V - qC_H - q \sum_{k\in E} U_k ,$$

which can be simplified as:

$$U_{\text{global}} = 8A_l + 4L + 2C_V + (1 - q)(2L + 4A_r + 4D + 4R + C_H + \sum_{k\in E} U_k). \quad (19)$$

Let t be the probability that the LAs of the called MU and calling unit are connected to the same LSTP. Since updating the location data table and the pointer table involves a simple access to a local memory, we assume that there are no additional costs to update or query this kind of table.

Scenarios 1 and 2 are applied when the called MU has a record in the location data table of the calling MU. The cost of Scenario 1, which is used when the called MU is found in its anchor LA, is given by:

$$\begin{aligned} S_1 &= t(4A_l + 2L + C_V) + (1 - t)(4A_l + 4L + 2R + 4D + C_V) \\ &= 4tA_l + 2tL + tC_V + 4A_l + 4L + 2R + 4D + C_V - 4tA_l - 4tL - 2tR - 4tD - tC_V \\ &= 4A_l + 4L + 2R + 4D + C_V - 2tL - 2tR - 4tD. \quad (20) \end{aligned}$$

When the called MU is not found in its anchor LA, Scenario 2 is applied. The cost of this scenario equals the cost of Scenario 1, plus the cost of passing over a pointer from the anchor LA of the MU to its current LA. This cost is given by:

$$S_2 = S_1 + 2A_l + L$$
$$= 6A_l + 5L + 2R + 4D + C_V - 2tL - 2tR - 4tD. \quad (21)$$

When the called MU does not have its location information stored in the location data table, Scenarios 3 and 4 are applied. The cost of Scenario 3, which is used when the called MU is found in its anchor LA, equals the cost of the location search in IS-41, given as:

$$S_3 = 4(A_l + L + A_r + D + R) + C_V + C_H. \quad (22)$$

When the called MU is not found in its anchor LA, Scenario 4 is applied. The cost of this scenario equals the cost of Scenario 3, plus the cost of passing over a pointer from the anchor LA of the MU to its current LA. As in Equation 21, this cost is:

$$S_4 = S_3 + 2A_l + L. \quad (23)$$

The total cost per time unit for locating an MU can be expressed as follows:

$$S_{\text{global}} = p(rS_1 + (1 - r)S_2) + (1 - p)(rS_3 + (1 - r)S_4). \quad (24)$$

Using Equations 21, 26, 27, 28, and 29, this cost can be rewritten as follows:

$$S_{\text{global}} = 6A_l + 5L + 4A_r + 4D + 4R + C_V + C_H - p[2t(L + R + 2D) + 4A_r + 2R + C_H]$$
$$- r(2A_l + L). \quad (25)$$

The total cost per time unit for location update and location search using the global architecture is obtained using Equations 16, 19, and 25.

$$C_{\text{global}} = m[8A_l + 4L + 2C_V + (1 - q)(2L + 4A_r + 4D + 4R + C_H + \sum_{k \in E} U_k)] +$$
$$1\{6A_l + 5L + 4A_r + 4D + 4R + C_V + C_H - p[2t(L + R + 2D) + 4A_r + 2R + C_H]$$
$$- r(2A_l + L)\} \quad (26)$$

For comparison purposes, we need to evaluate the costs of the original IS-41 scheme. We denote costs for various operations used in the IS-41 scheme as follows:

U_{IS41} : Cost for a location update operation

S_{IS41} : Cost for a location search operation

C_{IS41} : Total cost per time unit for location search and location update operations

The total cost per time unit for location update and location search under the IS-41 scheme is:

$$C_{is41} = mU_{is41} + 1S_{is41}, \tag{27}$$

where

$$U_{is41} = 4(A_l + L + A_r + D + R) + 2C_V + C_H, \text{ and}$$
$$S_{is41} = 4(A_l + L + A_r + D + R) + C_V + C_H.$$

Defining the relative cost of the global location management scheme as the ratio of the total cost per time unit for the global scheme to that of the IS-41 scheme, C_{global}/C_{is41}, we get as a function of the user's CMR:

$$\frac{C_{global}}{C_{is41}} = \frac{U_{global} + (\lambda/\mu)S_{global}}{U_{is41} + (\lambda/\mu)S_{is41}}. \tag{28}$$

Relation 28 uses the four probability terms: p, q, r, and t. Both p and r, which were defined above, can be used to classify users.

In order to quantify q and t, we assume that an LSTP consists of $x * x$ LAs arranged in a square, and each LA is itself a square. MUs are assumed to be uniformly distributed throughout the LSTP area and each MU exhibits the same arrival call rate at every VLR/MSC. Furthermore, each time an MU leaves an LA, one of the four sides is crossed with equal probability. Then, the probability that the MU's move is inter-LSTP is equal to the probability that the MU is in a border LA, multiplied by the probability that the MU's next move is to an LA belonging to a different LSTP. Define:

P_1 = Prob[MU lies in a border LA of the LSTP] = $4(x - 1)/(x * x)$

P_2 = Prob[MU's next move is to an LA belonging to a different LSTP] = 1/4

Þ Prob[MU's move is inter-LSTP] = $P_1 * P_2 = (x - 1)/(x * x)$.

Hence,

q = Prob[MU's move is intra-LSTP] = 1 - (x - 1)/(x * x).

Also, let us assume that all of the network SSPs are uniformly distributed among n LSTPs, and each SSP corresponds to a single LA. For example, take the case of the public, switched telephone network that includes 160 local access transport areas (LATAs) across the seven Regional Bell Operating Company (RBOC) regions (Bellcore, 1992), assuming one LSTP per LATA and the average number of LSTPs is 160/7, or 23 per region. Given that there are 1,250 SSPs per region, the number of SSPs per LSTP is 1250/23. Hence,

$$x = \sqrt{\frac{1250}{23}} \approx 7.4 \qquad \Rightarrow q \approx 0.88.$$

Under the conditions stated above, the probability t that both calling and called users are found in the same LSTP equals $1/n$ Þ t =1/23 = 0.043. Further quantitative results associated to the performance of the four scenarios can be found in Safa et al. (2002).

Conclusions and Future Work

In this chapter, we analyzed and proposed some mobility management models and schemes by taking into account their capability to reduce search and location update costs in wireless networks. The first model proposed is called the built-in memory model; it is based on the architecture of the IS-41 network and aims to reduce the HLR access overhead. The performance of this model was investigated by comparing it with the IS-41 scheme for different CMRs. Experimental results indicate that the proposed model is potentially beneficial for large classes of users and can yield substantial reductions in total user-location management costs, particularly for users who have a low CMR.

The built-in memory model appears promising when the higher elements of the network constitute the network performance bottleneck. The results show that the cost reduction obtained on the location update is very significant, while the extra costs paid to locate an MU simply amount to the costs of crossing a single pointer between two LAs. The built-in memory model was also compared with the forwarding pointers' scheme. The results show that this model consistently outperforms the forwarding pointers' strategy.

A second location management model to manage mobility in wireless communications systems was also proposed. According to this scheme, two tables are added to the IS-41 network architecture. A pointer table is added to each LA, and it tracks the MUs that moved out of this LA by setting a single pointer from this LA to the current LA. The location data table is located on an LSTP node and contains the data location of the MUs that are frequently called from the LAs connected to this LSTP. The results have shown that significant cost savings can be obtained compared with the IS-41 standard location management scheme, depending on the value of the MUs' CMR.

Finally, we presented the Mobile IP network architecture and mobility management algorithms in a real-time context. Compared to some conventional architecture and algorithms, the implementation of the architecture and algorithms produced better results for the location update and tunneling average times, as well as the CMR.

Many investigations are ongoing in real-time mobility management for Mobile IP networks. Such investigations address the implementation of real-time algorithms in real networks and suggest new algorithms and architectures. Also, since the current protocols are designed for micromobility, the global roaming area remains a highly challenging research domain.

References

Akyildiz, I. F., & Wang, W. (2002). A dynamic location management scheme for next generation multi-tier PCS system. *IEEE Transactions on Wireless Communications, 1*(1), 178-190.

Bellcore (1992). *Switching system requirements for interexchange carrier interconnection using integrated services digital network user part (ISDNUP)*(Tech. Ref. No.TR-NWT-000394).

Cayirci, E., & Akyildiz, I. F. (2002). User mobility pattern scheme for location update and paging in wireless systems. *IEEE Transactions on Mobile Computing, 1*(3), 236-247.

Cayirci, E., & Akyildiz, I. F. (2003). Optimal location area design to minimize registration signaling traffic in wireless systems. *IEEE Transactions on Mobile Computing, 2*(1), 76-85.

Chiussi, F. M., Khotimsky, D. A., & Krishnan, S. (2002). Mobility management in third-generation all-IP networks. *IEEE Communications Magazine, 40*(9), 124-135.

Diha, M., & Pierre, S. (2003). Architecture and algorithms for real-time mobility management in mobile IP networks. *Proceedings of the Second International Conference on Ad-Hoc, Mobile, and Wireless Networks (ADHOC-NOW)*, (pp. 49-59).

Escalle, P. G., Giner, V. C., & Oltra, J. M. (2002). Reducing location update and paging costs in a PCS network. *IEEE Wireless Communications, 1*(1), 200-209.

Fang, Y. (2003). Movement-based mobility management and trade off analysis for wireless mobile networks. *IEEE Transactions on Computers, 52*(6), 791-803.

Gallagher, M. D., & Randall R. A. (1997). *Mobile telecommunications networking with IS-41.* New York: McGraw-Hill.

Jain, R., & Lin, Y. B. (1995). An auxiliary user location strategy employing forwarding pointers to reduce network impacts of PCS. *Wireless Networks, 1*(2), 197-210.

Johnson, D. B., & Perkins, C. (2001). *Mobility support in IPv6* [Internet draft]. Internet Engineering Task Force. Retrieved from http://users.piuha.net/jarkko/publications/mipv6/drafts/mobilev6.html

Kwon, T. T., Gerla, M., & Das, S. (2002). Mobility management for VoIP service: Mobile IP vs. SIP. *IEEE Wireless Communications, 9*(5), 66-75.

Lee, Y. J., & Akyildiz, I. F. (2003). A new scheme for reducing link and signaling costs in mobile IP. *IEEE Transactions on Computers, 52*(6), 706-712.

Lin, Y. B., & Chen, Y. K. (2003). Reducing authentication signaling traffic in third-generation mobile network. *IEEE Transactions on Wireless Communications, 2*(3), 493-501.

Lin, Y. B., Cheng, H. Y., Cheng, Y.H., & Agrawal, P. (2002). Implementing automatic location update for follow-me database using VoIP and Bluetooth technologies. *IEEE Transactions on Computers, 51*(10), 1154-1168.

Lin, Y. B., Lee, P. C., & Chlamtac, I. (2002). Dynamic periodic location area update in mobile networks. *IEEE Transactions on Vehicular Technology, 51*(6), 1494-1501.

Mao, Z. (2002). An intra-LA location update strategy for reducing paging cost. *IEEE Communications Letters, 6*(8), 334-336.

Misra, A., Das, S., Dutta, A., McAuley, A., & Das, S. K. (2002). IDMP-based fast handoffs and paging in IP-based 4G mobile networks. *IEEE Communications Magazine, 40*(3), 138-145.

Mitrani, I. (1987). Modelling of computer and communication *system*. New York: Cambridge University Press.

Mouly, M., & Pautet, M. B. (1992). *The GSM system for mobile communications* (pp. 434-465). Telecom Pub: Alexandria, VA.

Safa, H., Pierre, S., & Conan, J. (2001). An efficient location management scheme for PCS networks. *Computer Communications, 24*(14), 1355-1369.

Safa, H., Pierre, S., & Conan, J. (2002). A built-in memory model for reducing location update costs in mobile wireless networks. *Computer Communications, 25*(14), 1343-1353.

Suh, B., Choi, J., & Kim, J. (2000). Design and performance analysis of hierarchical location management strategies for wireless mobile communication systems. *Computer Communcations Journal, 23*, 550-560.

TIA/EIA (1996). *Interim Standard IS-41-C*. Cellular Radio-Telecommunications Intersystem Operations. Retrieved 2004, from http://www.cdg.org/technology/cdma_technology/a_ross/Standards.asp

Wang, W., & Akyildiz, I. F. (2001). A new signaling protocol for intersystem roaming in next-generation wireless systems. *IEEE Journal on Selected Areas in Communications, 19*(10), 2040-2052.

Wong, V. W. S., & Leung, V. C. M. (2001). An adaptive distance-based location update algorithm for next-generation PCS networks. *IEEE Journal on Selected Areas in Communications, 19*(10), 1942-1952.

Wu, C. H., Lin, H. P., & Lan, L. S. (2002). A new analytic framework for dynamic mobility management of PCS networks. *IEEE Transactions on Mobile Computing, 1*(3), 208-220.

Xie, J., & Akyildiz, I. F. (2002). A novel distributed dynamic location management scheme for minimizing signaling costs in mobile IP. *IEEE Transactions on Mobile Computing, 1*(3), 163-175.

Zahariadis, T. B., Vaxevanakis, K. G., Tsantilas, C. P., Zervos, N. A., & Nikolaou, N. A. (2002). *Global roaming in next-generation networks. IEEE Communications Magazine, 40*(2), 145-151.

Section III

Network Support

Chapter VIII

Service Discovery in Wireless and Mobile Networks

Hitha Alex, University of Texas at Arlington, USA

Mohan Kumar, University of Texas at Arlington, USA

Behrooz A. Shirazi, University of Texas at Arlington, USA

Abstract

Service discovery is an important component of wireless and mobile network systems. An efficient service discovery mechanism would ensure high availability of services to users and applications, and high utilization of services. In this chapter, we discuss various issues and challenges facing the design and selection of a proper service discovery mechanism. This chapter also investigates service discovery mechanisms such as SLP, Jini, Salutation, and others, and assesses their suitability for applications in wireless and mobile environments.

Introduction

Mobile and wireless computing has evolved beyond the ability to wirelessly connect to read and browse the Web anywhere, anytime. Current trends involve exploiting local services, peers, and services in local and foreign networks with unknown infrastructures. For example, a mobile device should be able to use a printer in a new network or a personal digital assistant (PDA) should be able to use a faster Web cache service available in a new network. With the advent of location-based services and peer-to-peer computing, service discovery is now a critical middleware for the anywhere, anytime computing model adopted by mobile and pervasive computing networks (Helal, 2002). As the computing trends in wireless and mobile networks move more toward service–based, peer-to-peer computing, acquiring knowledge about available services poses great challenges due to the mobility of devices and the dynamic nature of the environment. The available services at any time and at any space may differ from those available at a different time and space. The anytime, anywhere computing model may also mean that mobile and wireless devices are not only consumers of services, but also providers of services. The number of services available in the network will continue to grow as more and more applications are developed day by day. An effective service discovery is therefore necessary to provide the knowledge of available services to the device and also to the networking infrastructure. Service advertisements and discovery can enable the pervasive space to dynamically change and evolve without major system reengineering. A service discovery scheme in wireless and mobile networks should (a) allow automatic discovery of services, (b) make new services in a network discoverable, (c) allow very little or no manual configuration of the device to make use of new services in a network or services in a new network, and (d) allow uninterrupted service provisioning.

In this chapter we will examine the important challenges for an effective service discovery model in wireless and mobile computing environments. In addition, we will investigate some of the existing service discovery models and their suitability to wireless and mobile networks. We will also investigate the future trends in the area of service discovery mechanisms.

Background

Services

A *service* is software that can perform specific function(s) on behalf of users and applications over a network. A *device* is any equipment that is used for computing, which includes conventional computers, small handhelds, specialty devices such as digital cameras, and printers. Most devices are represented on the network by one or more services; therefore, the term *service* is used in this chapter to represent both services and devices. *Web services* are a subset of services that can be accessed at an Internet universal resource identifier (URI).

Accomplishing an application task can be thought of as making use of a series of services. Mobile and wireless network devices should act as portals to receive computing services for the accomplishment of tasks, and also be able to provide services to other devices. Users of various mobile devices should be able to perform their day-to-day tasks without having to discover services. Efficient service discovery is an important component of any mobile and wireless networking infrastructure.

Service Discovery

Service discovery can be defined as the process of discovering locations of software entities or agents that can provide access to network resources such as devices, data, and services (Gibbins & Hall, 2001). This process should facilitate:

1. services to announce their presence to the network,
2. automatic discovery of local and remote services irrespective of the differences in the type of network and technology used,
3. automatic adaptation to mobile and sporadic availability,
4. services to describe their capabilities as well as query and understand the capabilities of other services, and
5. self-configuration without administrative intervention.

Therefore the primary goal of a good service discovery mechanism should be to make devices and networks smart so as to be aware of the available services.

Traditional computing environments are primarily serviced by static information services such as domain name systems (DNS; Mockapetris & Dunlap, 1988) and dynamic host configuration protocol (DHCP; Droms, 1997), which rely on static files or databases that are configured by system administrators. These services do not guarantee the availability of the objects registered with them. The semantics for searching are often very limited. There are no event generations when services register and deregister. All these factors listed above along with complexity in installation, configuration, and management of peripheral devices and the services make static information services unsuitable for a dynamic and mobile environment. Some directory services such as Lightweight Directory Access Protocol (LDAP) (Wahl, Howes, & Kille, 1997) and (Common Object Request Broker) CORBA ("CORBAservices," n.d.) can be used for service announcements and requests, but do not provide for spontaneous discovery. Discovery services can be built on top of these directory services.

Recent pervasive-computing project initiatives have identified service discovery as one of the major design components. For instance, the Intentional Naming System (INS; Adjie-Winoto, Schwartz, Balakrishnan, & Lilley, 1999) is an important component of the Oxygen project at Massachusetts Institute of Technology (MIT; Dertouzos, 1999). The Ninja project at UC, Berkeley (Gribble et al., 2001), employs secure wide-area service discovery (SDS; Hodes, Czerwinski, Zhao, Jospeh, & Katz, 2002) as the service discovery mechanism.

Figure 1. Service discovery with and without service utilization

The three phases of a service discovery process are

1. *Advertisements of services and their properties.* This is the process by which a service is made discoverable. New services that come into a new network need to make their services available to consumers in the network, requiring some form of advertising. Other circumstances under which advertisements may be required include state changes, service lifetime expiry, and so forth.
2. *Locating a service.* This is the process by which the location of a service is discovered, typically involving querying the network through a broadcast or a directory query that contains information about the service.
3. *Utilizing a service (optional).* This is the actual use of the services. Service discovery mechanisms may or may not involve actual utilization of the service as shown in Figure 1.

Challenges of Service Discovery in Wireless and Mobile Networks

Directory- or Peer-to-Peer-Based Mechanisms

Directory-type service discovery mechanisms make use of centralized or decentralized directories. A directory-type discovery mechanism allows automated discovery of services as mobile clients and services use this repository to search for or register services proactively or reactively. However, in an ad hoc or peer-to-peer networking environment, directory-centric service discovery may not be a cost-effective or even feasible solution. In such networks, services should negotiate one-on-one with each other, advertising their own capabilities and finding services that meet the needs they require. In a centralized directory model, the directory can become a bottleneck as the central point of failure. In a decentralized directory model, there will be more bandwidth consumption and consistency issues between different directories in the network. Peer-to-peer architecture suits well for a very small and dynamic environment when compared to a directory model, but suffers from immense bandwidth usage and very low scalability.

Some services will exist in both directory-centric and peer-to-peer environments. Therefore, ideally, a service discovery mechanism should provide for functioning equally well in both directory-based and peer-to-peer-based scenarios.

Service Description

A service description language is an important part of a service discovery mechanism as there is more than one participating entity in the process. Service description languages:

1. provide a communication language for participating entities of the service discovery process,
2. provide a common syntax and semantics for specifying services and service properties,
3. allow expressive service descriptions and query patterns, and
4. allow for filtering of service requests.

The service description language must provide services and clients with common semantics to establish communication and negotiate appropriate interfaces and parameters. In general, the services should advertise themselves, and clients need to query for these services. Due to the presence of hundreds of thousands of potential services in future networking environments, locating an appropriate service for a given task by a client can be a great challenge. The same service may be defined in different ways by different applications or clients. So in effect, it is the responsibility of the service discovery mechanism to map the appropriate definition to the location. The syntax of the service description language determines the *expressiveness* of this language, which in turn establishes the flexibility and semantic richness of the queries for a wide variety of devices and services. On the other hand, the complexity of the service description language can influence the precision of service advertisements, method of service information storage and searching of service information, and matching of a semantic-rich query to an item in the database. Web services description language (WSDL) is an example of an extensible markup-based language (XML) used to describe the network services businesses offer. In WSDL, language services are described as a set of endpoints operating on messages containing either document-oriented or procedure-oriented information. Also, most web services use XML to interact with each other.

Operating-System, Communication-Media, and Language Independence

In wireless and mobile computing environments, a wide variety of computing devices different from one another in their type, size, capacity, operating system, and communication protocols should be able to interoperate and communicate with each other seamlessly. This diversity should be transparent to the applications, entailing service discovery mechanisms to operate in heterogeneous environments well.

Ideally a service discovery mechanism should be able to operate independent of the operating system (especially small-device operating systems like Windows CE and Palm OS). It is also important that the service discovery be independent of the communication protocols (such as Internet Protocol [IP], Asynchronous Transfer Mode[ATM], infrared data association [IrDA], and Bluetooth) and communication media (such as wired, wireless, and Ethernet). The programming language used can also be a barrier, especially if the service discovery mechanism is dependent on the language-specific features such as remote method invocation (RMI) in Java. It will be ideal if the protocol can be implemented using at least the most widely used programming languages.

Scalability

As the range of the network and the number of clients and services in the network increase, the service discovery mechanism should scale well. In other words, the query time and resource consumption for the discovery process itself should grow more slowly than the range of the network and number of clients and services. Some factors that could affect the scalability of a service discovery mechanism are:

1. the difficulty in internetwork communication (due to firewalls, etc.),
2. the method of network layer transport (such as the use of multicast and broadcast),
3. the type of service information storage (centralized or decentralized),
4. the type of discovery mechanism (directory or peer-to-peer),
5. the method of query routing, and
6. the complexity of mapping an expressive query to an appropriate service (cost of search, update, and deletion of a directory).

In Gibbins and Hall (2001), the authors have examined the complexity and scalability characteristics of a number of commonly encountered architectures for service discovery. The scalability of a service discovery scheme determines its range or coverage in terms of network load or size. Most service discovery schemes that already exist or are under research are for small- to medium-sized networks and can be referred to as local-area service discovery. Discovery schemes that scale well to large-size networks such as the Internet can be referred to as wide-area service discovery. In SDS (Hodes et al., 2002), some attempts have been made to attain wide-area scalability through adaptive server-hierarchy management and lossy aggregation of service descriptions as they travel up the hierarchy. Little research is currently underway dealing with this challenge.

Security

Security is an important challenge in developing discovery protocols. Participating entities should be protected from malicious clients and services. Also, proper authentication and access control techniques are required to prevent unauthorized discovery and unrestricted access to resources. For example, one's laptop should not be allowed to print to anyone else's printer without proper permissions. Authentication requirements and the goal of automatic discovery by service discovery mechanisms are mutually conflicting issues. The strong point of SDS (Hodes et al., 2002) is the implementation of several security protocols and services. These protocols and services include public key authentication, one-way encrypted service announcements, secure RMI, certificates, and certificate authority structures.

Suitability for Dynamic Environments

Service discovery techniques should allow for highly dynamic updates that occur due to the mobility and failure of nodes and services in a network. This means, services appearing or disappearing should lead to immediate updates. *Eventing* is the asynchronous notification of significant conditions that can improve the responsiveness of the discovery mechanism to the dynamic nature of the environments. Examples of significant conditions include a needed service becoming available, a service leaving the network, or an important change in the state of a service (such as the printer running out of paper). A side effect of being responsive in a dynamic environment is the need for efficient garbage collection that expunges outdated information. Garbage collection prevents performance degradation due to clients trying to reach nonexistent services and services continuing to serve nonexistent clients. Leases are a popular garbage collection mechanism (e.g., JINI) where a service should periodically request a renewal to avoid expiration of a lease.

Functionality

The service discovery mechanism should respond appropriately to requests for different types of services. For example, in a printing service, clients may be occasionally interested in a very specific type of print quality (e.g., a high-resolution color laser printing) while typically basic services are requested. A client needing basic printing should be able to discover it without having to specify many details, and a client needing sophisticated services should be allowed to make more specific requests. Service discovery mechanisms should therefore be able to provide this functionality to its client applications. Service subtyping is one solution, where service descriptions and service information storage can accommodate service queries based on subtypes and attributes. Service subtyping provides other functionalities like service and attribute browsing.

Context awareness is another very important feature that a service discovery mechanism should have. It is often necessary to locate physically nearby resources. Other contextual information may include distance to the service, time, service load, and quality of service instances.

Cost

The costs of any service discovery technique can affect its widespread acceptance. The cost of a service discovery scheme can be either due to system or implementation costs. System costs include the burden on the available resources such as bandwidth, CPU, and power. Implementation expenses include direct and indirect costs. Direct costs include fees to procure and maintain the development tools and testing utilities. Indirect costs include implementation prerequisites, complexity, and code size.

Prerequisites

Prerequisites for the successful operation of a service discovery scheme not only affect the cost, but also restrict the breadth of implementation. Prerequisites may include use of static service information from DNSs (Mockapetris & Dunlap, 1988) and DHCP (Droms, 1997). Other examples of prerequisites are dependence on multicasting, and language-specific requirements such as JVM requirement for execution and RMI requirement for communication.

Maturity and Protocol Bridging

Stability of architecture is important in wireless and mobile computing environments. Nothing proves the validity and stability of architecture like an actual implementation. The effort that goes into actual implementation requires solving many hidden difficulties that were overlooked at design time.

In order to facilitate service discovery in wireless and mobile computing environments across various domains, heterogeneous devices, and communication protocols, either a single service discovery should dominate, or most commonly used technologies must be made interoperable. *Bridging* is one of the ways to achieve interoperability among different technologies. The inherent design characteristics will determine the ease with which a particular technology can be bridged with others in the market.

Ongoing Work in Service Discovery

In the last few years, considerable research has been done in the area of service discovery. Since these works were done by different organizations and institutions, their target environments, and therefore the design goals and features, differ. The salient features of some of these prominent service discovery research and standards will be discussed in this section.

Service Location Protocol

Service location protocol (SLP) is a service-discovery technology standard by the Internet Engineering Task Force (IETF; Kempf & Pierre, 1999). The entire protocol is described in "RFC 2608" (Guttman, Perkins, Veizades, & Day, 1999) and additional information about SLP can be found at http://www.srvloc.org. SLP is transparent to programming languages, although Application Program Interface (APIs) are currently defined for C and Java only. SLP is aimed at enabling applications to discover the existence, location, and configuration of needed services, and for services to advertise their availability in IP-based networks (Guttman, 1999). The architecture of SLP is given in Figure 2.

SLP consists of three main components.

1. *User agents (UAs).* These are software entities and agents that represent applications or clients that perform different user tasks. UAs perform the service discovery on behalf of clients or applications.
2. *Service agents (SAs).* These are software entities and agents that hold abstractions of attributes and the locations of services they represent, and also advertise on behalf of services.

Figure 2. Architecture of SLP

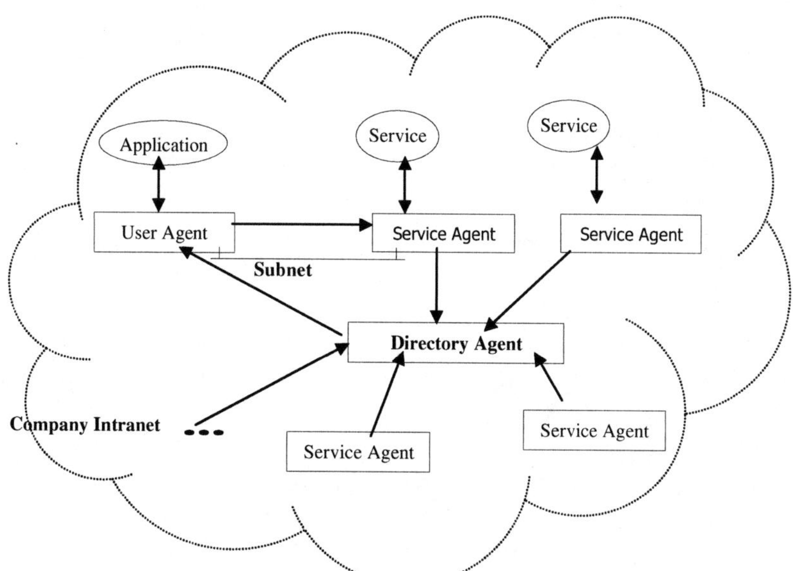

3. *Directory agents (DAs)*. These software entities and agents are responsible for the aggregation of service information received from SAs into repositories, and they respond to service requests from UAs.

SLP can function with or without DAs. The architecture where DAs are present follows a directory model for the service discovery. On the other hand, if no DAs are present, then SLP follows a peer-to-peer model for service discovery. In peer-to-peer-model SLP, UAs and SAs find each other through administratively scoped multicast. In directory-model SLP operation, clients and services need to first discover the DA(s) in the network. There are three different methods for DA discovery. In static discovery, SLP agents obtain the address of the DA(s) through DHCP (Droms, 1997). The role of a DHCP server is to distribute the address of DAs to hosts that request them. In active and passive discovery, multicast advertisements are used to discover the DA(s). In active discovery, UAs and SAs initiate the multicast request and DAs respond, whereas in passive discovery, DAs initiate the multicast and UAs and SAs get DA information.

SLP is a service discovery mechanism that is concerned only with the discovery, and not the utilization, of the discovered services. SLP provides the necessary procedures just enough to provide the clients with the information that they require to get in touch with the service they are looking for. Actual communication between clients and services that facilitate the service utilization is beyond the scope of SLP. There is no query forwarding by the DAs.

Service types are the fundamental naming conventions for services in SLP. Clients make their service queries through UAs by specifying the needed service types and, possibly, attributes. The specification of service type and attributes are provided using a "service URL" scheme discussed in "RFC 2609" (Guttman, Perkins, & Kempf, 1999). The service URL consists of all the necessary information such as type, location, and a service template. Service templates are documents registered with the Internet Assigned Numbers Authority (IANA) and which define the attributes of services, their default values, and their interpretation for a particular service type. SAs and UAs use the definitions in service templates to advertise and request a service type, respectively.

Other features of SLP include capability of service-type and attribute browsing, and the ability to deploy multiple DAs to enhance performance and robustness. In the case of multiple DAs, SAs should register with all the available DAs, and UAs need to query any of the DAs. In Zhao, Schulzrinne, and Guttman (2000), the authors propose mSLP, a fully meshed, peering DA architecture for SLP. In this architecture, peer DAs exchange the service registration information and keep the same consistent data, thus improving the reliability and scalability of SLP, Version 2.

SLP can operate in networks ranging from single LANs to networks under common administrations, also known as enterprise networks. These networks can have potentially tens of thousands of networked devices. However, due to the use of multicast and DHCP, which are not scalable to Internet, SLP's scalability is limited.

SLP has been implemented in many products by several leading industries like Sun Microsystems, Novell, IBM, Apple, Axis Communications, Lexmark, Madison River Technologies, and Hewlett-Packard (Guttman, 1999). The design of SLP is extensible and flexible, able to be bridged with other service discovery protocols and, although not implemented yet, research is ongoing in this direction.

Jini

Jini is a service discovery technology based on Java, developed by Sun Microsystems (Edwards, 2000; Newmarch, 2000; Oaks & Wong, 2000). Because of the platform-independent nature of Java, Jini can rely on mobile code to control services. Jini is a service discovery mechanism that is concerned with the discovery, as well as the utilization, of the services in primarily IP-based networks. Jini's architecture principle is similar to that of SLP with DAs. Jini operates only as a directory-style service discovery mechanism. Jini architecture is shown in Figure 3.

Jini consists of three main components.

1. *Lookup server (LUS).* This is a database for all services in the network. The Jini network should have at least one LUS. To join the Jini service discovery mechanism, every device and application in the network should place itself into a lookup table on the LUS. Besides pointers to services, lookup tables can also store Java-based program code for these services. This means that services may upload device drivers, interfaces, and other programs that help the clients access the service.
2. *Clients.* These are devices and applications that make use of Jini for service discovery. Clients can dynamically discover lookup services, search for interesting devices, and download these proxy objects.
3. *Services.* These are devices and applications that participate in the Jini service discovery mechanism. Upon initialization, Jini services download a proxy object to

Figure 3. Architecture of Jini

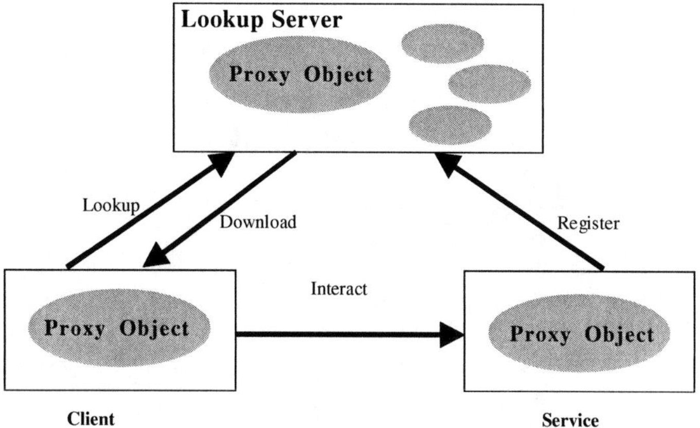

one or more LUSs to announce their availability to Jini entities. Methods in the proxy object allow clients to use the service without knowing anything about services a priori.

Jini relies heavily on the platform-independent nature of Java and the distributed computing support inherent in Java. Each Jini device is assumed to have a Java virtual machine (JVM) running on it. Jini relies on Java's object serialization, code downloading facilities, and RMI. Specification for Java RMI can be found at http://www.javasoft.com/rmi.

Since LUSs are required in the network for services to register with and clients to query, finding them in the network is critical for both clients and services. An LUS may be discovered in three different ways according to Jini specification. In a multicast request protocol, clients and services use multicast to discover the LUS in the network. In the second method, multicast announcement protocol, LUSs advertise their presence in the network through multicast to the clients and services. Lastly, in unicast discovery protocol, the location of an LUS is already known to clients and services through some other source such as a human operator.

When services register with one or more LUSs, the proxy object placed on the LUS will have a unique service ID, a set of already defined Java language types which is usually an interface and some attributes. Clients can query an LUS using the ID, Java language types, or the attributes. If there are multiple LUSs in the network, each LUS may be associated with a particular group and, hence, only the services associated with the group will be registered with that LUS. Then it is the responsibility of the client to query the appropriate LUS.

Jini, like SLP, can operate in networks ranging from single LANs to enterprise networks. These networks can have potentially tens of thousands of networked devices. However, due to the use of multicast, which is not scalable to Internet-sized networks, Jini is not scalable to the Internet. Although Jini implementations are not as widespread as SLP implementations, small industries are beginning to introduce Jini service discovery in various products. Links to some licensees of Jini technology can be found at http://www.sun.com/jini. Some research endeavors are underway to bridge Jini with other service discovery technologies. In Allard, Chinta, Gundala, & Richard (2003), the authors have designed and implemented a universal plug and play (UPnP) and Jini bridging framework. Jini-SLP-bridging efforts are described in Guttman and Kempf (1999).

Universal Plug and Play

UPnP is a Microsoft-developed service discovery technology aimed at enabling the advertisement, discovery, and control (use) of networked devices and services (Miller, Nixon, Tai, & Wood, 2001). UPnP can function both as a peer-to-peer-style service discovery mechanism and as a directory-mode mechanism in IP-based networks.

For service discovery, UPnP uses the Simple Service Discovery Protocol (SSDP) in IP-based networks. UPnP uses standard protocols such as Transport Control Protocol

Figure 4. Architecture of UPnP

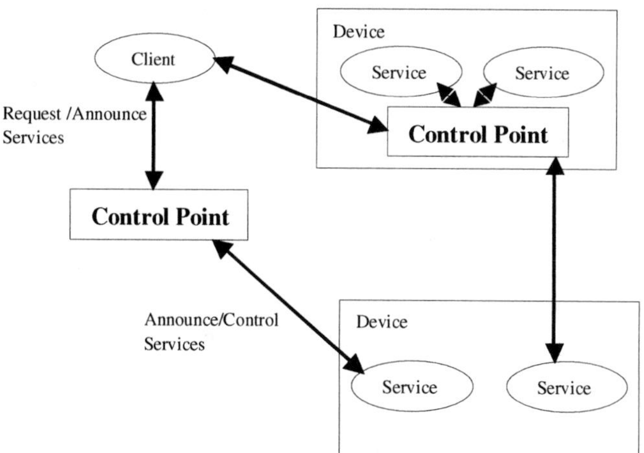

(TCP)/IP, Hyper Text Transfer Protocol (HTTP), and XML ("Extensible Markup Language," 2000) for communication and service description. UPnP is designed to be operating system, language, and communication media independent. The architecture of a UPnP network is as shown in Figure 4.

The main components that play a role in UPnP are services, clients, and *control points*. Services are the smallest units of control in the network. A control point in a UPnP network is a controller capable of discovering and controlling other devices. The control points discover the devices, and hence, all the services hosted by a device at one time. Also, control points retrieve service descriptions and invoke service actions on behalf of the clients.

The discovery process using SSDP is heavily dependent on HTTP over unicast and multicast. Every device on the network can install control points to perform truly as a peer-to-peer system. The discovery of control points is through multicast request or announcement. Devices that do not have a control point can use other control points in the network or depend solely on the multicast requests and announcement of services. Service announcements contain a service type and a URL for the service being advertised. Clients can access this Uniform Resource Locator (URL) using HTTP and invoke service actions through remote procedure calls (RPCs; Birrell & Nelson, 1984). HTTP is a very heavy-weight protocol, and hence the cost of using UPnP may be high for resource-restrained devices and networks. Also, it needs to be pointed out that although UPnP can work in a truly peer-to-peer model, to make use of full XML capabilities, at least one central Web server may be required somewhere.

UPnP is aimed at providing service discovery for small networks. The scalability is very limited, as it is more of a peer-to-peer technology and depends heavily on the multicast and HTTP. UPnP design is also flexible enough to be able to be bridged with other service discovery protocols, although this is not part of UPnP specification.

Salutation

Salutation is a service discovery mechanism developed by Salutation Consortium. Salutation Consortium is a collaboration of companies that make printers and similar devices such as HP, IBM, Xerox, and Canon, and these companies have implemented Salutation in many of their products. Additional information about the consortium can be found at http://www.salutation.org. Similar to SLP and UPnP, Salutation can also function as both a directory-style and peer-to-peer discovery mechanism. Salutation specification also includes the protocol for actual service invocation by the clients (*Salutation Architecture Specification*, 1999).

Salutation is a programming-language- and operating-system-independent service discovery mechanism. One important difference of Salutation from other service discovery technologies is that it is transport-protocol independent. However the network architecture is very similar to those of SLP and Jini.

Important components in a Salutation service discovery setup are clients, services, *Salutation managers* (SLMs), and *transport managers* (TMs). SLMs are the core of the architecture, just like the LUSs in Jini. SLMs provide a transport-independent interface to services and clients through the use of APIs that provide for service registration, service discovery, and optional service access function. Clients can access the service

Figure 5. Architecture of Salutation

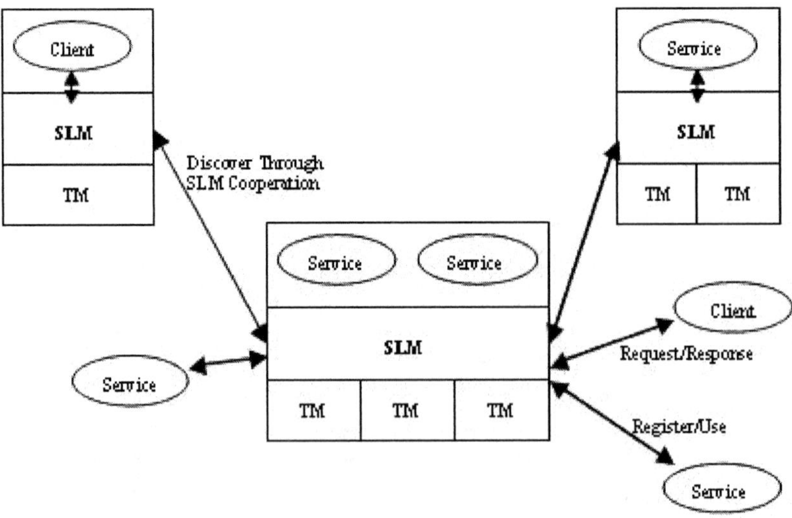

either through an SLM or through direct communication. Communication-protocol independence is achieved by the interface between SLMs and TMs. A TM is an entity dependent on the network transport it supports. There can be more than one TM for an SLM depending on how many transport protocols it supports. Clients and services can take the service of a local or remote SLM. The architecture of Salutation is depicted in Figure 5.

SLMs discover other SLMs in the network through broadcast RPCs. Also, clients and services discover SLMs in the network using broadcast RPCs. SLMs cooperate among themselves to interchange information about services registered with them upon a service request from a client. Salutation can function as a true peer-to-peer discovery model if every device in the network has an SLM on it, or solely through the use of broadcast RPCs by clients and services. Salutation defines a specific record format for describing and locating services and this format includes type and attributes.

Except for broadcast RPCs, Salutation is technology neutral. It can also be implemented with LDAP or any number of other directory services. Salutation design provides for its capability to be bridged with any other service discovery mechanism, including that for wireless technologies such as IrDA and Bluetooth (*Bluetooth Specification*, n.d.). Salutation is scalable only to enterprise-size networks, especially due to the dependence on broadcast RPCs for SLM discovery.

Figure 6. Architecture of INS/Twine

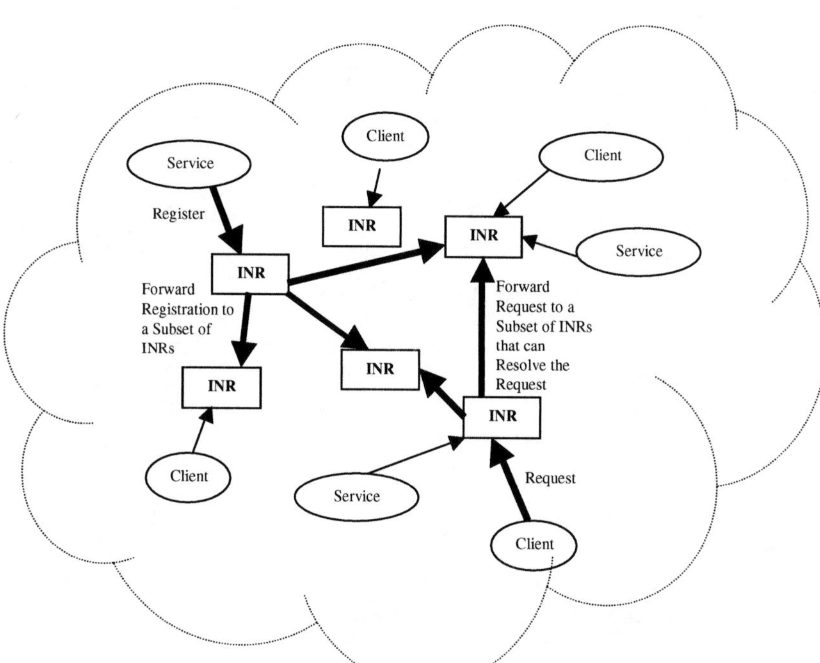

Intentional Naming System/Twine

INS is a research framework for service discovery designed at MIT. It is a naming system intended for expressive, robust, and responsive discovery of services and devices. This service discovery mechanism follows a decentralized directory model for discovery and provides the specification for service discovery as well as optional query and data forwarding to the service (Adjie-Winoto et al., 1999). INS is also the service discovery mechanism used by the Oxygen (http://oxygen.lcs.mit.edu) pervasive computing research project at MIT. INS/Twine is an improvement over INS in an effort to improve the scalability (Balazinska, Balakrishnan, & Karger, 2002). INS, as well as INS/Twine, does not depend on multicast, unlike all other service discovery technologies discussed, and can be considered as an operating-system- and programming-language-independent service discovery mechanism. INS has three main components that play a role in the service discovery: clients, services, and an application-level overlay network of intentional name resolvers (INRs). The architecture of INS is illustrated in Figure 6. An expressive service description, which gives adequate flexibility to clients in searching for an appropriate service, is one of the major design goals of INS. Fittingly, INS specifies an intentional naming scheme that uses name specifiers which are hierarchical arrangements of attribute value pairs. Every INR maintains a data structure called name tree, which is a tree superposition of all the name specifiers and the corresponding name info records. A name info record is a routing table for the next hop INRs and the IP addresses of the potential final destinations. A central authority called the domain space resolver (DSR) maintains information regarding the INRs in the domain. Clients and services need to communicate with only one INR. Services register with an INR with a name specifier and this is added to the name tree. This INR then uses application layer multicast to forward this information to every other INR in the network. Eventually every INR possesses information about every service in the network. So when a client queries an INR, the INR can do a lookup in its name tree and can either send the service information to the querying client (early binding) or can forward the query and data to the appropriate service on behalf of the querying client (late binding).

Although INS avoids the use of network layer multicast and hence avoids one bottleneck for scalability, the name tree lookup can still cause scalability issues as every INR has name tree entries for every service in the network. INS/Twine (Balazinska et al., 2002) is an effort to reduce this bottleneck as well. Authors claim INS/Twine as a peer-to-peer service discovery technology. However, the specification only provides for a cooperating network of INRs that resolves any service queries. In order to achieve a truly peer-to-peer mechanism, every device in the network should have an INR on them or be able to resolve the service queries on their own.

INS/Twine envisions wide-area scalability when compared to the enterprise-size scalability of INS. In order to achieve the wide-area scalability in INS/Twine, service registration is forwarded onto only a subset of all the INRs in the network. This subset of INRs is based on a hash table process algorithm to convert the name specifier to several hashing keys, and mapping these hashing keys to appropriate INRs in the network. Every INR should be able to implement this functionality for key mapping. So when a client queries for a service, the INR can map the query to the INRs that can resolve this query, and will then forward this query to the appropriate INRs. INRs resolving this query will send the response back to the initial INR to which the client queried. The remainder of the discovery procedure, naming tree organization, and lookup of INS/Twine remains the same as INS.

The scalability of INS/Twine will definitely be better than INS since the INRs contain information regarding only a subset of services in the network. However this improvement in the scalability may be at the cost of lesser fault tolerance and responsiveness as INS/Twine is more sensitive to INR failures when compared to INS. Although authors claim that INS/Twine can have Internet-size scalability, this is questionable due to the requirement of interdomain discovery in an Internet-scale network and needs to be further investigated.

Secure Service Discovery Service

SDS is a component in UC Berkeley's Ninja research project (Hodes et al., 2002). It is similar in many respects to other service discovery protocols discussed, especially IETF's SLP, but it features much stronger security, reliability, and scalability than those discussed so far. Like Jini, SDS is implemented in Java and depends on Java RMI. Unlike Jini, SDS uses XML for service descriptions. This is a powerful combination given the expressiveness of XML and portability of Java. SDS is a directory-style service discovery mechanism, with its local-area discovery techniques very similar to that of SLP.

The SDS model has clients, services, secure discovery service servers (SDSSs), certificate authorities, and capability managers. SDSSs are the counterparts of DAs in SLP. Clients search for and identify services by using the SDSSs to match XML-based service descriptions. Certificate authorities issue credentials that bind a service description to the service's public keys. The capability manager is a centralized service, similar to the certificate authority, which issues capabilities. Capabilities are like certificates, but they bind client names to service names rather than binding names to keys. A capability proves that a particular client is allowed to access a particular service. A service contacts a capability manager and lists all clients that are allowed to access it. The capability manager then creates and signs all necessary credentials and stores them for later distribution to the indicated clients. This mechanism allows for capability-based access control wherein the existence of a service can be hidden, in addition to allowing or disallowing access. However, access management is not very clean. Security protocols and services used in SDS are

1. public-key-authenticated server announcements, which assure authenticity of server announcements,
2. one-way encrypted service announcements through a combination of private and public key protocol, which assure privacy and authenticity of service descriptions,
3. secure RMI, which provides for two-way authenticated and encrypted remote-method invocation, and
4. certificates and certificate authority structure, which provide capability to authenticate all participants.

Like SLP, SDS also heavily depends on multicast for SDSS discovery among themselves as well as by clients and services. In Hodes et al. (2002), the authors discuss several ways to achieve wide-area scalability, which include name-specified mapping to neighbors, flooding, and query filtering. SDS adopted the query filtering, which is a hybrid of mapping and flooding on a hierarchical arrangement of SDSSs. More simply put, query filtering is filtered query flooding along the hierarchical interconnections. SDS anticipates many hierarchies to coexist and can be created dynamically. In hierarchies, root nodes are always a bottleneck and, hence, sensitivity to node failure will be very high. Also, when it comes to resolving multiple orthogonal attributes, the filtered query flooding will be more or less like query flooding. So in worst-case scenarios, the architecture will act like pure flooding for wide-area discovery, and this can hamper the scalability. Although the authors claim that SDS can have Internet-size scalability, this is questionable due to the requirement of interdomain discovery in an Internet-scale network and needs to be further investigated.

Other Service Discovery Mechanisms

Researchers have shown great interest in the field of service discovery. In addition to the schemes discussed so far, there are some more-recent activities in the area of service discovery. The Bluetooth service discovery protocol is intended for Bluetooth networks (*Bluetooth Specification*, n.d.). Bluetooth is a low-cost, low-power radio system designed as a cable replacement and short-range, personal-area networking technology. Bluetooth devices search to find other devices and services that lie within a 10-meter range of the transmitter. Bluetooth devices maintain sets of service records, each of which describes an available service. Service records consist entirely of descriptive attributes, describing the type of service, necessary protocols for communications, and so forth. To find a service, Bluetooth devices search the service records of other devices in the available space for the service of interest. Another service discovery endeavor is DEAPspace, which is a framework for interconnecting and discovering pervasive devices over a wireless medium (Hermann, Husemann, Moser, Nidd, Rohner, & Schade, 2000). While both Bluetooth and DEAPspace are designed for small networks, especially personal-area networks, it is feasible that these mechanisms can interoperate with other service discovery mechanisms intended for bigger networks. The authors in Castro, Greenstein, Muntz, Bisdikian, Kermani, and Papadopouli (2001) describe a service discovery scheme where clusters of directories dynamically organize into hierarchies to provide wide-area service discovery across domains.

Discussion on Existing Service Discovery Mechanisms

It is evident from the descriptions in the previous section about existing research in service discovery that many of the service discovery mechanisms logically operate in similar ways and have a number of features in common. A comparison of features of the service discovery mechanisms discussed in this chapter are summarized in Table 1. Directory-style discovery technologies are typically achieved through either a centralized or a distributed directory. In centralized mechanisms like SLP, JINI, and UPnP, if more than one directory is present in the network, there is no collaboration between these directories. On the other hand, in distributed mechanisms like Salutation, INS, and SDS, directories collaborate and communicate to provide service discovery. Peer-to-peer-style service discovery models are achieved either through the sole use of multicast or through the cooperation among the respective directory entities that need to be present on every device in the network. Every technology except INS uses multicast in some function of the discovery process. The basic mechanisms by which clients find services are advertisement and service request. Each technology uses one or both of these concepts. TCP/IP is one transport protocol all the discussed technologies depend on, although Salutation can work independent of transport protocol. Most of the technologies discussed do not have Internet-size scalability as one of the design goals and accordingly, these are scalable only up to enterprise-size networks, although these networks can be quite large and highly dispersed. Wide-area scalability is a design goal for both INS/Twine and SDS; however, across domains, Internet-size scalability is questionable.

Despite the logical similarity of these technologies, the implementations are diverse and therefore incompatible. Each model has its own model for key characteristics such as services, clients, service description, advertisement and discovery, interaction, and notification of events. One of the reasons for this heterogeneity is that these technologies were designed with different domains and applications in mind. These implementation differences may also be attributed to the reason for the difference in scalability of these technologies, although none of them may scale to Internet size.

Future Trends and Recommendations

In order to facilitate service discovery in wireless and mobile computing environments across various domains, heterogeneous devices, and communication protocols, either a single service discovery should dominate or the most commonly used technologies must be made interoperable. It is unlikely that all the diverse technologies will prevail in the end, and the probability will be guided by market share rather than technical merit.

Conclusion

An effective service discovery is necessary to provide the knowledge of available services to the devices and also to the networking infrastructure. We presented a discussion on the importance of service discovery and challenges in service discovery, and investigated various service discovery mechanisms.

We conclude that it is imperative to have a service discovery mechanism that is not a burden on the infrastructure and at the same time, highly available to devices and clients anytime, anywhere, regardless of device and network characteristics.

Table 1. Comparison of service discovery protocols

Feature	SLP	Jini	UPnP	Salutation	INS/Twine	SDS
Directory	Yes	Yes	Yes	Yes	Yes	Yes
Peer-to-Peer	Yes	No	Yes	Yes	Yes	No
Service Utilization	No	Yes	Yes	Yes	Yes	No
Service Description	Service Templates with Attributes	Attribute/Value Pairs	XML Based	Name/Value Pairs of Functional Units	Intentional Routing Trees	XML Based
Operating-System Independent	Yes	Yes	Yes	Yes	Yes	Yes
Transport-Protocol Independent	No	No	No	Yes	No	No
Programming-Language Independent	Yes	No	Yes	Yes	Yes	No
Scalability	Enterprise	Enterprise	Small Networks	Enterprise	Wide Area	Wide Area
Security	IP Dependent	Java Based	IP Based	Authentication	Not a Design Goal	Multiple Security Protocols and Services
Service Property Filtering	Yes	Yes	Yes	No	Yes	Yes
Prerequisites	TCP/IP, IP Multicasting, DHCP	Java VM and Java RMI, IP Multicasting	TCP/IP, RPC, HTTP, Unicast and Multicast	RPC	TCP/IP	TCP/IP, IP Multicasting, Java RMI
Implementation Examples	OpenSLP Novel Netware	Macromedia's Application Server—Jrun, NIST's Aroma Projector	UPnP Digital Media Receiver by Arcadyan Technology Corporation	IBM's NuOffice (Salutation Enhancements of Lotus Notes)	Research Project "Oxygen"	Research Project "Ninja"
Developer	IETF	Sun Microsystems	Microsoft	Salutation Consortium	MIT	UC, Berkeley

References

Adjie-Winoto, W., Schwartz, E., Balakrishnan, H., & Lilley, J. (1999). The design and implementation of an intentional naming system. *Proceedings of the ACM International Symposium on Operating Systems Principles (SOSP)*, (pp. 186-201).

Allard, J., Chinta, V., Gundala, S., & Richard, G. G., III. (2003). Jini meets UPnP: An architecture for Jini/UPnP interoperability. *Proceedings of the IEEE International Symposium on Applications and the Internet (SAINT)*, (pp. 268-275).

Balazinska, M. M., Balakrishnan, H., & Karger, D. (2002). INS/Twine: A scalable peer-to-peer architecture for intentional resource discovery. *Proceedings of the IEEE International Conference on Pervasive Computing (PerCom)*, (pp. 195-210).

Birrell, A. D., & Nelson, B. J. (1984). Implementing remote procedure calls. *ACM Transactions on Computer Systems, 2*, 39-59.

Bluetooth specification. (n.d.). Retrieved February 25, 2004, from http://www.bluetooth.org/spec/

Castro, P., Greenstein, B., Muntz, R., Bisdikian, C., Kermani, P., & Papadopouli, M. (2001). Locating application data across service discovery domains. *Proceedings of the Annual IEEE/ACM International Conference on Mobile Computing and Networking (MOBICOM)*, (pp. 28-42).

CORBAservices: Common object services specification. (n.d.). Object Management Group. Retrieved February 25, 2004, from ftp://ftp.omg.org/pub/.docs/formal/98-12-09.pdf/

Dertouzos, M. L. (1999, August). The future of computing. *Scientific American*, 52-55.

Droms, R. (1997). RFC 2131. *Dynamic host configuration protocol*. Retrieved February 25, 2004, from http://rfc-2131.rfc-index.com/rfc-2131.htm/

Edwards, W. K. (2000). *Core Jini*. Englewood Cliffs, NJ: Prentice Hall.

Exensible markup language (XML) 1.0 (2nd ed.). (2000). Retrieved February 25, 2004, from http://www.w3.org/TR/2000/REC-xml-20001006.pdf/

Gibbins, N., & Hall, W. (2001). Scalability issues for query routing service discovery. *Proceedings of the Second Workshop on Infrastructure for Agents, MAS and Scalable MAS*, (pp. 209-217).

Gribble, S. D., Welsh, M., Behren, R. Von, Brewer, E. A., Culler, D., Borisov, N., et al. (2001). The Ninja architecture for robust Internet-scale systems and services [Special Issue]. *Computer Networks, 35*(4), 473-497.

Guttman, E. (1999). Service location protocol: Automatic discovery of IP network services. *IEEE Internet Computing, 3*(4), 71-80.

Guttman, E., & Kempf, J. (1999). Automatic discovery of thin servers: SLP, Jini and the SLP-Jini bridge. *Proceedings of the 25th Annual Conference of the IEEE Industrial Electronic Society*, 722-727.

Guttman, E., Perkins, C., & Kempf, J. (1999). RFC 2609. *Service templates and service: Scheme.* Retrieved February 25, 2004, from http://rfc-2609.rfc-index.com/rfc-2609.htm/

Guttman, E., Perkins, C., Veizades, J., & Day, M. (1999). RFC 2608. *Service location protocol, version 2.* Retrieved February 25, 2004, from http://rfc-2608.rfc-index.com/rfc-2608.htm/

Helal, S. (2002). Standards for service discovery and delivery. *IEEE Pervasive Computing, 1*(3), 95-100.

Hermann, R., Husemann, D., Moser, M., Nidd, M., Rohner, C., & Schade, A. (2000). DEAPspace: Transient ad hoc networking of pervasive devices. *Proceedings of the ACM International Symposium on Mobile Ad Hoc Networking and Computing (MOBIHOC)*, (pp. 133-134).

Hodes, T. D., Czerwinski, S. E., Zhao, B. Y., Jospeh, A. D., & Katz, R. H. (2002, March). An architecture for secure wide-area service discovery. *ACM/Kluwer Wireless Networks, 8*(2-3), 313-230.

Infrared data association. (n.d.). Retrieved February 25, 2004, from http://www.irda.org/

Kempf, J., & Pierre, P. St. (1999). *Service location protocol for enterprise networks: Implementing and deploying a dynamic service finder.* New York: John Wiley & Sons.

Miller, B. A., Nixon, T., Tai, C., & Wood, M. D. (2001). Home networking with universal plug and play. *IEEE Communications Magazine, 39*(12), 104-109.

Mockapetris, P. V., & Dunlap, K. J. (1988). Development of domain name system. *Proceedings of the ACM International Conference on Applications, Technologies, Architectures, and Protocols for Computer Communication (SIGCOMM),* (pp. 123-133).

Newmarch, J. (2000). *A programmer's guide to Jini technology.* Berkley, CA: Apress L.P.

Oaks, S., & Wong, H. (2000). *Jini in a nutshell: A desktop quick reference.* Sebastopol, CA: O'Reilly & Associates, Incorporated.

Salutation architecture specification, version 2.0c. (1999). Retrieved February 25, 2004, from http://www.salutation.org/spec/Sa20e1a21.pdf/

Wahl, M., Howes, T., & Kille, S. (1997). RFC 2251. *Lightweight directory access protocol (v3).* Retrieved February 25, 2004, from http://rfc-2251.rfc-index.com/rfc-2251.htm/

Zhao, W., Schulzrinne, H., & Guttman, E. (2000). mSLP mesh-enhanced service location protocol. *IEEE International Conference on Computer Communications and Networks (ICCCN, 504-509).*

Chapter IX

Wireless Sensor Networks

Erdal Cayirci, Istanbul Technical University, Turkey

Abstract

A wireless sensor network is deployed either inside the phenomenon or very close to it. Unlike some existing sensing techniques, the position of sensor network nodes need not be engineered or predetermined. This allows random deployment in inaccessible terrains. On the other hand, this also means that sensor network protocols and algorithms must possess self-organizing capabilities. Another unique feature of sensor networks is the cooperative effort of sensor nodes. Sensor network nodes are fitted with an onboard processor. Instead of sending the raw data to the nodes responsible for the fusion, sensor network nodes use their processing abilities to locally carry out simple computations and transmit only the required and partially processed data. Realization of sensor networks requires wireless ad hoc networking techniques. In this chapter, we present a survey of protocols and algorithms proposed thus far for wireless sensor networks. Our aim is to provide a better understanding of the current research issues in this field. We also attempt an investigation into understanding design constraints and outline the use of certain tools to meet the design objectives.

Introduction

Advances in digital electronics, embedded systems, and wireless communications led the way to a new class of ad hoc networks, namely, wireless sensor networks (WSNs), that consist of sheer numbers of tiny nodes randomly deployed either inside the phenomenon or very close to it. Sensor network nodes are fitted with an onboard processor, and they can collaborate both in sensing and transferring the sensed data. WSNs have a wide range of potential applications, including security and surveillance, control, actuation and maintenance of complex systems, and fine-grain monitoring of indoor and outdoor environments. Some examples for these applications are explained below.

- **Military Applications:** WSNs can be an integral part of military command, control, communications, computers, intelligence, surveillance, reconnaissance, and targeting (C4ISRT) systems. The rapid deployment, self-organization, and fault-tolerance characteristics of sensor networks make them a very promising sensing technique for military C4ISRT. Since sensor networks are based on the dense deployment of disposable and low-cost sensor nodes, destruction of some nodes by hostile actions does not affect a military operation as much as the destruction of a traditional sensor. Some of the military applications are monitoring friendly forces, equipment, and ammunition; battlefield surveillance; reconnaissance of opposing forces and terrain; targeting; battle damage assessment; and nuclear, biological, and chemical attack detection and reconnaissance.
- **Environmental Applications:** Some environmental applications of sensor networks include tracking the movements of species, that is, habitat monitoring, monitoring environmental conditions that affect crops and livestock, irrigation, macro instruments for large-scale Earth monitoring and planetary exploration, and chemical and biological detection.
- **Commercial Applications:** The sensor networks are also applied in many commercial applications. Some of them are building virtual keyboards, managing inventory control, monitoring product quality, constructing smart office spaces, and environmental control in office buildings.

Sensor networks differ from conventional network systems in many aspects. WSNs usually involve a large number of spatially distributed, energy-constrained, self-configuring, and self-aware nodes. Furthermore, they tend to be autonomous and require a high degree of cooperation and adaptation to perform the desired coordinated tasks and networking functionalities. As such, they bring about new challenges and design considerations, which go much beyond conventional network systems. These design considerations, which are the reasons to develop new schemes and technologies rather than using available ad hoc networking technologies, can be summarized as follows (Akyildiz, Su, Sankarasubramaniam, & Cayirci, 2002).

- **Topology:** In WSNs, 100s to several 1,000s of sensor nodes (snodes) are densely deployed throughout the sensor field. The distance between two neighboring

Figure 1. Sensor networks topology

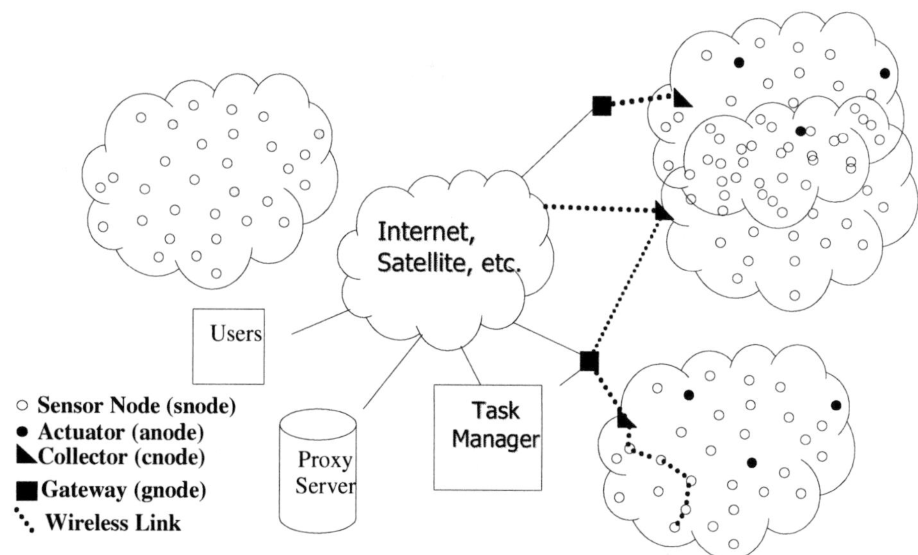

snodes is often limited to 10s of feet. The node deployment is usually done randomly by scattering nodes in the sensor field. In some applications, actuators (anodes) that control various devices can also be positioned within the sensor network. A collector node (cnode) that is often more capable than the other nodes in the field is also located either inside or close to the sensor field as shown in Figure 1. Cnodes, usually named sinks in the literature (Akyildiz et al., 2002), are responsible to collect the sensed data from snodes, and then serve the collected data to users. They are also responsible to start task disseminations in many applications. The sensed data by snodes are conveyed through the sensor network by multiple hops in an ad hoc manner, and gathered in cnodes that can be perceived as the interface between sensor networks and users. Multiple sensor networks can be integrated into a larger network through the Internet or direct links between either cnodes or gateways. In the future, an Internet user will presumably be able to query a sensor network that is located anywhere in the world or space.

Snodes may be statically deployed. However, device failure is a regular and common event due to energy depletion or destruction. It is also possible to have sensor networks with highly mobile nodes. Besides, snodes and the network experience varying task dynamics, and they may be target to a deliberate jamming. Therefore, sensor network topologies are prone to frequent changes after deployment.

- **Fault Tolerance:** Sensor networks should be able to sustain their functionalities without any interruption due to snode failures. Protocols and algorithms may be designed to address the level of fault tolerance required by the sensor network applications. The requirements of applications usually differ from each other. For example, the fault-tolerance requirements of a tactical sensor network can be considered much higher than those for a home application because snodes are prone to higher failure rates in tactical sensor networks, and the impact of sensor network failure in a tactical field can be much more important than the impact of a home sensor-network failure.

The differences in the requirements of various sensor network applications can be observed for almost every factor influencing the design of sensor networks. Moreover, trade-offs are generally required among these factors because there are stringent constraints related to them. Therefore a "one-size-fits-all" type, generic design is not possible for many tasks in sensor networks. Generally, different schemes are needed to fulfill the requirements of different applications.

- **Scalability:** The number of snodes deployed in a sensor field may reach millions in some applications. Moreover, the node densities may be as high as 20 nodes/m^3 in some applications. All schemes developed for sensor networks have to be scalable enough to fit with the node densities and numbers that are higher than all the other types of networks in the orders of magnitude.
- **Sensor Node Hardware:** An snode is made up of four basic components: sensing units, a processing unit, a transceiver unit, and a power unit. They may also have application–dependent, additional components such as a location-finding system, a power generator, and a mobilizer. Sensing units are usually composed of two subunits: sensors and analog-to-digital converters (ADCs). The analog signals produced by the sensors based on the observed phenomena or stimuli are converted to digital signals by the ADC, and then fed into the processing unit. Please note that an snode may be attached more than a single sensor. For example, a temperature and a humidity sensor can be attached to the same snode. The processing unit, which is generally associated with a small storage unit, manages the procedures that make an snode collaborate with the other nodes to carry out the assigned sensing tasks. A transceiver unit connects the node to the network. One of the most important components of an snode is the power unit. Power units may be supported with a power-scavenging tool such as solar cells. There are also other subunits, which are application dependent. Most of the sensor network routing techniques and sensing tasks require the knowledge of location with high accuracy. Thus, it is common that an snode has a location finding system. A mobilizer may sometimes be needed to move snodes when it is required to carry out the assigned tasks. All of these subunits may need to fit into a matchbox-sized module. The required size may be smaller than even a cubic centimeter in some applications.
- **Production Costs:** As stated in the scalability paragraph, a sensor network may contain millions of snodes. Therefore the cost of snodes has to be low in order for such a network to be feasible. The targeted node cost for some sensor network applications is less than \$1 while a Bluetooth node costs more than \$5. A Bluetooth node is delivered only for communications while an snode is for more, as explained in the sensor node hardware paragraph. This explains how challenging the targeted cost for snodes is.

- **Environment:** Snodes are densely deployed either very close to or directly inside the phenomenon to be observed. Therefore, they usually work unattended in remote geographic areas. They may be working in extremely harsh environments. They work under high pressure in the bottom of an ocean, in harsh environments such as debris or a battlefield, under extreme heat and cold such as in the nozzle of an aircraft engine or in arctic regions, and in an extremely noisy environments such as under intentional jamming.
- **Transmission Media:** In a multihop sensor network, nodes are linked by a wireless medium. These links can be formed by radio, infrared, or optical media. To enable global operation of these networks, the transmission medium must be available worldwide. One option for radio links is the use of industrial, scientific, and medical (ISM) bands, which offer license-free communication in most countries. Much of the current hardware for snodes is based upon radio frequency (RF) circuit design. Another possible mode for sensor networks is infrared. Infrared communication is license-free and robust to interference from electrical devices. Infrared-based transceivers are cheaper and easier to build. The main drawback of infrared is the requirement of a line of sight between sender and receiver. This makes infrared a reluctant choice for transmission medium in the sensor network scenario. Another interesting transmission scheme is introduced in Warneke, Liobewitz, and Pister (2001), where passive transmission using a corner-cube retroreflector (CCR) and active communication using a laser diode and steerable mirrors are examined. The unusual application requirements of sensor networks may make the choice of transmission media more challenging. For example, marine applications may require the use of the aqueous transmission medium.
- **Power Consumption:** The WSN nodes, being microelectronic devices, can only be equipped with a limited energy source. In some application scenarios, replenishment of power resources might be impossible. Sensor node lifetime, therefore, shows a strong dependence on battery lifetime. In a multihop ad hoc sensor network, each node plays the dual role of data originator and data router. The dysfunctioning of a few nodes can cause significant topological changes and might require rerouting of packets and reorganization of the network. Hence, power conservation and power management take on additional importance. In other mobile and ad hoc networks, power consumption has been an important design factor, but not the primary consideration, simply because power resources can be replaced by the user. In sensor networks, power efficiency is an important performance metric, directly influencing network lifetime.

  Power consumption in sensor networks can be divided into three domains: sensing, communication, and data processing. Sensing power varies with the nature of applications. Of the three domains, a node expends maximum energy in data communication. This involves both data transmission and reception. It can be shown that for short-range communication with low radiation power, transmission and reception energy costs are nearly the same. It is important that in this computation we not only consider the active power but also the start-up power consumption in the transceiver circuitry. The start-up time, being of the order of 100s of microseconds, makes the start-up power nonnegligible. Another important consideration related to data communications is about the path loss exponent λ. Due to the low-lying antennae, » is close to 4 in sensor networks. Therefore, routes that have more hops with shorter distances can be more power efficient compared to the routes that have fewer hops with longer distances.

Energy expenditure in data processing is much less compared to data communication. The example described in Pottie and Kaiser (2000) effectively illustrates this disparity. Assuming Rayleigh fading and fourth-power distance loss, the energy cost of transmitting 1Kb a distance of 100 m is approximately the same as that for executing 3 million instructions by a 100-million-instructions-per-second (MIPS)/W processor.

New schemes and protocols have been introduced to satisfy the constraints by these factors in the recent years. In the following sections we explore some of these schemes related to three layers of networking. In the next section, we explain application layer protocols for sensor networks. The protocols and schemes for transport layer are then discussed. Network-layer protocols are examined after that. Then we conclude our chapter. Please note that there is also need for data link and physical layer protocols for sensor networks. Although there are already some proposals related to these layers, we do not elaborate on them in this chapter.

Application Layer

Snodes can be used for continuous sensing, event detection, event identification, location sensing, and local control of anodes (Akyildiz et al., 2002). The concepts of the microsensing and wireless connection of these nodes promise many new application areas, for example, military, environment, health, home, commercial, space exploration, chemical processing, disaster relief, and so forth. Some of these application areas are described in the introduction. Application layer protocols needed to realize these applications are introduced below.

Application Layer Protocols for Wireless Sensor Networks

Although many application areas for WSNs are defined and proposed, potential application layer protocols for sensor networks remain largely unexplored. Two possible application layer protocols are explained below.

- **Sensor Management Protocol (SMP)** (Akyildiz et al., 2002)**:** System administrators interact with sensor networks by using SMP, which is a management protocol that provides the interfaces needed to perform tasks such as the following (Akyildiz et al.):

- introducing the rules related to data aggregation, attribute-based naming, and clustering to the snodes,
- exchanging data related to the location finding algorithms,
- time synchronization of the snodes,
- moving snodes,
- turning snodes on and off,
- querying the sensor network configuration and the status of nodes, and reconfiguring the sensor network, and
- authentication, key distribution, and security in data communications.

- **Sensor Query and Data Dissemination Protocol (SQDDP)** (Akyildiz et al., 2002): SQDDP provides user applications with interfaces to issue queries, respond to queries, and collect incoming replies. Note that these queries are generally not issued to particular nodes. Instead, attribute-based or location-based naming is preferred. For instance, "Find the locations of the nodes that sense a temperature higher than 70°F," is an attribute-based query. Similarly, "Find temperatures read by the nodes in Region A," is an example of location-based naming. Both attribute-based and location-based addressing schemes are explained in detail in the following sections.

Data Querying in Sensor Networks

One of the most challenging tasks in sensor networks is to synthesize the information requested by users from the available data measured or sensed by a large number of snodes. Since there are a sheer number of nodes with stringent energy constraints in a sensor network, it may not be feasible to fetch every reading of snodes for central processing. Instead, effective data querying and aggregation techniques are needed. In this section we focus on data querying in sensor networks.

Data queries in sensor networks can be continuous and periodical, continuous and event driven, or snapshot, that is, one-time queries. We can categorize sensor network queries also as aggregated or nonaggregated. Queries can also be complex or simple. Finally, queries for replicated data can be made. The users should be able to carry out any of these types of queries by using the data-querying scheme for sensor networks. One approach to realize this is to perceive a sensor network as a distributed database (Cayirci, 2003b).

Data Aggregation and Dilution by Modulus Addressing

In the data aggregation and dilution by modulus addressing (DADMA) scheme (Cayirci, 2003b), a sensor network is perceived as a distributed relational database composed of a single view that joins local tables located at snodes. Records in local tables are the measurements made upon a query arrival and consist of two fields, namely, *task* and *amplitude*. Since an snode may have more than one sensor attached to it, the task field, for example, temperature, humidity, and so forth, indicates the sensor that makes the

Figure 2. Sensor network as a distributed database

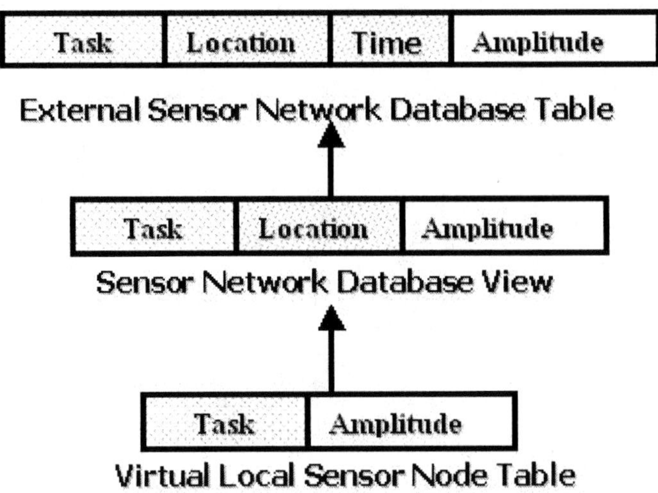

measurement. Snodes have limited memory capacity and they do not store the results of measurements. Therefore, the task field is the key field in the local tables created upon a query arrival. This perception of WSNs makes relational algebra practical to retrieve the sensed data without much memory requirement.

A sensor network database view (SNDV) can be created temporarily either at the cnode or at an external proxy server. An SNDV record has three fields, that is, *location*, task, and amplitude. While data is being retrieved from an snode, the sensed data is also joined with the location of the snode. Since multiple snodes may have the same type of sensors, that is, multiple sensors can carry out the same sensing task, the location and task fields become the key in an SNDV. For many WSN applications, the sensed data is needed to be associated with location data. For example, in target-tracking and intrusion-detection WSNs, sensed data is almost meaningless without relating it to a location. Therefore, location awareness of snodes is a requirement imposed by many WSN applications. There are a number of practical location-finding techniques reported for WSNs. We discuss them later in this chapter. If the location data is not available and not important for the application, the local identification field for the sensing snode replaces the location field.

It is also possible to maintain a database in a remote proxy server where the records obtained from queries, that is, the records at an SNDV, are stored after being joined with a *time* label. For example, a daemon can generate queries at specific time intervals, and insert the records in the SNDV resulting from these queries into the database after joining them with a time field. Note that each query results in a new SNDV where the results of the query are gathered temporarily.

In DADMA (Cayirci, 2003b) a query is started by a statement that has the structure given below. Note that the standard structured query language (SQL) notation is used in this statement except for the last field starting with the "based on" keyword.

Select [task, time, location, [distinct | all], amplitude,
[[avg | min | max | count | sum] (amplitude)]]
from [any , every , aggregate *m* , dilute *m*]
where [power available [<|>] *PA* |
location [in | not in] *RECT* |
t_{min} < time < t_{max} |
task = *t* |
amplitude [<|==|>] *a*]
group by task
based on [time limit = l_t | packet limit = l_p | resolution = *r* | region = *xy*]

A user can retrieve a subset of data fields available in an SNDV, and can aggregate amplitude data either by grouping data based on task and/or by using the *aggregate m* function given in Equation 1. Some of the snodes can also be excluded from a query by the *dilute m* function in Equation 2.

$$f(x) = x \; div \; m, \text{ and} \quad (1)$$

$$f(x) = (x / r) \; mod \; (m / r), \quad (2)$$

where

x is the grid location of a node relative to one of the axes,
r is the resolution in meters, and
m is the dilution or aggregation factor.

When the dilute *m* command is given by the user, every snode first uses Equation 2 to find out its location indexes in horizontal and vertical axes, and then compares these indexes with the region values *x* and *y* sent in the "based on" field of the query. If they match, the snode replies to the query. For example, the location indexes of an snode at location {46, 74} are {3, 1} for $m = 8$ and $r = 2$. Therefore, if the region value in the query is {3, 1}, this sensor should respond. Hence, only the snodes in $r \times r$-meter squares located in every m meters react to the query, and the others stay idle. This is a practical technique, especially when snodes are randomly deployed according to uniform distribution and the sensor network is monitoring environmental conditions such as temperature, humidity, and pressure.

For the same example, the indexes found out by using Equation 1 are {5, 9}. When the aggregate *m* command is received, the values measured by an snode are aggregated with the values measured by the other nodes having the same indexes. Hence, we can address the snodes at certain geographic locations, and aggregate data based on the location of nodes.

Querying Sensor Networks by Using Task Sets

The idea of task sets (TSs) is based on dividing a sensor field into subregions, and assigning a specified number of snodes in each subregion to every TS (Cimen, Cayirci, & Coskun, 2003). A viable option to define subregions is quadtree addressing, explained in the spatial addressing section. The number of nodes in each subregion varies because of the nonhomogenous distribution of nodes. Hence, the cost of querying the sensor field varies in different subregions. To balance this cost, forming TSs with a specific amount of nodes in each quadrant is proposed in Cimen et al. By TSs, users have an initiative to trade off between accuracy-reliability and communications costs. The number of nodes in a TS indicates the resolution of the data which can be collected by querying the TS. A higher number of nodes in a task set implies higher accuracy and reliability. On the other hand, more power is consumed as the number of nodes in a TS increases. The advantages of using TSs can be listed as follows.

- We can assign sensing tasks fairly among sensors in a subregion.
- We can trade off between accuracy-reliability and power consumption.
- We can design efficient load-balancing techniques and increase the lifetime of a sensor network.

Snodes can be assigned to TSs by using a central or distributed approach. In the central approach, a central node assigns nodes to the TSs. This central node can be the cnode or a remote node. In the distributed approach, snodes are self-organized into TSs. Users start TS formation by providing the following information:

- the number of TSs in each subregion,
- the maximum number of sensors in each TS, and
- the selection rules for TSs (e.g., sensor type, power available, random selection, etc.)

In the central approach, a central node (i.e., a cnode or remote node) assigns nodes to the TSs according to the parameters given by users, and notifies nodes about which TS

Figure 3. Assigning sensor nodes to task sets

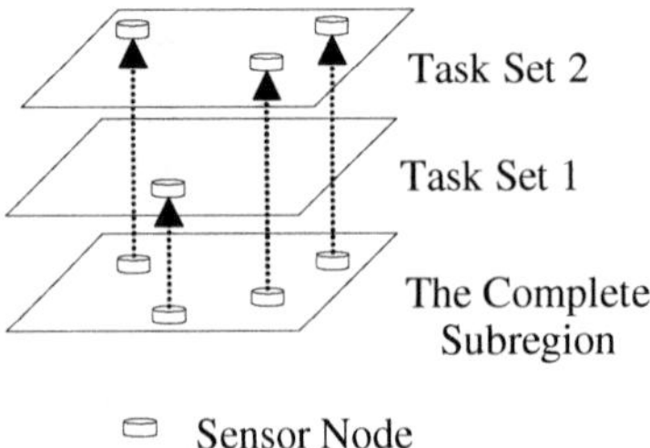

Table 1. An instance of a status table

Subregion Addresses	Sensor Type	Power Available	Task Set
1	1	0.95	2
1	1	0.98	1
1	1	0.93	2
1	1	0.96	2

they belong to. In the distributed approach, the forming-TS process starts after receiving the "form task sets" command from a user. Receiving this command, nodes prepare status data packets which consist of the fields given in Table 1 and broadcast it in their subregion, that is, quadrant. Nodes accept the status data packets from the other nodes in the same subregion and create status tables by using these status data packets. By using status tables and criterions for forming TSs, each node determines which TS it belongs to.

For example, TS1 can be specified as two nodes that have the highest power available in every subregion in a "form task set" command. Similarly, TS 2 can be specified as all nodes that are not in TS 1. After constructing a status table, a node can find out which TS it is in based on these specifications. In some cases, the status table of nodes in the same subregion may not be the same due to a hidden node problem. This may cause TSs to be formed slightly different than what a user specifies. However, this is negligible and can be corrected by central nodes as recognized in the reception of reports from TSs.

Snodes may be mobile and have limited power. They also may fail or be destroyed after being deployed. Therefore TSs may change in time and need to be updated. The trivial way to do this is to repeat the TS formation process with the same parameters. This can be done time based and synchronized by a central node, that is, the cnode or a remote node, by sending the "task set reformation" command at certain time intervals. In an alternative strategy, TS reformation can be started by snodes when the power available at an snode changes more than a threshold value, or an snode changes its subregion. When a "task set reformation" command is received, snodes run the same process as the one that they run at the "task set formation" command.

Other Schemes for Data Querying in Sensor Networks

In the active query forwarding in sensor networks (ACQUIRE) scheme (Sadagopan, Krishnamachari, & Helmy, 2003), each node that forwards a query tries to resolve it. If the node resolves the query, it does not repeat it, but sends the result back. Nodes collaborate with their n hop neighbors to resolve a query. The parameter n is named the look-ahead parameter. If a node cannot resolve a query after collaborating with n hop neighbors, it forwards it to another neighbor. When the look-ahead parameter n is 1, ACQUIRE performs as flooding in the worst case.

Mobility-assisted resolution of queries (MARQ) in large-scale mobile sensor networks (Helmy, 2003) makes use of the mobile snodes to collect data from the sensor network. In MARQ, every node has contacts that are some of the other snodes. When contacts move around, they interact with other nodes and collect data. Snodes collaborate with their contacts to resolve the queries.

The sensor query and tasking language (SQTL; Shen, Srisathapornphat, & Jaikaeo, 2001) is proposed as an application layer protocol that provides a scripting language. SQTL supports three types of events, which are defined by keywords *receive*, *every*, and *expire*. The receive keyword defines events generated by an snode when the snode receives a message, (every keyword defines events occurred periodically due to a timer time-out) and the expire keyword defines the events occurred when a timer is expired. If an snode receives a message that is intended for it and contains a script, the snode then executes the script.

Security in Sensor Networks

Security is one of the key issues, especially in tactical WSNs often deployed beyond the enemy lines. Various kinds of attacks against sensor networks are introduced in Karlof and Wagner (2003).

- **Spoofed, altered, or replayed routing information:** Routing information exchanged among nodes can be altered to create routing loops, attract or repel network traffic, extend or shorten the routes, generate false error messages, partition the network, increase the end-to-end latency, and so forth.
- **Selective forwarding:** Malicious nodes may refuse to forward certain messages, and drop them.
- **Sinkhole attacks:** A malicious node can be made very attractive to surrounding nodes with respect to the routing algorithm.
- **Wormholes:** An adversary tunnels messages received in one part of the network and replays them in a different part. Hence, the traffic in the network can be forwarded to an unintended receiver.
- **"Hello" flood attack:** A malicious node may broadcast routing or other information with large-enough transmission power and convince every node in the network that the adversary is its neighbor.
- **Acknowledgment spoofing:** An adversary can spoof link-layer acknowledgments for overheard packets addressed to neighboring nodes.
- **Sybil attack:** A single node presents multiple identities to other nodes in the network. This reduces the effectiveness of fault-tolerant schemes and poses a significant threat to geographic routing protocols.

The security protocols for sensor networks (SPINS) scheme is one of the security schemes proposed for sensor networks (Perrig, Szewczyk, Wen, Culler, & Tygar, 2001). In SPINS, two secure building blocks are used: the Secure Network Encryption Protocol

(SNEP) and μ TESLA (the micro version of the Timed, Efficient, Streaming, Loss-Tolerant Authentication Protocol). SNEP provides data confidentiality, two-party data authentication, and data freshness. It is based on a shared counter instead of randomization for semantic security. Hence, it has low overhead. μ TESLA provides authenticated streaming broadcast. It assumes that nodes are loosely time synchronized. An authentication key is broadcast by the node that makes broadcasts with specified time intervals. The current key is the function of the next key. Therefore, when every node knows the initial key, they can authenticate the central node every time that it broadcasts the new key.

A denial of service (DoS) attack is any event that diminishes the network capacity to perform its expected function. Each layer's vulnerabilities and defense mechanisms to DoS attacks are discussed for a typical sensor network in Wood and Stankovic (2002). Security, network bandwidth, and power consumption in sensor networks are discussed in Chen, Cui, Wen, and Woo (2000), where two applications have been implemented: target tracking and light sensing. Routing security in WSNs is introduced in Karlof and Wagner (2003).

The quarantine-region notion introduced in Sancak, Cayirci, Coskun, and Levi (2004) is an approach to reduce the cost of data security schemes in sensor networks. In this scheme, when the cnode finds out that there is an attack, it broadcasts a defend message. When an snode receives a defend message, it does not relay unauthenticated messages during a time period t_q. If it receives an unauthenticated message during t_q, it first requests from the last hop node of the message for authentication. If the last hop node fails in authentication, the node assumes that it is in the quarantine region and does not relay any data messages unless it is successfully authenticated, and it transmits its messages authenticated. As shown in Figure 4, the authentication region is the region where the transmissions of the malicious node can be received.

Figure 4 shows a sensor field where a quarantine region is indicated by the shaded area. The nodes 3, 4, 7, and 8, and an antinode are in the quarantine region, therefore they have to send authenticated messages. When a sensor has a message to send, it first checks if it is quarantined. If so, it sends the message authenticated. Nodes do not relay any unauthenticated messages when they are quarantined. For example, Nodes 3 and 4 do

Figure 4. Quarantine region

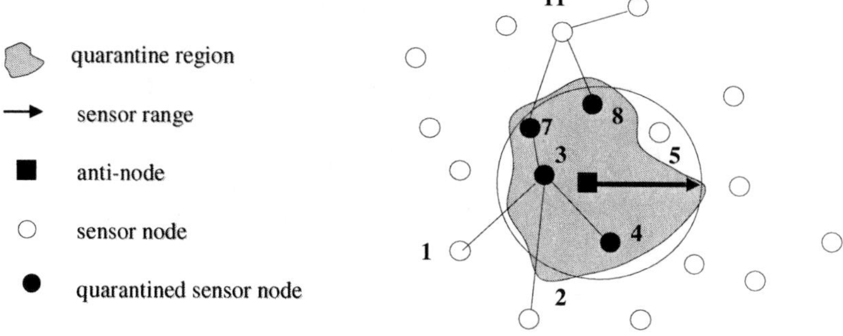

not relay messages coming from the antinode if they are not authenticated. Similarly, Node 7 does not relay unauthenticated messages coming from Nodes 3 and 4. Node 11 receives authenticated messages from Nodes 7 and 8, and transmits them unauthenticated. Nodes outside the quarantine regions do not need authentication to transmit a message, even if the message was an originally authenticated message coming from a quarantine region, unless their messages go through a quarantine region. For example, when Node 3 receives an unauthenticated message from Node 1 or 2, it first requests for authentication from the node that sends the message, and relays the message only after successful authentication.

Snodes determine when they can be out of the set of quarantined nodes. If a quarantined snode does not detect an unsuccessful authentication attempt during the quarantine period t_q, it switches back to not-quarantined mode. Snodes start a quarantine period every time they detect an unsuccessful authentication attempt. When a sensor is out of the quarantined set, it sends its messages unauthenticated unless the relaying node requests authentication, and it relays also unauthenticated messages. Snodes stay out of the quarantined set until they receive a defend message from the cnode.

Localization in Sensor Networks

Localization (Bulusu, Heideman, & Estrin, 2000; Doherty, Pister, & Ghaoui, 2001; Erdogan, Cayirci, & Coskun, 2003; Nasipuri & Li, 2002; Niculescu & Nath, 2003; Patwari, Hero, Perkins, Correal, O'Dea, 2003; Saverese, Rabaey, & Beutel, 2001; Savvides, Han, & Srivastava, 2003; Savvides, Park, & Srivastava, 2002) is one of the key issues in sensor networks because sensed data are almost meaningless without associating them with location data, in many applications. Moreover, localization may be needed for some tasks such as spatial addressing of the nodes and geographical routing. Similar to many other aspects related to sensor networks, the required level of localization accuracy differs from one application to another. For example, it may be enough to indicate in which room the measurement is carried out in one application. On the other hand, an accuracy level up to centimeters may be required in another application.

The first option for node localization is the Global Positioning System (GPS). However, GPS is not always a viable option for sensor networks. Nodes may be located in places where signals coming from satellites are not received with the required strength. In addition to this, GPS modules may be too expensive to attach every node in some applications. Therefore, GPS-less techniques are important for sensor networks.

GPS-less techniques are generally based on multilateration by using either the estimated distance or angle from beacon nodes. When the angle-based multilateration technique is used, the intersection point of the lines drawn from beacon nodes at the estimated directions gives the location of the node as shown in Figure 5a. In distance-based multilateration, the intersection of the circles that have the related beacon nodes at their centers and the radius equal to the distances from the beacon nodes is the estimated location of the node as shown in Figure 5b.

The distance from a beacon node can be estimated by using one of the following techniques: received signal strength (RSS), time of arrival (TOA), and time difference of

Figure 5. Multilateration

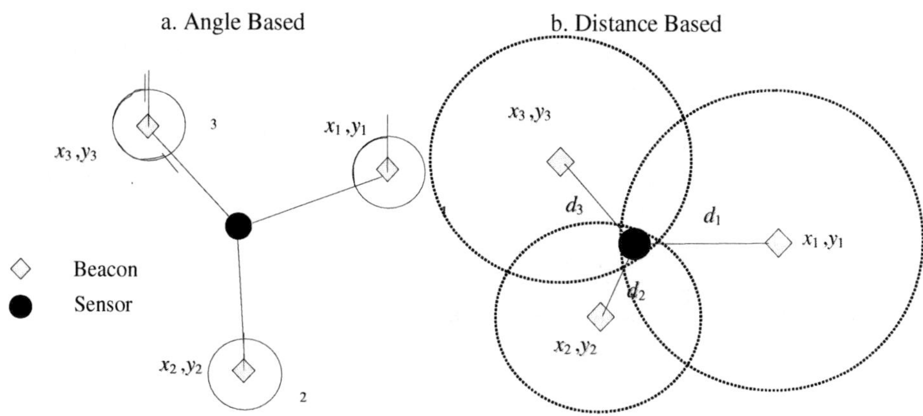

arrival (TDOA). The technique for estimating the direction of a beacon node is named angle of arrival (AOA). All these techniques have pros and cons. In RSS, a node knows the location of the beacons and the strength of the signals transmitted by them. Then it estimates the distance of the beacons by using a propagation model and received signal strengths. The results may not be highly accurate due to multipath effects, other impairments such as shadowing and scattering, and non-line-of-sight conditions. In TOA or TDOA, the node is time synchronized with the beacon nodes. It knows the location of the beacons together with the transmission time of the signals. When the node also knows the reception time, it takes a simple computation to find out the distance of the beacons based on the propagation speed of the signal. The results obtained by TOA or TDOA may also be impaired due to multipath effects and non-line-of-sight conditions. Moreover, the propagation speed of the RF signals is too high for sensor networks where the distances between nodes are only limited to a few meters in most of the cases. Therefore, ultrasound signals that have lower propagation speed may be preferred for this technique. The AOA technique is based on the usage of special antenna configurations. It may also be inaccurate due to multipath effects, non-line-of-sight conditions, and other sources of impairments in the wireless medium.

There are three approaches to carry out node-localization computations in sensor networks: centralized, distributed, and locally centralized (Savvides et al., 2003). In the centralized approach, all measurements are sent to a central node by snodes. The central node finds out the locations of the nodes by using these measurements, then disseminates the results. Since snodes have limited computational power and memory space, this may be a viable option for some applications. Moreover, in some applications, snodes may not need localization information, but the central node that carries out some tasks such as route optimization, optimal sensor-field-coverage computations, spatial data aggregation, and so forth, may need localization data. Also, the centralized approach may perform better for collaborative multilateration explained below. In the distributed approach, nodes find out their locations themselves. Clusters where a central node for each cluster computes the locations of the nodes in the cluster are established in the locally centralized approach.

In collaborative multilateration (Savvides et al., 2003), snodes collaborate with the other snodes for localization when they do not receive signals from enough numbers of beacons. For example, two snodes that can receive signals from two beacons can collaborate to alleviate the lack of the third beacon as shown in Figure 6. The basic idea is to have at least *n* equations to estimate *n* variables. In this approach, the solution uniqueness for the equations is required. The details about collaborative multilateration can be found in Savvides et al.

One very simple approach for node localization is the sectoral sweeper (SS) scheme (Erdogan et al., 2003). Although the resolution of the SS scheme is not as high as the other techniques explained in this section, it is simple enough to be implemented without any additional hardware or software components in the snodes. Moreover, the resolution of the SS scheme is high enough for many sensor network applications. The basic idea in the SS scheme is based on task dissemination by using directional antennae. Each task is also associated with minimum and maximum RSS values and a unique task identification. When an snode reports for a task, the task identification implies also a specific region. Please note that the borders of the task region cannot be very well defined, but a little amorphous as shown in Figure 7 due to multipath and non-line-of-sight effects. Creating overlapping task regions for the same task can enhance the resolution of the SS scheme. When an snode reports for multiple tasks, the intersecting area of the reported task regions is the location of the node. This intersection area is smaller than a task region. Hence location estimation with higher resolution can be achieved.

Time Synchronization in Sensor Networks

Time synchronization is also important for sensor networks not only because of the requirements by the protocols in various layers such as medium access control, that is,

Figure 6. Collaborative multilateration

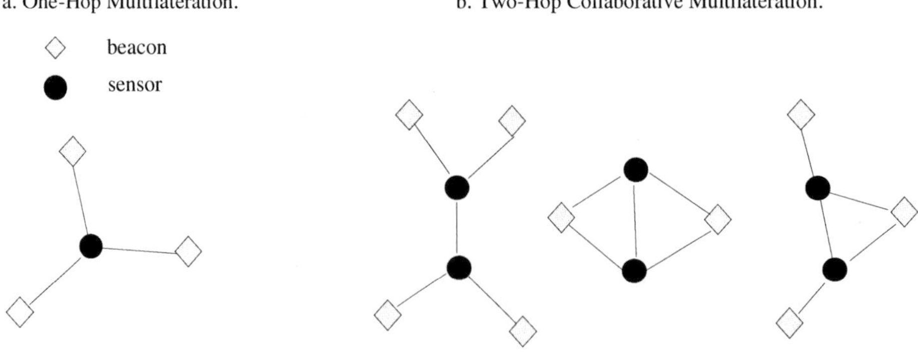

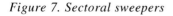

Figure 7. Sectoral sweepers

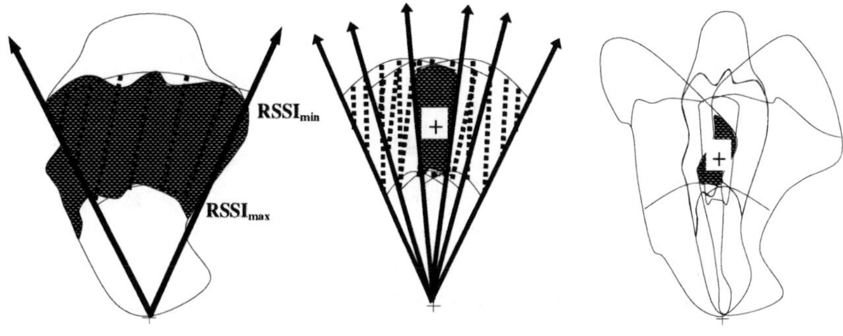

scheduling, and network layer, that is, routing and aggregation, but also sensed data are often needed to be related with a time. However, time synchronization is a more challenging task in sensor networks compared to the other ad hoc networking technologies due to the factors explained in the introduction. We can list the factors influencing time synchronization in large systems as follows (Elson, Girod, & Estrin, 2002; Levine, 1999; Mills, 1994, 1998; Su & Akyildiz, 2003).

- **Temperature:** Temperature variations during the day may cause the clock to speed up or down (a few microseconds per day).
- **Phase noise:** Access fluctuation can occur at the hardware interface, response variation of the operating system can cause interruptions, there can be a jitter in delay, and so forth.
- **Frequency noise:** The frequency spectrum of a crystal has large sidebands on adjacent frequencies.
- **Asymmetric delay:** The delay of a path may be different for each direction.
- **Clock glitches:** Hardware or software anomalies may cause sudden jumps in time.

Time-synchronization algorithms for sensor networks can be categorized into three broad classes. The first category is centralized time synchronization, where snodes are synchronized to a central timeserver. Network time protocol (Mills, 1994) falls in this category. The reference broadcast synchronization scheme (Elson et al., 2002) is an example of the distributed approach where the time is translated hop by hop throughout the network. In the third approach, a sensor network self-organizes for time synchronization (Su & Akyildiz, 2003). Nodes are synchronized within clusters, then clusters are synchronized with each other. This is a scalable scheme for sensor networks that may have thousands of nodes.

Transport Layer

The ultimate goal of a sensor network is the detection of specified events of interest occurred in a sensor field. Since the detection ranges of snodes often overlap, the same event is usually reported by multiple snodes. The sheer number of snodes, the environmental characteristics of sensor fields, and power limitation of the nodes may pose frequent, unexpected loss of data packets. Apart from occasional losses, all packets that report the same event information may be lost in some cases. Therefore, an event may be completely lost although multiple snodes report it. To overcome this problem, new end-to-end event transfer schemes are needed. In this chapter, we introduce a new group of end-to-end, reliable event transfer schemes. We focus on the problem of reporting the detected events by snodes to the cnode. Since there is not much work carried out on end-to-end flow-control and congestion-control issues, we do not discuss them in this chapter.

To the best of our knowledge, there has been a limited number of works on the design of an efficient, reliable transport protocol for sensor networks. The reliable multisegment transport (RMST; Stann & Wagner, 2003) scheme is one of these works, and it is designed to provide end-to-end, reliable data packet transfer for directed diffusion in a network-layer protocol, explained in the following section. RMST is a selective negative-acknowledgment-based (NACK) protocol that has two modes: caching mode and noncaching mode. In the caching mode, a number of nodes along a reinforced path, that is, the path that the directed diffusion protocol uses to convey the data to the cnode, are assigned as RMST nodes. Each RMST node caches the fragments of a flow. Watchdog timers are maintained for each flow. When a fragment is not received before the timer expires, a NACK is sent backward in the reinforced path. The first RMST node that has the required fragment along the path retransmits the fragment. The cnode acts as the last RMST node, and it becomes the only RMST node in the noncaching mode.

The pump-slowly, fetch-quickly (PSFQ) scheme (Wan, Campbell, & Krishnamurty, 2003) is similar to RMST (Stann & Wagner, 2003). PSFQ comprises three functions: message relaying (pump operation), relay-initiated error recovery (fetch operation), and selective status reporting (report operation). Every intermediate node maintains a data cache in PSFQ. A node that receives a packet checks its content against its local cache and discards any duplicates. If the received packet is new, the time to live (TTL) field in the packet is decremented. If the TTL field is higher than 0 after being decremented, and there is no gap in the packet sequence numbers, the packet is scheduled to be forwarded. The packets are delayed a random period between T_{min} and T_{max}, and then relayed. A node goes to fetch mode once a sequence-number gap is detected. The node in fetch mode requests the retransmission of lost packets from neighboring nodes.

PSFQ and RMST schemes are designed to enhance end-to-end data-packet-transfer reliability. The event-to-sink reliable transport (ESRT; Sankarasubramaniam, Akan, & Akyildiz, 2003) protocol is the first transport layer protocol that focuses on end-to-end reliable event transfer (EERET) in wireless sensor networks. In ESRT, reliable event transfer is not guaranteed, but increased by controlling the event-reporting frequencies of snodes. ESRT does not fit the requirements of many sensor network applications, either. Therefore, a new set of end-to-end reliable event-transfer protocols is proposed in Tezcan, Cayirci, and Caglayan (2003).

The main design issues of the EERET schemes in Tezcan et al. (2003) are the collective-cooperative paradigm and energy efficiency. The EERET schemes comply with the protocols at the lower layers and aim to increase reliability of event delivery with minimum energy expenditure. Characteristics of these schemes are summarized as follows.

- Designed for sensor networks, considering their characteristics such as scalability, hardware constraints, and power consumption.
- Simple and lightweight.
- Designed to provide different classes of reliability requirements to different applications.
- Independent from underlying network protocols.
- Compatible with existing applications.

In the reliable event transfer approach, an event is defined as the critical data generated by snodes. An event can be classified as critical or not by the application. The EERET schemes are designed for the delivery of these critical data packets. In most cases, the same critical data are generated by more than one snode because snodes are usually densely deployed in WSNs. We emphasize that an event is successfully transferred to the cnode when at least one packet reporting the event is received by the cnode.

For example, Nodes a, b, c, and d detect the same event illustrated in Figure 8. Since all three nodes can generate data packets reporting this event, the end-to-end transfer of the event can be succeeded even if one of them is received by the cnode. Reliability can be implemented in the transport layer, which has been traditionally concerned as an end-to-end issue. Another alternative is using medium access control (MAC) layer reliability, which is a hop-by-hop mechanism, where intermediate nodes take the responsibility for

Figure 8. End-to-end reliable event transfer

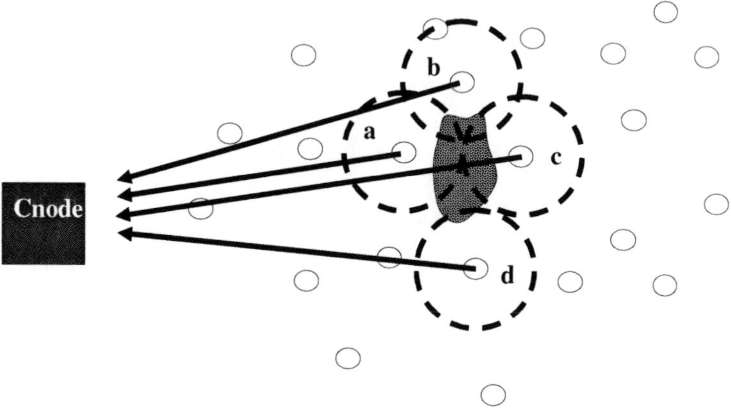

loss detection and recovery. Since the aim is to transfer the event to the cnode successfully rather than transferring each sensor report, hop-by-hop error recovery will bring extra unacceptable cost due to the large number of snodes.

The EERET schemes are categorize into two groups: nonacknowledgment-based (NoACK) and acknowledgment-based (ACK) schemes. In NoACK-based schemes, three methods, namely, event-reporting frequency, node density, and implicit acknowledgment, can be given. For the second group, another three schemes, that is, selective acknowledgment, enforced acknowledgment, and blanket acknowledgment, can be listed.

The next two sections explain the NoACK-based schemes and ACK-based schemes. NoACK-based schemes are a collection of alternative methods of increasing reliability without waiting for end-to-end acknowledgment. In contrast, ACK-based schemes make use of acknowledgment, but not as we use in connection-oriented, end-to-end protocols. Acknowledgment is used in different ways to provide reliability.

Nonacknowledgment-Based Schemes

In this section, the alternative schemes where snodes do not wait for an end-to-end acknowledgment are presented.

Implicit Acknowledgment

One method for reliable event transfer in sensor networks is implicit acknowledgment. Implicit acknowledgment makes use of the broadcast characteristic of the wireless channel. When a data packet sent by an snode is repeated by its gradients, that is, the gradient nodes repeat their packets to convey them to the cnode, the successful reception of the packet is also implied. Since it does not need a separate acknowledgment packet, its additional overhead is only due to listening to the media for a time interval.

One may argue that this is not an end-to-end scheme but a hop-by-hop technique. This is correct. However, this technique increases the end-to-end, reliable event-transfer rate.

Event-Reporting Frequency

This scheme is used by the ESRT protocol (Sankarasubramaniam et al., 2003), which is based on the event-to-cnode reliability model. The level of event-delivery reliability is controlled by an increasing or decreasing event-reporting frequency. As the reporting frequency increases, the number of packets sent by an snode increases. This decreases the probability that the reported event is lost. However, it also incurs additional power consumption.

One other point is that the reporting frequency can be increased until a certain point, beyond which the reliability drops. This is because the network is unable to handle the increased injection of data packets, and data packets are dropped due to congestion. The details about the scheme can be found in Sankarasubramaniam et al. (2003).

Node Density

In sensor networks, there are usually multiple nodes that have overlapping sensing regions. Hence, it is possible that multiple nodes collaborate to detect the same event. The number of nodes that report the same event has an impact on the end-to-end event-transfer reliability. As the number of snodes in critical regions or the number of nodes involved in reporting an event is managed using a network management protocol, the EERET rate can also be controlled. Higher EERET rates can be achieved by increasing the number of nodes involved in a sensing task. The TS concept (Cimen et al., 2003) explained previously can be used to manage the number of nodes involved in a sensing task. Therefore, we already have some practical techniques that can be used for node-density-based end-to-end event reliability control.

Acknowledgment-Based Schemes

Although the acknowledgment mechanism is a traditional way of achieving end-to-end reliability, it may not be viable for many WSN applications due to the following reasons.

- Most of the WSN applications have very stringent energy constraints. Therefore, the overhead of the acknowledgment packets may not be justifiable.
- Since some of the events reported by snodes may not be as critical as some others, generating acknowledgment for all packets received may incur unnecessary cost.
- Since many snodes may report the same event, acknowledging all of them by a single acknowledgment may be more effective.

Selective Acknowledgment

Since WSN consists of thousands of snodes, which are densely deployed, waiting for an acknowledgment for each data packet is inappropriate (Tezcan et al., 2003). Instead, each snode may activate the acknowledgment mechanism when it detects critical data. There are various ways to determine whether a data packet carries critical data or not. One approach is using a threshold value. Snodes and the cnode come to an agreement on a threshold value before deployment that depends on the application. Then, an snode decides whether the measurement is critical or not by comparing the agreed threshold value. For example, a temperature sensor may report the temperature periodically. Unless the reported temperature does not change by a threshold value, the reported data can be accepted as not critical. When a change of more than the threshold value is observed, this can be accepted as critical. In other cases, lost data can be obtained by interpolation.

Snodes wait for acknowledgments only when critical data is reported. On the other end, each data packet received by the cnode is also compared with the threshold value and categorized as critical or not. If a critical data packet is received, an acknowledgment packet is sent to the snode immediately. If the snode does not receive an acknowledgment packet during a predetermined time-out period, it retransmits the packet. In this regard, such an acknowledgment mechanism controls the event transfer with minimum overhead on the traffic, and increases reliability.

Enforced Acknowledgment

In enforced acknowledgment (Tezcan et al., 2003), the basic idea is almost the same as in selective acknowledgment. The difference is that the cnode does not compute whether the received data packet carries critical data or not. Instead, the snode computes this before sending the packet, and marks the packet if it carries critical data. The cnode sends back an acknowledgment when it receives a packet marked as a packet that carries critical data.

Blanket Acknowledgment

Multiple snodes reporting the same event may be acknowledged by a single acknowledgment packet (Tezcan et al., 2003). A single acknowledgment packet can be broadcast to all snodes reporting the same event by using a technique similar to sectoral sweepers. For example, the blanket-acknowledgment scheme can be used in the sensor networks for a disaster-relief operation where snodes are responsible to report live humans trapped under rubble. When the cnode acknowledges the presence of a live human under the rubble, snodes do not need to worry whether their report is acknowledged or not.

Blanket acknowledgment can also be used in conjunction with selective and enforced acknowledgment to broadcast the acknowledgment packets.

Network Layer

The network layer is one of the most challenging layers in sensor networks because all of the factors influencing sensor networks have an impact on network-layer issues, such as addressing, data aggregation, and routing.

Addressing in Sensor Networks

The unique features and application requirements of WSNs (Akyildiz et al., 2002) pose new challenges in node addressing. Fixed and universal addressing of snodes is not a viable option for many sensor network applications. Therefore, attribute-based naming or local identification of snodes has thus far been considered to address a specific sensor for various purposes such as node management, data querying, data aggregation, and routing. We categorize the node-addressing techniques applicable for sensor networks as follows (Cayirci, 2003a).

- **Attribute-based naming and datacentric routing:** Attribute-based naming (Intanagonwiwat, Govindan, & Estrin, 2000) is one of the earliest techniques used for node addressing in WSNs. In this technique, nodes that measure a certain amplitude for a specified attribute are called, for example, "nodes that measure a temperature more than 35°C."
- **Spatial addressing:** Spatial addressing is especially useful in applications such as intrusion detection and target tracking, where queries are mainly based on node locations. It is also needed for spatial data aggregation and geographic routing schemes.
  - **Polygonal addressing:** In this technique, the border of a region is defined, and then the nodes inside, outside, or in a buffer zone that has a certain depth along this border are queried. Borders of the region can be detected by a distributed edge-detection (Chintalapudi & Govindan, 2003) algorithm, or can be specified by giving a series of geographical locations.
  - **Sectoral sweepers** (Erdogan et al., 2003): Spatial regions can also be specified by using a directional antenna, where snodes that receive a signal within a certain range of received signal strength respond to a query.
  - **Quadtree- or Octree-based addressing** (Cimen et al., 2003): Quadtrees and octrees can be used to partition the sensor field into quadrants or octants, and then these quadrants or octants are queried. Since a location-aware node can easily find out which quadrant or octant it is in for a given tree depth, nodes in the queried partitions respond. Various query patterns that can be created by this approach are examined in Cimen et al.
  - **Modulus addressing** (Cayirci, 2003b): Hash functions can also be used to include or exclude snodes in queries.
- **Using local identifications for nodes, and mapping the destination of the user queries to the local identifications:** In this scheme, every node is addressed by a local identification in the sensor field. The destination of incoming task packets or queries is mapped to these local identifications by the intermediate nodes such as cnodes or proxy servers. Users may indicate a destination by using either an attribute-based naming or spatial addressing scheme. A gateway node maps this address to a local identification.
- **Address reuse:** Address reuse is especially useful for MAC layer addressing of snodes. The same addresses can be assigned to multiple nodes as long as this does not cause a conflict. A distributed protocol for address reuse is proposed in Schurgers, Kulkarni, and Srivastava (2002). In this protocol, a node first broadcasts a "hello" message. The nodes that hear this message reply with an "info" message where they declare their local addresses. The node that broadcasts the "hello" message randomly selects an address not used by its neighbors by using the data in the "info" messages. If there is a conflict due to a hidden node problem, the first node that detects this conflict sends a "conflict" message to both of the nodes, and one of them changes its address.

Data Aggregation in Sensor Networks

In sensor networks, data aggregation (Boulis, Ganerival, & Srivastava, 2003; Cayirci, 2003b; Zhao, Govindan, & Estrin, 2003) is important due to two reasons. First, it is needed for tasks such as data fusion, data association, and data aggregation. Second, it is perceived as a way to reduce the communication overhead. Therefore, there has been considerable research effort on data aggregation that can be classified as follows.

- **Temporal or spatial aggregation:** Data can be aggregated time based or location based. For example, the temperature readings for every hour or temperature readings for various regions in a sensor field can be averaged. Also, a hybrid-approach combination of time- and location-based aggregation can be used.
- **Snapshot or periodical aggregation:** Data aggregation can be made snapshot, that is, one time, on the receipt of a query. Alternatively, temporarily aggregated data can be reported periodically.
- **Centralized or distributed aggregation:** A central node can gather and then aggregate data, or data can be aggregated while being conveyed through a sensor network. A hybrid approach is also possible where clusters are set, and a node in each cluster aggregates the data from the cluster.
- **Early or late aggregation:** Data can be aggregated at the earliest opportunity, or aggregation of data may not be allowed before a certain number of hops so as not to hinder the collaboration among the neighboring nodes.

Routing in Sensor Networks

Routing in sensor networks has attracted many researchers as a new research field recently. Although many of them were designed based on a critical assumption that snodes are not mobile, one can find a routing protocol that can fulfill the requirements of almost any sensor network application. The routing protocols can be broadly classified as datacentric, hierarchical, or location based (Akkaya & Younis, 2004).

Datacentric Routing Protocols for Sensor Networks

Flooding, gossiping, rumor routing (Braginsky & Estrin, 2002), sensor protocols for information via negotiation (SPIN; Heinzelman, Kulik, & Balakrishan, 1999), sequential assignment routing (SAR; Sohrabi, Gao, Ailawadhi, & Pottie, 2000), directed diffusion (Intanagonwiwat et al., 2000), energy-aware routing (Shah & Rabaey, 2002), threshold-sensitive and energy-efficient sensor network (TEEN; Manjeshwar & Agrawal, 2002), and constraint anisotropic diffusion routing (CADR; Chu, Hausecker, & Zhao, 2002) are examples of the protocols that fall in this category. In this section, we explain two of these protocols.

SPIN is based on the advertisement of data available in snodes. When an snode has data to send, it broadcasts an advertisement (ADV) packet. The nodes interested in this data

Figure 9. SPIN protocol

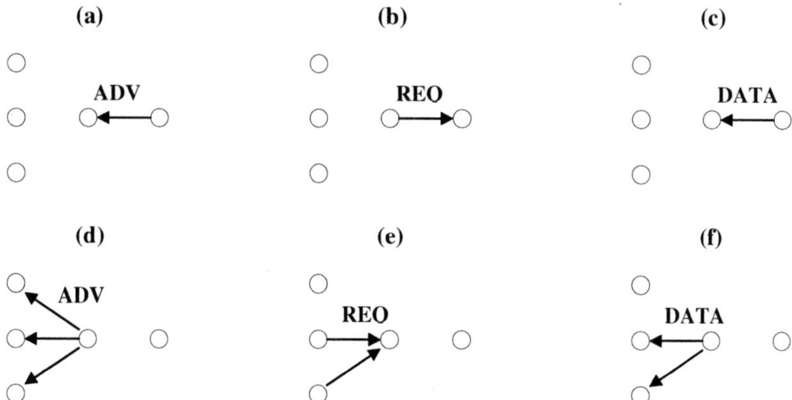

reply back by a request (REQ) packet. Then the snode disseminates the data to the interested nodes by using data (DATA) packets. When a node receives data, it also broadcasts an ADV packet and relays DATA packets to the nodes that send REQ packets. Hence, the data is delivered to every node that may have an interest. This process is shown in Figure 9.

In SPIN, the routing process is stimulated by snodes. Another approach, namely, directed diffusion, is cnode oriented. In directed diffusion, the cnode floods a task throughout the sensor network. While the task is being flooded, snodes record the nodes which send the task to them as their gradient, and hence, the alternative paths from snodes to the cnode is established. When there is data to send to the cnode, this is forwarded to the gradients. One of the paths established is reinforced by the cnode. After that point, the packets are not forwarded to all of the gradients but to the gradient in the reinforced path. Directed diffusion is illustrated in Figure 10.

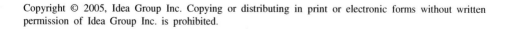

Hierarchical Routing Protocols for Sensor Networks

The low-energy, adaptive clustering hierarchy (LEACH; Heinzelman, Chandrakasan, & Balakrishnan, 2000), power-efficient gathering in sensor information systems (PEGASIS;

Figure 10. Directed-diffusion protocol

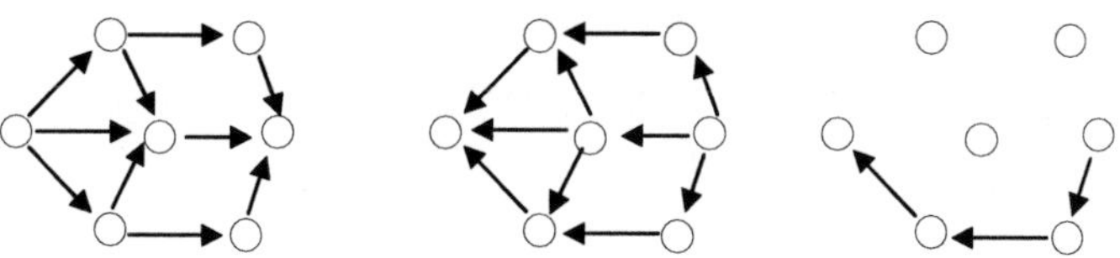

Lindsey & Raghavendra, 2002), and the self-organizing protocol (Subramanian & Katz, 2000) are in this category. These techniques tackle the scalability factor by clustering nodes for routing. For example, in LEACH, any snode can elect itself as a cluster head at anytime with a certain probability. Snodes access the network through the cluster head, which requires minimum energy to reach.

Location-Based Routing Protocols for Sensor Networks

Location-based algorithms such as the minimum-energy communication network (MECN) and small MECN (SMECN; Li & Halpern, 2001), and Geographic Adaptive Fidelity (GAF; Yu, Heideman, & Estrin, 2001) make routing decisions based on geographic locations of snodes. In SMECN, it is assumed that the exact locations of snodes are known. Based on these locations, a sensor network is represented as a graph. Then the subgraph that connects all nodes with minimum energy cost is computed by using a graph theoretic approach.

Conclusion

WSNs constitute a very challenging yet equally important venue for research in many disciplines. Factors such as scalability, fault tolerance, operating environment, production cost, hardware constraints, network topology, transmission media, and power considerations influence the design of protocols and algorithms needed for sensor networks. In this chapter, we explain these factors, and then examine the proposed schemes that fulfill the constraints related to them. We focus on application-, transport-, and network-layer protocols, and summarize the results of the research reported in the literature for these layers.

References

Akkaya, K, & Younis, M. (2004). A survey on routing protocols for wireless sensor networks. *Ad Hoc Networks* (to appear).

Akyildiz, I. F., Su, W., Sankarasubramaniam, Y., & Cayirci, E. (2002). A survey on sensor networks. *IEEE Communications Magazine*, 38, 102-114.

Boulis, A., Ganerival, S., & Srivastava, M. B. (2003). Aggregation in sensor networks: An energy accuracy tradeoff. *Proceedings of IEEE International Workshop on Sensor Network Protocols and Applications (SNPA)*, (pp. 128-138).

Braginsky, D., & Estrin, D. (2002). Rumor routing algorithm for sensor networks. *Proceedings of the ACM Workshop on Sensor Networks and Applications (WSNA)*, (pp. 22-30).

Bulusu, N., Heideman, J., & Estrin, D. (2000). GPS-less low cost outdoor localization for very small devices. *IEEE Personal Communication, 7*, 28-34.

Cayirci, E. (2003a). Addressing in wireless sensor networks. *Proceedings of COST-NSF Workshop on Exchanges and Trends in Networking (Nextworking)*, Crete.

Cayirci, E. (2003b). Data aggregation and dilution by modulus addressing in WSNs. *IEEE Communications Letters*, 7 (8), 355-357.

Chen, M., Cui, W., Wen, V., & Woo, A. (2000). Security and deployment issues in a sensor network.

Chintalapudi, K. K., & Govindan, R. (2003). Localized edge detection in sensor fields.

Chu, M., Hausecker, H., & Zhao, F. (2002). Scalable information-driven sensor querying and routing for ad hoc heterogeneous sensor networks. *International Journal of High Performance Computing Applications, 16*(3), 90-110.

Cimen, C., Cayirci, E., & Coskun, V. (2003). Querying sensor fields by using quadtree based dynamic clusters and task sets. *Proceedings of the IEEE Military Communications Conference (MILCOM)*, Boston.

Doherty, L., Pister, K. S. J., & Ghaoui, L. E. (2001). Convex position estimation in wireless sensor networks. *Proceedings of the IEEE International Conference on Computer Communications (INFOCOM).*

Elson, J., Girod, L., & Estrin, D. (2002). Fine grained network time synchronization using reference broadcasts. *Proceedings of the ACM International Symposium on Operating Systems Design and Implementation (OSDI).*

Erdogan, A., Cayirci, E., & Coskun, V. (2003). Sectoral sweepers for sensor node management and location estimation in ad hoc sensor networks. *Proceedings of the IEEE Military Communications Conference (MILCOM).*

Heinzelman, W. R., Chandrakasan, A., & Balakrishnan, H. (2000). Energy-efficient communication protocol for wireless microsensor networks. *Proceedings of the IEEE Hawaii International Conference on System Sciences*, (pp. 1-10).

Heinzelman, W. R., Kulik, J., & Balakrishan, H. (1999). Adaptive protocols for information dissemination in wireless sensor networks. *Proceedings of the IEEE/ACM International Conference on Mobile Computing and Communications (MOBICOM)*, (pp. 174-185).

Helmy, A. (2003). Mobility-assisted resolution of queries in large-scale mobile sensor networks [Special issue]. *Computer Network, 43*(4), 437-458.

Intanagonwiwat, C., Govindan, R., & Estrin, D. (2000). Directed diffusion: A scalable and robust communication paradigm for sensor networks. *Proceedings of the IEEE/ACM International Conference on Mobile Computing and Communications (MOBICOM),* Atlanta.

Karlof, C., & Wagner, D. (2003). Secure routing in wireless sensor networks: Attacks and countermeasures. *Proceedings of IEEE International Workshop on Sensor Network Protocols and Applications (SNPA),* Alaska, (pp. 113-127).

Levine, J. (1999). Time synchronization over the Internet using an adaptive frequency locked loop. *IEEE Transactions on Ultrasonics, Ferroelectronics, and Frequency Control*, 888-896.

Li, L., & Halpern, J. Y. (2001). Minimum-energy mobile wireless networks revisited. *Proceedings of IEEE International Conference on Communications (ICC).*

Lindsey, S., & Raghavendra, C. S. (2002). PEGASIS: Power efficient gathering in sensor information systems. *Proceedings of the IEEE Aerospace Conference.*

Manjeshwar, A., & Agrawal, D. P. (2002). TEEN: A protocol for enhanced efficiency in wireless sensor networks. *Proceedings of the IEEE Wireless Communication and Networking Conference (WCNC).*

Mills, D. L. (1994). *Internet time synchronization: The network time protocol. Global States and Time in Distributed Systems.* IEEE Computer Society Press.

Mills, D. L. (1998). Adaptive hybrid clock discipline algorithm for the network time protocol. *IEEE Transactions on Networking*, 6 (5), 505-514.

Nasipuri, A., & Li, K. (2002). A directionality based location discovery scheme for wireless sensor networks. *Proceedings of the ACM Workshop on Wireless Sensor Networks and Applications (WSNA)*, (pp. 105-111).

Niculescu, D., & Nath, B. (2003). Localized positioning in ad hoc networks. *Proceedings of IEEE International Workshop on Sensor Network Protocols and Applications (SNPA)*, 42-50.

Patwari, N., Hero, A.O., Perkins, M., Correal, N.S., & O'Dea, R.J. (2003). Relative location estimation in wireless sensor networks. *IEEE Transactions on Signal Processing.*

Perrig, A., Szewczyk, R., Wen, V., Culler, D., & Tygar, J. D. (2001). SPINS: Security protocols for sensor networks. *Proceedings of the IEEE/ACM International Conference on Mobile Computing and Communications (MOBICOM)*, (pp. 189-199).

Pottie, G. J., & Kaiser, W. J. (2000). Wireless integrated network sensors. *Communications of the ACM*, *43*(5), 551-558.

Sadagopan, N., Krishnamachari, B., & Helmy, A. (2003). The acquire mechanism for efficient querying in sensor networks. *Ad Hoc Networks.*

Sancak, S., Cayirci, E., Coskun, V., & Levi, A. (2004). Sensor wars: Detect and defend against spam attacks in sensor networks. *Proceedings of the IEEE International Conference on Communications (ICC),* Paris.

Sankarasubramaniam, Y., Akan, O. B., & Akyildiz, I. F. (2003). ESRT: Event-to-sink reliable transport in wireless sensor networks. *Proceedings of the IEEE/ACM International Conference on Mobile Computing and Communications (MOBICOM),* Annapolis.

Saverese, C., Rabaey, J., & Beutel, J. (2001). Locationing in distributed ad hoc wireless sensor networks. *Proceedings of the IEEE International Conference on Acoustics, Speech and Signal Processing (ICASSP),* Salt Lake City.

Savvides, A., Han, C., & Srivastava, M. (2003). Dynamic fine grained localization in ad-hoc networks of sensors. *Proceedings of the IEEE/ACM International Conference on Mobile Computing and Communications (MOBICOM).*

Savvides, A., Park, H., & Srivastava, M. (2002). The bits and flops of the n-hop multilateration primitive for node localization problems. *Proceedings of the ACM*

Workshop on Wireless Sensor Networks and Applications (WSNA), (pp. 105-111).

Schurgers, C., Kulkarni, G., & Srivastava, M. B. (2002). Distributed on-demand address assignment in wireless sensor networks. *IEEE Transactions on Parallel and Distributed Systems*, *13*(10), 1056-1065.

Shah, R., & Rabaey, J. (2002). Energy aware routing for low energy ad hoc sensor networks. *Proceedings of the IEEE Wireless Communication and Networking Conference (WCNC).*

Shen, C., Srisathapornphat, C., & Jaikaeo, C. (2001). Sensor information networking architecture and applications. *IEEE Personal Communications*, 52-59.

Sohrabi, K., Gao, J., Ailawadhi, V., & Pottie, G. J. (2000). Protocols for self organization of a wireless sensor network. *IEEE Personal Communications*, 16-27.

Stann, F., & Wagner, J. (2003). RMST: Reliable data transport in sensor networks. *Proceedings of the IEEE International Workshop on Sensor Network Protocols and Applications (SNPA)*, 102-112.

Su, W., & Akyildiz, I .F. (2003). *Time diffusion synchronization protocol for sensor networks* (Tech. Rep. No.). Georgia Tech.

Subramanian, L., & Katz, R. (2000). An architecture for building self configurable systems. *Proceedings IEEE/ACM Workshop on Mobile Ad Hoc Networking and Computing (MOBIHOC).*

Tezcan, N., Cayirci, E., & Caglayan, U. (2003). End-to-end reliable event transfer in wireless sensor networks. *Proceedings of the IEEE International Symposium on Personal, Indoor and Mobile Radio Communications (PIMRC 2004).*

Wan, C.-Y., Campbell, A. T., & Krishnamurty, L. (2003). PSFQ: A reliable transport protocol for wireless sensor networks. *Proceedings of the ACM Wireless Sensor Networks and Applications (WSNA)*, (pp. 1-11).

Warneke, B., Liobewitz, B., & Pister, K. S. J. (2001). Smart dust: Communicating with a cubic millimeter computer. *IEEE Computer*, *34* (1), 2-9.

Wood, A. D., & Stankovic, J. A. (2002). Denial of service in sensor networks. *IEEE Computer*, 54-62.

Yu, Y., Heideman, J., & Estrin, D. (2001). Geography-informed energy conservation for ad hoc routing. *Proceedings of IEEE/ACM International Conference on Mobile Computing and Communications (MOBICOM).*

Zhao, J., Govindan, R., & Estrin, D. (2003). Computing aggregates for monitoring wireless sensor networks. *Proceedings of the IEEE International Workshop on Sensor Network Protocols and Applications (SNPA)*, (pp. 139-149).

Chapter X

Connection Admission Control in Wireless Systems

Tuna Tugcu, Georgia Institute of Technology, USA

Abstract

Connection Admission Control (CAC) is the process that decides which connection requests are admitted to the system and allocated resources. CAC in wireless networks differs from wireline networks due to mobility and scarcity of wireless resources, and the physical properties of the radio channels. In this chapter, the basic issues in CAC for wireless systems are discussed in the context of resource management and trade-off between blocking and dropping rates. Though it is not among the topics of this chapter, quality of service (QoS) provisioning is also briefly mentioned due to its relationship with CAC. Following the discussion of the common and different points of CAC in both wireline and wireless systems, admission control in next-generation wireless systems is explained.

Introduction

Connection Admission Control (CAC) is the heart of a telecommunications system since it both determines the system's throughput and affects user satisfaction. During the past few decades, research has focused on improving CAC in wireline networks to increase throughput without impairing user satisfaction. However, the mobility of users, the shortage of wireless resources, and the physical properties of the radio channels in wireless systems add new dimensions to the complexity of the problem. Since the scarce radio resources constitute the bottleneck in wireless systems, CAC schemes for wireless systems generally focus on the management of the radio resources rather than the resources in the wireline portion of the system.

Efficient management of the scarce radio resources in wireless systems is vital to the overall performance of the system. The CAC process, which follows the paging process, is a crucial part of resource management. It is the CAC scheme that decides which connection requests are admitted into the system and granted resources. The CAC scheme must consider the impacts of accepting requests on the other connections in the same and surrounding cells. These impacts result from both the interference caused by the new connections and the possible future handovers to the surrounding cells.

The CAC schemes in wireless systems differ from their counterparts in wireline networks by taking care of user mobility. A wireless system must be able to convey the connection of an active user (i.e., a user with an ongoing connection) from one cell to the other when the user moves between cells. To avoid service interruption during such handover operations, the system needs to employ prioritization or reservation schemes to keep connection dropping and blocking rates at reasonable levels. Once a connection request is admitted, the system must do its best to keep the connection alive.

CAC also plays a significant role in providing quality of service (QoS). Future telecommunication systems like 3G and Next-Generation Wireless Systems (NGWS) aim at providing integrated services such as voice, high-bandwidth data, and multimedia with QoS support. CAC must consider QoS requirements during connection set-up and handovers. In the case of NGWS, QoS requirements must be translated between the subsystems in the case of vertical handoffs. CAC needs to work hand in hand with resource management to achieve this task.

In this chapter, the basic issues in CAC for wireless systems are discussed. To provide a thorough discussion of admission control, we start by addressing the issues common to CAC in both wireline and wireless systems. Then, we focus on issues that are specific to wireless systems. This discussion is followed by admission control in NGWS. We finally conclude by summarizing the important points in admission control.

CAC and Resource Management

The wireless spectrum constitutes the bottleneck in wireless systems. CAC is the process of deciding which connection requests are admitted into the system and allocated these scarce resources. Efficient management of the radio resources determines the upper limit on the performance of the overall system. To achieve greater efficiency, the topics of wireless spectrum assignment to the cells, admission of new and handover connections into the system, and QoS provisioning should be considered together.

Resource Management

Link capacity in the backbone of a wireless system is typically abundant. Even if a link in the backbone does not have enough capacity, its capacity can easily be increased. However, this is not the case for the radio links. The capacity of the radio link is dictated by the radio spectrum allocated to the wireless system. Coexistence of multiple generations of wireless systems limits the frequency band available for new wireless systems, making the radio spectrum the scarcest resource in the network. It is the task of CAC and resource management schemes to administer this scarce resource efficiently to maximize user satisfaction and network throughput.

Resource management, FDMA-based systems can be classified into three categories according to the assignment of channels to cells: *fixed channel allocation schemes* (FCA), *dynamic channel allocation schemes* (DCA), and *hybrid channel allocation schemes* (HCA; Katzela & Naghshineh, 1996). FCA schemes partition the available spectrum into frequency bands, and assign the frequency bands to cells according to a reuse criterion. Though the FCA schemes are simple to implement, they do not adapt to fluctuations in traffic. In DCA schemes, channels are allocated dynamically from a pool according to need. DCA schemes adapt to changing traffic patterns better than FCA schemes, but their complex, centralized mechanisms are the major drawbacks. Furthermore, their performance deteriorates under high traffic-load conditions. HCA combines FCA and DCA schemes to overcome the problems in both schemes. On the other hand, in CDMA systems, frequency planning is not a problem, but inner- and outer-cell interference is the key issue in admission control (Jeon & Jeong, 2002; Liu & Zarki, 1994; Shin, Cho, & Sung, 1999; Tugcu & Ersoy, 2001).

Types of Failure in Connection: Blocking and Dropping

In wireline networks, the CAC scheme aims at reducing the connection blocking rate, the ratio of rejected connection requests. This is achieved by maintaining a conservative approach in the allocation of resources to connections. Once a connection is admitted, the network ensures that the connection gets enough resources.

The situation becomes more complex in the case of wireless systems due to the mobility of the users. In addition to connection blocking rate, CAC in wireless systems must also consider the *connection dropping rate* (also called *forced connection-termination rate*), the ratio of ongoing connections that are forcefully terminated without the will of the users due to insufficient radio resources in the new cell during handover.

From the perspective of the mobile user, the most tangible criteria that determine user satisfaction are connection blocking and dropping rates. It is widely accepted that connection dropping is considered to be more annoying than connection blocking (Lin & Chlamtac, 2001; Luo, Thng, & Zhuang, 1999; Tugcu & Ersoy, 2001). In order to reduce the connection dropping rate, CAC must consider the *pressure from surrounding cells* while making the admission decision for a new connection request. In other words, during the admission decision for a new call, CAC must put aside some resources for possible handover calls from the surrounding cells. Therefore, the CAC scheme running in a cell cannot work independent of the surrounding cells.

Prioritization and queuing are two direct solutions for lowering the connection dropping rate (Senarath & Everitt, 1995). In prioritization, handover connections are given higher priority than new connections. However, this approach results in the starvation of new connections in cells intersected by highways. Since the number of handover requests to these cells will be very high, stationary users in such cells will not be able to get enough radio resources. Queuing, on the other hand, is not practical since the time required to complete connection establishment or handover is very short compared to connection duration.

The connection dropping rate can be lowered by reserving some resources, called *guard channels*, specifically for handover connections. Since the guard channels are reserved for handoff calls, new call requests will not be granted if all channels except the guard channels are busy. Therefore, determining the optimum number of guard channels is a crucial issue for the system performance. If the number of guard channels is too high, many new call attempts will fail although there are free channels. On the contrary, if the number of guard channels is too low, many handoff events will fail resulting in a high forced call-termination rate. Thus, there is a trade-off between connection blocking and connection dropping. The decision for the optimum number of guard channels is both time and space dependent. The number of guard channels depends on the location of the cell and random events like traffic congestion, accidents, or festivals. Therefore, assigning a predetermined number of guard channels to each cell increases dropping rates in some cells and blocking rates in others.

To adjust the number of guard channels dynamically to cope with varying connection traffic, channel reservation is applied. Channels are reserved only in the cells on the future path of the subscriber. Since the path that the subscriber will follow is independent of the planning of the spectral resources, it is not possible to exactly know the set of cells on his or her way. However, a good estimation for this set can be made by considering the fact that subscribers move toward a destination instead of making random moves. Therefore, the path that a subscriber follows is the concatenation of multiple line segments toward the destination. A reservation area may be formed by considering this fact. Though not guaranteed, it will be very likely that the subscriber will remain in this reservation area in the near future. A generic algorithm for CAC with reservations is given in Figure 1.

In the literature, most of the previous work on guard channels assigns a fixed number of guard channels to each cell (Gavish & Sridhar, 1997; Katzela & Naghshineh, 1996; Liu & Zarki, 1994; Shin et al., 1999). As stated above, a fixed number of guard channels is vulnerable to fluctuations in new call generation and handoff rates. Methods for a variable number of guard channels have been proposed in Hou, Fang, and Akansu (2000), Kim, Lee, Lee, Kim, and Mukherjee (1999), Levine, Akyildiz, and Naghshineh (1997), Ma, Han, and Trivedi (2002), Oliviera, Kim, and Suda (1998), and Tugcu and Ersoy (2001). The work in Levine et al. and Oliviera et al. is for TDMA and FDMA systems. The air interface is not specified in Kim et al.. Only the work in Hou et al., Ma et al., and Tugcu and Ersoy propose a variable number of guard channels for CDMA systems. In Lee and Cho (2000),

Figure 1. Generic CAC with reservations

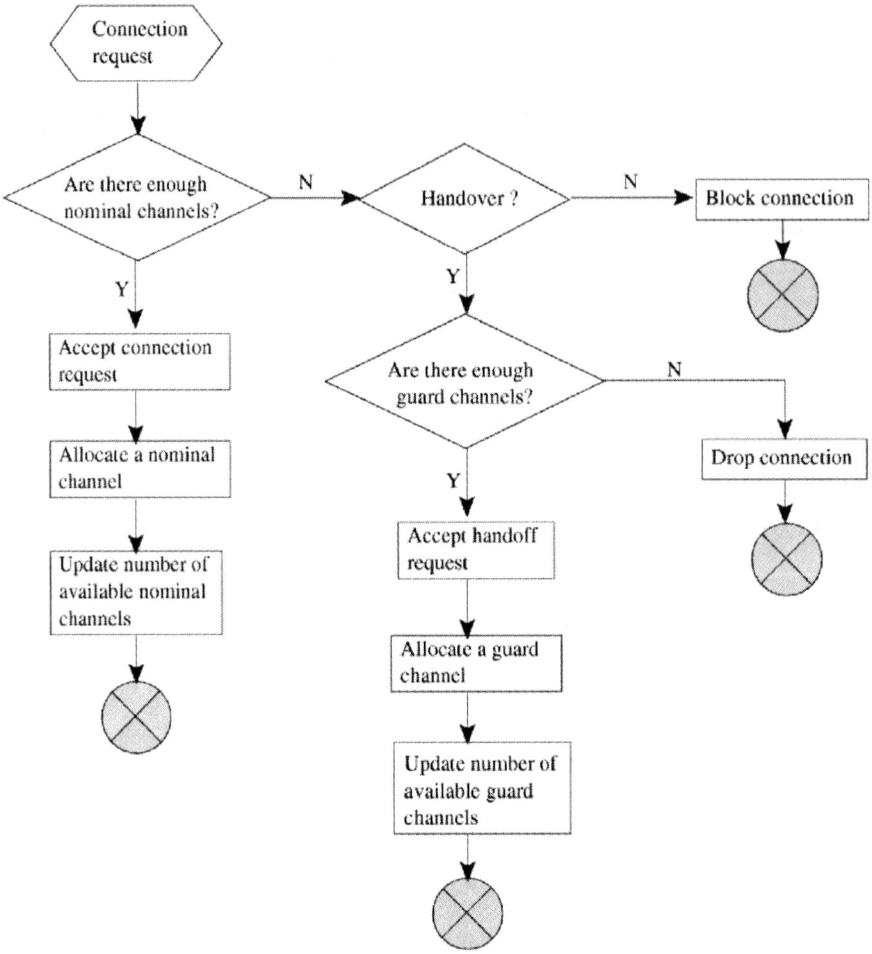

a scheme that borrows channels from stationary calls participating in handoffs for handoff requests by moving mobiles is proposed.

Considering QoS in Admission Control

CAC tries to maximize the number of connections admitted into the system while maintaining the QoS requirements of existing connections. Therefore, CAC must avoid congestions in advance and come up with a Boolean decision to accept or reject.

Applications may require QoS in different ways.

- **Quantitative (parameterized) QoS requirements:** To provide the guaranteed service required by this category of applications, the system should enforce explicit resource reservation in CAC. Strict traffic shaping and control together with efficient scheduling should be employed. The most-commonly used attributes and characteristics are packet size, data rate, maximum delay, delay variance, and acknowledgement policy.

- **Qualitative QoS requirements:** This category of applications does not require hard guarantees as opposed to quantitative QoS. The traffic for these applications is served according to differentiated services principles. The service classes are prioritized over each other, but no per-flow guarantees are provided.
- **Best effort:** These applications do not require any QoS service at all. The system may reserve a limited amount of bandwidth, determined by the system policy, to ensure these applications do not starve.

On the mobile terminal side, the flow of data from the application to the networks is as follows. If guaranteed service is required, the *resource reservation entity* in the mobile terminal communicates with its peer in the wireless access node (typically the base station or the access point) to set up the path and make necessary reservations. The application passes all data packets to the *classification entity*, which places the packets into the appropriate queue. The packets are retrieved from the queues by the *packet scheduler entity* and handed to the *channel access entity*, which transmits the packets on the wireless medium according to the basics of the access technology used. If guaranteed service is not required, the resource reservation step above is skipped.

Connections are generally defined using the following QoS attributes:

- Attributes related to bandwidth:
  - Average data rate
  - Peak data rate
  - Minimum acceptable data rate
  - Maximum burst size (maximum number of bits sent at the peak rate)
- Attributes related to delay:
  - Maximum delay
  - Maximum jitter (variation in delay)
- Attributes related to reliability of the connection:
  - Bit Error Rate (BER) or Frame Error Rate (FER)
  - Maximum loss ratio (ratio of undelivered frames to received frames)

CAC has three basic components: traffic descriptors, admission criteria, and measurements (Figure 2). A traffic descriptor is a set of parameters that characterize a traffic source. A typical traffic descriptor is a *token bucket*, composed of a token fill rate r and a token bucket size b. A source described by such a token bucket will transmit at most $r \times t + b$ bytes. Admission criteria are the rules used by the CAC scheme to make the decision. CAC must consider the effect of the new connection on the existing connections. To make the decision, CAC needs an accurate measure of the amount of congestion

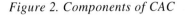

Figure 2. Components of CAC

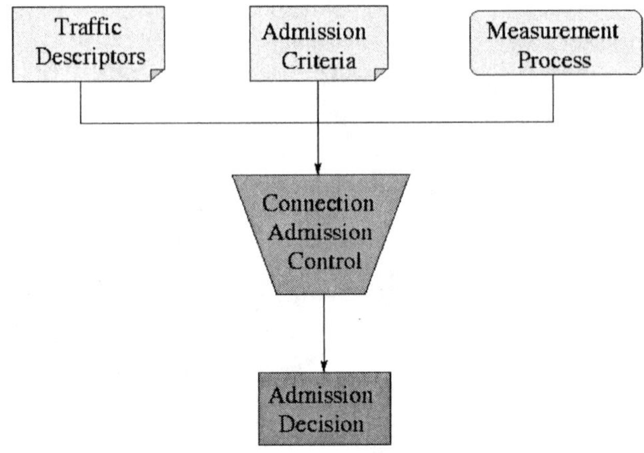

and the amount of resources in the network. Measurements can be made using a time window, point samples, or exponential averaging (Floyd, 1996; Jamin, Danzig, Shenker, & Zhang, 1997; Tse & Grossglauser, 1997).

For each connection request, CAC considers the network measures (such as multipath, path loss, and interference) and the traffic characteristics, along with the QoS requirements of all existing connections, to make the acceptance decision. As mentioned before, handoff connections must be prioritized over new connections to lower connection dropping rate. If prioritization is provided by means of reservations, the resources required by the reservations should also be considered.

Once the request has been accepted, the system must monitor the traffic sent over the connection with a *policer* and ensure that the connection does not violate the QoS specifications. On the mobile terminal side, the data must be pushed into the system using a *traffic shaper* so that frames are not dropped by the policer in the network. Flows from multiple queues are placed in frames and transmitted over the physical channel using a *scheduler*. If the quality of the radio resources deteriorates, QoS should also be degraded gradually. In the case of multimedia communications, feedback must be provided to the encoder to reduce the offered data rate.

A Look into the Future: Next-Generation Wireless Systems

The aim of NGWS is to provide high-bandwidth access anytime and anywhere. Various types of services, including multimedia with different levels of QoS requirements, will be provided independent of the location and speed of the user. Though existing wireless systems are also designed with these objectives in mind, they fail to satisfy all of the requirements *simultaneously* due to constraints like global coverage, indoor and outdoor communications, and frequent handoffs. WLAN, PCS, satellite systems, and their future generations together with new wireless systems like 4G Mobile are candidates as subsystems (Figure 3). These systems and the new technologies to come will serve collaboratively as subsystems in NGWS in order to provide high-bandwidth access everywhere.

Since the service areas of the subsystems overlap, the mobile terminals will have access to multiple subsystems simultaneously. Though a mobile terminal has access to multiple subsystems simultaneously, NGWS must select one of the subsystems for connection. Among the accessible subsystems, one subsystem that can accommodate the connection request will be selected subject to service class and user preferences.

The selection of the subsystem is a critical factor in the performance of the overall NGWS. Trivial solutions like selecting the subsystem with best signal level will result in the accumulation of connections in some of the subsystems. Such an accumulation will cause blocking of service for mobiles that cannot access lightly loaded subsystems, resulting in higher outage rate, lower throughput, and unstable service throughout the NGWS (Tugcu & Vainstein, 2003).

Let b_i^s denote the ith access node (e.g., base station, etc.) of subsystem s and c_i^s denote the capacity of b_i^s. Each access node b_i^s periodically transmits its load information, l^S_i, to all of its neighbors, and also keeps record of their loads. We denote the recorded value of l^t_j at access node b_i^s with l''_j. Since the load information is exchanged between only a few access nodes in the vicinity and the information exchange is performed over abundant wired links, this overhead is negligible. Also, l''_j may not be exactly up to date, but since load in a cell does not fluctuate wildly, l''_j will be reasonably close to l^t_j. We denote the new load of b_i^s if request rq is accepted with $\hat{l}_i^s(rq)$. We also denote the load of b_j^t, based on the recorded value l''_j, if rq is accepted with $\hat{l}_j^{\prime t}(rq)$.

With each connection or handoff request rq, we associate an ordered list of reachable access nodes in which ordering criteria is the user's preferences for the class of rq. We denote the ordered list of access nodes specified in request rq as $\mathcal{L}_{ac}(rq)$. For outgoing connection set-up and handoff requests, a mobile terminal, MT, sends the request to the first access node, b_i^S, in $\mathcal{L}_{ac}(rq)$. However, for incoming connections, the initiator (caller) is a remote node that is not aware of the subsystems accessible by MT, the availability of the resources in the subsystems, and the user preferences for MT. Furthermore, the paging process, which precedes connection establishment, need not be done through

Figure 3. NGWS architecture

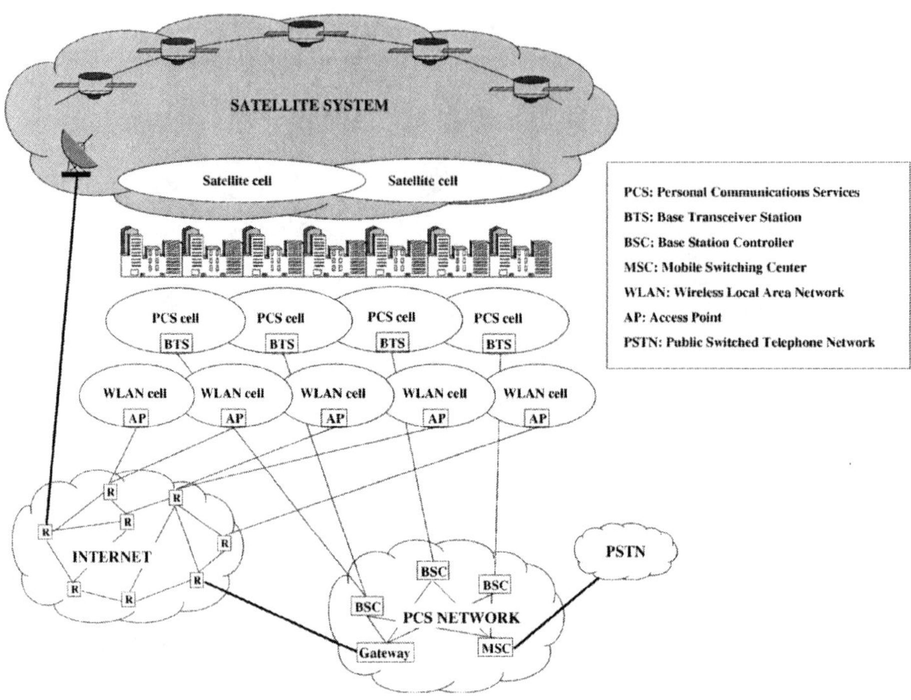

the subsystem over which the connection will be established. Therefore, in the paging reply message, *MT* specifies $\mathcal{L}_{ac}(rq)$ to be used in connection admission. Then, the initiator sends the connection establishment request to the first access node, b_i^S, in $\mathcal{L}_{ac}(rq)$. Since $\mathcal{L}_{ac}(rq)$ contains the identifiers of a few access nodes, its overhead is negligible. The CAC algorithm for NGWS is depicted in Figure 4.

Conclusion

CAC is the process that determines which requests are admitted into the system and granted resources. The decision made by the CAC scheme is important for both the user and the operator. The user immediately experiences the impacts by a connection termination. The operator suffers from lower utilization and customer satisfaction.

A general admission control scheme makes its decision based on the traffic descriptors and available network resources. In wireless systems, CAC also tries to minimize dropping and blocking rates. To achieve this goal, techniques such as reservations, guard channels, and queuing can be used. The admission decision is generally made in

Figure 4. CAC algorithm for NGWS

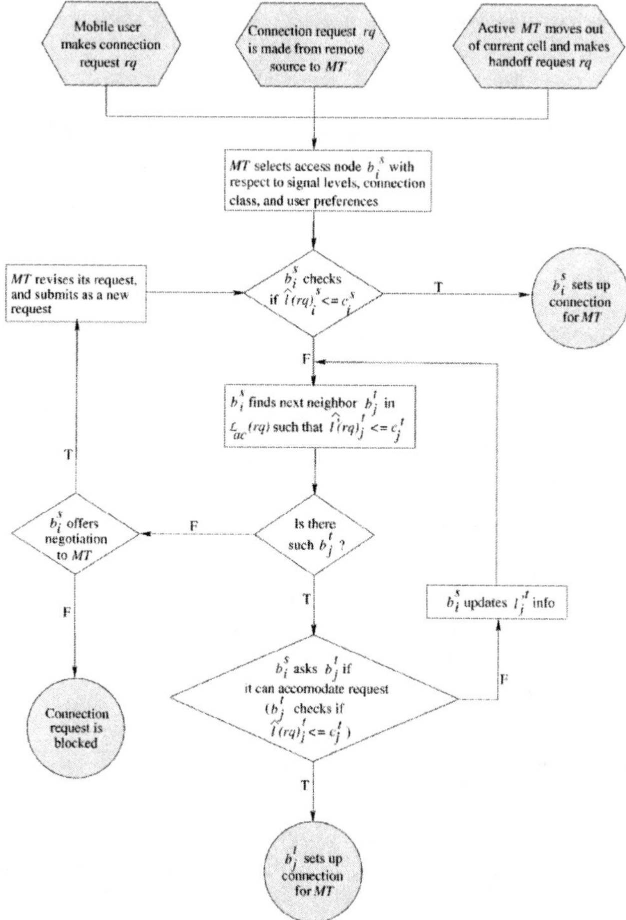

the backbone with the assistance of the mobile terminal, but it is also possible to move the burden from air interface to the backbone network.

References

Acampora, A. S., & Naghshineh, M. (1994). Control and quality of service provisioning in high-speed microcellular networks. *IEEE Personal Communications*, *1*(2), 36-43.

Floyd, S. (1996). Comments on "Measurement-based admissions controlled-load service." *ACM Computer Communications Review*.

Gavish, B., & Sridhar, S. (1997). Threshold priority policy for channel assignment in cellular networks. *IEEE Transactions on Computers*, *46*(3), 367-370.

Goodman, D. J. (1997). *Wireless personal communications systems*. MA: Addison-Wesley.

Hong, D., & Rappaport, S. (1986). Traffic modeling and performance analysis for cellular mobile radio telephone systems with prioritized and nonprioritized handoff procedures. *IEEE Transactions on Vehicular Technology*, *35*(3), 77-92.

Hou, J., Fang, Y., & Akansu, A. N. (2000). Mobility-based channel reservation scheme for wireless mobile networks. *Proceedings of the IEEE Wireless Communications and Networking Conference (WCNC)*, (pp. 527-531).

Jamin, S., Danzig, P., Shenker, S., & Zhang, L. (1997). A measurement-based admission control algorithm for integrated services packet networks. *IEEE/ACM Transactions on Networking*, *5*(1), 56-70.

Jeon, W. S., & Jeong, D. G. (2002). Call admission control for CDMA mobile communications systems supporting multimedia services. *IEEE Transactions on Wireless Communications*, *1*(4), 649-659.

Katzela, I., & Naghshineh, M. (1996). Channel assignment schemes for cellular mobile telecommunication systems: A comprehensive survey. *IEEE Personal Communications*, *3*(3), 10-31.

Kim, Y. C., Lee, D. E., Lee, B. J., Kim, Y. S., & Mukherjee, B. (1999). Dynamic channel reservation based on mobility in wireless ATM networks. *IEEE Communications Magazine*, *37*(11), 47-51.

Lee, D.-J., & Cho, D.-H. (2000). Performance analysis of channel-borrowing handoff scheme based on user mobility in CDMA cellular systems. *IEEE Transactions on Vehicular Technology*, *49*(6), 2276-2285.

Levine, D. A., Akyildiz, I. F., & Naghshineh, M. (1997). A resource estimation and call admission algorithm for wireless multimedia networks using the shadow cluster concept. *IEEE/ACM Transactions on Networking*, *5*(1), 1-12.

Lin, Y.-B., & Chlamtac, I. (2001). *Wireless and mobile communications*. New York: Wiley Computer Publishing.

Liu, Z., & Zarki, M. E. (1994). SIR-based call admission control for DS-CDMA cellular systems. *IEEE Journal on Selected Areas in Communications*, *12*(4), 638-644.

Luo, X., Thng, I., & Zhuang, W. (1999). A dynamic channel pre-reservation scheme for handoffs with GoS guarantee in mobile networks. *Proceedings of the Fourth IEEE Symposium on Computers and Communications (ISCC)*.

Ma, Y., Han, J. J., & Trivedi, K. S. (2002). Call admission control for reducing dropped calls in CDMA cellular systems. *Computer Communications*, *25*(7), 689-699.

Oliviera, C., Kim, J. B., & Suda, T. (1998). An adaptive bandwidth reservation scheme for high-speed multimedia wireless networks. *IEEE Journal on Selected Areas in Communications*, *16*(6), 858-873.

Posner, C., & Guerin, R. (1985). Traffic policies in cellular radio that minimize blocking of handoffs. *Proceedings of the 11th International Teletraffic Congress (ITC)*.

Qiu, X., Chawla, K., Chuang, J., & Sollenberger, N. (2001). Network-assisted resource management for wireless data networks. *IEEE Journal on Selected Areas in Communications, 19*(7), 1222-1234.

Senarath, G., & Everitt, D. (1995). Performance of handover priority and queuing systems under different handover request strategies for microcellular mobile communication systems. *Proceedings of the IEEE Vehicular Technology Conference (VTC)*, 897-901.

Shin, S. M., Cho, C.-H., & Sung, D. K. (1999). Interference-based channel assignment for DS-CDMA cellular systems. *IEEE Transactions on Vehicular Technology, 48*(1), 233-239.

Sutivong, A., & Peha, J. M. (1997). Novel heuristics for call admission control in cellular systems. *Proceedings of the Sixth IEEE International Conference on Universal Personal Communications (ICUPC)*, 129-133.

Tripathi, N. D., Reed, J. H., & Vanlandingham, H. F. (1998). Handoff in cellular systems. *IEEE Personal Communications, 5*(6), 26-37.

Tse, D., & Grossglauser, M. (1997). Measurement-based call admission control: Analysis and simulation. *Proceedings of the IEEE International Conference on Computer Communications (INFOCOM)*, (pp. 981-989).

Tugcu, T., & Ersoy, C. (2001). Resource management in DS-CDMA cellular systems using the reservation area concept. *Proceedings of the Fourth European Personal Mobile Communications Conference (EPMCC).*

Tugcu, T., & Vainstein, F. (2003). Mathematical foundations of resource management in next generation wireless systems. *IEEE International Symposium on Personal, Indoor and Mobile Radio Communications (PIMRC).*

Section IV

Location-Based Services

Chapter XI

Indexing Mobile Objects: An Overview of Contemporary Solutions

Panayiotis Bozanis, University of Thessaly, Greece

Abstract

Mobile computing emerged as a new application area due to recent advances in communication and positioning technology. As David Lomet (2002) notices, a substantial part of the conducted work refers to keeping track of the position of moving objects (automobiles, people, etc.) at any point in time. This information is very critical for decision making, and, since objects' locations may change with relatively high frequency, this calls for providing fast access to object location information, thus rendering the indexing of moving objects a very interesting as well as crucial part of the area. In this chapter we present an overview on advances made in databases during the last few years in the area of mobile object indexing, and discuss issues that remain open or, probably, are interesting for related applications.

Introduction

During the last years, a significant increase in the volume and the diversity of the data which are stored in database management systems has happened. Among them, spatio-temporal data is one of the most fast developing categories. This phenomenon can be easily explained since there is a flurry of application development concerning continuously evolving spatial objects in several areas. To name a few, mobile communication systems, military equipment in (digital) battlefields, air traffic, taxis, truck and boat fleets, and natural phenomena (e.g., hurricanes) all generate data whose spatial components are constantly changing.

In the standard database context, data remains unchanged unless an update is explicitly stated; for example, the phone number in an employee's record remains the same unless it is explicitly updated. If this assumption was employed to continuously moving objects, then highly frequent updates should be performed. Otherwise, the database would be inaccurate and thus query outputs would be obsolete and unreliable.

In order to capture continuous movement and, additionally, spare unnecessary updates, it is widely accepted to store moving object positions as time-dependent functions, which results in updates triggered only by function parameter changes. For example, when objects follow linear movement, the parameters could be the position and the velocity vector of each object at the particular time the function (and therefore the object) is registered to the database. Usually, the moving objects are considered responsible for updating the database about alterations of their movement.

The following paragraphs present a comprehensive review on the various indexing proposals for accommodating moving objects in database systems, so that complex queries about their location in the past, the present, or the future can be served. The more elementary problem of location management, which asks for storing and querying the location of mobile objects based on the underlying network architecture, is surveyed in Pitoura and Samaras (2001). The works of Agarwal, Guibas, et al. (2002), Wolfson (2002), and Lomet (2002) discuss various aspects of modeling and manipulating motion, while the "lower level" subject of organizing (indexing) data for efficient broadcasting in wireless mobile computing is treated in Chen, Wu, and Yu (2003) and Shivakumar and Venkatasubramanian (1996); the interested reader could consult all these references for a wider introduction.

Definitions and Background

The indexes developed to accommodate moving objects can be classified into two broad categories:

(a) those optimizing queries about past states of movement, the so-called *historical queries*, and

(b) those designed to answer queries about future positions of the moving objects, which are termed *future* or *predictive queries*.

This categorization is not strict since there are structures enabling queries about both past and future positions. One can also group the indexes based upon whether the object trajectories are indexed or the objects themselves. However, in this study, we have chosen the classification according to whether the proposals are *practical* ones (that is, they have been actually implemented and experimentally investigated in realistic environments) or *theoretical* ones, which aim mainly at indicating the inherent complexity of the problem since their adoption to real applications is problematic because of the hidden constants or the involvement they exhibit, and, thus, their use is avoided.

Moving Object Representation

Since the size or the shape of a moving object is unimportant compared to the significance of its position as time evolves, its representation is directly related only to time and location. On the other hand, the way this spatiotemporal information is registered depends on the kind of the processing needed, namely, the postprocessing of recorded data and the exploration of current and future data location.

In the first case, object trajectories must be maintained. This means that one has the complete knowledge of location at any past time instance. Obviously, this is impossible, not only for storage limitations but also due to the nature of the underlying application; for instance, communication frameworks, like Global Positioning System (GPS) equipment, generate discrete location data. Therefore, trajectories must be calculated by interpolating the sampled locations. Since linear interpolation is widely accepted for this task, each trajectory eventually turns into a polyline, that is, a sequence of connected line segments, in three-dimensional space. In conclusion, the problem of indexing trajectories reduces to indexing semantically related line segments. As we will see in the sequel, this extra information requires cautious extensions to spatial database indexes. The interested reader could find more about spatial indexes in Gaede and Günther (1998).

In the second case, the object position must be considered as a time function $x(t)$. Employing x, the application administrating the database can anticipate the future object locations, given that objects report any change in x parameter(s). Usually, $x(t)$ is modeled as a linear function of time, $x(t) = x(t_{ref}) + v(t - t_{ref})$, specified by two parameters: (a) the *reference position* $x(t_{ref})$ at a specific time t_{ref} and (b) the *velocity vector* v. This kind of representation is characterized as *parametric*, and its parameters define a dual to the (real) time-location space framework, which, as we elaborate later, was exploited for the development of new indexing methods.

Query Types

Location-Based Queries

The queries of this category can be further subdivided into *range queries* and *proximity* or *nearest-neighbor* (NN) ones. There are three different classes of range search queries: (a) *time-slice* or *snapshot query* (r, t), which specifies a hyper-rectangle r located at a time instant t, and asks for all moving objects that will be contained in r at that time, (b) *window query* (r, t), which requests reporting all objects crossing the hyper-rectangle r during the time interval $t = [t_s, t_e]$, and (c) *moving query* (r_1, r_2, t), where $t = [t_s, t_e]$, which specifies a $(d + 1)$-dimensional trapezoid τ by connecting r_1 at t_s and r_2 at t_e, and inquires all objects that will pass through τ.

On the other hand, a proximity query asks for specifying the nearest moving objects to a given location (spatial point) at time instance t or during a time interval t. There is also a generalization which asks for the *k nearest neighbors* (kNN). Sometimes, *reversed nearest neighbor* (RNN) searching is required, which asks for all objects having a given one as their nearest neighbor.

Trajectory-Based Queries

This type of queries concerns: (a) the topology of the trajectories, that is, information about the semantics of the movement, for example, objects entering, leaving, crossing, and bypassing a region during a given time instance or interval (for example, "Find all mobile phones entering a particular cell between 2 p.m. and 5 p.m. today."), and (b) derived or navigational information, for example, traveled distance, covered area, and velocity (for instance, "Report all objects whose traveled distance between 2 p.m. and 5 p.m. three days ago was smaller than 60 km/h.").

Continuous Queries

This kind of queries stands as a quite natural enhancement of the location-based ones by considering that the query range or point is also moving, for example, "Find all my nearest restaurants as I drive towards the current direction for the next 5 minutes." Despite its conceptual simplicity, these questions are not straightforward at all since they are equivalent to constantly posing location-based queries. As we will exhibit, current solutions capitalize on the semantics of the movement to achieve efficient processing.

Soundness-Enriched Queries

The members of this category are distinguished from the simple location-based counterparts since they demand answers enriched with validity-temporal or spatial-information.

Specifically, they additionally specify the future time t that the result expires and the change that will occur at time t. For instance, when one asks, "Report the nearest pharmacy to my position as I am moving now," the database will return the nearest pharmacy ID i, the future time t that i ceases to be the closest, and a new pharmacy i' that would be the next nearest at time t. Alternatively, the database can return a validity region r around the query position within which the answer remains valid. Returning to the previous example, i will be accompanied by the region that covers as the nearest.

This class of queries aims at reducing subsequent queries. To be explanatory, consider the scenario of a moving user posing a query to a server. The query and the server responses are delivered via a wireless network. Based on the fact that the mobility makes the validity of the answer highly volatile, the extra transmitted information could spare network bandwidth since the user will release new inquiries only when it is absolutely necessary.

Practical Indexes

The indexes of this category can be roughly divided into two subcategories according to the time dimension: All these structures that are able to answer queries about the present and the future belong to the first group, while in the second one, one can find indexes which accommodate historical spatiotemporal data, and so, their main purpose is to respond to inquires about the past. We will see that this taxonomy is not strict-the structure in Sun, Papadias, Tao, and Liu (2004) indicates this fact.

Querying the Present and the Future

The members of this grouping can be further classified based on three types of supporting inquiries: range, nearest neighbor, and soundness enriched.

Structures Supporting Range Queries

Tayeb, Ulusoy, and Wolfson (1998) presented one of the earliest works on indexing mobile objects. Using linear functions to approximate object movement, they reduced the problem to indexing lines in the *xt*-plane. The transformed data set is stored in a periodically generated bucket Point Region (PR) quadtree (Samet, 1990), which stores a line segment in every quadrant of the underlying space that it crosses, partially or fully. The authors provided algorithms for answering snapshot and window queries. In short, this index suffers from the need of continuous rebuilding and the rather high space requirements.

Kollios, Gunopoulos, and Tsotras (1999b) suggested solutions for one- and two-dimensional range searching: When the objects are moving on the line, then either they could be indexed in the *xt*-plane using standard spatial solutions like R-trees (Guttman,

1984), or the problem could be solved by simplex range searching in the two-dimensional dual space. In the last case, when Hough-X dual space (Jagadish, 1990) is used, the employment of a dynamic version of external partition trees (Agarwal, Arge, Erickson, Franciosa, & Vitter, 2000) guarantees linear space complexity and $O(n^{1/2+\varepsilon} + k)$ worst-case query performance. On the other hand, the alternative of employing Hough-Y dual space (Jagadish, 1990) permits a practical approximation algorithm with linear space and expected logarithmic query time by segmenting the plane into c horizontal stripes and indexing them with c independent B+-trees in the same way Kollios, Gunopoulos, and Tsotras (1999a) did. When one accepts exact answers for a specified time period T, then the authors provide a solution of linear space complexity and worst-case logarithmic query performance by storing the relative ordering of the objects. Specifically, since between two crosses of objects the relative order of objects remains unchanged, they discover all objects crossing during T and store the orderings in an external version of persistent lists (Driscoll, Sarnak, Sleator, & Tarjan, 1989). Finally, Kollios et al. considered also (a) the extension to a 1.5-dimensional space (that is, movement into routes) by, first, indexing the trajectories and then, on each trajectory, the moving points, and (b) the two-dimensional version of the problem by mapping each trajectory to a point in a four-dimensional space and indexing the resulting points with external partition trees.

On the other hand, Chon, Agrawal, and El Abbadi (2001) conducted preliminary investigation in which two parameters are sufficient for indexing moving objects. Adopting the convention that object movement can be described by four independent parameters, namely the velocity, the starting time, and the starting and ending positions of the movement, they examined all six different combinations when only two parameters can vary, the other two being constant, and concluded that the most promising dimensionality reduction results if one varies only the starting time and the ending position. The validity of the conclusion depends on performance study conducted using the SS-tree (White & Jain, 1996) as the underlying spatial index, and comparing the results with the dual space transformation of Kollios et al. (1999b).

In Šaltenis, Jensen, Leutenegger, and Lopez (2000), TPR-tree, a time-parameterized (TP) version of R*-trees (Beckmann, Kriegel, Schneider, & Seeger, 1990) for objects moving with constant velocities in one-, two-, and three-dimensional space, was introduced, which became the de facto spatial index for future queries. Therefore, we will be more analytical in our discussion. So, TPR-tree can be defined as a balanced, multiway tree. The moving points are accommodated in the leaf nodes while the internal nodes store pairs of a pointer to a subtree T and a TP bounding rectangle (TPBR) R_T, augmented with a velocity vector in such a way, it can bound the positions of all moving objects or other bounding rectangles in T. The TPBRs are defined in a conservative manner: If t_{ref} is the reference time and S is the set of all involved objects, then in each dimension x_i, the lower bound is set to be the minimum x_i-coordinate value in S at time t_{ref}, moving with the minimum observed velocity in S, and the upper bound is defined to be the maximum x_i-coordinate value in S at time t_{ref}, moving with the maximum observed velocity in S. Please notice that TPR-tree indexes the future trajectories of moving points as infinite lines. TPBRs never shrink and, since they may mistakenly grow, the authors suggested algorithms that keep the index tuned for H time units; after that, a global reorganization (or rebuilding) of the structure is necessary. Toward this end, the insertion and bulk-loading algorithms of the R*-tree are generalized so that their respective objective functions are time parameterized in the following way. Let A be an objective function and $A(t)$ its generalized counterpart with the involved metrics, like perimeters or overlap, being time dependent. Then, they minimize the integral

$$\int_{t_{\mathrm{ref}}}^{t_{\mathrm{ref}}+H} A(t)dt.$$

Three types of queries are supported: (a) time-slice queries, (b) window queries, and (c) moving queries. Type a calls for calculating the bounding rectangles of the index at time t before the intersection test is performed. Types b and c are served based on a very simple observation: the extents of a moving TPBR and the moving query should intersect in each dimension at a time point. The authors provide analytical formulae that compute such time intervals, thus avoiding the time-expensive, generic polyhedron intersection tests. In conclusion, one could say that the TPR-tree is very practical and, in the sense of avoiding the usage of dual space or the reduction of higher dimensional spaces, is a straightforward solution tested for uniformly generated one-, two-, and three-dimensional workloads and various values of the validation parameter H.

TPR-Tree algorithms extended in Benetis, Jensen, Karθiauskas, and Šaltenis (2002) so that they could answer two-dimensional NN and RNN queries for a query point q during a time interval t. For the first case, they employ the standard approach of prune and depth-first search in R-tree-like indexes, like, for example, (Roussopoulos, Kelly, & Vincent, 1995) maintaining a list of intervals whose union equals to t, each associated with a point (or points) which is (are) the closest to q among the examined so far. The authors provide a metric M so that parts of the tree, on the average closest to q, are to be visited first; in this way, TPBRs with no chance of enclosing closer than the current closest set of descendant points are early pruned. The RNN queries are served based on the following fact: If we divide the space around the query point q into six equal sectors s_i by straight lines intersected at q, then there exist *at most* six RNN points, at most, one in each s_i. So, we first find the NN point(s) of q for each sector, which is a candidate for being the RNN point of q. If there exist two or more of them in a sector, then there are no tentative RNN points for q. Then each candidate point is checked for whether it has q as the NN. In order to spare disk access, all RNN candidate points are tested in one traversal of the index.

Šaltenis and Jensen (2002) proposed R^{EXP}-tree for indexing the current and anticipated future positions of moving objects, based on the assumption that objects' positions expire after a time period; this is quite natural to the context of location-based services where objects that have not reported their position within a given time period t_{exp} are assumed to be uninterested in the service, and so they are declared expired and are removed from the data set. Also, the authors introduced a new parameter, the query window length W, which is an upper bound on how far from their issue time queries are expected to refer to. This extra knowledge of the expiration time is used to derive better (tighter) TPBRs. In the one-dimensional case, let S be the set of involved moving objects (points or intervals), t_{exp} be the maximum expiration time observed in S, and $h = \min\{H, t_{\mathrm{exp}}\}$. Then it is proven that the best TPBR enclosing S is the trapezoid with upper and lower bounds containing the edges of the convex hull of S, which intersect the line $t =$

$t_{ref} + h/2$. Extending this observation to the d-dimensional case, the authors suggested either (straightforward) independent computation in each dimension, or the processing, in random order, of each dimension so that the computation in the ith dimension considers the processing made for the first i - 1 dimensions. The latter solution produces tighter TPBRs based on analytical formulas. As for the update operations, R^{EXP}-tree uses integrals of the form

$$\int_{t_{ref}}^{t_{ref}+\min\{H, t_{exp}\}} A(t)dt,$$

while employing a lazy removal of expired objects (points or TPBRs) only after an update operation discovers an expired entry. Compared to TPR-trees, where the objects are assumed not to expire, R^{EXP}-trees exhibit better experimental performance on artificially generated index workloads of factor 2 or more without any degradation of the update time.

Procopiuc, Agarwal, and Har-Peled (2002) introduced the Spatiotemporal Self-Adjusting R-tree (STAR) for two-dimensional moving points as an improvement over TPR-trees (Šaltenis et al., 2000) which self-adjusts whenever the query performance deteriorates without user interference; actually, the user specifies the parameter determining the quality or space consumption and self-adjusting time or query performance trade-offs. Using the result of Agarwal and Har-Peled (2001) for approximating the extent of moving points, they store the Minimum Bounding Rectangles (MBRs) as sequences (to be accurate, chains) of points so that every MBR, at any time instance, is described by interpolating along two axes. Additionally, employing a priority queue, they "refresh" the approximations as time evolves in order to keep them valid, and redistribute the children of a node v when the children of v overlap too much. In order to exhibit the advantages of the STAR technique, the authors provide experimental evidence on both synthetic and realistic data sets, following the method used for experimentation in Šaltenis et al. (2000). Their main findings can be summarized as follows: (a) A speedup of 2-3 with respect to TPR-tree was achieved, (b) the deterioration of the scheme over time was proven to be not too much, and (c) the proposed approximations and heuristics actually work well. Overall, one can argue that the STAR proposal is a good example of how one can incorporate theoretical (geometrical) results into a practical mobile index in order to achieve good query time performance.

TPR-trees were recently improved in Tao, Papadias, and Sun (2003) by introducing TPR*-trees, which exhibit new insertion and deletion algorithms. In order to achieve this, the authors suggested a cost model for the original TPR-tree which emphasizes the factors that influence its performance. To be more specific, Tao et al. observed that the probability that a node is intersected by a query window q during a time interval t_q depends on the area $A_{SR}(o', t_q)$ swept by its extension-according to q characteristics-on each axis MBR o' during t_q. This metric is quite different from the integral metrics of TPR-trees, which do not differentiate between static and moving MBRs of the same area. So, the main goal of TPR*-tree's update algorithms is minimizing the quantity

$$C(q) = \sum_{\forall \text{node with MBR } o} A_{SR}(o', t_q).$$

In order to make the quantity work for every query q, the authors suggest optimizing TPR*-trees for the static point query q_0 for the time horizon of H time units, a choice which is fully justified by a thorough experimentation. Additionally, the insertion and deletion algorithms employ some more-elaborated decision making processes for insertion path selection, node reinsertion, and children node redistribution, which pay off a lot in terms of search performance. Finally, Tao et al. conducted extensive experiments that proved the superiority of TPR*-trees over the TPR-trees under all conditions; the average query cost is almost five times less and the average update cost is nearly constant while the TPR*-tree remains effective as time evolves. In conclusion, based on the previous discussion, the TPR*-tree can be characterized as the "state-of-the-art" index for serving range queries about the present and the future on moving objects. This fact, none the less, does not invalidate the practicality of the TPR-trees when one affords some administration cost on index maintenance in order to be able to answer two-dimensional NN and RNN queries for a query point q during a time interval t.

Structures Supporting Nearest Neighbor Queries

The members of this subcategory are additionally distinguished as follows.

Simple Proximity Indexes

Kollios et al. (1999a) suggested practical solutions for dealing with the (conceptually) simple problem of locating the nearest moving neighbor of a static query in the plane. The authors present performance studies for (a) indexing in the *xt*-plane, which is equivalent to indexing line segments or lines (i.e., the trajectories of the objects) with standard spatial indexes, like, for example, R-trees (Guttman, 1984) or R*-trees (Beckmann et al., 1990), so that one has to find the closest trajectory, and (b) segmenting the plane into c horizontal stripes and indexing in the dual Hough-Y space, with the employment of c B+-trees (Comer, 1979), one for each of the c horizontal stripes. The generality of the approach also permits finding the NN (a) within a specified time interval-since the query point becomes a line segment, and (b) for restricted data movement in fixed line segments (routes) or, as it is called, movement in the 1.5-dimensional space; one first indexes line segments in spatial indexes, and, after locating the nearest route, then employs an NN search on the objects of the identified route. In short, this work, besides being the first one, can be characterized as preliminary since the cases studied are very restrictive.

On the other hand, Aggarwal and Agrawal (2003) introduced methods for indexing a special case of moving objects with nonlinear trajectories in arbitrary dimensions. Specifically, the functional representation $F(\Theta, t)$ of the trajectory, where $\Theta = (\theta_1, \theta_2, \ldots, \theta_k)$ is its associated parametric representation, is said to satisfy the convex hull property when, for any set of n trajectories $F(\Theta^1, t), F(\Theta^2, t), \ldots, F(\Theta^n, t)$, the following holds: If Q' lies inside the convex hull of $\Theta^1, \Theta^2, \ldots, \Theta^n$, then $F(\Theta', t)$ lies inside the convex hull of $F(\Theta^1, t), F(\Theta^2, t), \ldots, F(\Theta^n, t)$. Although the convex hull property is a property of the particular parametric representation of a trajectory and not a property of the trajectory itself, the authors demonstrated that very interesting categories of movement satisfy the property: d-dimensional trajectories with constant velocity, d-dimensional parabolic trajectories, elliptic orbits, and trajectories that accept approximate Tailor expansion. Since the convex hull property relates the locality in parametric space to the locality of the positions of objects, Aggarwal and Agrawal suggested the indexing of parametric representations of their trajectories using common multidimensional indexes like R*-trees (Beckman et al., 1990). Actually, they exhibited their method for solving the NN search problem by introducing a branch-and-bound, best-first algorithm for pruning and searching the underlying index which was investigated for linear and parabolic trajectories in three- and two-dimensional space, respectively. In conclusion, this work is quite interesting as being the first one extending, in a nontrivial way, the class of indexed trajectories for NN inquiries.

Continuous Proximity Indexes

Song and Roussopoulos (2001b) studied the kNN problem for a moving query point and static (i.e., not moving) data points, suggesting algorithms that extend static kNN ones, which is equivalent to the continuous kNN problem. The intuition behind the solutions presented there is that, when the query point moves to a new position, then some part of the previous answer must also belong to the new one. So, a series of conditions were proved that state when the previous answer set, or a part of it, remains valid as the query point moves to a new location, which search bound is appropriate to reinitiate the branch and bound at the new position, and how a prefetched-in-main-memory NNs set of cardinality $M > k$ is changing with query movement. The proposed algorithms were experimentally examined, and their simplicity renders them easy to implement, a fact that one should consider when using off-the-shelf R-trees.

Continuous NN queries were treated from a different perspective in Ishikawa, Kitagawa, and Kawashima (2002). Based on the observation that ellipsoid areas around moving query points are better than circular ones when the movement is conducted in a semiconstrained manner-for example, consider cars moving on city roads-they proposed an incremental ellipsoid query-generation algorithm which is built up on current, past, and future trajectory positions, and carefully selected metrics. The authors suggested indexing the static data set with a "standard" spatial structure, like, for example, R*-trees, but they do not reuse the previous answer for evaluating the next query position; instead, they search the spatial index from scratch, anticipating that page caching will spare some I/O cost. In a nutshell, this method needs elaborate tuning to be effective since the incremental update procedures are quite costly.

Continuous kNN search in a static data set has been studied by Tao, Papadias, and Shen (2002). They actually followed the approach of Tao and Papadias (2002) which, as we will see in the sequel, is based on influence points and TP NN search. However, the authors tried to avoid the performance of an NN search from scratch each time the result expires. Instead, they suggest algorithms for evaluating the answers within a single traversal of the input-point set. Assuming the accommodation of the input set in an R-tree $\mathcal{T}$, the authors observed that when the query point is moving on a line segment ℓ, the result consists of a list of l of NN points p_i; p_is partition ℓ into a number of disjoint subsegments ℓ_i, each one of them having p_i as the NN. So, Tao et al. proposed a bound-and-bound traversal of T that employs heuristic node-pruning rules which capitalize on the fact that a new point belongs to l as long as it "cancels" a current member(s) of l. The authors also provided conditions for guiding the search while pruning nodes during continuous kNN, and suggested cost models for node access estimation in case of uniform distributions; for the general case, they recommend the use of histograms. The extensive experimental evaluation of the methods strongly suggests the usefulness of the proposal when, of course, the data set is a static one.

Raptopoulou, Papadopoulos, and Manolopoulos (2003) proposed an algorithm for answering continuous kNN queries on moving points stored in a TPR-tree. Since the squared Euclidean distance between two moving points is described by a parabolic function of time, the authors observed that the kNN points during a time period t_q can be determined by the k levels of the arrangement of the squared distance functions of moving points with respect to the moving query during t_q. So they suggested a two-phase algorithm. First, the underlying TPR-tree is searched in a depth-first manner according to the minimum distance metric:

$$mindist(x, y) = \sqrt{\sum_j | x_j - y_j |^2}$$

between the moving query and the bounding rectangles at a starting time of t_q. When m ($m \geq k$) moving points are located, then, by considering their arrangement, the kNN points for the entire t_q are determined. Next, the second phase commences, which retraverses the TPR-tree, employing two pruning heuristics, and refines the answer as long as relevant, "promising" subtree branches still exist. The conducted experiments advocate the applicability of the method in case one has adopted the TPR-tree for indexing moving points and needs to serve continuous kNN query requests.

Finally, Iwerks, Samet, and Smith (2003) presented algorithms for answering continuous kNN queries on a constantly moving point set whose members can change either location or velocity. The proposed methods refer to static query points, but they can be applied to moving ones as well. Their solution capitalizes on the observation that answering continuous queries about objects within distance D to the query point is much easier to maintain than a kNN one. So, the authors suggested one to first filter points with a continuous within-distance query in order to reduce the number of points considered for the kNN query. By calculating the time instances when points change their distance to the query point or when points change their order with the current kNN point, and by ranking these two kinds of events with a priority queue, Iwerks et al. presented algorithms for query maintenance through the proper use of the queue, which is experimentally investigated. In general, the approach is quite interesting, extending naturally the repertoire of mobile NN queries; but, as Iwerks et al. noticed, further research is needed for fine-tuning.

Structures Supporting Soundness-Enriched Inquires

This class of indexes returns, along with the answer to either range or proximity queries, either temporal or spatial validity information, which specifies the conditions under which the answer remains in effect. In general, the solutions presented so far are relatively new, but also quite promising for further development and research.

Time-Parameterized Queries

Tao and Papadias (2002) introduced TP queries which return (a) the objects that satisfy the (spatial) conditions of the query, (b) the expiration time t_{exp} of the validity of the answer, and (c) the change that invalidates the answer at time t_{exp}. TP queries are very useful in contexts like location-based applications, air-traffic control, and so forth. In order to solve the problem, the authors introduced the concept of the influence time t^o_{inf} associated with each moving object *o*, which indicates the time *o* influences the result. In this way, finding the expiration time of the answer reduces to discovering the minimum influence time of all objects which, in turn, is equivalent to NN search with distance metric being the influence time. The above observation is valid for the treatment of window, kNN, and join queries. In the first case, the influence time of an object or its MBR equals the minimum intersection time along all dimensions. The authors provided also analytical formulae for evaluating the t^o_{inf} of an object *o*, which is used in a standard branch-and-bound traversal of the index accommodating the object set. The same rationale was also followed for the kNN queries; however, due to relatively increased complexity of the intersection conditions, simpler approaches were proposed which, none the less, behave excellent in practice. Now, in the join case, the influence time is defined for pairs of objects (o_1, o_2) as the minimum time when o_1 and o_2 either stop or start satisfying the join conditions. Concluding their work, Tao and Papadias also exhibited how TP queries can be used to answer continuous *spatiotemporal* and *earliest event* queries. The first type refers to the case when a time interval is defined during which one should provide results as they are generated. It follows promptly that this can be served by continuous TP queries at times when the current result expires. The second type asks for the evaluation of the earliest time in the future a specified event could take place; for example, in the scene of moving point objects and query point *q*, find the first time *q* catches a point. By surrounding *q* with a time-varying radius cycle, this query reduces to TP by evaluating the earliest time the circle contains a point, which, in turn, equals to determining the smallest radius of such a circle.

Validity Queries

In Zheng and Lee (2001), the problem of enabling the mobile clients to determine the validity of query results based on their current location was considered. This approach aims at reducing the network traffic by reducing the number of client queries: The server, additionally to the query result, returns a validity region *r* within which there is no alteration of the answer, and so, clients can issue subsequent queries only when they leave *r*. Please note that this treatment departs from the "standard" future prediction queries where the time, and *not* the query location, defines the answer. This view permits the following approach: Since the data set is static, its Voronoi diagram is constructed. This diagram is used to locate the NN of a moving query with respect to its current position by determining the Voronoi cell *c* which encloses it. Additionally, the maximum circle around a query point which does not cross any bounding edge of *c* is calculated and returned to the user as the safer lower bound of the validity of the query result.

Zhang, Zhu, Papadias, Tao, and Lee (2003) also dealt with validity NN, kNN, and window queries. For the first two cases, observing that the validity region is the Voronoi and order-k Voronoi cell, respectively, they suggested algorithms that implicitly calculate the respective cell-that is, without calculating the whole Voronoi or the order-k Voronoi diagram-by, first, finding the NN(s), and then issuing TP NN or kNN queries (Tao & Papadias, 2002) for locating the points defining the border of the cell. The third case of the validity region of a window query is reduced to finding the maximal rectangle around the focus (center) of the window, where the result remains unchanged, which is then sharpened by removing the parts of it that would force the query containing other points not in the reported answer. Zhang et al. proved that these calculations require posing one standard window query, one "holey" window query, and just a few main-memory TP window queries.

Querying the Past

Historical databases on moving objects accommodate spatiotemporal information which can be processed to either *report* or *enumerate* all objects satisfying certain spatial and temporal conditions. In the first case, the corresponding indexes are characterized as *reporting*, while, in the second one, they are termed as *aggregating* or *enumerating*. Here we must note that an aggregating index does not simply count objects; it must be able to provide whatever summarized data, like, for example, average statistics, the related application needs.

Reporting Indexes

In Pfoser, Jensen, and Theodoridis (2000), spatiotemporal queries on mobile object trajectories without capabilities in future prediction queries are treated. The work is based on the observation that the simple use of R-tree (Guttman, 1984), and therefore the employment of MBR approximations, for storing the line segments of trajectories just leave a lot of "dead" space. They proposed two solutions, the STR-tree and the TB-tree. The first one consists of an extension of R-trees so that lines belonging to the same trajectory are kept close together while the main dimension is time. This is accomplished by trying to store new segments to the leaf accommodating their predecessors, and, when this is impossible, it performs leaf split, putting together more recent segments into a new node only if there exists a vacant predecessor, at most, p - 1 levels up; otherwise, new segments are inserted to leaves lying to the right of the search path. This modification permits the combined queries since, at the first stage, the execution of the R-tree range search algorithm identifies the segment trajectories belonging to the initial range and, at the second stage, recursive range queries with endpoints of discovered segments retrieve the final answer.

On the other hand, TB-tree departs from the underlying rationale of R-trees, which assumes independency among geometries, and focuses on spatiotemporal queries. For this, it cuts every trajectory into pieces, each piece is stored in a leaf, and the various leaves are connected to form a linearly linked list. In order to insert a new segment, one first locates the leaf accommodating its predecessor as STR-tree does. If there does not exist any space, then a new leaf is created and becomes the rightmost one of the index.

This approach means that a trajectory is allocated to a set of leaves, and the leaves are organized in a tree hierarchy so that combined queries in the beginning locate the relevant leaves, and then follow list pointers as long as the second range constraints are met. Experimental evaluation of the index proved its efficiency compared to the R-tree. Further tuning of the TB-tree performance in case of it being constrained by the presence of infrastructure, like roads, lakes, and so forth, was achieved in Pfoser and Jensen (2001) by suggesting segmentation of the original query range in a set of subranges so that dead space can be eliminated.

The case of indexing moving objects with potential changing extended for historical queries without future prediction capability (snapshot and small-range queries) was treated in Kollios, Gunopoulos, Tsotras, Delis, and Hadjieleftheriou (2001) and Hadjieleftheriou, Kollios, Tsotras, and Gunopoulos (2002). The authors followed the approach of approximating the object movement with an MBR, but, in order to eliminate empty space and overlap, they suggested the usage of artificial splits and insertion of the pieces into a partially persistent R-tree (Kumar, Tsotras, & Faloutsos, 1998). The first work refers to the case of linear functions of time. It presents a greedy algorithm which, given the preferred number of splits is proportional to the number of objects, calculates split points which minimize the overall empty space, capitalizing on the monotonicity of such a kind of objects, which guarantees that the savings in empty space decrease as the number of splits increases. This property does not apply to the case of general time functions, so the second paper suggested (a) an optimal, straightforward, dynamic programming algorithm, quadratic to the number of splits, and (b) two greedy solutions of linear-to-the-number-of-splits and subquadratic-to-the-number-of-objects complexity, aiming at finding the next object to split according to the maximum possible volume reduction.

The treatment of mobile objects in a spatiotemporal context was also addressed in Porkaew, Lazaridis, and Mehrotta (2001). Assuming the use of the "ubiquitous" R-tree (Guttman, 1984) as the underlying index, the authors proposed algorithms and intersection conditions for various combinations of spatial and temporal range, NN, and kNN queries when the trajectories are approximated by MBRs, and when the movement is described in location-velocity, multidimensional, parametric space.

Song and Roussopoulos (2001a, 2003), based on the fact that indexing static objects has been extensively investigated, suggested the zoning-based, index-updating police. The space is partitioned into regions, the so-called *zones*. Each zone is related to the objects it contains through a bucket. In that way, objects trigger updates only when they move into a different zone while range queries are reduced to querying the buckets whose zones intersect the query region. The generality of the approach gives the freedom to choose any index for bucket manipulation while the authors introduced a new index, the SEB-tree, for indexing the objects that left the bucket. SEB-tree exploits the fact that "absent" objects can be represented by two-dimensional points with increasing coordinate values. This property permits the segmentation of the plane into rectangular regions indexed by a B-tree.

Lazaridis, Porkaew, and Mehrotra (2002), assuming the approximation of object trajectories by MBRs which are stored in R-trees, presented algorithms for moving window queries with known (predictive) and unknown (nonpredictive) moving patterns. In short, the authors suggested the use of a priority queue and specific intersection conditions which guide and prune the search of the underlying R-tree in a way similar to the one in Roussopoulos et al. (1995) that serves standard kNN queries in static points.

Aggregating or Enumerating Indexes

Papadias, Tao, Kalnis, and Zhang (2002) introduced indexes for answering aggregate queries in spatiotemporal databases, that is, queries that ask for summarized data over two-dimensional regions satisfying specific spatiotemporal conditions. The first index, aggregate RB-tree (aRB-tree), refers to the case of fixed spatial dimensions where one needs to maintain historical summary data. The aRB-tree consists of the generalization of aggregate R-trees (aR-trees) of Papadias, Kalnis, Zhang, and Tao (2002) to three-dimensional space (two spatial dimensions and one time dimension). The main block is an R-tree indexing spatial regions in the form of MBRs. Each MBR region stores accumulated data for the entire region and historical summarized data for the time dimension in a B-tree. In this way, the spatial part of the query guides the search using the R-tree part, and, in each intersecting node, the corresponding B-tree is accessed for the temporal part. When the regions are dynamic, the authors proposed two solutions. The first one, aggregate historical RB-tree (aHRB-tree), uses the node copying technique (Driscoll et al., 1989) for the main R-tree in order to avoid unnecessary replication of unaffected regions. The second one, the aggregate 3DRB-tree (a3DRB-tree), can be used when all region changes are known in advance; in such case, each version of a region is stored in the form of a three-dimensional box as a distinct entry in a 3DR-tree (Papadias, Kalnis, et al., 2002), resulting, thus, in space reduction compared to the aHRB-tree.

The shortcoming of the above work is that it does not support distinct counting of the objects; if an object remains in the query region for several time stamps during the query interval, then it will also participate in the result several times. As the distinct counting property is very crucial in many decision-making queries, like, for example, traffic analysis and mobile phone users' statistics, Tao, Kollios, Considine, Li, and Papadias (2004) suggested algorithms that do not use direct counting of the objects, as the volume of the data or legal issues about personal data may prohibit this approach. Instead, they are using FM sketches (Flajolet & Martin, 1985), as one nowadays finds in many data stream processing algorithms. So, they replace the summary B-tree of the aRB-tree proposal with a "sketch" B-tree, where the aggregate information of the leaves is maintained in an FM structure, while each intermediate node sketch is formed by or forming the sketches of all the sketches in its subtree. In this way, the resultant structure becomes an *approximate*, distinct counting index of bounded probability failure, while the search algorithm of aRB-tree remains the same. The authors actually proposed three heuristic rules that exploit the pruning capabilities of sketches additively to the spatial and temporal conditions of the query. Additionally, they discussed how their scheme can be extended to support distinct sum queries, how their sketch B-tree can be applied to a hierarchy of increasing-resolution regular grids, and in which way sketches can be used in mining spatiotemporal association rules.

A first approach to support approximate queries about the past, the present, and the future was recently reported in Sun et al. (2004). In order to achieve their goals, the authors presented an adaptive multidimensional histogram (AMH) for answering present-time queries. AMH is based on a regular grid partition of the plane into cells. Each cell maintains its frequency, that is, the number of enclosing objects. The cells are distributed into a (upper bounded) number of buckets with the aid of a binary partition tree. Each bucket contains its area, average, and average squared frequency of the respective cells. Employing continuous bucket merging and splitting during idle CPU cycles, AMH reorganizes itself, in an interruptive way, to follow the data distribution and, thus, it can successfully serve present-time queries by retrieving the buckets intersected by the query window. The outdated bucket versions are stored in a main-memory index for answering historical queries, which is either a packed B-tree or a three-dimensional R-tree, depending upon the update and query performance trade-off one accepts. The historical index gradually emigrates to disk. Finally, based on the observation that overall data distribution varies slowly and smoothly despite the abrupt nature of the individual objects' velocities, the authors applied exponential smoothing, a time-series forecasting method, for serving queries about the future with the use of recent history data.

Theoretical Solutions

Basch, Guibas, and Hershberger (1999) provided a theoretical framework for dealing with moving objects by introducing the concept of *kinetic framework*, which refers to storing only a combinatorial instance of the data set at any time. That is, in spite of the continuous character of the movement, the data structure changes only on certain discrete *kinetic events* which depend strictly on combinatorial properties. The crux of the framework is that since the way the points move is known, one can calculate when kinetic events will take place and manipulate them with a priority queue, the so-called *event queue*.

A series of theoretical, computational-geometry main-memory results followed the above paradigm. To name a few; Agarwal, Gao, and Guibas (2002) proposed algorithms for "kinetizing" kd-trees; Karavelas and Guibas (2001) coped with kinetic spanners; Czumaj and Sohler (2001) approximate kinetic data structures like binary search trees, range trees, and heaps; Kaplan, Tarjan, and Tsioutsiouliklis (2001) designed kinetic heaps; Agarwal and Har-Peled (2001) presented approximate algorithms for maintaining descriptors of the extent of moving points; Agarwal, Guibas, Murali, and Vitter (2000) considered the cylindrical binary space partitions; Agarwal, Basch, de Berg, Guibas, and Hershberger (2000) proved lower bounds of kinetic planar subdivisions; and Agarwal, Eppstein, Guibas, and Henzinger (1998) studied kinetic minimum spanning trees. In the following lines, we will present secondary-memory theoretical solutions, that is, theoretical indexing schemes for data residing in a database.

Indexes Supporting Range Queries

Agarwal, Arge, and Erickson (2003) proposed four mainly theoretical indexing schemes for moving points in the plane. The first one improves upon the approach of Kollios et al. (1999b); instead of mapping points to four-dimensional space and using one-level partition trees, they project the points to the *xt*-planes and to *yt*-planes, employ transformations to dual *xt*-planes and to *yt*-planes, and the resulting point sets are stored to a two-level external partition tree so that the answer set is included in two stripes of the dual planes. This solution has linear space and $O(n^{1/2+\varepsilon} + k)$, *k* output size, and $O(\log^2 n)$ amortized update complexity. The structure can also serve range queries for a time interval *t* with the same complexity bounds. The authors also suggested two indexes based on the kinetic framework of Basch et al. (1999) for queries referring to the present time or arriving in a strict chronological order. In case of one-dimensional linear moving points, they stored the points ordered across the line in a kinetic B-tree which is updated only when two points interchange positions; these moments are handled by an external priority queue. The solution has linear space and logarithmic query and update complexity, and it can be combined with partition trees to achieve a trade-off between the query time and the number of kinetic events. In order to extend the approach to two dimensions, the authors stored the points into a kinetic version of range tree whose function is governed by a global event queue and two kinetic B-trees, one for each coordinate. This approach results in increases in space and in update time by an $O(\log_B n / \log_B \log_B n)$ factor. The fourth solution offers approximate results to NN searching by replacing the Euclidean metric with a polyhedral one, whose unit ball is a regular *m*-gon, and storing the points in a three-level quasi-linear space index with $O(n^{1/2+\varepsilon}/\sqrt{\delta})$ query time, $0 < \varepsilon$, $\delta < 1$. The first two levels are partition trees on dual *xt*- and *yt*-planes while the third one consists of lists containing the lower envelope of the trajectories.

Agarwal, Arge, and Vahrenhold (2001) provide kinetic solutions for answering window and moving window queries for one- and two-dimensional moving points with time complexities that depend on the number of events $\phi(t_q)$ that occurred between the current time t_{now} and the query time t_q. For the case of moving points on the line, the *yt*-plane is segmented into a logarithmic number of vertical strips or slabs so that the number of events in slab *i* is $O(nB^i)$, *B* being the page capacity, and, in each slab, the part of the arrangement that belongs to it is represented by $O(n/B^i)$ carefully selected *y*-ordered levels which are stored into a persistent B-tree. In this way, a window query with $nB^{i-1} < \phi(t_q) \leq nB^i$ reduces to first, locating the respective window, and then searching the right version of the B-tree in $O(\log_B n + k/B + B^{i-1})$ time, *k* being the output size. The update cost of an event is bounded by $O(\log^3 n)$ and the space complexity is $O(n/B \log_B n)$. The authors extended their approach to the case of moving points in the plane. They first divided the *xyt*-space into a logarithmic number of slabs along the time axis so that the number of events in slab *i* is $O(nB^i)$. In each slab, the projections of the arrangement in the *xt*- and *yt*-planes are represented in an analogy to the one-dimensional case manner, that is, by carefully selected levels of arrangements. This construction results in $O(\log^3 n)$ update cost of an event, $O(n/B \log_B n)$ space complexity, and

$$O(\sqrt{n / B^i}\,(B^{i-1} + \log_B n) + k / B)$$

moving window query cost.

Indexes Supporting Soundness-Enriched Queries

Tao, Mamoulis, and Papadias (2003) presented the first theoretical bounds on validity information of various types of range query, or of NN-search query answers which refer to their expiration time *t* and their change at time *t* to remain valid. More specifically, when the query's length and movement are chosen from a constant number of combinations and the point set is static, the query cost is logarithmic and the space is linear by subdividing the plane into disjoint areas which are stored in a persistent B-tree. When the point set is static, the query's length is arbitrary, but the movement is axis parallel; then, by combining a primary B-tree of logarithmic fan-out with an external priority search tree (Arge, Samoladas, & Vitter, 1999), the query cost is $O(\log_B^2(n/B)/\log_B\log_B(n/B))$ and the space is $O(n/B\log_B(n/B)/\log_B\log_B(n/B))$. For the general case of static point set and queries with arbitrary length and movement, a slightly modified version of partition trees (Agarwal, Arge, et al., 2000) permits linear space and $O((n/B)^{1/2+\varepsilon})$ query cost, whereas a double transformation to the slope-rank space allows the application of simplified external range trees so that a validity query costs $O(\log_B^2(n/B)/\log_B\log_B(n/B))$ time and $O(n^2/B\log_B(n/B)/\log_B\log_B(n/B))$ space. When the data points are dynamic whereas the query is static, the authors proposed plane sweeping during which the ordering of the trajectories is stored in a persistent, aggregate B-tree which guarantees logarithmic query cost and $O(n^2/B\log_B(n/B))$ space. The general case of both dynamic data points and query is only considered in one-dimensional space by accommodating the cells of the arrangement currently intersected by the query (the zone) in a B-tree and maintaining it with an external priority queue, resulting in linear space consumption and logarithmic query time. On the other hand, for the NN search queries, when the points are statically lying in the plane, the careful use of Voronoi diagrams solves the problem with linear space and logarithmic query cost, whereas, in case of moving points on the line, the storage of the zone permits linear space and logarithmic query complexity.

Conclusion and Future Direction

A broad category of applications, emerging from the technological advances that occurred in the last years in telecommunications and hardware, constitute the framework known as mobile computing. A key issue in such a framework is the fast and on-time access to a constantly changing data set. This fact indicates the crucial role of indexing moving objects. In this study we provided a comprehensive review of contemporary solutions in this area of research from the database (application) perspective, which, none the less, presents some very interesting topics for further consideration. First, it would be very helpful of the indexes to provide results that capture the uncertainty associated with the location of moving objects due to network delays and the continuous character of motion. Trajcevski, Wolson, Zhang, and Chamberlain (2002) deal with the issue of modeling and querying about uncertainty, but, there is a lot to be done. It would be also very interesting to efficiently cope with nonlinear trajectories since the scope and the range of indexable moving objects will be significantly extended. Another appealing subject, especially for extending mobile application capabilities, is the design of indexing structures capable of serving "mixed" queries concerning the past and the future of movement; the approach in Sun et al. (2004) can be considered as a first attempt toward this end. The incremental valuation of validity queries is very intriguing, as well. Finally, from an engineering perspective, it would be very helpful (a) to test all indexes with real data sets, as, until now, every experimental investigation is conducted with "semireal" ones, where the movement component is actually generated, and (b) to design efficient updating algorithms for the indexes, different from the usual "deletion and reinsertion" practice, accepting perhaps a trade-off between either the query time or the accuracy of the result and the update time.

References

Aggarwal, C. C., & Agrawal, D. (2003). On nearest neighbor indexing of nonlinear trajectories. In *Proceedings of the 22nd ACM SIGACT-SIGMOD-SIGART Symposium on Principles of Database Systems,* June 9 -12, 2003, San Diego, California, USA (pp. 252-259). New York: ACM.

Agarwal, P. K., & Har-Peled, S. (2001). Maintaining approximate extent measures of moving points. In *Proceedings of the 12th Annual Symposium on Discrete Algorithms,* January 7-9, 2001, Washington, DC (pp. 148-157). New York: ACM/ SIAM.

Agarwal, P. K., Arge, L., & Erickson, J. (2003). Indexing moving points. *Journal of Computer and System Sciences, 66* (1), 207-243.

Agarwal, P. K., Arge, L., & Vahrenhold, J. (2001). Time responsive external data structures for moving points. In F. K. H. A. Dehne, J.-R. Sack, & R. Tamassia (Eds.), *Algorithms and Data Structures, Seventh International Workshop, WADS 2001, Providence, RI, USA, August 8-10, 2001, Proceedings, LNCS 2125* (pp. 50-61). Berlin, Germany: Springer.

Agarwal, P. K., Arge, L., Erickson, J., Franciosa, P. G., & Vitter, J. S. (2000). Efficient searching with linear constraints. *Journal of Computer and System Sciences, 61*(2), 194-216.

Agarwal, P. K., Basch, J., de Berg, M., Guibas, L. J., & Hershberger, J. (2000). Lower bounds for kinetic planar subdivisions. *Discrete & Computational Geometry, 24*(4), 721-733.

Agarwal, P. K., Eppstein, D., Guibas, L. J., & Henzinger, M. R. (1998). Parametric and kinetic minimum spanning trees. In *39th Annual Symposium on Foundations of Computer Science, FOCS '98,* November 8-11, 1998, Palo Alto, California, USA (pp. 596-605). New York: IEEE Computer Society.

Agarwal, P. K., Gao, J., & Guibas, L. J. (2002). Time responsive external data structures for moving points. In R. H. Mohring & R. Raman (Eds.), *Algorithms: ESA 2002, 10th Annual European Symposium,* Rome, Italy, September 17-21, 2002, Proceedings, LNCS 2461 (pp. 5-16). Berlin, Germany: Springer.

Agarwal, P. K., Guibas, L. J., Edelsbrunner, H., Erickson, J., Isard, M., Har-Peled, S., et al. (2002). Algorithmic issues in modeling motion. *ACM Computing Surveys, 34*(4), 550-572.

Agarwal, P. K., Guibas, L. J., Murali, T. M., & Vitter, J. S. (2000). Cylindrical static and kinetic binary space partitions. *Computational Geometry, 16*(2), 103-127.

Arge, L., Samoladas, V., & Vitter, J. S. (1999). On two-dimensional indexability and optimal range search indexing. *Proceedings of the 18th ACM SIGACT-SIGMOD-SIGART Symposium on Principles of Database Systems,* May 3-June 2, 1999, Philadelphia, Pennsylvania (pp. 346-357). New York: ACM Press.

Basch, J., Guibas, L. J., & Hershberger, J. (1999). Data structures for mobile data. *Journal of Algorithms, 31*(1), 1-28.

Beckmann, N., Kriegel, H.-P., Schneider, R., & Seeger, B. (1990). The R*-tree: An efficient and robust access method for points and rectangles. In H. Garcia-Molina & H. V. Jagadish (Eds.), *Proceedings of the 1990 ACM SIGMOD International Conference on Management of Data,* Atlantic City, NJ, May 23-25, 1990 (pp. 322-331). New York: ACM.

Benetis, R., Jensen, C. S., Karθiauskas, G., & Šaltenis, S. (2002). Nearest neighbor and reverse nearest neighbor queries for moving objects. In M. A. Nascimento, M. T. Ozsu, & O. R. Zaiane (Eds.), *International Database Engineering & Applications Symposium, IDEAS'02,* July 17-19, 2002, Edmonton, Canada, Proceedings (pp. 44-53). New York: IEEE Computer Society.

Chen, M.-S., Wu, K.-L., & Yu, P. S. (2003, January/February). Optimizing index allocation for sequential data broadcasting in wireless mobile computing. *IEEE Transactions on Knowledge and Data Engineering, 15*(1), 161-173.

Chon, H. D., Agrawal, D., & El Abbadi, A. (2001). Storage and retrieval of moving objects. In K.-L. Tan, M. J. Franklin, & J. C. S. Lui (Eds.), *Mobile Data Management, Second International Conference, MDM 2001,* Hong Kong, China, January 8-10, 2001, Proceedings, LNCS 1987 (pp. 173-184). Berlin, Germany: Springer.

Comer, D. (1979). The ubiquitous B-tree. *ACM Computing Surveys, 11*(2), 121-137.

Czumaj, A., & Sohler, C. (2001). Soft kinetic data structures. In *Proceedings of the 12th Annual Symposium on Discrete Algorithms,* January 7-9, 2001, Washington, DC (pp. 865-872). New York: ACM/SIAM.

Driscoll, J. R., Sarnak, N., Sleator, D. D., & Tarjan, R. E. (1989). Making data structures persistent. *Journal of Computer and System Sciences, 38*(1), 86-124.

Flajolet, P., & Martin, G. N. (1985). Probabilistic counting algorithms for data base applications. *Journal of Computer and System Sciences, 31*(2), 182-209.

Gaede, V., & Günther, O. (1998). Multidimensional access methods. *ACM Computing Surveys, 30*(2), 170-231.

Guttman, A. (1984). R-trees: A dynamic index structure for spatial searching. In B. Yormark (Ed.), *SIGMOD'84, Proceedings of Annual Meeting,* Boston, Massachusetts, June 18-21, 1984 (pp. 47-57). New York: ACM Press.

Hadjieleftheriou, M., Kollios, G., Tsotras, V. J., & Gunopoulos, D. (2002). Efficient indexing of spatiotemporal objects. In C. S. Jensen, K. G. Jeffery, J. Pokorny, S. Šaltenis, E. Bertino, K. Bohm, & M. Jarke (Eds.), *Advances in Database Technology EDBT 2002, Eighth International Conference on Extending Database Technology,* Prague, Czech Republic, March 25-27, Proceedings, LNCS 2287 (pp. 251-268). Berlin, Germany: Springer.

Ishikawa, Y., Kitagawa, H., & Kawashima, T. (2002). Continual neighborhood tracking for moving objects using adaptive distances. In M. A. Nascimento, M. T. Ozsu, & O. R. Zaiane (Eds.), *International Database Engineering & Applications Symposium, IDEAS'02,* July 17-19, 2002, Edmonton, Canada, Proceedings (pp. 54-63). New York: IEEE Computer Society.

Iwerks, G. S., Samet, H., & Smith, K. (2003). Continuous *k*-nearest neighbor queries for continuously moving points with updates. In J. C. Freytag, P. C. Lockemann, S. Abiteboul, M. J. Carey, P. G. Selinger, & A. Heuer (Eds.), *VLDB 2003, Proceedings of 29th International Conference on Very Large Data Bases,* September 9-12, 2003, Berlin, Germany (pp. 512-523). St. Louis, MO: Morgan Kaufmann.

Jagadish, H. V. (1990). On indexing line segments. In *VLDB 1990, Proceedings of 16th International Conference on Very Large Data Bases,* August 1990, Brisbane, Queensland, Australia (pp. 614-625). St. Louis, MO: Morgan Kaufmann.

Kaplan, H., Tarjan, R. E., & Tsioutsiouliklis, K. (2001). Faster kinetic heaps and their use in broadcast scheduling. In *Proceedings of the 12th Annual Symposium on Discrete Algorithms,* January 7-9, 2001, Washington, DC, (pp. 836-844). New York: ACM/SIAM.

Karavelas, M. I., & Guibas, L. J. (2001). Static and kinetic geometric spanners with applications. In *Proceedings of the 12th Annual Symposium on Discrete Algorithms,* January 7-9, 2001, Washington, DC (pp. 168-176). New York: ACM/SIAM.

Kollios, G., Gunopoulos, D., & Tsotras, V. J, (1999a). Nearest neighbor queries in a mobile environment. In M. H. Bohlen, C. S. Jensen, & M. Scholl (Eds.), *Spatio-Temporal Database Management, International Workshop STDBM'99,* Edinburgh, Scotland, September 10-11, 1999, Proceedings, LNCS, 1678 (pp. 119-134). Berlin, Germany: Springer.

Kollios, G., Gunopoulos, D., & Tsotras, V. J. (1999b). On indexing mobile objects. In *Proceedings of the 18th ACM SIGACT-SIGMOD-SIGART Symposium on Principles of Database Systems,* May 31-June 2, 1999, Philadelphia, Pennsylvania (pp. 261-272). New York: ACM Press.

Kollios, G., Gunopoulos, D., Tsotras, V. J., Delis, A., & Hadjieleftheriou, M. (2001). Indexing animated objects using spatio-temporal access methods. *IEEE Transactions on Knowledge and Data Engineering, 13*(5), 742-777.

Kumar, A., Tsotras, V. J., & Faloutsos, C. (1998). Designing access methods for bitemporal databases. *IEEE Transactions on Knowledge and Data Engineering, 10*(1), 1-20.

Lazaridis, I., Porkaew, K., & Mehrotra, S. (2002). Dynamic queries over mobile objects. In C. S. Jensen, K. G. Jeffery, J. Pokorny, S. Šaltenis, E. Bertino, K. Bohm, & M. Jarke

(Eds.), *Advances in Database Technology EDBT 2002, Eighth International Conference on Extending Database Technology,* Prague, Czech Republic, March 25-27, Proceedings, LNCS 2287 (pp. 269-286). Berlin, Germany: Springer.

Lomet, D. (2002). Letter from the editor-in-chief [Special issue]. *Bulletin of the Technical Committee on Data Engineering, 25*(2), 1.

Papadias, D., Kalnis, P., Zhang, J., & Tao, Y. (2001). Efficient OLAP operations in spatial data warehouses. In C. S. Jensen, M. Schneider, B. Seeger, & V. J. Tsotras (Eds.), *Advances in Spatial and Temporal Databases, Seventh International Symposium, SSTD 2001,* Redondo Beach, California, USA, July 12-15, 2001, Proceedings, LNCS 2121 (pp. 443-459). Berlin, Germany: Springer.

Papadias, D., Tao, Y., Kalnis, P., & Zhang, J. (2002). Indexing spatio-temporal data warehouses. In *Proceedings of the 18th International Conference on Data Engineering,* February 26- March 1, 2002, San Jose, California, USA (pp. 166-175). New York: IEEE Computer Society.

Pfoser, D., & Jensen, C. S. (2001). Querying the trajectories of on-line mobile objects. In *Proceedings of the Second ACM International Workshop on Data Engineering for Wireless and Mobile Access, MobiDE,* May 20, 2001, Santa Barbara, California, USA (pp. 66-73). New York: ACM.

Pfoser, D., Jensen, C. S., & Theodoridis, Y. (2000). Novel approaches to the indexing of moving object trajectories. In A. El Abbadi, M. L. Brodie, S. Chakravarthy, U. Dayal, N. Kamel, G. Schlageter, & K.-Y. Whang (Eds.), *VLDB 2000, Proceedings of 26th International Conference on Very Large Data Bases,* September 10-14, 2000, Cairo, Egypt (pp. 395-406). St. Louis, MO: Morgan Kaufmann.

Pitoura, E., & Samaras, G. (2001, July/August). Locating objects in mobile computing. *IEEE Transactions on Knowledge and Data Engineering, 13*(4), 571-592.

Porkaew, K., Lazaridis, I., & Mehrotta, S. (2001). Querying mobile objects in spatio-temporal databases. In C. S. Jensen, M. Schneider, B. Seeger, & V. J. Tsotras (Eds.), *Advances in Spatial and Temporal Databases, Seventh International Symposium, SSTD 2001,* Redondo Beach, California, USA, July 12-15, 2001, *Proceedings, LNCS 2121* (pp. 59-78). Berlin, Germany: Springer.

Procopiuc, C. M., Agarwal, P. K., & Har-Peled, S. (2002). STAR-tree: An efficient self-adjusting index for moving objects. In D. M. Mount & C. Stein (Eds.), *Algorithm Engineering and Experiments, Fourth International Workshop, ALENEX 2002,* San Francisco, California, USA, January 4-5, 2002, Revised Papers, LNCS 2409 (pp. 178-193). Berlin, Germany: Springer.

Raptopoulou, K., Papadopoulos, A., & Manolopoulos, Y. (2003). Fast nearest-neighbor query processing in moving-objects databases. *GeoInformatica, 7*(2), 113-137.

Roussopoulos, N., Kelly, S., & Vincent, F. (1995). Nearest neighbor queries. In M. J. Carey & D. A. Schneider (Eds.), *Proceedings of the 1995 ACM SIGMOD International Conference on Management of Data,* San Jose, California, May 22-25, 1995 (pp. 71-79). New York: ACM Press.

Šaltenis, S., & Jensen, C. S. (2002). Indexing of moving objects for location-based services. In *Proceedings of the 18th International Conference on Data Engineer-*

ing, February 26-March 1, 2002, San Jose, CA (pp. 463-472). New York: IEEE Computer Society.

Šaltenis, S., Jensen, C. S., Leutenegger, S. T., & Lopez, M. A. (2000). Indexing the positions of continuously moving objects. In W. Chen, J. F. Naughton, & P. A. Bernstein (Eds.), *Proceedings of the 2000 ACM SIGMOD International Conference on Management of Data,* May 16-18, 2000, Dallas, Texas, USA (pp. 331-342). New York: ACM.

Samet, H. (1990). *The design and analysis of spatial data structures.* Reading, MA: Addison-Wesley.

Shivakumar, N., & Venkatasubramanian, S. (1996). Efficient indexing for broadcast based wireless systems. *Mobile Networks and Applications, 1*, 433-446.

Song, Z., & Roussopoulos, N. (2001a). Hashing moving objects. In K.-L. Tan, M. J. Franklin, & J. C. S. Lui (Eds.), *Mobile Data Management, Second International Conference, MDM 2001,* Hong Kong, China, January 8-10, 2001, Proceedings, LNCS 1987 (pp. 161-172). Berlin, Germany: Springer.

Song, Z., & Roussopoulos, N. (2001b). *k*-nearest neighbor for moving query point. In C. S. Jensen, M. Schneider, B. Seeger, & V. J. Tsotras (Eds.), *Advances in Spatial and Temporal Databases, Seventh International Symposium, SSTD 2001,* Redondo Beach, CA, USA, July 12-15, 2001, Proceedings, LNCS 2121 (pp. 79-96). Berlin, Germany: Springer.

Song, Z., & Roussopoulos, N. (2003). *k*-nearest neighbor for moving query point. In M.-S. Chen, P. K. Chrysanthis, M. Sloman, & A. B. Zaslavsky (Eds.), *Mobile Data Management, Fourth International Conference, MDM 2003,* Melbourne, Australia, January 21-24, 2003, Proceedings, LNCS 2574 (pp. 340-344). Berlin, Germany: Springer.

Sun, J., Papadias, D., Tao, Y., & Liu, B. (2004). *Querying about the past, the present and the future in spatio-temporal databases.* Paper accepted for presentation at the 20th IEEE International Conference on Data Engineering (ICDE), Boston, MA, March 30-April 2, 2004. Retrieved October 10, 2003, from http://www.cs.ust.hk/~dimitris/publications.html

Tao, Y., & Papadias, D. (2002). Time-parameterized queries in spatio-temporal databases. In M. J. Franklin, B. Moon, & A. Ailamaki (Eds.), *Proceedings of the 2002 ACM SIGMOD International Conference on Management of Data, Madison,* Wisconsin, June 3-6, 2002 (pp. 334-345). New York: ACM.

Tao, Y., Kollios, G., Considine, J., Li, F., & Papadias, D. (2004). *Spatio-temporal aggregation using sketches.* Paper accepted for presentation at the 20th IEEE International Conference on Data Engineering (ICDE), Boston, MA, March 30-April 2, 2004. Retrieved October 10, 2003, from http://www.cs.ust.hk/~dimitris/publications.html

Tao, Y., Mamoulis, N., & Papadias, D. (2003). Validity information retrieval for spatio-temporal queries: Theoretical performance bounds. In T. Hadzilacos, Y. Manolopoulos, & J. F. Roddick (Eds.), *Advances in Spatial and Temporal Databases, Eighth International Symposium, SSTD 2003,* Santorini Island, Greece, July 24-27, 2003, Proceedings, LNCS 2750 (pp. 159-178). Berlin, Germany: Springer.

Tao, Y., Papadias, D., & Shen, Q. (2002). Continuous nearest neighbor search. In *VLDB 2002, Proceedings of 28th International Conference on Very Large Data Bases,* August 20-23, 2002, Hong Kong, China (pp. 287-298). St. Louis, MO: Morgan Kaufmann.

Tao, Y., Papadias, D., & Sun, Q. (2003). The TPR*-tree: An optimized spatio-temporal access method for predictive queries. In J. C. Freytag, P. C. Lockemann, S. Abiteboul, M. J. Carey, P. G. Selinger, & A. Heuer (Eds.), *VLDB 2003, Proceedings of 29th International Conference on Very Large Data Bases,* September 9-12, 2003, Berlin, Germany (pp. 790-801). St. Louis, MO: Morgan Kaufmann.

Tayeb, J., Ulusoy, Ö., & Wolfson, O. (1998). A quadtree-based dynamic attribute indexing method. *The Computer Journal, 41*(3), 185-200.

Trajcevski, G., Wolson, O., Zhang, F., & Chamberlain, S. (2002). The geometry of uncertainty in moving objects databases. In C. S. Jensen, K. G. Jeffery, J. Pokorny, S. Šaltenis, E. Bertino, K. Bohm, & M. Jarke (Eds.), *Advances in Database Technology EDBT 2002, Eighth International Conference on Extending Database Technology,* Prague, Czech Republic, March 25-27, Proceedings, LNCS 2287 (pp. 233-250). Berlin, Germany: Springer.

White, D. A., & Jain, R. (1996). Similarity indexing with the ss-tree. In S. Y. W. Su (Ed.), *Proceedings of the 12th International Conference on Data Engineering,* February 26-March 1, 1996, New Orleans, Louisiana (pp. 516-523). New York: IEEE Computer Society.

Wolfson, O. (2002). Moving objects information management: The database challenge. In A. Y. Halevy & A. Gal (Eds.), *Next Generation Information Technologies and Systems, Fifth International Workshop, NGITS 2002,* Caesarea, Israel, June 24-25, 2002, Proceedings, LNCS 2382 (pp. 75-89). Berlin, Germany: Springer.

Zhang, J., Zhu, M., Papadias, D., Tao, Y., & Lee, D. L. (2003). Location-based spatial queries. In A. Y. Halevy, Z. G. Ives, & A.-H. Doan (Eds.), *Proceedings of the 2003 ACM SIGMOD International Conference on Management of Data,* San Diego, California, USA, June 9-12, 2003 (pp. 443-454). New York: ACM.

Zheng, B., & Lee, D. L. (2001). Semantic caching in location-dependent query processing. In C. S. Jensen, M. Schneider, B. Seeger, & V. J. Tsotras (Eds.), *Advances in Spatial and Temporal Databases, Seventh International Symposium, SSTD 2001, Redondo Beach, CA, USA,* July 12-15, 2001, Proceedings, LNCS 2121 (pp. 97-113). Berlin, Germany: Springer.

Chapter XII

Extensible Platform for Location-Based Services Deployment and Provisioning

Manos Spanoudakis, University of Athens, Greece

Angelos Batistakis, University of Athens, Greece

Ioannis Priggouris, University of Athens, Greece

Anastasios Ioannidis, University of Athens, Greece

Stathes Hadjiefthymiades, University of Athens, Greece

Lazaros Merakos, University of Athens, Greece

Abstract

Location-based services can be considered the most rapidly expanding field of the mobile communications sector. The proliferation of mobile-wireless Internet, the constantly increasing use of handheld, mobile devices and position-tracking technologies, and the emergence of mobile computing prepared the grounds for the introduction of this new type of services with an impressively large application domain and use range. The combination of position-fixing mechanisms with location-dependent, geographical information can offer truly customised, personal communication services through the mobile phone or other type of devices. In this chapter, motivated by the technology advances in the aforementioned areas, we present a generic platform for delivering Location-based services (LBSs) to the nomadic user. The platform features a modular architecture, which can be easily extended. Although the overall architecture of the platform is discussed, the focus is on the technical specifications, the design, the functionality, and the prototype implementation of its central component, the kernel. The kernel is responsible for coordinating communication with the various pluggable components in order to provide the full range of operations involved in the LBS delivery chain (i.e., from initial deployment to invocation, execution, and delivery of results).

Introduction

During the early '90s, the introduction of the Internet signaled the beginning of a new era in the sectors of telecommunications and networking. The idea of a whole world interconnected in a global network accessible at anytime by anyone (La Porta, Sabnani, & Gitlin, 1996) had such allurement that it gave a tremendous boost to the Internet right from the beginning of its existence.

Nowadays, over 10 years after this landmark point, things are once more changing in warping speed. A new concept, that of "wireless Internet" has emerged and constantly gains significant support. The rapid evolution of wireless and mobile telephony during the last decade has played a significant role in the evolution of the Internet towards its wireless or even mobile form. However, wireless Internet is still in its infancy, and many issues remain to be addressed in order to form a competitive solution in the area of telecommunications and a strong complement to the wireline infrastructure.

One of the most crucial points in the adoption of every new technology is what it has to offer to its potential users. Practically, people will not be interested in the technology itself, no matter how advanced it may be, but will care for the new services introduced by the technology, as this is what they actually use and pay for. Location-based services (LBSs) is an example of new-generation services that can be developed and offered within this wireless Internet framework along with the traditional Internet services that will also be ported to the wireless domain. An important aspect for providing new services is to develop means and tools that will assist both their creation and provisioning but other aspects as well, such as management and billing.

This chapter presents a new middleware and service-provisioning platform which facilitates location-based-services provisioning toward the public with minimum effort from the involved actors (e.g., service provider, mobile operator, etc.) and with no impact to the standardized wireless Internet framework. Moreover, the platform adopts an open architecture so that it can accommodate future technologies as well as be integrated in telecommunication infrastructures yet to come.

The rest of the chapter is structured as follows. First, we provide a brief overview of research and achievements in technologies strongly coupled with the concept of LBS provisioning. The next section presents a detailed description of the PoLoS architecture. The functionality of the various software components comprising the platform is extensively discussed. Subsequently, we introduce the service specification language to be used for writing new services that will exploit the designated middleware in order to be deployed and delivered to end users. The last two sections address issues regarding to the future of LBS systems and present our conclusions summing up the presented architecture.

Related Research

Faced with an increasingly difficult challenge in growing both average revenue per user (ARPU) and number of subscribers, wireless operators and their partners are developing a host of new products, services, and business models based on data services. LBSs, which can be characterized as the most rapidly expanded field of the mobile communications sector, are a key part of this portfolio, although complex LBSs are still a few years away from introduction. According to In-Stat/MDR (2003), the revenue opportunity for wireless carriers only will rise from $5 million in 2002 to more than $167 million in 2006. Other research conducted by Ovum, a large European consulting firm on telecommunications, software, and IT services, claims that the LBS market will generate over $5 billion for European operators only by the end of 2005 (Pawsey, Green, Dineen, & Munoz Mendez-Villamil, 2002). Such estimates justify the increased interest of all key players in the LBS chain (e.g., telecom operators, content providers, service providers, etc.) regarding the development of advanced solutions that can boost the market. The rapid proliferation of mobile-wireless Internet, the constantly increasing use of handheld, mobile devices and position-tracking technologies, and the emergence of mobile computing prepared the grounds for the introduction of this new type of applications with an impressively large application domain and use range. The combination of position-fixing mechanisms with location-based, geographically dependent information can offer truly customised personal communication services through mobile phones or other types of devices.

The term LBSs is used to describe a family of services that depend on the knowledge of the geographic location of mobile stations. Some of the most prominent LBS are:

- Navigation, travel, and tourist information services depending on the user's actual location.
- End user assistance services. This category includes low-usage services designed to provide end users with safety networks for difficult situations such as roadside automotive assistance and emergency services.
- Monitored person location, which includes data for health care, emergency calls, and so forth.
- Third-party tracking services for both corporate and consumer markets. These services can be the basis for more advanced ones such as fleet management, asset tracking, and people finding.
- Triggered services, which are servicesthat are automatically initiated whenever a user enters a predetermined area or on a periodic time basis. Examples include location-based billing and advertising services.

LBSs cover a whole range of user needs from emergencies (such as the Federal Communications Commission [FCC] E911 mandate for wireless emergency calls) to amusement (e.g., friend finder and Point of Interest [POI]s). In order for these applications to be made available to the public, a synergy of different, yet complementary, technologies and architectures is required. These technologies will be overviewed in the next paragraphs.

Positioning Technology

Positioning technology is a key point in the LBS context. During the last years, research in this area has been impressive. A plethora of solutions (Hightower & Borriello, 2001) that provide an estimation of the user's present location have emerged, each with distinct characteristics and capabilities, and are hence applicable to different circumstances. All solutions can be divided into two major categories: the "terminalcentric," where the mobile terminal itself determines its position even in the absence of a mobile network, and the "networkcentric," where the position of the terminal is determined by the network infrastructure. However, a broader categorization based on the necessary underlying infrastructure follows.

Satellite Based

The Global Positioning System (GPS) is the prevalent solution in this category. A constellation of satellites (space segment) transmits information enabling a receiver (earth segment) to determine its position. Location determination uses the timed difference of arrival of satellite signals and is performed either entirely within the mobile unit or within the network providing 10-to-40-meter accuracy. It renders the user totally independent and, in principle, allows access to any LBS from third-party service providers. The second method, known as Differential GPS (Parkinson & Enge, 1996), can potentially provide location accuracy at the meter and submeter level by using infrastructure-based assistance.

Advanced research has been performed during the recent years in this area and, already, two new systems are emerging. The European Space Agency (ESA) works on Galileo (Benedicto, Dinwiddy, Gatti, Lucas, & Lugert, 2000), which, according to its specification, will deliver positioning services of high accuracy (4 to 10 m). On the other hand, the Russian Federation Government in coordination with the Russian Space Forces has launched GLObal Navigation Satellite System [GLONASS] (http://www.glonass-center.ru), which aims to provide significant benefits to the civil user community through a variety of applications offering, however, significantly lower performance (accuracy in the range of 57 to 70 meters).

The major drawbacks of the satellite-based solutions are their inadequacy to support positioning in indoor environments in conjunction with the need for costly and power-consuming handheld devices. A hybrid "network-assisted" approach, called the Assisted Global Positioning System (A-GPS), enhances GPS capabilities by using fixed GPS receivers to fetch data that can complement the readings of the terminal (Djuknic & Richton, 2001), thus alleviating the above-mentioned deficiencies.

Terrestrial-Infrastructure Based

Most of the solutions available for location determination through terrestrial infrastructure adopt the networkcentric approach (Lopes, Villier, & Ludden, 1999; Ludden & Lopes, 2000). The simplest example of a networkcentric positioning technique is the cell-ID method. Cellular networks have a built-in capability to identify the cell where a specific mobile terminal is located with an appropriate level of accuracy. This capability is an inherent part of mobility management. In addition to this cell-based, coarse approach for determining a terminal's location, a number of radio-based techniques have been developed such as time of arrival (TOA) and enhanced-observed time difference (E-OTD; Drane, Macnaughtan, & Scott, 1998).

Almost all major mobile operators have already incorporated in their networks support for location services by providing a gateway mobile location center (GMLC). A GMLC is a special network node with location-retrieval capabilities, specified by the Third-Generation Partnership Project (3GPP) initiative (*3GPP*, 2002). The GMLC is the first node that external-location service clients can access in a GSM public-land mobile network (PLMN). Terrestrial-infrastructure-based solutions are generally less accurate than their satellite counterparts, however, they are widely used as they do not require expensive additions to the terminal devices (i.e., GPS receivers).

Moreover, significant progress has been achieved lately in the area of Wireless Local Area Network (WLAN)-based location technology, typically used in indoor environments (Castro, Chiu, Kremenek, & Muntz, 2001). The problem of positioning in these environments has been solved with ad hoc solutions, such as Microsoft's RADAR system (Bahl & Padmanabhan, 2000), due to the actual lack of standardization for location-data acquisition from WLAN devices. Results from the Internet Engineering Task Force (IETF) Geopriv Workgroup (Cuellar, Morris, Mulligan, Peterson, Polk, 2002) are expected on this issue, but also the Location Interoperability Forum Mobile Location Protocol (LIF MLP)—considering a possible third Generation (3G)-WLAN interworking—seems to be suitable for the purpose of standardizing a WLAN-based location solution.

Geographic Information System

Geographic information systems (GISs) represent a comprehensive database related to spatial information with an integrated set of tools for querying, analysing, and displaying data. Currently, the integration of GIS technologies with position-fixing devices has proven a powerful decision-making tool that can be exploited in scientific investigations, resource management, and development planning. Sophisticated GIS databases may be derived from censuses, surveys, maps in conventional and digital form, and satellite imagery. The importance of the spatial dimension in large-scale analyses for problems of environmental monitoring, regional development, and land-use planning is widely recognized. Advanced GIS technologies present the spatial aspect as well as the visual representation (3D models) of reality that allow for immediate reactions in a number of situations (e.g., natural disasters, fire fighting, traffic conjunction, etc.), and provide navigational and accessibility information to and from different locations.

The current GIS technology is developing toward the following principal directions.

- The Open GIS Consortium (OGC) open standards specifications (http://www.ogc.org).
- The adoption of an Relational Database Management System (RDBMS) as a common, central database for features (geometry and attributes).
- The development of the extended markup language (XML) and, specifically, the geographical markup language (GML) prototypes for data exchange between applications.

LBS Platforms

Many corporate software vendors, prompted by both the boom of mobile Internet and the lack of ready solutions that could handle the different aspects pertaining to the LBS provisioning chain, developed software tools and middleware platforms for handling both the creation and delivery of such services. Most of these platforms offer a robust environment, which, to some extent, can be used for developing services. The vast majority of LBS platforms are built using Java enterprise technologies, and most of them comply with industry standards through open interfaces.

Platforms such as Celltick's Interactive Broadcast platform (http://www.celltick.com) that supports only the Short Messaging Service (SMS) and Wireless Application Protocol (WAP) interfaces for Global System for Mobile Communications (GSM)/General Packet Radio Service (GPRS), or Ericsson's Mobile Positioning System (http://www.ericsson.com) are enhanced position-tracking solutions for second Generation (2G) or 3G mobile networks. Kivera's Location Engine (http://www.kivera.com) does not implement a built-in interface with network operators for positioning, and thus can be considered a spatial data provider platform and not an autonomous LBS platform. The most advanced platforms available today cater to all standard interfaces used by mobile phones (SMS, Multimedia Messaging Service [MMS], WAP) as well as HyperText Markup Language (HTML) for 2.5G and 3G HTML-enabled mobile phones and personal digital assistants (PDAs). In terms of positioning technologies, the ones supported are mostly those interfacing with gateway-mobile-location-center-compliant (GMLC) GSM/Universal Mobile Telecommunications System (UMTS) networks. LocatioNet's (http://www.locationet.com) and Cellpoint's MLS/MLB (http://www.cellpoint.com) platforms cover all the aforementioned features together with a set of off-the-shelf applications, while LocatioNet comes with a high-performance GIS engine. Webraska (http://www.webraska.com) and ESRI (http://www.esri.com), in their platforms, add support for GPS-enabled handheld devices. Telenity's Canvas Location-Enabling Server (http://www.telenity.com), apart from supporting standardized XML interfaces with positioning and GIS servers, facilitates service creation through a service creation environment (SCE). The result is a service in the service-creation markup language (SCML) format that is deployed in the platform similarly to the PoLoS deployment procedure discussed below. Deployment of LBSs through Web services or Java Application Programming Interface (API)s is supported by Autodesk's LocationLogic (http://www.autodesk.com), whilst Openwave with Location Studio (http://www.openwave.com) provides Web service interfaces for external applications to request location-specific information from GIS and positioning components, acting more like a mediator than an integrated platform hosting application. Finally, in the world of local wireless networks, PanGo provides the Proximity Platform (http://www.pangonetworks.com) for deploying and delivering LBS for 802.11-based WLANs, while Appear Networks follows a different approach with its Provisioning Server (http://www.appearnetworks.com) that focuses on fast and effortless delivery of all types of mobile applications (including location-aware applications) for local execution on the wireless device.

The above discussion clearly shows that despite the variety of platforms that exist today, there is no integrated solution that covers all aspects of the LBS provisioning paradigm, that is, from the specification of a new service to the actual deployment and delivery to end users. Most of the available platforms focus on the process of deploying and delivering the service, and neglect issues such as service-logic creation and specification. Others do consider these latter issues, but restrict their capabilities to a limited subset of the full LBS spectrum. What is equally important is the clear separation of all the existing platforms to those designed for outdoor "macro" environments and those focusing solely on the delivery of indoor "micro" applications. The first family of platforms cannot be considered as a candidate for location-aware applications in indoor environments (such as exhibitions, museums, etc.) due to the lack of positioning components interfacing with indoor positioning technologies such as WLAN or Bluetooth, which greatly differ from their GSM/UMTS counterparts. Similarly, platforms operating over local wireless networks satisfy only indoor location services' needs. Hence, it is straightforward that delivering an all-purpose, integrated LBS platform is still an open issue. The PoLoS platform appears as the sole candidate for transparently bridging indoor and outdoor environments, placing them under a unified system with its built-in support for cellular, GPS, and WLAN positioning technologies.

Service Languages

Recently, numerous service languages have appeared, having the XML standard as their syntactical basis. Languages such as Call Processing Language [CPL] (Lennox & Schulzrinne, 2000), VoiceXML/Call Control eXtended Markup Language (CCXML), and Service Control Mark-up Language [SCML] (Bakker & Jain, 2002) have been specified with an XML-based syntax and are used in providing next-generation telephony services. The use of XML for expressing procedural languages has both benefits and drawbacks. Element nesting used in procedural languages is directly supported by XML. However, serial interpretation is not a core feature of XML and implicit rules must be established, such as the ordering of tags and content. Fortunately, both the document object model (DOM) and Simple API for XML (SAX) interfaces, the commonly used XML processing tools, are guaranteed to provide the XML elements in the order of their appearance. Another feature of the XML representation is that it can easily be presented in a visual, instead of the traditional textual, form. This allows the development of visual tools for easy service building.

PoLoS Platform Architecture

The PoLoS platform was built upon state-of-the-art technologies in the areas of information mobility, distributed mobile computing, and GISs. Its objective is to provide a single, homogeneous and expandable middleware platform for end-to-end provision of the aforementioned functionality.

The PoLoS platform provides the full functionality needed by both the service operator, which will deploy the service, as well as the end user, who will take advantage of the deployed service. To achieve this, PoLoS features a component-based architecture as illustrated in Figure 1. Each component within this modular architecture has a specific and clearly defined functionality, which is further elaborated below. The clean separation between these functional entities is another design goal of the PoLoS architecture. The autonomy granted to each component through this architectural approach makes possible the operation of the platform in distributed environments (i.e., different nodes and organizations hosting different components), thus rendering it scalable and enhancing not only its reconfigurability and reusability of each component separately, but also its business versatility. It also fits quite well into a wide range of business models involving the management of components by different parties and actors.

Figure 1 clearly shows that the heart of the PoLoS architecture is the component named kernel. This is the entity which hosts the logic of the LBS, handles all execution issues, and coordinates the communication of the latter with the various peripheral components. Five main interfaces of the kernel are identified toward the following entities.

- the positioning component (POS),
- the GIS component,

Figure 1. The PoLoS overall architecture

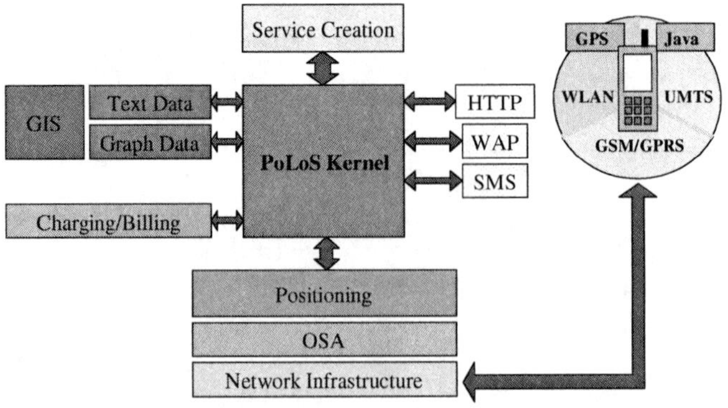

- the interfaces component,
- the SCE, and
- the charging and billing component

All components interact with the kernel using distributed architectures in order to allow them to reside in different locations. It is very important that these well-defined interfaces are all built upon standard technologies, making peripheral components easily replaceable by others implementing the desired functionality. Therefore, all peripheral components can be regarded as plug-ins to the kernel. This approach renders the platform capable to adapt to future advances of the positioning and GIS technology, or to accommodate new transport protocols apart from the existing ones (Hypertext Transfer Protocol [HTTP], WAP, SMS).

PoLoS Kernel

The kernel component can be considered the heart of the PoLoS platform. Its role is twofold. It serves as the run-time environment for all LBSs offered to the end user. Moreover, it acts as the coordinating entity for all external components. In order to run the deployed services and to fulfill requests posted for them, a variety of peripheral components need to be invoked.

The kernel itself comprises distinct modules, each one having a specific role during the life cycle of a service (from its creation, to deployment and execution). This modular design allows for the easy, centralized management of the provided services as well as for a variety of different invocation modes. At the moment, apart from the standard request-response model, the platform inherently supports internally scheduled invocations on a time- or event-basis according to the push paradigm. The five modules that collectively form the kernel are depicted in Figure 2 and are the service deployer, the service invocation module (SIM), the service execution environment (SEE), the scheduler, and the service registry.

Each module exposes a well-defined set of interfaces for interoperating with other modules or external components, as well as for management purposes.

The implementation of the kernel is heavily based on Java technologies, and more specifically, on the Java 2 Enterprise Edition (J2EE) framework (*J2EE*, 2003) and the enterprise JavaBeans (EJB) specification (*Enterprise JavaBeans 2.1*, 2003). Initial development was done on the Jboss application server (http://www.jboss.org), an open-source environment. Nevertheless, provision has been made to ensure compatibility with commercial application servers such as BEA's Weblogic and IBM's WebSphere. Furthermore, every module complies with the Java management extensions (JMX) framework (*JMX Framework*, 2003) in order to be externally manageable by an administrating entity.

Figure 2. Kernel's internal architecture

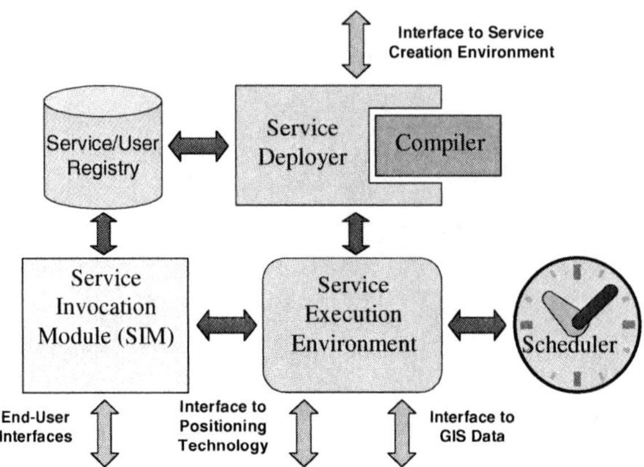

Figure 3. Service deployer's phased processing

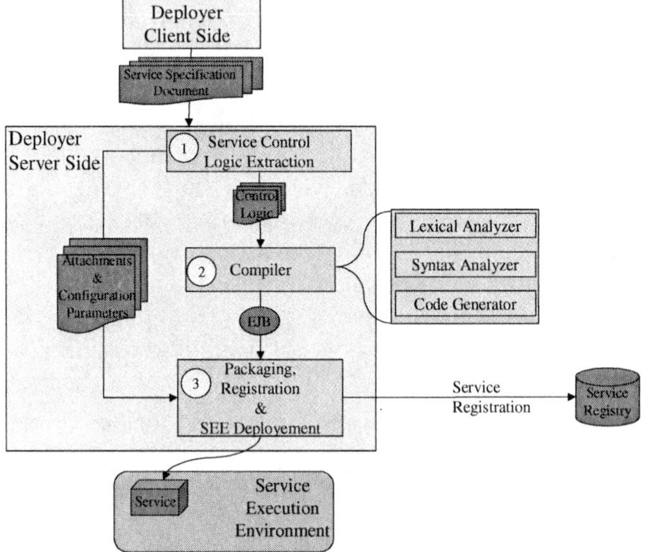

Service Deployer

The service deployer is the entry point for all new services entering the kernel. As shown in Figure 3, the deployer consists of two distinct parts, the client and the server side. The client side resides externally to the Kernel (usually at the service creator's premises) and uses Web services to communicate with the server side. The former is usually invoked from external sources (e.g., the SCE component) in order to inject new services into the kernel.

A service specification document describing the new service is sent through a Simple Object Access Protocol [SOAP] (Graham, Davis, Simeonov, Daniels, Brittenham, Nakamura, Fremantle, Koenig, Zentner, 2002) message to the Web service provider side of the deployer. The communications channel between the SCE and the kernel has been secured using transport layer security and secure sockets layer (TLS/SSL). Therefore, the client application has to present a digital certificate in order to be authenticated and authorized, and it is up to the private key infrastructure (PKI) scheme used by the operator and service provider to accept any kind of certificates issued by third-party-trusted certificate authorities (Johner, Fujiwara, Yeung, Stephanou, Witmore 2000).

Initially, the document is checked to verify the integrity of the received data. Subsequently, a three-phase procedure is followed as explained below. As stated, the initial form of the service comes in an XML-based language. An initial processing is performed in order to locate and extract the service-control logic part contained in the service specification document, which will feed the compiler. In the second phase, the compiler is invoked and performs lexical and syntactical analysis on the extracted part in order to produce the executable form of the service. Since the design was based on open technologies, the current implementation of the compiler generates java bytecode (in the form of EJBs). The service parameters should be extracted from the XML document and saved in a service-related repository in order to be available when the service is invoked. This repository resides within the SEE. The initial document may contain attachments that can range from HTML and Jave Server Pages (JSP) pages to compiled modules (e.g., java classes, midlets), and their role is to enhance visually the delivered service or provide sophisticated functionality that cannot be covered by the specified language. Finally in the third phase, the compiled service, along with its attachments, is packaged, and a request for deployment inside the SEE of the kernel is performed. Although this part of the deployer's functionality (deployment request) depends on the application server used (i.e., Jboss), it can be easily adapted to support any other application server. If the request completes successfully, the new service is registered within the service registry. By default, a newly deployed service cannot be invoked. It is the responsibility of the platform administrator to decide that a service should be considered as ready to be invoked. This functionality allows conditional billing setup or other per-service, additional parameterizations to take place by the platform operator before actual provision of the service.

Service Invocation Module

The SIM is the front-end mechanism used to receive requests within the kernel. It consists of the components shown in Figure 4.

The dispatcher acts as the entry point to the PoLoS kernel for all incoming requests irrespective of their source. Such service requests can be originated either from the external environment (e.g., terminal device) and forwarded to the SIM via the interfaces component (see Figure 2), or from within the kernel. The latter ones originate from the scheduler as well as from a running service (to be explained below). It is implemented as an EJB, making possible remote access to the kernel since the interfaces component is

Figure 4. Service invocation module's architecture

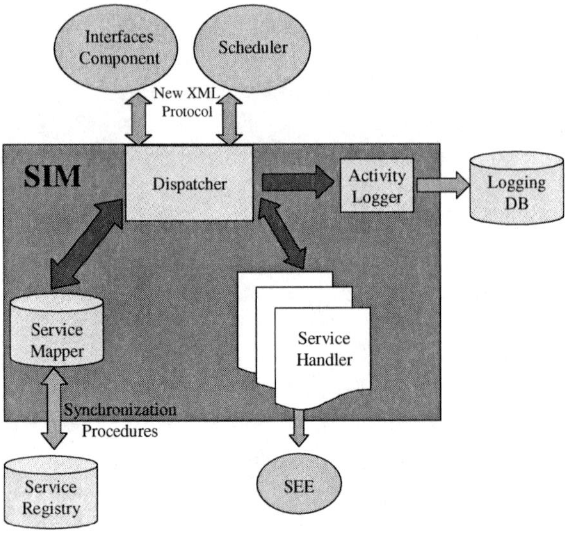

external to the kernel and they can be deployed in different servers. The dispatcher communicates synchronously with the interfaces component by exchanging information modeled using a new, specially designed, XML-based protocol. Subsequently, it retrieves the name that the requested service is bound to inside the kernel SEE by consulting the service mapper. The service mapper is an object that replicates, in memory, the bindings between the names that the services expose to PoLoS users and those used internally to identify the service instances. This information is generated by the deployer and is kept synchronized with that of the persistent service registry using transactional techniques.

The actual communication with the SEE is carried out by the service handlers. After each successful binding, the dispatcher, in order to process the request, retrieves a service handler instance from a pool. The service handler incorporates a generic, transparent mechanism for invoking all possible services. Such a mechanism is independent of the invoked service itself and the necessary parameters. The latter, as well as the returned results, are encapsulated in an XML stream. Moreover, the service handler is responsible for checking the security context of the invocation request. Such a checking is implemented using the Java Authentication and Authorization Service (JAAS) security framework as supported by any EJB-2.0-compliant container. Details on PoLoS authentication schemes are provided in subsequent paragraphs. Finally, logging of service activity, which constitutes the basis for the charging and billing system, is performed by the activity logger component of the SIM. The implementation uses standard, lightweight log4j calls (Gulcu, 2003), thus achieving the minimum delay interference of the logging procedure in the service execution cycle.

Service Execution Environment

The SEE serves as the run-time context for all deployed services. It is based on an EJB container, therefore providing the platform with features and advantages inherent to the EJB technology, such as portability, scalability, efficiency, instance pooling, and so forth. The executable code of the service arrives from the deployer module in bytecode format compliant with the EJB 2.0 specification. Only the corresponding entry methods specified in the service-control-language (SCL) document of the service are available to the SIM for accessing the service functionality. In order to achieve independence of the container's default deployment mechanism, the platform stores metainformation for each service in the service registry. This information will be used for redeploying services in case the platform abruptly terminates operation. Redeployment is possible through the deployer's management interface. This approach was adopted in order to have full control of the deployment procedure rather than rely on the default container's functionality.

During service execution, the SEE is responsible for coordinating its interaction with any external entity. Such an entity could be either the POS component (for retrieving user's location information) or the GIS component (for information requested by the user). Since POS and GIS are components external to the PoLoS kernel, and therefore can reside in a different environment, they are not directly addressed by the SEE. For each component there exists a relevant handler in order to achieve the communication.

Communication with the GIS component is implemented using Web services over a secure HTTP connection as shown in Figure 5. The Web services client resides inside the SEE and is invoked using an EJB interface, whilst the server is part of the GIS component. Adoption of this approach makes the kernel fully compatible with any possible commercial GIS system that can expose its functionality as a Web service. The POS component, on the other hand, is a pure java plug-in and is accessed through standard Remote Method Invocation (RMI)/Internet Inter-ORB Protocol (IIOP) mecha-

Figure 5. GIS component client-server communication

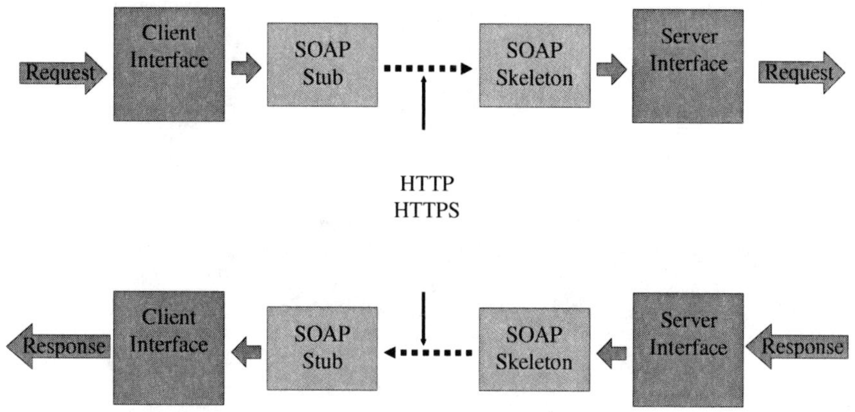

nisms (Roman, 2002). The aforementioned invocation pattern is depicted in Figure 6. As mentioned above (SIM presentation), the PoLoS kernel allows combining existing services in order to implement a new service. This is achieved by allowing a service executing in the SEE to address another service through the SIM, thus making it a potential initiator of communication with the SIM.

Another feature of the SEE is the service debugging capability. The service logs execution traces upon request. Such logs can be used for debugging purposes by the service creator. The service creator can initiate a debug session for a specific instance of the service in order to read these traces from the SCE environment through a Web services interface provided by the kernel.

Scheduler

The role of the scheduler is to implement the logic required to support the automatic execution of service instances within the platform (Tsetsos, Sekkas, Priggouris, & Hadjiefthymiades, 2004). The scheduler component supports the triggered execution of services and the implementation of the push service paradigm (Cheverst, Mitchell, & Davies, 2002). It features built-in support for both event- and time-based service execution, and a variety of scheduling schemes are provided:

- one-time scheduling,
- infinite, periodic scheduling, and
- periodic scheduling within a fixed interval, and so forth.

Figure 6. Service-external components invocation pattern

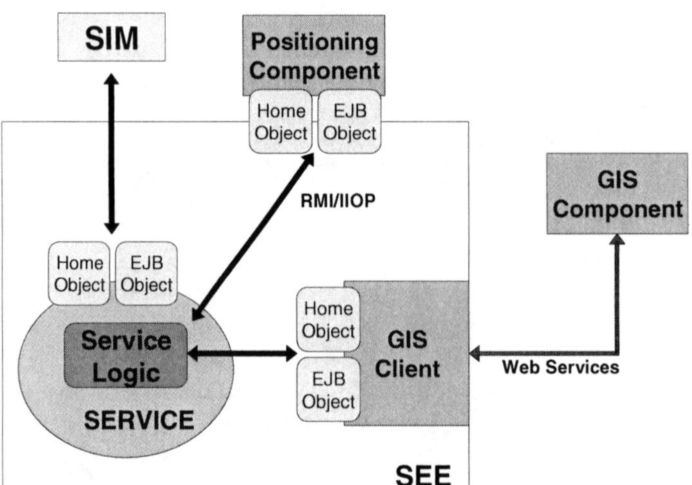

Figure 7. Scheduler's internal architecture

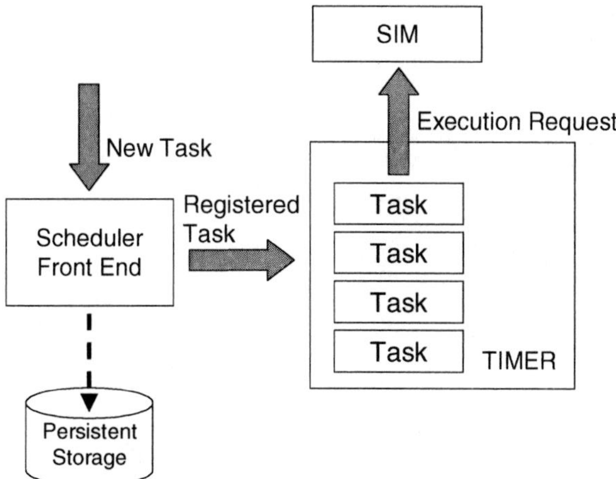

The scheduler also provides means for delivering location events coming from the POS component to the service (e.g., the user entered a mall or a building block) with the minimum possible delay. This is part of the functionality provided by the Open Services Architecture (OSA)/Parlay mobility specification (Parlay-ETSI, 2001), and is also particularly interesting in scenarios of personalised location-based advertising.

A very important design goal is to ensure that scheduled tasks will always be executed even after a system failure. To achieve the objective of maximum reliability, the scheduler features internal mechanisms for persistently storing information pertaining to each scheduled task in order to cope with potential platform shutdowns. Upon reinitialization of the scheduler, all stored tasks that have not yet expired are reloaded into the timer for execution. Only tasks that are marked for future and/or multiple executions are saved in the storage and are removed when their time frame is depleted.

The internal architecture of the scheduler (as shown in Figure 7) is based on the EJB specification and the RMI framework.

Service and User Registry

The service registry is a storage area that contains information for all services deployed in the platform. When a new service is successfully deployed (or redeployed), the deployer updates the service registry with the appropriate information. Each registered service has numerous attributes, some of which are significant for its execution, while others are for informational purposes only. The most important attributes are summarized in Table 1.

Registered services have three possible states (Figure 8): started, stopped, and stopping. Stopped services are services which have been successfully deployed but are not

Table 1. Service attributes

Attribute	Description
ID	Unique identifier within the platform
Status	STARTED|STOPPED|STOPPING
JNDIName	Name used to bind service ID with the name used internally by the kernel (JNDI)
AccessControl	Whether service is free or not
Roles	Names of the roles associated with the service

Table 2. User attributes

Attribute	Description
ID	Unique identifier within the platform
Username, Password	Used only when a request cannot be identified directly by other means
GSMDevices (IPDevices)	A set of Mobile Station ISDNs (IPs) identifying the user using SMS, WAP (HTTP)

available to end users and do not have running instances within the SEE. This is the default state for all newly deployed services. In order for the service to be accessible by end users, the administrator must explicitly start it (using the administration console), thus setting its state to started. Finally, stopping services are services that are in the process of being stopped. They do not accept requests but have running instances inside the SEE. After the completion of those instances, the service moves to the stopped state. The service mapper can be considered as the in-memory counterpart of the service registry for started services only.

Along with the services, the same registry contains information pertaining to registered users (Table 2). This information is structured in such a way that, together with the service-specific information, it implements the role-based policies upon which the authentication and authorization schema is built (Priggouris, Hadjiefthymiades, & Merakos, 2004). A user must be registered in the platform in order to have access to services other than those that are freely available. Every service accessible to registered users has a number of roles associated with it, each having a different set of execution privileges on this service. The attributes of roles are generated according to the information contained in the service description (EJB's deployment descriptor, in this case). Moreover, a registered user is assigned to one or more of the above-mentioned

Figure 8. Registered service state diagram

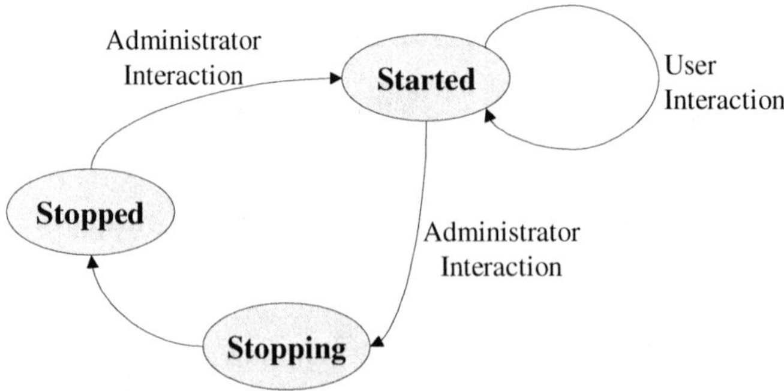

roles for each service he or she has access to. These roles are retrieved from the registry upon the user's admission to the platform and are used for controlling every service invocation.

The service and user registry is implemented using directory services, thus allowing for the fast retrieval of the stored data.

Peripheral Components

The term peripheral components is attributed to the relative location of these components within the platform in relation to the kernel. Each component has clear functionality within the platform and features a well-defined API (implemented on top of open protocols such as HTTP and RMI) for interfacing with the kernel. This approach makes each component autonomous and independent from the rest of the platform. Existing components can be removed at anytime and new ones, implementing similar functionality, can be plugged in without impact on the operation of the platform. The only requirement is that the new versions must implement the same interfaces. For this reason, each component supports extended management capabilities available through an appropriate Web management console. A more-detailed presentation of the functionality and implementation of each component follows in subsequent paragraphs.

Interfaces Component

The interfaces component is responsible for the communication of the platform with external entities (i.e., the user's terminal equipment). This communication is carried out transparently, irrespective of the underlying transport technology, which is used as bearer of the user's request. The current release of the platform supports the three most widely established protocol suites, namely, SMS, WAP, and HTTP. Nevertheless, its modular architecture makes it expandable, allowing it to accommodate additional sets of application protocols in the future. The interfaces component communicates directly with the SIM of the kernel using a new XML-based protocol specially designed for this purpose.

The design of the component, depicted in Figure 9, shows clearly the existence of special front-end modules dedicated to different protocols. The main task of these modules is to translate an incoming request, conveyed in a protocol other than HTTP, to HTTP. This request is then fed to a generic servlet (GServlet) which processes it (decapsulates the spatial, security, and session information), forwards the information using the XML-based protocol to the SIM, and subsequently handles the outbound flow. This output stream has to be formatted to the requested protocol. The aforementioned GServlet resides inside an HTTP server (and servlet engine) that can be considered to be the boundary between the kernel and the interfaces component. The architecture of each of the protocol-specific components is presented.

The WAP module consists of two submodules, one for WAP pull requests (generated by a user's terminal) and one for outgoing push messages coming from the kernel's

Figure 9. Interfaces component design

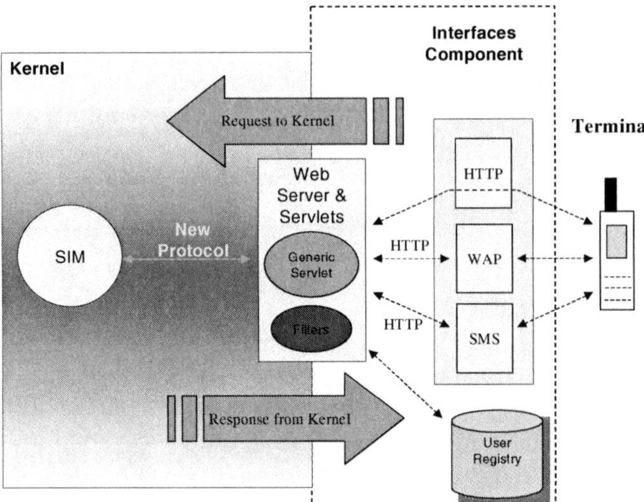

scheduler. The former receives the HTTP-encapsulated requests from the WAP gateway, while the latter has to encapsulate data coming from the service into a WAP push message as specified in the Push Access Protocol (PAP; *Push access protocol specification*, 2001) by the Open Mobile Alliance (OMA), and then forward it to the push proxy gateway for delivery.

The SMS module communicates with the SMS gateway (SMSC) of the network operator using the standard OSA interface. The technology of choice for implementing this interface was Web services and SOAP, as recommended by the Parlay Group (Lozinski, 2003). The component inside the SMS module that handles the interaction with the SMSG is the SMS OSA wrapper, as shown in Figure 10. It acts as the Web services provider for incoming requests and client for outgoing requests, having a different instance for every distinct SMSC. Moreover, it has to translate the asynchronous interfaces provided by OSA to synchronous kernel invocations and vice versa.

Apart from serving as the means for communication between the platform and the user's terminal, the interfaces component provides a first level of access control by intercepting a set of filters between an incoming HTTP request and the GServlet. The functionality of these filters will be elaborated later in the chapter in the context of the overall security framework discussion. Finally, the interfaces component is responsible for extracting location information from the request's header arriving from GPS-enabled clients.

Figure 10. SMS wrapper high-level architecture

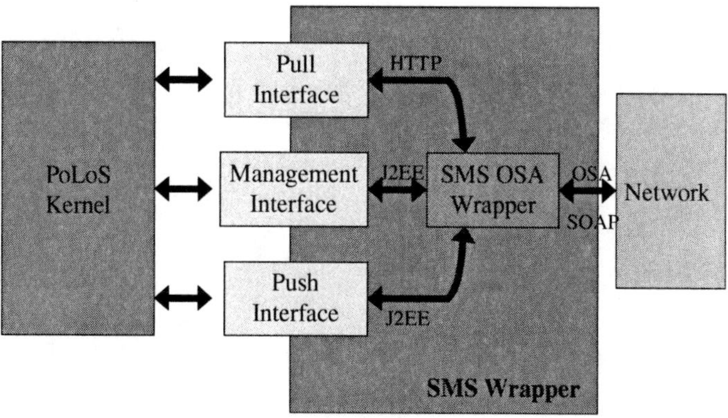

Positioning Component

The POS is responsible for providing a unified interface for location information to services. A major requirement for an LBS system is positioning technique independence. Therefore, POS integrates the underlying, major positioning technologies, acting as a layer of abstraction for the kernel. Using this approach, all complexities inherent to the location-tracking techniques, as well as their diversifications, are hidden from the kernel, which is presented with a simple, high-level and well-defined interface. Currently, the component supports location retrieval from standard 2G and 3G networks, WLAN, and GPS-enabled terminals. The latter is possible through a special software agent running inside the mobile terminal for pushing location data to the platform. Despite the existence of standards for location services in cellular networks, in WLANs the current status is complicated. For this reason a new framework, which provides positioning of objects to LBS applications transparently of the different indoor positioning mechanisms, is designed. It resembles the positioning architecture used by GSM/UMTS networks. It incorporates a novel entity, named the generic WLAN location center (GWLC), which similarly to GMLC, is the entry point for all location requests arriving to the WLAN network (Papazafeiropoulos, Prigouris, Marias, Hadjiefthymiades, & Merakos, 2003). For the aforementioned network infrastructures, communication is achieved through specific wrapper modules and standard OSA/Parlay open interfaces (Moerdijk & Klostermann, 2003).

The POS component discriminates between four types of positioning requests. These are:

- request response (RR), where a service asks for a user's position,
- periodic request (PR), that is actually a periodically activated RR through the kernel's scheduler,

Figure 11. Positioning component modular architecture

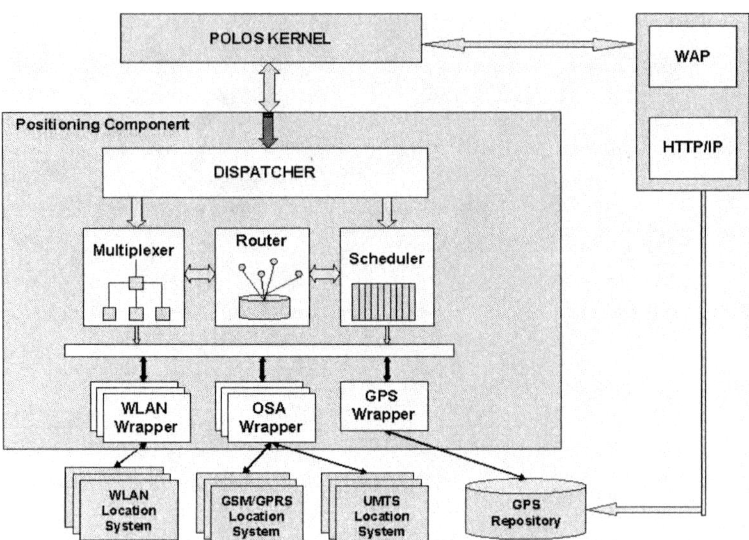

- event-driven request (ER), which resembles a network alert for cases where a user enters a specified area, and
- generic request (GR), which is a variant of RR type with the difference that a request is forwarded to all underlying networks that the user is registered with.

The modules comprising the POS component, illustrated in Figure 11, are the dispatcher, the router, the quality of service (QoS) scheduler, the multiplexer, and the wrappers. They are implemented using EJB and standard RMI technologies, and have the following functionality.

Location requests from the kernel are routed by the dispatcher to the appropriate submodule for further processing. If the request is of type ER, it is processed by the multiplexer. Its responsibility is to multiplex requests coming from the same user terminal in order to end up with a single request registration within a certain network for each user, reducing the load at the network for providing location information. All other types of requests are handled by the scheduler. The scheduler works asynchronously using priority queuing mechanisms that satisfy the requirement for differentiated treatment based on QoS. Requests are once again forwarded to the appropriate wrapper. The router encapsulates the logic for resolving the IP address of the appropriate network's gateway based on the incoming user's address (network prefix). This mapping is stored in an internal registry and can be easily administered through a management interface.

The most sophisticated submodule of the POS component is the set of wrappers. Three types of wrappers exist:

- GMLC wrappers for communicating with the GMLC of GSM/UMTS networks,
- WLAN wrappers for interfacing with the GWLC entity of WLANs, and
- GPS wrappers.

Each wrapper instance is bound to a specific network operator. It embodies a set of message queues for incoming and outgoing traffic, and communicates with the actual network exposing OSA functions as Web services. Concerning GPS positioning, the respective wrapper consults the GPS repository of the PoLoS platform to retrieve the coordinates. Filling this repository with data is part of the interfaces component's functionality.

GIS Component

The GIS Component provides a running service with geographical information retrieved from deployed GIS repositories. As users are interested in LBSs usable in different countries or regions, the GIS component will potentially interact with multiple, different GIS repositories (for different countries and different systems). Together with the lack of standardization in GIS data and systems, this imposes the need to support the use of each country's coordinate reference system. For example, in Greece, the Greek geodetic system (EGSA87) can be used, whereas France can use the Lambert geo reference system, thus assuring the easy integration of existing and evolving GIS data by "local" content providers in the platform and the provision of GIS services in different countries. The role of the GIS component, besides the retrieval of textual and graphical content, is to format it in such a way as to create representations (e.g., maps) of the requested area. The GIS component is capable of supporting the following types of services:

- indoor and outdoor localization service,
- indoor and outdoor navigation service,
- outdoor proximity service for finding POIs, and
- outdoor geocoding and reverse geocoding service for finding street details based on provided coordinates or the coordinates of a given place, accordingly.

Its implementation uses the client-server model with the client residing within the kernel while the server is found within the component and has access to the GIS repository. Communication between the two parts is achieved using Web services/SOAP protocol. Such an approach allows the two involved parties to be fully independent of each other. This is a very important aspect since they can be operated in different nodes, using different implementation technologies, which fits perfectly with the scenario and business model of LBS provisioning by a consortium of organizations.

In order to provide generic functionality to the PoLoS platform the GIS repository uses ESRI shapefiles as its digital geographic data type. A shapefile is a data file format introduced by ESRI (http://www.esri.com) for storing geographic features in vector

Figure 12. GIS component architecture

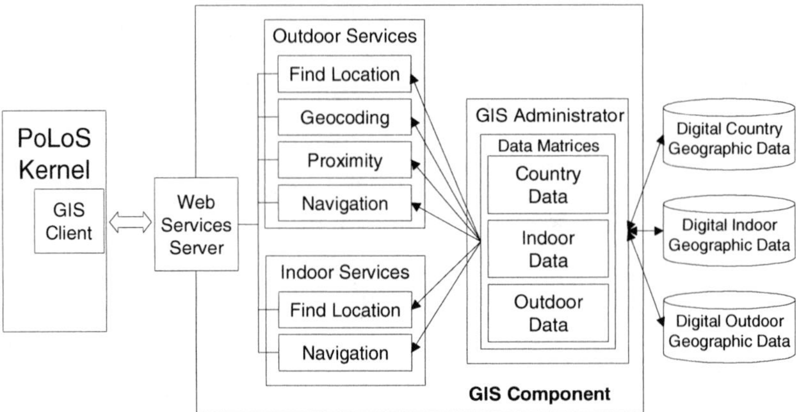

format. This means that map features are represented as sets of *x* and *y* coordinates. The choice of the specific data format is by no means restrictive for the implementation of the GIS algorithms since data coming from repositories in different formats can be easily converted using standard transformation-conversion software. Metainformation on the content supplied by the component is kept in the GIS administrator data matrix (Figure 12).

Service Creation Environment

A major advantage of PoLoS over other similar platforms is that it comes with full support for LBS creation. The SCE developed for this reason is not just an Software Development Kit (SDK) or API, but instead it is an advanced integrated development environment (IDE) simplifying the process of creating and deploying a service. It is structured around the service specification language (the XML-based scripting language that will be discussed in the last paragraph) and caters to both easy creation and deployment of services. It is based on the Eclipse framework, an open-source IDE supported by numerous companies (e.g., IBM, Borland, SAP, etc.).

The main goals of such an environment are to simplify the editing of the service script as well as to support easy and secure deployment of the script, along with possible attachments, to the execution engine.

There are two types of editors available in the SCE: a textual and a visual editor. The textual editor, besides standard text-processing facilities such as search-replace and undo-redo actions, provides the service creator with more advanced features such as syntax highlighting. Moreover, a distinct pane for managing variables used throughout

Figure 13. SCE: Textual editor perspective

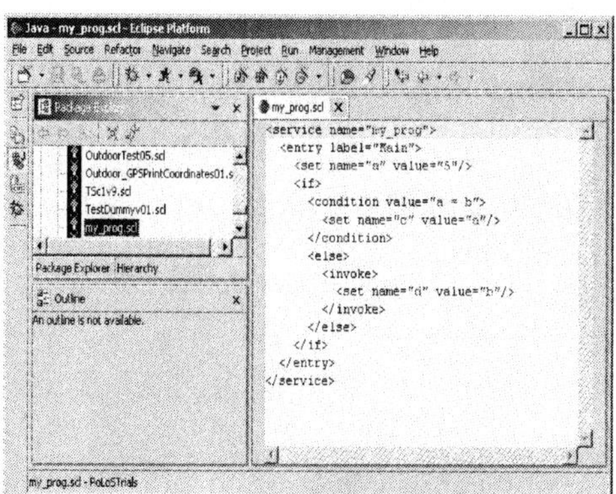

Figure 14. SCE: Visual editor perspective

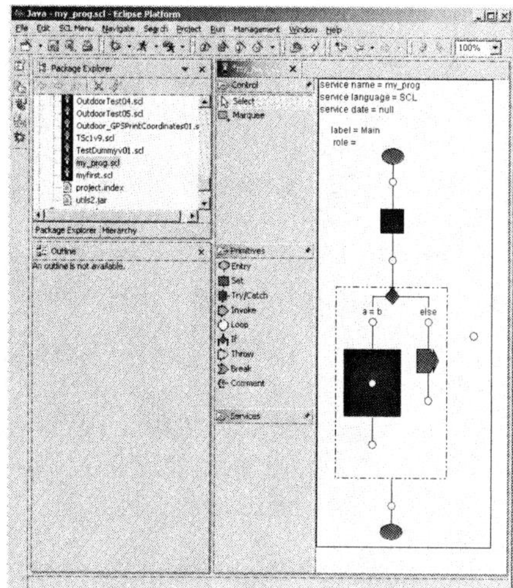

the script is supplied for improving the creator's productivity. Another pane is used for displaying all information pertaining to attachments. A screen dump of the textual editor is shown in Figure 13.

The alternative of the visual editor has capabilities found in all modern visual programming environments. The visual editor gives the developer the ability to construct services through a block-by-block approach, and visible information flows between blocks simply by dragging and dropping components from a pallet. A component can also be an existing service, thus allowing easy service aggregation. Furthermore, switching between visual and textual perspective (always kept synchronized) is supported throughout the entire service development process. A screen dump of the visual editor is shown in Figure 14.

A user-friendly menu is available for deploying services. This functionality is compatible with the functionality provided by the deployer client module. It consists of functions other than deployment for retrieving all deployed services with their status, and for verifying whether a service is well formed according to the language specifications without actually deploying it.

Communication with the deployer within the kernel is achieved using Web services/SOAP. The crucial issue of secure communication between the service creator and the platform host was resolved not only by exchanging data over HypterText Transfer Protocol Secure (HTTPS) using SSL/TLS, but also by forcing the creator to present a digital client certificate for his or her authentication to the PoLoS platform.

Finally, the SCE serves as client to the service debugger offered by the SEE. Using the appropriate SCE menu, the user can initialise a debug session, view service log traces in a special log window, and save them in a file if necessary.

Charging and Billing Component

Charging and billing are essential operations for the successful deployment of a telecommunication services platform in a commercial environment. The PoLoS platform architecture has integrated capabilities for keeping track of system and user activities in order to perform such operations. The flexibility provided by the charging and billing component is such that various existing charging models can be accommodated (Priggouris et al., 2003; Tsalgatidou & Pitoura, 2001). Actors involved in the LBS provisioning chain cooperate on the basis of specific business relationships. End users, service providers, and network operators will have to charge or be charged, possibly following different charging models for each case.

The PoLoS platform delivers charging and billing functionality through extensive and detailed, per-component logging. Logging is performed by each component separately in a database. Functionality is enhanced through a set of management interfaces based on customisable views for each involved actor. Therefore, the network operator who might be interested in charging the LBS provider for the location information retrieved through the OSA interface, that meets a predefined QoS, will need to have access to the POS component's logs. Similarly, if the service provider has decided to charge its users on a per-transaction basis it can consult the service activity logs generated by the kernel. Also, complex combinations of usage logs can result in very accurate and adaptable billing models.

PoLoS Security Framework

As discussed above, PoLoS caters to both free and secure service invocation. The latter is supported by addressing the following two issues:

- authentication, which secures access to the platform, and
- authorization, which guarantees that only users with certain rights and privileges can access specific services.

PoLoS incorporates extensive security features building upon a two-level authentication infrastructure (Figure 15), hence making the platform attractive for different business models. Both authentication and authorization are handled by creating a security context for each user accessing the platform. The invocation of a service instance takes place inside this context, which allows users to access services with different roles, and therefore, with different privileges. Creation of the context is achieved through the use of the JAAS framework in conjunction with the EJB 2.0 role-based security specification. The authentication of a user request takes place against the user and service registry and results in the establishment of a new security context for the designated user. Such context contains all specific roles assigned to the user for all services and characterizes the user for all subsequent interactions with the platform. Roles are service specific.

The first level of security consists of performing access control on a platform or service level based on the address of the incoming request. It is implemented in the interfaces

Figure 15. Secure invocation framework

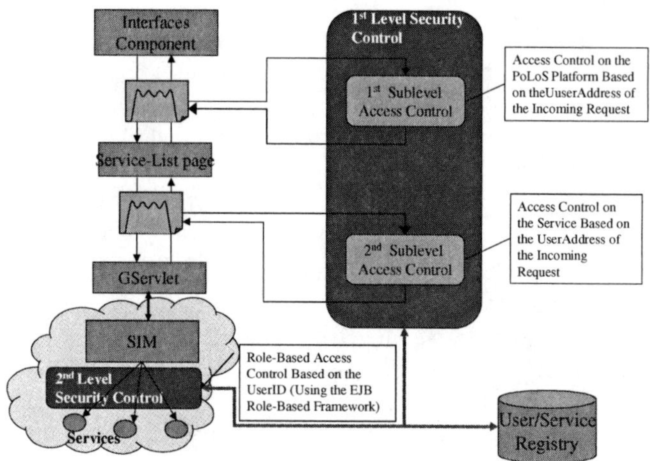

component by filtering the HTTP request before it reaches the GServlet. The service registry of the platform has been enriched with information that allows predetermined address ranges and networks to have access to a specific service. Thus, a service that is provided to a specific network operator can be configured to be accessed only by users served by this operator. This feature is quite attractive since PoLoS could be operated by a value-added service provider, which is not bound to a specific network operator but could possibly want to provide a service, which will be accessed exclusively by users attached to the designated network.

Moreover, similar flexibility in terms of allowing access to a specific group of users has been implemented on the platform level. This means that, if needed, the administrator of the platform could restrict or allow access to the platform (i.e., for all available services) to a specific set of users. This scenario seems attractive in the case where PoLoS is hosted by a specific network operator, which aims to provide a subset of services only to its users.

Once the request passes successfully the security control of the first level, authentication and role-based access control, as defined by the EJB 2.0 specification, are invoked. This second-level security control takes place inside the SIM using the JAAS framework, as already discussed.

Service Control Language

One of the main objectives of the PoLoS project was the definition of a new scripting language for specifying LBSs. Although one might argue that building a new language would add only complexity and confusion to the process of service creation, its implementation was considered advantageous for several reasons. First, the nature of general-purpose scripting languages is such that it does not fit the very specific needs of LBSs. They are too generic, providing capabilities that are not compliant with the semantics of an LBS. Furthermore, they lack support for abstractions commonly found in the context of these applications. The majority of services tend to have rather simple internal business logic and data manipulation requirements, but make extended use of predetermined components of the platform (e.g., for retrieving the location of a requestor). Thus, it is clear that the role of such a language is to efficiently orchestrate the operation of different platform components in order to provide the desired service.

The SCL designed is considered an optimal solution for the PoLoS platform. Its syntax is based on XML, a widespread notation in the telecommunications field, providing great flexibility and extensibility. At the same time, it caters to strict semantics and automatic verification of structure correctness. SCL does not include any kind of input or output filtering and processing, therefore it is kept independent from interface characteristics. Normally, the internal processing logic together with the complex operations carried out by the invoked platform components are sufficient for specifying most of the services. In the case where this is inadequate, the complementary exploitation of additional native-language components, such as class libraries, HTML and JSP pages, or downloadable applets and midlets, can provide the desired functionality. The SCL can channel information flow to such components with no additional burden or alteration of the generic execution model. A service specification file can be created using the SCE without this being mandatory. It can, as well, be created in any American Standard Code for Information Interchange (ASCII) editor with the aid of the deployer's client utility for adding attachments. An example of a service specification language file is presented in Figure 16. The service described here simply retrieves the location of the user issuing the request and returns a textual representation of it.

SCL Specification

The SCL is the part of a service specification language document placed between the <service> tags. The rest of the document includes configuration settings that can be either service-wide parameters and constants or service dependencies (<prerequisites> tag) and attachments, as explained above. In the example of Figure 16, this attachment is a binary jar file encoded in Base64 format. The next paragraphs will concentrate on the SCL part of the document.

Data Structures

Just like any general-purpose language, SCL supports three main categories of data structures, namely, variables, arrays, and records. SCL incorporates a nonstandard approach, combining these categories under a single naming scheme supported by XML's nesting facilities. The following two definitions are identical:

Figure 16. A service specification language document

```
<?xml version="1.0" encoding="UTF-8"?>
<PolosXML>
  <service lang="SCL" name="GetMyLocation" date="10/09/2003">
    <entry label="main">
      <set name="userpos">
        <invoke component="POS">
          <set name="userid" value="parameters.userid"/>
        </invoke>
      </set>
      <set name="useraddress">
        <invoke component="GIS">
          <set name="coordinates.north" value="userpos.coordinates.north"/>
          <set name="coordinates.east" value="userpos.coordinates.east"/>
        </invoke>
      </set>
      <set name="result.location" value="useraddress.textposition"/>
    </entry>
  </service>
  <configuration>
    <prerequisites>
      <service name="aserviceid"/>
        <library name="org.polos.alibraryid"/>
    </prerequisites>
    <parameters>
      <set name="language" value="en"/>
      <set name="en">
        <set name="Text1" value="Hello"/>
        <set name="Text2" value=" World"/>
      </set>
    </parameters>
    <description>
      This is a sample service for retrieving my location
    </description>s
    </config>
  </configuration>
  <include type="jar">
    <identity name="jce2.jar" encoding="Base64" date="2002/11/07 14:30:24"/>
    <content>
      e3wUAAJ8IAAAVAAAATU.....NEpDRUpBU1MuRFNBM2jimM3G== <!-Base64 encoding -->
    </content>
  </include>
</PolosXML>
```

<set name="en"><set name="Text1" value="Hello"/><set name="Text2" value="World"/></set>

and

<set name="en.Text1" value="Hello"/><set name="en.Text2" value="World"/>.
Such nesting of variables can continue in arbitrary depth.

Data Repositories

The term repository is used as a generic, multipurpose storage area. It can range from the session information kept in a Web server to the return parameters. It follows exactly the same notation scheme as the one described above for variables. Provision was made for certain specific repositories for configuration purposes, service or session-wide variables, and for input and output parameters.

In the example shown in Figure 16, the input parameters repository is accessed in the following line:

<set name="userid" value="parameters.userid"/>.

Data Types and Expressions

The SCL adopts the same approach as most contemporary scripting languages for data typing. A variable's type is deduced at run time from the context it is used in and the operations that are performed on it. The language discriminates between three data types, which are the following.

- Numbers: Numbers in SCL are integers. An implementation may choose to limit the range of values for numbers for practical reasons, although a minimum size of 32 bits is supported.
- Strings: Strings in SCL are similar to strings in other languages. There are no language limitations to the maximum size of a string. Strings support the unicode character set.
- Booleans: Boolean expressions are actually numbers. Any non-zero number corresponds to a boolean "true," while zero corresponds to "false."

Communication with other Services and Components

Services are accessed by external entities using the notion of entry points. Each service has a number of named entry points, which are similar to the procedures exposed as the service's interface. All code in SCL resides under an entry point. If an entry is not specified, the entry point "main" is used. The single entry point in the previous example is defined as follows:

<entry label="main"> ...code goes here... </entry>.

Invoking other components is achieved using the special <invoke> tag as in the case of the POS component presented below. There, a single value named userid is passed as the invocation parameter:

<invoke component="POS"><set name="userid" value="parameters.userid"/></invoke>.

A service can invoke another service using the above tag and the attributes "service" and "entry" instead of the "component" attribute:

<invoke service="ServiceX" entry="EntryPointX">...</invoke>.

Finally, Java methods found in classes included using the <include> tag can be invoked by using the "method" attribute and the desired class.

Execution Flow Control

The service language offers the capability of controlling the flow of service execution through the use of the conditional execution and loop constructs. The mechanism offering conditional execution is the <if> tag. The structure of the <if> tag is as follows:

<if><condition expr="...">...code goes here...</condition><condition expr="...">...code goes here...</condition> </if>.

The first expression that evaluates to "true" is used to execute the code just like in any other language.

Looping is supported via the <loop> tag. This causes an infinite iteration of the included commands until the <break/> tag is encountered, in which case the execution resumes after the </loop> tag of the innermost loop:

<loop>...code...<if><condition expr="..."><break/></if>...code...</loop>.

Future Trends

Certain issues have to be addressed before LBSs become the elusive "killer applications" that mobile operators envisage. First of all, the advent of third-generation mobile terminals will add a new perspective with high-quality visual presentation of data that clients are willing to pay for. These capabilities are at the moment available mostly in expensive GPS-enabled PDAs. Moreover, LBS will be easier adopted by the customers once the focus moves from traditional pull applications to more elaborate push applications where the user can easily, and with minimal effort, select when and what kind of information is to be provided. However, the real boost for LBS is likely to come with the interoperability of the operators at the national and international levels. It is obvious that business applications such as fleet tracking will certainly require roaming of users, which inevitably leads to the cooperation of different operators under the same platform. Interoperability can only be achieved using standards and open APIs. For this reason, standardization organizations, telecommunications companies, and research institutes formed the Open Location Services Initiative (OpenLS). OpenLS is devoted to standardizing the interfaces between the components comprising an LBS system in order to accelerate the availability of location services. The PoLoS platform is built upon industry standards and, thus, can be considered ideal for supporting multiple operators and environments. Moreover, it is our intention to extend the PoLoS platform to deliver a reference LBS system covering the widest possible spectrum of applications and environments.

Conclusion

In this chapter we discussed issues related to the design and development of the PoLoS middleware and service provisioning platform for LBSs. The PoLoS platform is an integrated, Java-based software enabling end-to-end delivery of LBSs that builds upon industry standards in wireless, mobile, and GIS technologies. Its modular infrastructure provides all necessary core elements that guarantee easy development, deployment, provision, and management of new and compelling location services. The design of the platform adopts the following orientation, deriving from the general requirements of an LBS system involving actors like users, operators, and service providers.

- **Portability:** The platform is independent of specific hardware and operating systems.
- **Reusability:** It is a generic platform, decoupled from service-logic basic components, and thus is unaffected by dynamic, new service introduction
- **Independence from underlying technologies:** The PoLoS platform has the following features.
- It is not bound to specific network technologies. It covers both outdoor and indoor environments (GSM/GPRS, UMTS, and WLAN) and can support an arbitrary number of location techniques by simply adding the appropriate new interfaces.
- It is not coupled with specific GIS technologies.
- It uses open interfaces toward the GIS, the network, and end-user devices.
- **Support for many operation paradigms:** namely, server push, client pull, and event scheduling
- **Flexible service handling:** Service logic is fully covered by the platform while the service itself is specified through the SCE.
- **Roaming across different infrastructures**: Users are free to move from an outdoor to an indoor WLAN environment.
- **Generic system infrastructure tied with a generic business model:** There is a clear separation between service creation, service provisioning, and the network,

allowing each function to be handled by independent (yet cooperating) organizations.

- **Security:** Authentication and authorization using role-based mechanisms is inherently supported, as well as secure network communication using TLS/SSL.

The platform can provide a user's positioning information from multiple sources such as terrestrial position fixing, WLAN, and GPS. Its unique feature is that it has built-in support for retrieving location from indoor WLAN networks, and even for allowing roaming between outdoor and indoor environments. It can be accessed by a wide range of mobile terminals over the most commonly used protocol suites (SMS, WAP, and HTTP). Also, key concepts of the component architecture that make the platform flexible and extensible to future needs were presented in the chapter. From a business perspective, privacy, security, and charging and billing functionality are significant factors for the establishment of such a system. The PoLoS platform fulfills these requirements by introducing novel approaches based on state-of-the-art technologies. Finally, a new XML-based SCL was presented with the purpose of assisting in the task of service creation on the platform. The syntax of the language, along with its unique characteristics, was also presented in detail.

Acknowledgments

The work presented here has been performed in the framework of the project IST-2001-35283 "PoLoS," which is partly funded by the European Community and the Swiss BBW (Bundesamt fur Bildung und Wissenschaft). The authors would like to acknowledge the contributions of their colleagues from CSEM, ALCATEL SEL AG, Telefonica I+D, Epsilon Consulting Ltd., INTRACOM SA, EPSILON SA, and the University of Athens.

References

3GPP: Technical specification group services and systems aspects; network architecture (Release 5). (2002-03) TS 23.002 V5.6.0. Retrieved May 2004 from http://www.3gpp.org

Bahl, P., & Padmanabhan, V. (2000). RADAR: An in-building RF-based user location and tracking system. *Proceedings of the IEEE International Conference on Computer Communications (INFOCOM)*, (pp. 775-784).

Bakker, J.-L., & Jain, R. (2002). Next generation service creation using XML scripting languages. *Proceedings of the IEEE International Conference on Communications (ICC)*, New York.

Benedicto, J., Dinwiddy, S. E., Gatti, G., Lucas, R., & Lugert, M. (2000). *GALILEO: Satellite system design and technology developments*. European Space Agency.

Castro, P., Chiu, P., Kremenek, T., & Muntz, R. (2001). A probabilistic room location service for wireless networked environments. *Proceedings of the International Conference on Ubiquitous Computing (UBICOMP)*, (pp. 18-34).

Cheverst, K., Mitchell, K., & Davies, N. (2002). Exploring context-aware information push. *Personal and Ubiquitous Computing, 6*(4), 276-281.

Cuellar, J., Morris, J. B., et al. (2004, February). *Geopriv requirements* [RFC 3693]. Retreived April 2004 from http://www.ietf.org/rfc/rfc3693.txt

Djuknic, G. M., & Richton, R. E. (2001). Geolocation and assisted GPS. *IEEE Computer, 34*(2), 123-125.

Drane, C., Macnaughtan, M., & Scott, C. (1998). Positioning GSM telephones. *IEEE Communications Magazine, 36*(4), 46-54.

Enterprise JavaBeans 2.1 specification (Public draft). (2003) Sun Microsystems, Inc. Retrieved April 2004 from http://java.sun.com/products/ejb/

Graham, S., Davis, D., Simeonov, S., Daniels, G., Brittenham, P., Nakamura, Y., Fremantle, P., Koenig, D., & Zentner, C. (2002). *Building web services with SOAP, XML and UDDI*. Sams Pub.

Gulcu, C. (2003). *The complete log4j manual*. Lausanne, Switzerland: L QOS.ch.

Hightower, J., & Borriello, G. (2001). Location systems for ubiquitous computing. *IEEE Computer, 34*(8), 57-66.

In-Stat/MDR. (2003). *Location-based services: Finding their place in the market* (Rep. No. IN0300357WI).

J2EE platform specification (Proposed final draft no. 3). (2003). Sun Microsystems. Retrieved November 2003 from http://java.sun.com/j2ee/j2ee-1_4-pfd3-spec.pdf

Java Management Extensions (JMX) 1.2 Specification, Sun Microsystems. Retrieved September 2004 from http://java.sun.com/products/JavaManagement

Johner, H., Fujiwara, S., Yeung, A., Stephanou, A., & Whitmore, J.(2000). *Deploying a public key infrastructure*. IBM Red Books. Retrieved April 2004 from http://www.redbooks.ibm.com/redbooks/pdfs/sg245512.pdf

La Porta, T. F., Sabnani, K. K., & Gitlin, R. D. (1996). Challenges for nomadic computing: Mobility management and wireless communications. *ACM Journal of Nomadic Computing, 1*(1), 3-16.

Lennox, J., & Schulzrinne, H. (2000). *Call processing language framework and requirements*. RFC 2824. Retrieved on March 2004 from http://www.ietf.org/rfc/rfc2824.txt

Lopes, L., Villier, E., & Ludden, B. (1999). GSM standards activity on location. *Proceedings of IEEE Colloquium on Novel Methods of Location and Tracking of Cellular Mobiles and Their System Applications,* London.

Lozinski, Z. (2003). *Parlay/OSA: A new way to create wireless services*. IEC Mobile Wireless Data. Retrieved June 2003 from http://www.parlay.org/specs/library/IEC_Wireless_A_ New_ Way_to_Create_Wireless_Services.pdf

Ludden, B., & Lopes, L. (2000). Cellular based location technologies for UMTS: A comparison between IPDL and TA-IPDL. *Proceedings of IEEE VTC2000*.

Moerdijk, A.-J., & Klostermann, L. (2003). Opening the networks with Parlay/OSA: Standards and aspects behind the APIs. *IEEE Network Magazine, 17*(3), 58-64

Papazafeiropoulos, G., Prigouris, N., Marias, I., Hadjiefthymiades, S., & Merakos, L. (2003). Retrieving position from indoor WLANs through GWLC. *Proceedings of the IST Mobile & Wireless Communications Summit*, (pp. 199-204).

Parkinson, B. W., & Enge, P. K. (1996). *Differential GPS*. In B. W. Parkinson & J. J. Spilker, Jr. (Eds.), *Global Positioning System: Theory and applications* (Vol. 2, pp. 3-50). Washington, DC: American Institute of Aeronautics and Astronautics.

Parlay-ETSI. (2001). Open Service Access API. Part 6: Mobility SCF. ES 210915-6 v.1.1.1. Retrieved May 2004 from http://www.parlay.org/specs/

Pawsey, C., Green, J., Dineen, R., & Munoz Mendez-Villamil, M. (2002, February). *Ovum forecasts: Global wireless markets 2002-2006*. Ovum.

Priggouris, G., Hadjiefthymiades, S., & Merakos, L. (2004). An XML framework for multi-level access control in the enterprise domain. *Proceedings of Sixth International Conference on Enterprise Information Systems (ICEIS)*, Porto.

Priggouris, G., Hadjiefthymiades, S., Tsetsos, B., Sekkas, O., Papazafeiropoulos, G., Priggouris, N., Duspiva, M., Lopez-Aladros, R., Kassapoglou-Faist, C., Peinado, J.L.M., Iacovides, E., Karagiozidis, M., & Alevropoulos, C. (2003, August). PoLoS process methods and process re-engineering methodologies. *PoLoS Deliverable* D-121. Retrieved March 2004 from http://www.polos.org/

Roman, E. (2002). *Mastering enterprise Javabeans* (2nd ed.). London: John Wiley & Sons.

Tsalgatidou, A., & Pitoura, E. (2001). Business models and transactions in mobile electronic commerce: Requirements and properties [Special issue]. *Computer Networks, 37*(2), 221-236.

Tsetsos, V., Sekkas, O., Priggouris, G., & Hadjiefthymiades, S. (2004). A component-based scheduling architecture for the enterprise domain. *Proceedings of Seventh IEEE International Symposium on Object-Oriented Real-Time Distributed Computing (ISORC)*, Vienna, Austria.

WAP Push architectural Overview. (2001). WAP Forum. Retrieved September 2004 from *http://www.openmobilealliance.org/tech/affiliates/wap/wap-250-pusharchoverview-20010703-a.pdf*

Chapter XIII

Location-Dependent Query Processing Benchmark

Ayse Yasemin Seydim, Central Bank of the Republic of Turkey, Turkey

Margaret H. Dunham, Southern Methodist University, USA

Abstract

Benchmarks define techniques which can be followed to determine the effectiveness of a given software or hardware design. Ever since the development of the Wisconsin Benchmark and subsequent transaction-processing (TPC) benchmarks, there has been a concensus and general acceptance of these performance comparison tools. However, these benchmarks are not sufficient to determine the performance of mobile-based applications. For example, these traditional benchmarks ignore some of the important wireless-mobile features such as location-dependent queries and movement of the mobile host. In this chapter we examine the issues needed for the development of such a mobile query benchmark. In particular, we focus on queries which involve location-dependent features. We first examine the unique aspects of this mobile architecture which impact any benchmark design, and then propose a benchmark suitable for it.

Introduction

A benchmark defines a common testing criterium which facilitates the comparison of two systems. Benchmarks for traditional database systems have been thoroughly examined with several different types proposed (Gray,1993). Typically, a database benchmark consists of three features.

- *Queries:* These are typically simplistic versions of real-life queries which could be executed to evaluate the necessary components for query processing.
- *Data:* Data may be artificial data which are created to represent typical data in that domain. Definitions of data requirements need to be sufficiently abstract to ensure use in any possible system.
- *Execution Guidelines:* Execution guidelines indicate specifically how the benchmark is to be executed, what performance metrics are to be used, and how these metrics are to be generated. These guidelines should be applicable for evaluating a real implementation, a test bed, a prototype, or even a simulation of a proposed implementation.

Although these are the basic features for a benchmarking tool, generally there is no specific definition provided for the networking or connectivity characteristics of the environment. Today, mobile computing has become necessary for the applications to serve the mobile (and also stationary) users who want to be able to process from anywhere, anytime. Existing benchmarking tools, though, are simply not adequate for such a mobile computing environment. Data, queries, and execution guidelines are not directly applicable to the environment and do not include the architectural and connectivity issues that are specific to mobility. There is no "typical" application for mobile computing, moreover, a debit-credit banking application does not seem to be a reasonable choice for a mobile environment. It is obvious that a major reason the existing benchmarks are inadequate is because the mobility aspect is completely ignored. For example, if a query is requested from a mobile unit (MU), the way to test the movement of the MU should be specified in the benchmark. In this chapter we present the guidelines for a mobile computing benchmark.

We propose a benchmark in which queries are requested from the MU and executed at a node in the fixed wired network, a usual case for a mobile user. Although different types of queries should be included in a mobile computing benchmark, those that highlight the uniqueness of that environment are crucial. Such database queries are those whose results depend on the requester's location, *location-dependent queries* (LDQs; Seydim, Dunham, & Kumar, 2001a, 2001b). Examples include, "Where is the closest hotel?" and, "What Italian restaurants are within five miles?" Data used in the benchmark must include data which contain location components, *location-dependent data* (LDD; Dunham & Kumar, 1998; Ren & Dunham, 2000). Any proposed mobile computing benchmark should support not only the LDQs but also more traditional types as well. Moreover, the benchmark must characterize typical applications with the mobility characteristics. Therefore, a location-dependent benchmark should include the following features.

- *Queries:* Queries should be representative of all types of queries which can be stated in a mobile computing environment including (but not limited to) LDQs.
- *Data:* Data may be artificial, that is, created to represent any LDD.
- *Mobile Unit Behavior:* A method to abstract MU behavior should be specified. Mobility patterns can be based on real-life behaviors as well as simulated trajectories.
- *Execution Guidelines:* Similar to traditional benchmarking guidelines, execution should be targeted to the wireless environment with queries requested from an MU and processed at a fixed data server.

Note that we are only concerned with retrieval queries and do not explicitly address updating and specific hardware features of any implementation. Guidelines of a location-dependent benchmark have been established; data invalidation, client caching, and so forth should be evaluated with the implementation details for a targeted implementation. They are independent of the query performance benchmark requirements. We view the benchmark requirements as applicable to any updating scheme or implementation framework.

This chapter is organized as follows. First we provide an overview of the mobile computing environment and the query processing supported in this environment. Then we summarize the related benchmark work and give the guidelines of the location-dependent benchmark. A test bed designed to apply the benchmark to our previously proposed middleware architecture, the *location dependent services model* (LDSM; Seydim, 2003), is given as a starting point for further implementation. Finally, we summarize the benchmark study.

Mobile Computing Query-Processing Framework

We provide a benchmark for a traditional mobile computing architecture. Here we assume that the wireless connection is provided by a wireless service provider like AT&T, and a third party vendor or the same wireless provider, called a location service, determines the location of the mobile object. We assume that database queries are requested by users located at MUs. MUs are connected to content providers through wireless links provided by wireless service providers. The exact architecture is not crucial for use of this benchmark as long as the queries are submitted to content providers from MUs. The requirements, then are illustrated below:

MU « Wireless Provider (with Location Service) « Content Provider.

The architecture described is not dependent on any content service provider, wireless service, or location service. Location of the mobile object should be bound to the query in order to be answered. This study assumes location-dependent content is on fixed hosts, that is, an infrastructured, wireless mobile environment. Processing in this LDQ benchmark is not dependent on the mobile-wireless system architecture. Study assumes the location is provided in any granularity by any provider. The architecture, location-dependent services manager, is discussed in a previous study (Seydim et al., 2001a). Base stations and so forth are considered as part of the wireless service provider. The LDSM architecture is suggested to be independent of any content service provider, wireless service, or location service. Location of the mobile object should be bound to the query in order to be answered. It assumes location-dependent content is on fixed hosts.

The unique features of the queries in this mobile environment are captured by the benchmark. We briefly summarize some of these important features in the following paragraphs. Interested readers are referred to the literature for further information (Seydim et al., 2001a, 2001b).

Queries asked in the mobile environment may have a slightly different structure and format from traditional database queries. Location-related queries are queries which have at least one location-related attribute or location-related predicate. Location relatedness and LDQs are discussed in another study in detail (Seydim et al., 2001b). For example, the query, "What are the names and addresses of the restaurants within five miles?" seeks for the restaurants within five miles of the current position of the query issuer. In order to provide the answer to the query, first the application has to know the location of the issuer. The query can later be bound to this location. Thus, location dependence in queries implies that the information asked is related to a location, but the location is not explicitly known when the query is asked. In addition, the granularity of the query location could be at any level. Common granularities include latitude-longitude, cell ID, zip code, and so forth.

Obviously, there will be location-related and non-location-related attributes in any query. To answer any location-based request while the user is moving, it may become necessary to identify location-related attributes. Sometimes the statement of the query will not have any location-related attribute, but the way it is stated will have an implication that the query issuer's current position is involved in the selection criteria. If the location attribute is implied in the query, which is called an LDQ, the implied location-related attribute has to be added to the query and the query is processed on location-dependent data. Once the user's location is known, the query becomes *location aware*, with an explicit indication of this special location attribute. Therefore, a query including any location-related attribute in its predicates is called a *location-aware query* (LAQ). An LAQ produces the same result set independent of the place it is issued. This is not the case for an LDQ. Traditional database queries which do not involve location data are referred to here as *non-locationrelated queries* (NLR-Qs). Any mobile computing benchmark must support all three types of queries.

When an LDQ is issued, the location with which it is to be processed must be determined. We assume that the query is bound to a location referred to as the *binding location* (BL). A traditional location services manager or the Global Positioning System (GPS) may bind the location to latitude and longitude. In addition, when the query is processed at the data provider, it will be processed using the location supported in the database. This location granularity at the content provider is referred to as the *data location* (DL). The DL may be at a completely different granualarity than the BL. Thus, the benchmark needs to examine the effectiveness of processing queries with any BL or DL combination. Conversion of the BL to DL to facilitate effective processing of an LDQ should be supported by the wireless database environment and needs to be tested in any proposed benchmark. This process of converting from BL to DL is referred to as *location leveling* (Seydim, 2003). In this research, we do not consider implementation details of the LDD, how they are stored, or how location leveling is performed. Furthermore, we assume the query location does not change until the results are returned to the user.

Querying location-dependent information in the mobile environment has been an important research area. LDD is data whose values change with location. In Dunham and Kumar (1998), LDD have been viewed as spatial replicas depending on a data region where they are included. Query-processing approaches based on physical organization of data and location binding are discussed in Kumar and Dunham (1998). In line with these studies, a formal model has been presented in Ren and Dunham (2000) to describe the mobility of objects. Some research has concentrated on querying location data and has examined problems related to the frequent updates thereof (Imielinski & Badrinath, 2002; Pitoura & Bhargava, 1994). Another group of work concentrates not only on the moving object's location, but also on a time constraint (Sistla, Wolfson, Chamberlain, & Dao, 1997; Wolfson, Xu, Chamberlain, & Jiang, 1998). Direction and speed of the mobile object are included in this spatiotemporal model similar to Liu's, Bahl's, and Chlamtac's work (1998; Liu et al. suggested location prediction in wireless networks also using a cell-based location model.). The queries in those Moving Object Database (MOD) applications carry the application or business logic within their classification. In our work, we assume that a mobile object's location is provided by a location service provider at some granularity. More on related research on LDQ processing can be found in Seydim (2003). Recently there has been much research which has looked at predicting future locations for MUs (Aljadhai & Znati, 2001; Liu et al., 1998; Liu & Gerald, 1995). These works are related to our research in that we assume a query has to be bound to a location. This may be either that obtained by a location service (our assumption) or may be a future location based on predictions of user movement.

Related Work

The benchmarking of database systems is recognized as extremely difficult due to the breadth of application domains, types of database systems, and hardware alternatives involved. Some benchmarks have tested the functionality of the systems, and some of them have targeted the evaluation of the performance. One of the first database system benchmarks was the Wisconsin Benchmark which targeted relational database systems (Bitton, DeWitt, & Turbyfill, 1983; DeWitt, 1993). Predefined queries are used for testing the major components of the database system. There have been many successors to this approach which improved on it. Current accepted TPC benchmarks are mainly based on determining the transaction-per-second (TPS) processing rate for sets of queries (Gray, 1993; Serlin, 1993). They are loosely based on debit and credit banking applications.

Although location-dependent applications are seen as special cases of spatial applications, most of the spatial data properties do not seem to be necessary. In fact, spatial benchmarking requires vast amounts of data, complex operators, different indexes, and visualization testing. The SEQUOIA 2000 spatial-database storage benchmark (Stonebraker, Frew, Gardels, & Meredith, 1993) is the first benchmark that tested massive data with complex data types and sophisticated searching. It is basically a storage benchmark that uses real data sets with raster and vector representations. Performance has been evaluated by the total elapsed time and the retail price of the hardware, which does not seem to be a constraint today. Spatial join is not a part of the SEQUOIA benchmark where instead, spatial selection, that is, windowing, has been used.

On the other hand, spatial join operations are evaluated by Gunther (1993), and Hoel and Samet (1995). A hierarchical tree structure based on containment is used for spatial join in Gunther , where each node is a spatial object. A model to compare different strategies using this tree structure has been defined, and parameter and variables are given. In Hoel and Samet, the spatial join operation is evaluated using R-tree, R*-tree, R+-tree, and the PM Random (PMR) quadtree in an application domain with line segment data. Time to build the data structure and time to perform the spatial join are measured and evaluated.

The performance of spatial join algorithms are also compared in Gunther, Oria, Picouet, Saglio, and Scholl1(998), which used both artificial and real data sets. Sets of rectangles can be produced with a Web-based generator and real data is taken from the SEQUOIA 2000 storage benchmark. Sample size and selectivity of the join predicate are the main parameters. Not as specific as the others, Gurret, Manolopoulos, Papadopoulos, and Rigaux (1999) aim to provide a generalized approach where all kinds of spatial access methods and related query-evaluation techniques can be performed under a common framework. As Paton, Williams, Dietrich, Liew, Dinn, and Patrick (2000) claim, due to the lack of a standard, a widely used spatial database benchmark, a general one for vector spatial databases with 26 queries, is another effort in spatial benchmarking.

Although some of the queries in these benchmark studies may be used for a location-dependent applications framework, a more simplified benchmark with mobility characteristics is much better to serve the purpose.

We are aware of only one other work which examines benchmarks for location queries (Theodoridis, 2003). Theodoridis has proposed ten queries to test targeting spatiotemporal data and MOD queries. He proposes a precise set of data (Human, Building, Road) that is different from ours (Hotel, Restaurant). The movement of MUs is captured in his approach by the Human data explicitly encorporating the route of a user's movement. In our approach, however, we view the movement of users as a integral part of the benchmark, just as data and queries. Thus, we view this as separate from the data rather than being encorporated into it. While Theodoridis included two operations associated with initializing and updating the databases, we have followed the approach used in previous database benchmarking techniques that assume this is part of the preparation needed to run the benchmark. Thus we view how these are done as being up to the user. While Theodoridis suggests ten queries, we provide a general query model (Figure 3) and 29 basic queries. Our queries are categorized based on type: general (no location-related data), location aware, location dependent, and location queries. We thus frame our queries based on earlier LDQ frameworks—not ad hoc as seems to be used in Theodoridis' work. There are several aspects which we investigate as part of our benchmark that Theodoridis ignores. We discuss how geographic data can be generated based on data distributions. We discuss selectivity of data. We define basic mobility patterns. In general, we feel that our work provides a more general approach to LDQ benchmarking. However, we also feel that Theodoridis' work in MOD queries is more extensive than ours. Theodoridis' work also examines indexes to be used to efficiently process MOD queries.

Benchmark Guidelines

This section discusses the major guidelines of the proposed location-dependent benchmark. This benchmark breaks the location-dependent application testing into distinct components and gives a direction for further implementations. Descriptions of the database relations that form the benchmark are given, queries, their types, and how they are going to be used are discussed, and we give mobility-behavior details. Execution guidelines and the metrics to evaluate the system are also presented.

Benchmark Data

The benchmark for location-dependent applications should involve database relations which will give different results depending on the location attributes in the query predicates. The LDQs issued are usually read-only queries that ask for information dependent on the mobile query issuer's location. One may notice that most cited types of LDQs are for hotel and restaurant information. Thus, it would make sense to use this data as the benchmark data. Initially, we study read-only queries and leave the update

Figure 1. Location hierarchy for the proposed benchmark

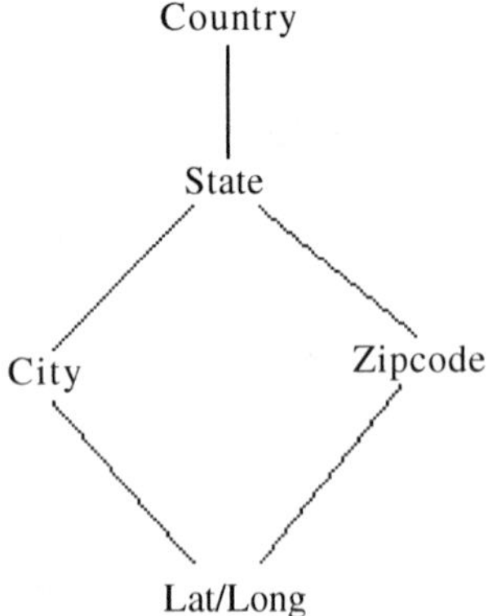

transactions to a further study. Thus, only selection and projection operations are defined.

It is necessary to have different location granularities defined in the data to represent an actual implementation. Rather than using complex data types (i.e., spatial), such as line and polygon data definitions, one can use the inherent relationships between the location concepts. In our benchmark we assume that data comes from hotel and restaurant domains. Containment relationships are defined as shown in the location concept hierarchy in Figure 1.

The schemas for the Hotels and Restaurants relations are shown in Table 1. Since this is a location-dependent benchmark, it is necessary to have at least one location-related attribute in each relation, and those need to be in different granularities. Hotels and Restaurants relations have unique keys which are the ID numbers, and they are also accessed by some nonunique keys which are location related. The two relations have different candidate (secondary) keys in order to test the difference in location granularities. We assume that the Hotels relation has a nonunique key of City, and the Restaurants relation has Zip Code. For simplicity, the length of the latitude and longitude attributes are chosen similarly to the most commonly used TIGER (topological, integrated, geographic-encoded referencing) data (*TIGER*, 1999) of the United States Postal Service. Specification of these attributes can be integrated with the ones given in the mobile location protocol specification of the *Location Interoperability Forum* (2000).

Data Generation

Data values for the Hotels and Restaurants relations may be obtained from real database relations or artificially generated ones. Here, it is important that location-related attributes exist in the data and in the concept hierarchy. All LDD points lie in a *workspace* or *test area* which is defined as a rectangular area represented by upper-right-corner and

Table 1. Schemas for hotels and restaurants

Attribute	Type	Relation
Restaurant/Hotel ID	integer(5), NLR, Unique	Both
Restaurant/Hotel Name	char(60), NLR	Both
Address	char(45), LR	Both
City	char(20), LR, Nonunique	Hotel
State Code	char(2), LR	Both
Zip Code	char(10), LR, Nonunique	Restaurant
Country Code	integer(3), LR	Both
Latitude	integer(9), LR	Both
Longitude	integer(10), LR	Both
Occupancy	integer(5), NLR	Hotel
Number of Tables	integer(5), NLR	Restaurant
Other	char(50), NLR	Both

lower-left-corner coordinate values in a Cartesian coordinate system. The lower-left-corner point is assumed to be at the origin (0, 0) so it is enough to give maximum x and maximum y values to define the test area.

While generating the artificial data, some points can be chosen in the workspace to emulate the hypothetical city centers. The latitude and longitude x and y coordinate values should be created for the hotel and restaurant point data. Figure 2 shows the center points with *c*s and other point data (x, y) pairs with dots. These data points can be distributed randomly in the workspace with any type of distribution, such as a Gaussian distribution. With Gaussian distribution, points will be scattered more densely towards each area center. For the hotels having restaurants, randomly chosen hotel point data can be added to restaurant point data.

Although the benchmark assumes an artificial workspace, this two-dimensional space could be overlayed on a real area (such as a state) to provide a more realistic approach. If the benchmark is to be executed in a real, live mobile computing environment, this artificial workspace concept cannot be implemented precisely, but it is a reasonable alternative when used in simulations or test beds.

Values of the attributes are also assigned randomly from a finite set of values. The value of City, State, Zip Code, and Country Code are all from set-valued domains. For instance, we can use the set CityTX = {Austin, Dallas, Houston, ...} for cities in Texas. These values can be real values, therefore, we are not restricting the assignments with a predefined set of values that is given in the benchmark.

Selectivity of Data

The result of a query depends on the selection criteria: sometimes all records will be requested, sometimes only the records in one area. Therefore, the number of records that is returned is going to change. The number of records returned has been defined as the selectivity factor of the attributes of relations (O'Neil, 1993). For example, if hotel data is equally distributed in a test area with four city regions, then the selectivity factor for a city will be 25% of the records. Then, for 100 hotel data points, 25 will be in City 1, 25

Figure 2. Data distributions in a test area

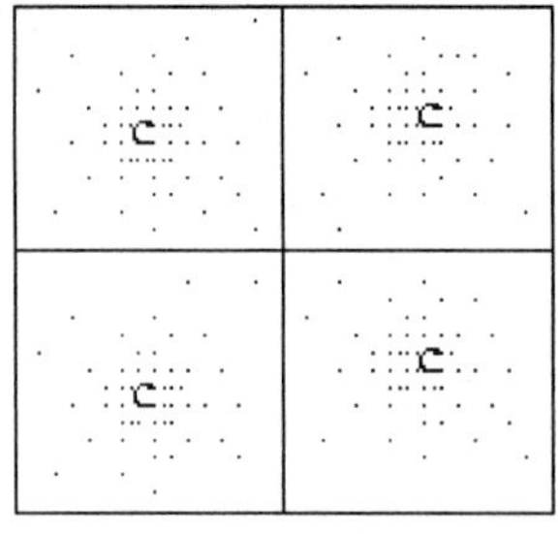

(a) Cities with selectivity 25%

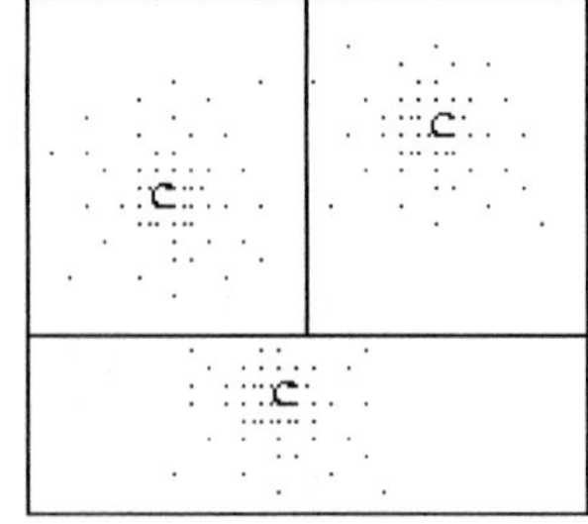

(b) Zipcoded data with selectivity 35%, 35% and 30%

will be in City 2, and so forth. The distribution of Hotel and Restaurant data points depending on the key location attribute (City, Zip Code, State, Country) shows the selectivity of the queries. Figure 2 illustrates the example of data distribution in two different test area partitionings. In Figure 2a, four cities are defined with equal numerical distribution, that is, the selectivity of City attribute value is 25% each. In Figure 2b, there are three centers where selectivity of the attribute Zipcode is 35%, 35%, and 30% for three different zip codes. Note that this may not represent a real-life implementation.

The selectivity defined here is for the location-related attributes given in the query, where the query is location aware. When mobility is the case, a user's current location value might be needed in the query, which in turn might bring up the location-granularity mismatch (LGM) problem. When there is an LGM problem, the BL will be different from the candidate (secondary) key of the data set. Therefore, selectivity after location binding (e.g., Zip Code) and the selectivity after translation to the DL (e.g., City) may be different.

Queries in the Benchmark

The application for the benchmark should include LDQs, LAQs, and NLR-Qs issued for hotels and restaurants databases. The benchmark queries are a mix of these three basic types. The number of relations and predicates used in the query determines the query as a *simple* (S), *Multirelation* (M), or *Compound* (C) query. We also differentiate those in the basic query set to be used in the evaluations.

Benchmark queries are created randomly for each run of the experiment. An SQL-like query model is shown in Figure 3. There are basically five parts in the model, which are used to create a query for the benchmark. Parts of the query model are explained as follows:

Figure 3. Benchmark query model

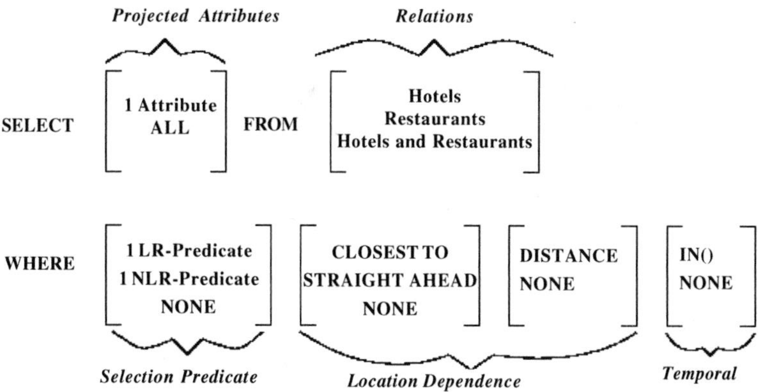

- *Projected Attributes:* To determine the attributes to be shown to the user, either all or one attribute must be chosen. A projection operation is implied.
- *Relations:* Attributes are selected from one relation, either Hotels or Restaurants, or both. If there are two target relations with the same projection attributes and selection predicates, the query is an M query. Thus, the same query has to be issued for the second relation, too.
- *Selection Predicate:* The selection predicate might include a location-related predicate, a non-location-related predicate, or nothing at all. A location-related predicate is associated with a location-related attribute, whereas a non-location-related predicate is related to a traditional attribute.
- *Location Dependence:* The location-dependence portion of the query model contains two operators, CLOSEST TO and STRAIGHT AHEAD, with a DISTANCE keyword. These are defined as location-related operators, where they imply a window within which the MU stays. The CLOSEST TO operator implies a circular area within a specified distance, where STRAIGHT AHEAD refers to a rectangular area in the direction of movement. These operator meanings are pictorially shown in Figure 4. Unless a DISTANCE is specified, the value of the radius of the circular (or the longer side rectangular) area should be set to a default distance. If additional operators are going to be used by a tester, the meaning of the operators should be defined explicitly.
- *Temporal:* A time parameter can also be defined in the last part of the query model. This parameter is used in testing the prediction queries (Sistla et al., 1997), which refers to a future location. When the keyword IN is used, the time value specified is added to the current time stamp to estimate the predicted location of the mobile user. Otherwise, the current time is assumed.

By using the query model, the following basic types of queries are designed for the benchmark.

- *Traditional or NLR-Q:* SELECT Occupancy FROM Hotels
- *LAQ:* LAQ has at least one location-related predicate (SELECT Name FROM Restaurants WHERE Zip Code = "12345").

Figure 4. Location-related operator meanings

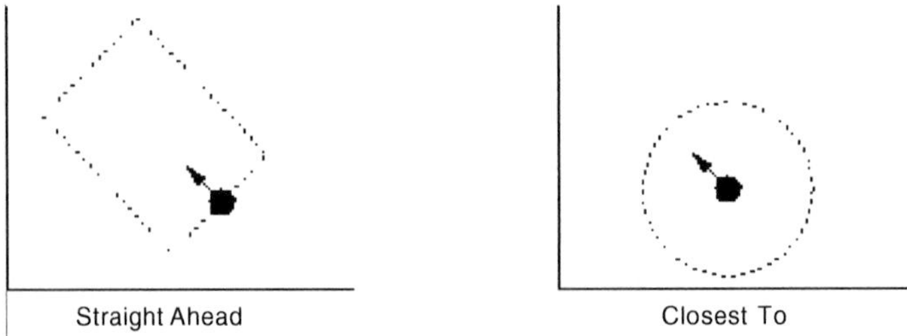

- *LDQ:* LDQ requires location binding and has special location-related operators (SELECT all FROM Hotels WHERE CLOSEST TO (five miles)).
- *Prediction Query (PQ):* PQ is an LDQ which is issued for a future time (SELECT all FROM Hotels WHERE CLOSEST TO (5 miles) IN (10 min)).

Basic query sets are created using the defined query model. These *base query set*s for each type are shown in Table 2. This table also includes the type of queries to guide the tester. Note that the Hotels and Restaurants relations are location-related relations since they have at least one location-related attribute in their schemas. Therefore, when we select all attributes from any of these relations, as in queries Q21 and Q22, then the queries can be viewed as LAQs. The BL granularity is also given to evaluate the results. It is shown in italics if location leveling is required.

Another type of query is a *continuous query* (CQ), an LDQ that is issued continuously during the movement of a mobile user. The same query is issued again after a period of time, which means queries overlap, that is, the overlapping frequency between queries is not zero. The rate of two consecutive queries being the same is represented by overlap frequency, namely, the OverlapRate. OverlapRate can be given as a benchmark parameter and at least one of the LDQs can be used in the CQ test. On the other hand, a query which is not a part of this query model but used during location binding is *location query* (LQ). LQ is issued to obtain current MU position from a location service, for example, "SELECT CurrentLocation." We have also added an LQ to the query set to use in location binding.

Mobility Behavior Description

In exploring the query processing performance in location-dependent applications, modeling the MU behavior becomes necessary. There are two behavioral components: *connectivity* and *modeling of movement.* First, planned and unplanned disconnections can be modeled in the benchmark. In this study, however, we do not explicitly model disconnections and we assume the MU is connected all along its path after activation.

Secondly, there are three dimensions in movement modeling: speed, direction, and movement pattern. At any time, an MU will have a direction and a driving speed towards that direction. When a mobile user is going along a path, its direction and average speed will probably not change. Therefore, if a location service knows the current location of the MU, its speed, and direction, then it can estimate the new location depending on its movement pattern.

In general, moving objects follow predefined paths, the routes to specific places. Movement patterns for each mobile user can be modeled in the reference region within which all the movement occurs. If we are going from our house to work, we normally drive on the routes we have taken in the past. Every path consists of small sections of road which can be treated as straight line segments. Depending on his or her mobility pattern, a user can drive from one place to another by using only one line segment, and/or come back to his original place with the same segment. He or she can also stop at intermediate

Table 2. Benchmark queries

Query Set	#	Query	BL
Non-Location-Related Queries	Q11	SELECT Occupancy FROM Hotels WHERE HotelID = 12345	N/A
	Q12	SELECT NumberofTables FROM Restaurants WHERE RestaurantID = 67890	N/A
Location-Aware Queries	Q21	SELECT all FROM Hotels	N/A
	Q22	SELECT all FROM Restaurants	N/A
	Q23	SELECT all FROM Hotels WHERE City = "Dallas"	N/A
	Q24	SELECT all FROM Restaurants WHERE Zip Code = "75275-0000"	N/A
	Q25	SELECT Zip Code FROM Hotels WHERE City = "Richardson"	N/A
	Q26	SELECT Zip Code FROM Restaurants WHERE City = "Richardson"	N/A
	Q27	SELECT all FROM Hotels and Restaurants WHERE City = "Richardson"	N/A
Location-Dependent Queries	Q31	SELECT all FROM Hotels WHERE CLOSEST TO DISTANCE(5 miles)	Zip Code
	Q32	SELECT all FROM Hotels WHERE CLOSEST TO DISTANCE(5 miles)	City
	Q33	SELECT all FROM Restaurants WHERE CLOSEST TO DISTANCE(5 miles)	City
	Q34	SELECT all FROM Restaurants WHERE CLOSEST TO DISTANCE(5 miles)	Zip Code
	Q35	SELECT all FROM Hotels and Restaurants WHERE CLOSEST TO DISTANCE(5 miles)	Zip Code
	Q41	SELECT all FROM Hotels STRAIGHT AHEAD DISTANCE(10 miles)	City
	Q42	SELECT all FROM Hotels STRAIGHT AHEAD DISTANCE(10 miles)	Zip Code
	Q43	SELECT all FROM Restaurants STRAIGHT AHEAD DISTANCE(10 miles)	Zip Code
	Q44	SELECT all FROM Restaurants STRAIGHT AHEAD DISTANCE(10 miles)	City
	Q45	SELECT all FROM Hotels and Restaurants WHERE STRAIGHT AHEAD DISTANCE()	Cell
Compound Location-Dependent Queries	Q51	SELECT all FROM Hotels WHERE Name "Marriot" and CLOSEST TO DISTANCE(5 miles)	Zip Code
	Q52	SELECT all FROM Restaurants WHERE Name "Wendy's" and CLOSEST TO DISTANCE(5 miles)	City
	Q53	SELECT all FROM Hotels WHERE Zip Code = "75275-0000" and CLOSEST TO DISTANCE(5 miles)	City
	Q54	SELECT all FROM Restaurants WHERE City = "Dallas" and CLOSEST TO DISTANCE(5 miles)	Zip Code
Prediction Location-Dependent Queries	Q61	SELECT all FROM Hotels WHERE CLOSEST TO DISTANCE() IN(2 min)	Zip Code
	Q62	SELECT all FROM Restaurants WHERE CLOSEST TO DISTANCE() IN(2 min)	City
	Q63	SELECT all FROM Hotels and Restaurants WHERE CLOSEST TO DISTANCE() IN(2 min)	Zip Code
Join Location-Dependent Queries	Q71	SELECT all FROM Hotels, Restaurants WHERE Hotels.Zip Code = Restaurants.Zip Code and CLOSEST TO DISTANCE(5 miles)	Zip Code
	Q72	SELECT all FROM Hotels, Restaurants WHERE Hotels.Address = Restaurants.Address and CLOSEST TO DISTANCE(5 miles)	City
Location Query	Q81	SELECT CurrentLocation	Any

points. As a result, we view the movement pattern or path of a mobile user to be one of the three types: *one way*, *round trip*, or *random*.

While one-way and random-trip paths both have different source and destination locations, a round-trip path has the same source and destination point. A random-trip path can be considered to be constructed by more than one one-way trips. Each distinct path can be given a path identification number, namely, path ID. We view the workspace in two dimensions and model the location of the mobile unit in a two-dimensional space.

Figure 5 shows an example for each movement type in a test area. We summarize the movement patterns with two randomly generated points, s and t, as follows.

- *One Way:* The MU moves from point s to t at a uniform speed (e.g., MU1, PATH1, one line segment).
- *Round Trip:* The MU moves from point s at a uniform speed to point s_1, waits a certain time at point s_1, then moves from point s_1 to point $s = t$ at a uniform speed (e.g., MU2, PATH2, two line segments).

- *Random:* The MU moves in a set of one-way segments. There are *n* segments where endpoints are labeled as s_i, where 1 *d•i d•n*. The entire set of segments defines the movement pattern. Thus, we have the following segments: <s, s_1>, <s_1, s_2>, ..., <s_{n-1}, s_n> = <s_{n-1}, t>.

The ending point of segment *i* is the starting point of segment *i* + 1 (e.g., MU3, PATH3, five line segments). There is a waiting time between the segment activations, and it may be different at each intermediate point. We assume the available paths are predefined and numbered in the location service component. This approach is realistic since street maps imply predefined paths from one destination to another.

Location Service and Path Generation

No matter how it is implemented, a location service calculates and returns the location of a mobile user for the specific time asked. The MU's assigned path ID, the ID of the MU, the starting time for that drive, and the time for the requested location have to be provided for this purpose. In return, the location service estimates the position by calculating the distance from the source point by using the movement starting time and the required time (query time). Initially, the granularity of the location to be bound to the issued query is chosen as geographic coordinates—latitude and longitude. When the MU reaches its final destination according to the given time (which must be greater than the total travel time), the same *x*, *y* coordinates are returned. This can also be used for testing queries issued by stationary users.

Figure 5. Mobility pattern types

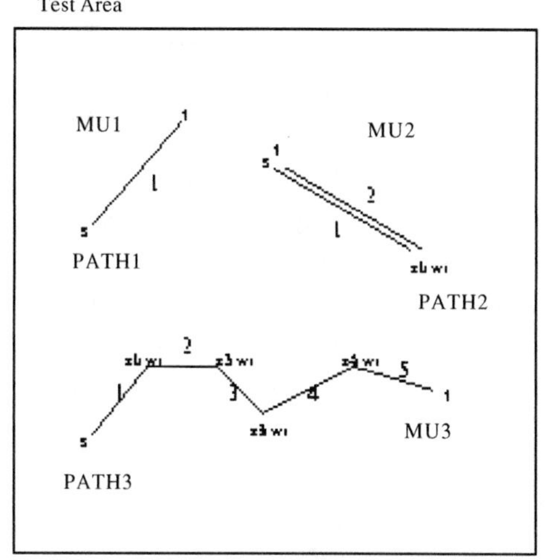

s: source point
sh: intermediate points
t: target point
w: random waiting time

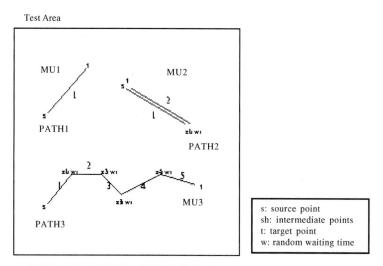

MU1 One-Way, MU2: Round Trip, MU3 Random

The location service provides a table lookup calculation for the position of a mobile user at a given time. A real location database can be used for the location determination, but this approach is simple and serves the purposes to begin with. A more realistic way of location estimation can be performed in real time with the help of a wireless network or the wireless device (GPS) for a location service. For simplicity, a third-party location service is assumed to provide location in latitude and longitude granularity in a certain path for this initial version of the benchmark. However, the intent here should be to obtain a location independent from a location service of any kind.

In a simulated movement environment, MU movements are generated according to the predefined paths. However, in a real-world implementation, a mobile device will be required to move in one of these patterns. In this case, the entire movement pattern of the MU causes some concerns as this is difficult to duplicate from one implementation to another. Moreover, the use of this benchmark with real-world MU movement needs future research.

A given path ID can be used by any MU at any time, and there is a speed limit associated with each path. The speed for the paths are chosen randomly within a range and associated with the path number. The basic assumption here is the mobile user keeps an average speed within this limit. For each set of paths, the maximum number of users in the mobile environment determines the number of paths for each type of movement. That is, for each type of movement, the number of paths is equal to the maximum number of mobile users. For instance, if the benchmark accommodates 100 separate mobile users, there will be 100 one-way, 100 round-trip, and 100 random-movement paths created in one path set before the benchmark is executed. The same path can be used by more than one mobile user as in real life.

During path generation, every path has a source and destination point which are chosen randomly in the workspace. The limits of the test area are given as the path generation parameter along with the maximum number of MUs.

Execution Guidelines

The basic benchmark queries can be tested in an architecture consisting of any wireless network that is connected to a fixed network, where the data servers reside. Queries are issued from an MU and the data sets are stored on fixed host servers. The general benchmark architecture is shown in Figure 6. The mobility of the users and the location service is simulated with the predefined communication costs.

It is obvious that there is a need for preprocessing or translation for processing an LDQ. After the query (or MU) is bound to a location, the new query is sent to the related database server. Note that there may be LGM between the LDQ stated and the data location granularity. However, there is no restriction on the implementation of location binding and query translation.

Figure 6. Benchmark architecture

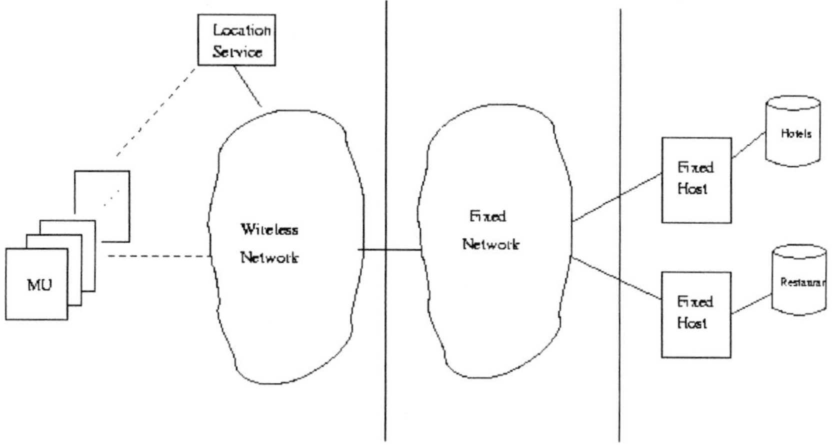

Benchmark Activities and Execution

Mobile users and their movement patterns (paths) constitute the *activity sets* of the benchmark execution. In fact, the MUID-PathID pairs correspond to movement patterns of each MU in their profiles. Each MU, then, submits queries from a *query set* along the travel during a run. Therefore, a set of MUID-PathID assignment pairs are to be generated before the benchmark starts. These activity sets are saved to create the same inputs to the system when needed.

To generate an activity set, the distribution of the travel types (one way, round trip, random) are given to create the corresponding percentages for each type. The number of MUs, the MU IDs' range, the path group ID from which to choose the paths, and the percentages of each movement type are the basic activity set generation parameters, but one can add more to it depending on the implementation.

The file name of the activity set can be created by adding the number of the activity set to a predefined name. At least three activity sets for each path group should be created during benchmark preparation process. The set of queries are chosen from the base query sets (at least one from each group) and put together in a query set file. The total time period to run the queries, mean arrival time between queries, distribution of each query type, and the overlap frequency between consecutive queries (OverlapRate) are needed to prepare a query set file.

After preparation of the data, the benchmark is executed with the activity set, query set, and the path group file under the given parameters. These parameters include the number of queries in each query set, the maximum number of MUs, mean activation time between the MU for each activity set, the random number seeds, mean arrival time of queries, overlap percentage between consecutive queries, and benchmark run period. We summarize the input to an execution as

- benchmark workspace definition (at least 400x400 units),
- area (Zip Code regions, City boundaries) definitions (overlaying areas on workspace),
- Hotels data set (at least two City centers, 50 to 150 hotel point data),
- Restaurants data set (at least two Zip Code area centers, 100 to 200 points),
- path group files (at least two path groups),
- activity set files (at least three MUID-PathID pair files for each path group), and
- query sets (at least three sets with different distributions of query types, including the base query set in each).

A universal time counter should be set before the execution. MUs are to be activated (simulated) at random points using a uniform distribution with the given mean activation time from the activity set. Each MU should issue queries from the query set with the mean query issue time interval. MUs continue to move until reaching their destinations. Queries continue to be issued until all the queries have finished. This may mean that some queries are actually issued when the corresponding MU has finished its movement and is at its destination point. After all MUs are stopped and all queries are executed, benchmark reports should be prepared from history files.

Benchmark Metrics

The main purpose of a benchmark is to provide tools to compare and evaluate the performance and functionality of different implementations of the same model. The same set of queries are to be tested and assessed by using metrics. In the location-dependent benchmark, performance measurements include the response time and the quality metrics as precision, recall, and false positives of the results. In a mobile benchmark, focus should be on the evaluation of the query processing within the network at least for location determination and location translation, and on the functionality provided by the system. Response time that includes preprocessing time for handling location mismatches is important to compare similar architectures. Therefore, we explicitly present this period in the response time calculations. Additionally, with the functionality testing, the quality of the location translation to solve the mismatch problem should be gathered. Precision and recall of the conversions are among these quality assessment metrics.

In this benchmark setting, *response time* (RT) for each query is the first metric to consider among the performance evaluation criteria. RT is calculated using all of the following parts:

- any processing time at MU after issue of the query, M_p,
- transfer time from MU to network, M_c,
- time in network before query reaches to communication interface, N_{p1},
- transfer time of the query from network to fixed host, F_c,
- processing time (access time) for the query results, F_p,
- transfer time of results from fixed host to network, F_c*n,

- time within network until the query results arrive to the MU, $N_{p2} * n$,
- transfer time from network to MU, $M_c * n$,
- time to display the results to the user, M_d.

Here, n is the number of records that is returned as the result of the query. Therefore, there is a factor n depending on the number of resulting records. Thus, the total RT for one query can be written as:

$$RT = M_p + M_c + N_{p1} + F_c + F_p + (F_c * n) + (N_{p2} * n) + (M_c * n) + M_d. \quad (1)$$

For a traditional S query, there will be no processing cost on the MU within the network so M_p, N_{p1}, and N_{p2} will be almost zero. However, there might be some processing cost for the LDQs since they might need location binding and query translation before being processed in the data server. The processing cost of LDQs, either on the MU or in the network, depends on different implementations, so it is left flexible to add in this evaluation. Communication cost from MU to network, M_c, is to be set to a fixed value, but if the tests are run from a wired (stationary) terminal, this value is updated. RT is also affected by the resulting number of records that is determined by the selectivity factor of the attributes mentioned earlier. Note that in a real implementation or even simulation, RT can be easily calculated; it does not have to be done in this piecewise fashion.

Not only does an architecture need to provide a basic framework for location-dependent applications, but also it should provide the best answers to the queries in these. The response of the query Q31 will be a set of hotels. Different implementations of the query may obtain different results. This could be based on how and when the MU is bound to a location. It may also be based on how CLOSEST is calculated and what location granularities are used. The quality of the answers depends on the BL and DL granularities and the processing provided. If there is no granularity mismatch for an application, then we can definitely conclude that the best quality of the results are provided by the system. Otherwise, if BL is either a fine or coarse granularity, the data accessed will give better results if DL is closer to that fine or coarse granularity. Base query set tables also have the BL granularities to evaluate the semantic differences. Quality is to be evaluated by precision, recall, false positives, and missing data values. In the benchmark, since we have modeled a location service that estimates the location of the MU as a latitude and longitude pair, precision will be always less than one. These metrics are to be defined in the detailed guidelines for comparing systems.

A Benchmark Test Bed Design

A benchmark test bed including a proposed middleware architecture has been designed as a basis for future implementation (Seydim, 2003). This proposed middleware is called the *location dependent services manager* (LDSM). In this section, the major architectural view and the test bed design for the benchmark are presented. The execution and implementation of the proposed test bed is general enough to support any test bed environment involving a middleware layer which supports the initial evaluation and analysis of a query (including location leveling) prior to submitting to a content provider.

Test Bed System Model

The system components including the middleware with the query translation steps and the location leveling processing should be tested. The system model, *System Under Test* (SUT), is designed as a mobile client, middleware, server environment and is shown in Figure 7. In this setting, content providers are hypothetically settled on fixed hosts, and the middleware resides on an application server. Numerous mobile users are to be simulated and the corresponding movement patterns are to be assigned.

In SUT, queries should be created in advance by a query generator depending on the frequency between queries and the probability of each property existing in a query. The whole time period for the experiments or the number of queries will be given by the test programmer. As required by the benchmark, two content databases, Hotels and Restaurants, should be defined to reside on different content providers. After the query is sent to the middleware, the need for location binding should be determined and, if needed, the location request should be transferred to the location service.

Depending on the path assigned to the mobile user, which should be prepared in the activity set generation phase, the location service returns the corresponding current or future location of the given MU. LDSM performs the location leveling and its other functions to process and route the query. The components for the test bed are summarized in Figure 8.

Figure 7. System model

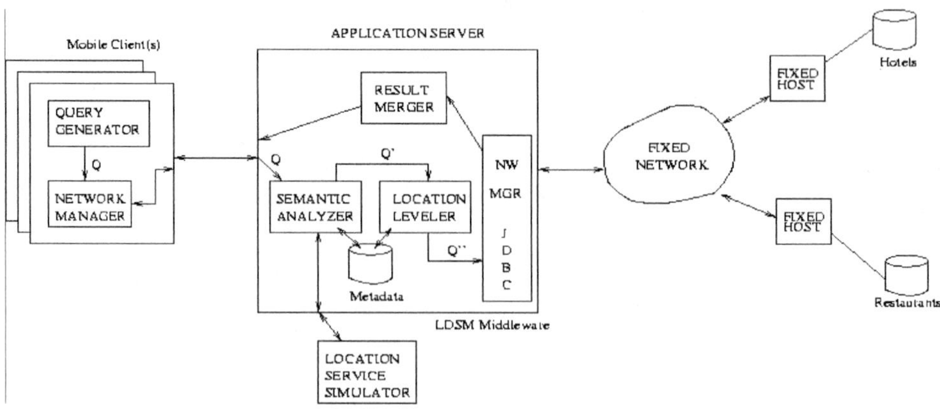

Figure 8. Benchmark execution

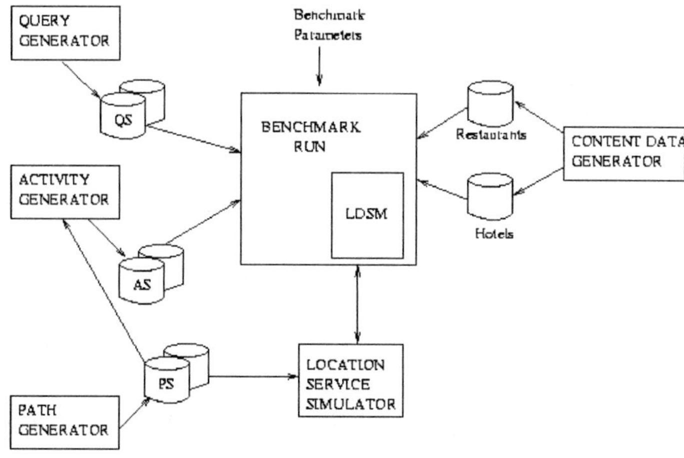

Suggested Test Environment

The suggested test environment is designed to be implemented on the SMU Engineering Network. A realization scenario for each component of the benchmark framework can be given as follows.

- *Mobile Client:* In this basic design of the test bed, mobility is simulated on the same machine as the middleware runs. The number of MUs should given as the benchmark run parameter. Each client submits one query at a time from the pregenerated query sets and waits for the response. There is no queuing system or transaction management installed in the mobile clients. The activity sets should be generated a priori to determine the mobile users and movement patterns tuples for the run of simulation.
- *Middleware:* One or more processing units to accommodate the middleware software can be used in the benchmark. Here, only one processing unit is considered. Application is to be developed on a desktop computer running Microsoft Windows 2000 with a system of a minimum 256 MB RAM, 20GB hard disk, and Pentium III 1Ghz.
- *Database Servers:* Servers can be any processing unit that will keep the data sets. Sample data sets (Hotels and Restaurants) are created on workstations with at least 512 MB RAM running a Unix operating system, where Oracle 8.1.5 DataBase Management System (DBMS) is installed.
- *Network:* Fixed network connections of any type can be used. Local-area network connection in SMU is with a 100 MB Ethernet cable from the application server to the fixed backbone network.

Figure 9. Movement path generator

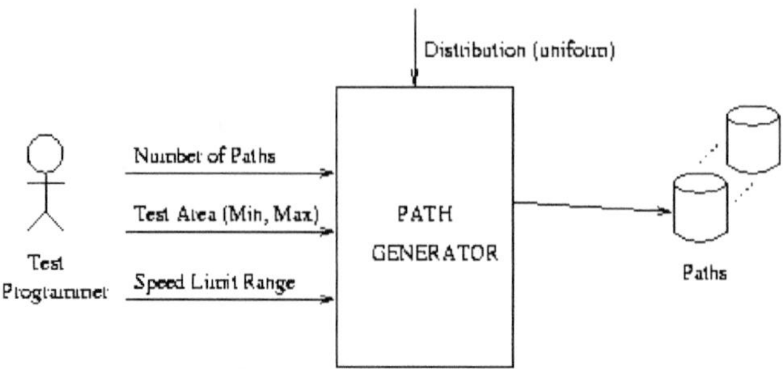

- *Benchmark/LDSM Software:* Any programming language can be used for the implementation of middleware and the benchmark tests. The necessary database connection drivers should be available. For the location service and the interface to data servers, Java programming language Java Development Kit (JDK) 1.3.1. and Oracle Java Database Connectivity (JDBC) native driver are used.

Data and Queries

Hotels and Restaurants data are entirely dependent on their location, which we have shown in their schemas in Table 1. As a start, four sample data sets are found on the Internet for Hotels and Restaurants, namely, Super 8 motel, Sheraton, Jack in the Box, and Wendy's for Texas.

When executing the benchmark, the set of queries from the base set should be specified. To accomplish this, the following characteristics of the queries should be given as input for each execution:

- time period to run the queries,
- mean arrival time of queries,
- the probabilities and distributions of each query type, and
- the overlap percentage between consecutive queries (OverlapRate).

Query sets will be created in the implementation of the benchmark.

Mobility Behavior

The path generation component is to be implemented in Java. Considering every path consists of one or more line segments, a path can be described by the following attributes:

- path number,
- segment's number,

- starting point coordinates,
- ending point coordinates,
- speed, and
- delay—the number of seconds at the intermediate point for round-trip and random types of movements.

The path generator component is summarized in Figure 9. For providing diversity, at least two path groups can be generated during the benchmark preparation process. The group to be used for that run of the benchmark will be chosen during activity generation.

The location service component is implemented in Java. The designed messaging between the middleware and the location service is shown in Figure 10.

Activity Sets

A set of mobile-user-path assignments should be generated before the benchmark starts. These activity sets will be saved in data files for reuse. The activity generator is the component for this generation.

For each activity set, the distribution of the travel types (one way, round trip, random) should be given to create the corresponding percentages for each type. The number of the activity set, the number of MUs for that activity set, the starting identification number of the MU, the path group number, and the trip distributions will be given as input to the activity generation. The file name of the activity set can be created by adding the number of the activity set to a predefined name. At least three activity sets for each path group should be used in test bed execution.

Distribution of travel types may play a significant role during the benchmark run in case of testing of a caching or prefetching strategy. For the specific middleware tests, there is no need to concentrate on the travel type percentages, but to create diversity, different percentage groups can be assigned in each activity set generation. The basic input to the generation process are shown in Figure 11.

Figure 10. Location service messaging

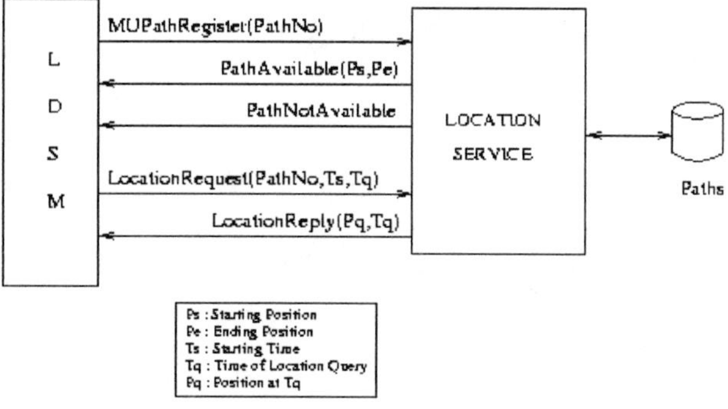

Figure 11. Activity generator

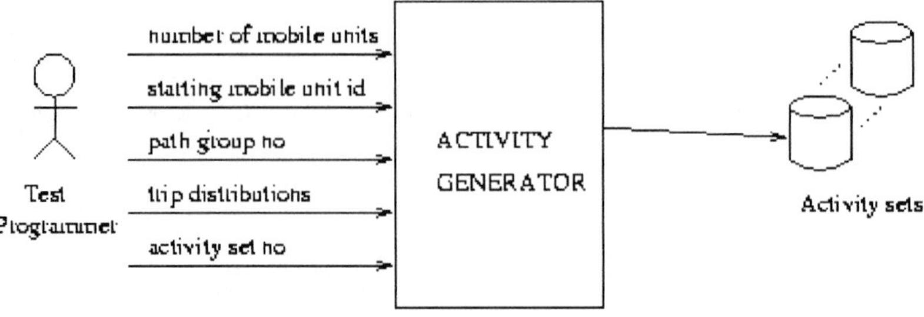

Test Bed Execution Procedure

After the preparation process for the runs, when the data files, path files, activity set files, and query set files are ready, the benchmark can be executed. Parameters give the mean activation time between the MUs, mean query issue time, and the random number seeds. Steps in the *benchmark run* can be stated as follows.

1. Set time counter for the run.
2. Start the MUs with the activation time intervals.
   (a) Each MU should issue queries with the mean query issue time intervals (random).
   (b) Stop the MU if the queries are finished or it reached the destination point.
3. After all MUs are stopped, prepare reports from the history files.

Steps in the *MU activation* can be summarized as follows.

1. MU starts.
2. MU path registers.
3. If the path is not an existing path, MU becomes zombie.
4. Otherwise, write to history and active MU files.
5. Issue a query from the query set.
   (a) Perform semantic analysis.
   (b) If LDQ, ask Location(Path, Ts, Tq).
   (c) Wait for query reply.
   (d) Wait for ThinkTime.
   (e) If not finished or not at destination, go to Step 5.
   (f) If Pq is equal to Pt, then deactivation may start after query reply comes or, if Ps = Pt, deactivation starts.

After the benchmark run, reports should be generated from history files.

Conclusion

The guidelines given in this chapter are preliminary benchmark considerations in a mobile environment. The proposed benchmark includes the following components.

- Data:
    - Hotels and Restaurants relations
    - Location hierarchy for location granularities
    - Workspace or test area with city centers concept
- Queries:
    - LDQ, LAQ, NLR-Q
    - S, M, C
    - LGM tested
- Mobility Behavior (Movement)
    - One way, round trip, random
- Execution Guidelines
    - Performance Metrics:
        - RT
        - Quality
    - Activity sets
    - Query sets
    - Path groups

Connectivity of the mobile device is another area that should be modeled in the benchmark. Mobile devices have limitations compared to fixed devices. Due to limited bandwidth, wireless network conditions, and energy limitations, mobile devices are more prone to disconnections. For planned disconnections, frequent voluntary disconnections are considered to be *sleeper* types of users, and infrequent voluntary disconnections are *workaholic* types of users (Barbara & Imielinski,1994). If we have the same number of involuntary disconnections, the total disconnections should be larger for sleepers than workaholics. Planned and unplanned disconnections can be modeled in the benchmark, which will also impact query processing and the location leveling. This benchmark can be extended to include connectivity rates.

There should be some metrics to evaluate how the location mismatches are solved. The amount or convergence to expected values will be a measure of quality. Therefore, precision, recall, and other metrics should be considered for different granularity graphs. We leave the details of these performance metrics to future refinements of the benchmark. These are the preliminary guidelines targeted to the mobile environment. Only the location service, path generation component, and the data server interface component are implemented.

There are other areas which require further examination. In the future, we hope to be able to incorporate some of the aspects from Yannis Theodoridis' excellent benchmark work to create a benchmark approach more comprehensive than either of our separate works. In the future, results of studies involving real data are hoped to be detailed.

Acknowledgments

We would like to thank Mark Fontenot for his design and implementation of the LDSM test bed. We also acknowledge work done by SMU undergraduate CSE students in implementing an LDSM simulation visualization tool. We are particularly indebted to Vijay Kumar for earlier collaboration on many of these issues. Finally, we would like to thank the anonymous reviewers for providing constructive comments which have produced a better work.

This material is based upon work supported by the National Science Foundation under Grant No. IIS-9979458.

References

Aljadhai, A., & Znati, T. (2001). Predictive mobility support for qos provisioning in mobile wireless environments. *IEEE Journal on Selected Areas in Communicaitons, 10*, 1915-1930.

Barbara, D., & Imielinski, T. (1994). Sleepers and workaholics: Caching strategies in mobile environments. *Proceedings of ACM SIGMOD International Conference on Management of Data,* (pp. 1-12).

Bitton, D. J., DeWitt, D., & Turbyfill, C. (1983). Benchmarking database systems: A systematic approach. *Proceedings of Ninth International Conference on Very Large Data Bases, VLDB'83,* (pp. 8-19).

DeWitt, D. J. (1993). The Wisconsin benchmark: Past, present, and future. In J. Gray (Ed.), *The benchmark handbook for Database and Transaction Systems*, (2nd edition, pp. 269-315). Morgan Kaufmann.

Dunham, M. H., & Kumar, V. (1998). Location dependent data and its management in mobile databases. In R. Wagner (Ed.), *Proceedings of Ninth International Workshop on Database and Expert System Applications, DEXA'98* (pp. 414-419). Vienna, Austria: IEEE Computer Society.

Gray, J. (Ed.). (1993). *The benchmark handbook for database and transaction processing systems (2nd ed.).* San Mateo, CA: Morgan Kaufmann Publishers.

Gunther, O. (1993). Efficient computation of spatial joins. *Proceedings of Ninth International Conference on Data Engineering (ICDE '93),* (pp. 50-59).

Gunther, O., Oria, V., Picouet, P., Saglio, J. M., & Scholl, M. (1998). Benchmarking spatial joins à la carte. *Proceedings of International Conference on Scientific and Statistical Database Management (SSDBM'98),* (pp. 32-41).

Gurret, C., Manolopoulos, Y., Papadopoulos, A., & Rigaux, P. (1999). The BASIS system: A benchmarking approach for spatial index structures. *Proceedings of Spatio-Temporal Database Management, International Workshop, STDBM'99,* Edinburgh, Scotland, (pp. 152-170).

Hoel, E. G., & Samet, H. (1995). Benchmarking spatial join operations with spatial output. *Proceedings of 21st International Conference on Very Large Data Bases,* VLDB'94, (pp. 606-618).

Imielinski, T., & Badrinath, B. R. (2002). Wireless graffiti: Data, data everywhere matters. *Proceedings of 28th International Conference on Very Large Data Bases,* VLDB'02, Hong Kong, China.

Kumar, V., & Dunham, M. H. (1998). *Defining location data dependency, transaction mobility and commitment* (Tech. Rep. No. 98-CSE-01). Dallas, TX: Southern Methodist University.

Liu, T., Bahl, P., & Chlamtac, I. (1998). Mobility modeling, location tracking, and trajectory prediction in wireless (ATM) networks. *IEEE Journal on Selected Areas in Communicaitons, 6,* 922-936.

Liu, Y., & Gerald, Q. (1995). Efficient mobility managment support for wireless data services. *Proceedings of the 45th IEEE Vehicular Technology Conference,* Chicago.

Location Interoperability Forum (LIF). (2000). Retrieved May 11, 2004, from the Location Interoperability Forum Web site: http://www.locationforum.org

O'Neil, P. E. (1993) The set query benchmark. In J. Gray (Ed.), *The benchmark handbook for database and transaction systems* (pp. 361-395). Morgan Kaufmann.

Paton, N. W., Williams, M. H., Dietrich, K., Liew, O., Dinn, A., & Patrick, A. (2000). VESPA: A benchmark for vector spatial detabases. *Proceedings of Advances in Databases, 17th British National Conference on Databases, BNCOD 17,* Exeter, UK, July 3-5 (pp. 81-101).

Pitoura, E., & Bhargava, B. (1994). Building information systems for mobile environments. *Proceedings of Third International Conference on Information and Knowledge Management, CIKM'94,* (pp. 371-378).

Ren, Q., & Dunham, M. H. (2000). Using semantic caching to manage location dependent data in mobile computing. *Proceedings of Sixth Annual International Conference on Mobile Computing and Networking, MobiCom 2000,* (pp. 210-221).

Serlin, O. (1993). The history of DebitCredit and the TPC. In J. Gray (Ed.), *The benchmark handbook for database and transaction systems* (pp. 21-40). Morgan Kaufmann.

Seydim, A. Y. (2003) *Location dependent query processing in mobile environments* (Tech. Rep. No.). Doctoral dissertation. Dallas, TX: Southern Methodist University, Department of Computer Science and Engineering.

Seydim, A. Y., Dunham, M. H., & Kumar, V. (2001a). An architecture for location dependent query processing. *Proceedings of Fourth International Workshop on Mobility in Databases and Distributed Systems (MDDS'01),* Munich, Germany.

Seydim, A. Y., Dunham, M. H., & Kumar, V. (2001b). Location dependent query processing. In S. Banerjee, P. K. Chrysanthis, & E. Pitoura (Eds.), *Proceedings of Second ACM International Workshop on Data Engineering for Mobile and Wireless Access, MobiDE'01* (pp .47-53). Santa Barbara, California: ACM.

Sistla, A. P., Wolfson, O., Chamberlain, S., & Dao, S. (1997). Modeling and querying moving objects. *Proceedings of 13th International Conference on Data Engineering (ICDE'97),* (pp. 422-432).

Stonebraker, M., Frew, J., Gardels, K., & Meredith, J. (1993). The SEQUOIA 2000 storage benchmark. *Proceedings of ACM 1993 SIGMOD International Conference on Management of Data,* (pp. 2-11).

Theodoridis, Y. (2003). Ten benchmark database queries for location-based services. *The Computer Journal, 46* (4), 713-725.

TIGER: Topological integrated geographic encoded referencing/zip-zone improvement plan technical guide. (1999). United States Postal Service. Retrieved May 11, 2004, from the USPS National Customer Support Center Web site: http://www.usps.com/ncsc/addressmgmt/tiger_print.htm

Wolfson, O., Xu, B., Chamberlain, S., & Jiang, L. (1998). Moving objects databases: Issues and solutions. *Proceedings of International Conference on Scientific and Statistical Database Management, (SSDBM'98),* (pp. 111-122).

Chapter XIV

Location-Dependent Data Access and Queries

Gyula Rabai, Budapest University of Technology and Economics, Hungary

Sandor Imre, Budapest University of Technology and Economics, Hungary

Abstract

Location-dependent data access technology will create a new family of services in mobile telecommunication systems of the near future. This technology is based on positioning, communication over various protocols, and spatial databases. Location-dependent applications will be used in many fields such as emergency services, information services, and financial services. This chapter intends to present the infrastructure that makes it possible to create such services, and provides an insight on key application development issues. The focus is on location information access as the latest achievements of this research field are presented.

Introduction

In the past decade, mobile communication technology has gone through a major evolution. As a result of this, today we have cellular radio networks that cover a large percent of the earth and serve millions of customers. These radio networks are capable of providing voice and data access to mobile devices moving between different locations. This infrastructure can be exploited to a further extent if location-based services, such as location-dependent data access, location-dependent queries, automatic reconfigurations, and location-based push services, are introduced.

Those of us who are interested in this technology or would like to develop services to exploit this opportunity must have a clear understanding of how location-based data access works. It is also important to see what tools are available and how they can be used to create advanced location-based applications. In the following text, the theoretical background and some of the most advanced tools are presented, as positioning techniques, various data access technologies, spatial database systems, and spatial queries are discussed. Some of the most interesting research achievements and development activities carried out in a European Fifth Framework project, Configurable Radio with Advanced Software Technology (CAST; Madani et al., 2000), are also outlined. The reader of the chapter will get insight on how location-based data access technology was used in this project and what role it will play in Fourth-Generation Wireless (4G) wireless networks. During the discussion, the focus is on location-based data access and queries.

In the first part of this chapter, an overview of positioning technologies and the description methods for location-based data are presented. In the second part, data access methods for retrieving location information are discussed, including location-based information retrieval using the Wireless Application Protocol (WAP), data access with the help of J2ME, and mobile execution environment Classmark 3 (MExE CM 3). In the third part, the discussion is about the extension of the SQL language to support the querying and displaying of location information. In this part, many useful examples are presented. In the fourth section, along with a set of example applications, the system developed in the frames of the CAST project is introduced as we present how it is possible to dynamically update the configuration of wireless terminals using location-based queries. In the final part, some open issues of the research associated with location-based data access and queries will be presented, and the effects of future improvements to the wireless infrastructure will be predicted.

Basics of Location-Based Data Access and Queries

In the field of mobile computing, location-based data access and location-based information retrieval are very hot topics. Information providers see new opportunities in providing information either by push technologies or by user-initiated queries based on the location of the user. New technologies are being developed for location-dependent data access to fulfill the needs. Among these technologies, some deal with positioning, some with retrieving location information from the terminal or the network, and others are focused on the location information storage and management. The reader must understand what kind of location information is available, how it is represented, and how it can get to the information provider.

The geographic location of a mobile terminal can be identified by its position on the surface of the earth. To address each location, two-dimensional or three-dimensional geographic positioning can be used. Two-dimensional geographic positioning offers latitude and longitude information. Three-dimensional geographic positioning adds altitude to the two above. Two-dimensional positioning is often easier to handle, and in most cases, it is sufficient to solve a given task. In the next part, the focus is on two-dimensional positioning, thus the latitude and longitude coordinates for a given location are represented as a two-dimensional vector, <latitude, longitude>, where longitude ranges from -180 (west) to 180 (east), and latitude ranges from -90 (south) to 90 (north). Using this representation, < 47.30, 19.05> is an example of the geographic coordinates of Budapest, Hungary. To address the whole surface of the earth with the precision of 0.05 km, nine bytes are necessary if single precision floating-point numbers are used (Imielinski & Navas, 1999). Two-dimensional geographic positioning makes it possible to represent distances between any two points and allow us to use various geometric objects to represent locations and areas. The simplest objects we can use are (a) a point, (b) a circle (center point and radius), and (c) a polygon ($point_1$, $point_2$, ..., $point_n$), where each vertex is represented using geographic coordinates.

In an application based on location-dependent data access, mobile clients located in a region can be addressed by a polygon drawn around the region. The polygon is translated into geographical coordinates and can be used to look up relevant information. Polygons can also represent things like outlines of cities, districts, flood plains, or boundaries of a shopping area. Polygons can be combined to form complex representations. A polygon within another polygon might geographically represent a restricted area for driving inside a city.

Different geometry types can have different meanings. While a polygon can be used to represent a geographic area, a circle can be a measure of precision. For example, if the location of a mobile device is represented by a circle, the geographical coordinates of the center point would show the highest probability of the mobile device's location, and the radius of the circle would represent the level of precision the positioning technology could provide.

If geometry types are described by a set of longitude and latitude coordinates, the *vector* model is used in the application. Alternatively we can choose a fundamentally different type of geographic model: the *raster* model. In the raster model the information about points, lines, and polygons are not encoded and stored as a collection of coordinate pairs. A location database based on the raster model comprises a collection of grid cells like a scanned map. The raster model is very useful for describing continuously varying features, such as accessibility cost for a hospital, while the vector model is a good choice for describing discreet objects or areas. Typical analytical operations, such as proximity analysis and overlay analysis, can be performed on both models.

Location-based applications in wireless networks often work with moving objects. If a car moving on the road issues a query for available gas stations on the way ahead, the returned information needs to be different from the one that was issued from a fix location. In the first case, we are talking about information retrieval based on moving reference objects, and in the second case, a query based on stationary reference objects is discussed.

Determining the Position of a Mobile Terminal by Network

In a wireless network, the location of a mobile terminal can be determined by different network entities and can be represented using two-dimensional geographic positioning. It is important to see what the major differences are when location information is provided by the mobile terminal, and how it is possible to handle location information that was determined by the network itself.

One of the most obvious ways to identify the location of a mobile device in a cellular radio network is to identify the cell the terminal is using. This information is always present since all calls, messages, and data packets must be routed to the relevant base stations, and position updates are always performed when handover takes place. However, the level of precision of this location information is often not sufficient for mobile applications. Most of the time a high level of accuracy is needed. Determining the position of the user to a fine level of accuracy is not an easy task. To see how it can be done, we examine the possibilities in a Global System for Mobile Communication (GSM) network.

In GSM the user's location can be determined from data that is inherently present in the network. Although some of this data is not originally intended for mobile positioning, it is able to give enough clues to heuristically locate a mobile phone to an acceptable level of accuracy. This information consists of network parameters such as the serving-cell identity, timing advance (TA), and neighboring cell measurements (Bajada, 2003). By identifying the serving cell the mobile phone is using, we can give a brief estimate of its location. The level of accuracy in this case depends on the size of the cell, which might vary from a few hundred meters to several kilometers.

A better approach to determine the position is based on TA. In a GSM network, each mobile terminal transmits data in bursts to the nearest base station. To make sure each transmission burst from the mobile terminal arrives at the base station in the correct time slot, a measurement of the round-trip delay is made and a TA value is calculated. This method can be used to determine the distance of the mobile terminal from the base station in steps of approximately 550 m (Figure 1). Unfortunately the distance from the base station itself is not fit for most applications since it only provides a circular area where the mobile terminal is located. If advanced antenna technology is used, the angle of

Figure 1. Positioning using timing advance

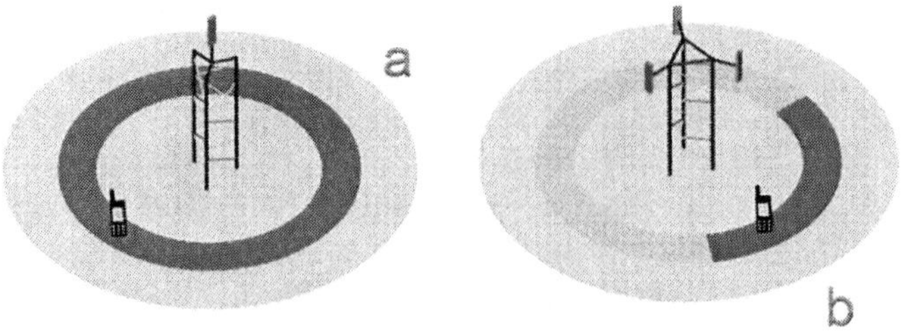

Figure 2. Circular E-OTD

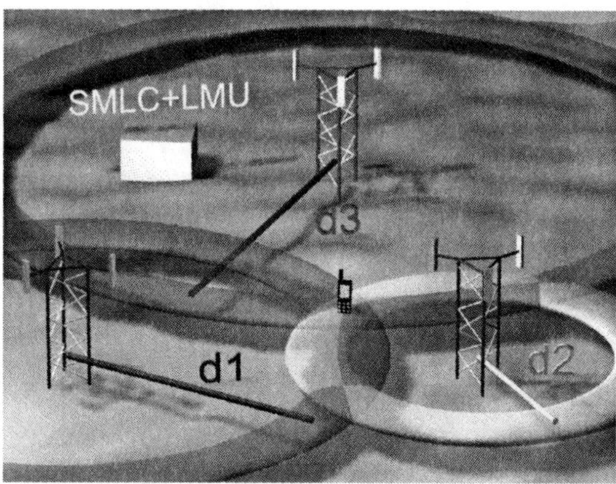

arrival of the transmission can also be determined. This way, a slice of the circle is selected, which is much better. Alternatively, it is possible to use TA information with more than one base station at the same time (Figure 2). The technology used for this purpose is called enhanced observed time difference (E-OTD; Chamberlain, 1995; Lohman, Stoltzfus, Benson, Martin, & Cardenas, 1983). E-OTD is used in several ways. One of the most common is circular E-OTD, where circles are set up and their intersection gives a quite precise result.

Location Information Provided by Mobile Terminal

In May 2000, the U.S. government stopped jamming the signals from Global Positioning System (GPS) satellites for use in civilian applications, dramatically improving the accuracy of GPS-based location data to five to 50 m. Thanks to this decision, smart mobile devices, which are equipped with GPS receivers, are now capable of determining their position very accurately. From these devices, location-based queries can be issued very easily because the location information is available in the handset at any time at an acceptable level of precision. If a mobile device equipped with a GPS receiver uses location-based services, only a data communication channel is needed between the device and the information provider. Therefore, in this scenario it is very easy to create location-based applications. On the other hand, there are some disadvantages to this approach. GPS receivers can only be used outdoors and building a GPS receiver into each handset increases the cost of the terminals. If the location is determined and sent to the information provider by the mobile terminal, the information is not available on the server side and location-based push services are hard to create. Another problem is concerning security. Once we rely on location information sent by a client, the information provider cannot trust the information, which limits the number of possible applications.

To sum it up, location information can be determined using different technologies and the service provider who is using this information must have an access method to get this information.

Location-Dependent Information Retrieval

Access methods to retrieve location information highly depend on the application architecture used in the wireless network. In some scenarios, client-server, peer-to-peer, and stand-alone applications require location data. To be able to handle the different application architectures, MExE is used.

Mobile Execution Environment

MExE provides a standardized execution environment for mobile stations and server-side operators, and an ability to negotiate supported capabilities between the communicating parties. With the MExE technology, applications can be developed independent of any platform. MExE has recommended application architectures described by classmarks. The most popular architectures today are Classmark 1 (CM 1), Classmark 2 (CM 2), and CM 3. MExE CM 1 is the WAP environment. This environment uses a client-server approach similar to the World Wide Web (WWW) to enable mobile devices to retrieve any information located on the Internet. The mobile devices are equipped with a relatively simple Internet browser, the WAP microbrowser, that can be used to download content. The MExE CM 2 and CM 3 application technologies are similar in the way they both rely on applications downloaded into mobile terminals. The difference is that in CM 3, we allow applications to provide services for mobile clients allowing peer-to-peer solutions. Personal Java and its successor, J2ME CLDC MIDP, comply with the requirements of the MExE CM 3 specifications. These specifications cover requirements on user-interface customization, user-profile management, service management (discovery, configuration, control, and suspension), quality-of-service support, and security.

As one can see, there are many options, and the method for data access is highly dependent on the MExE. Today the most widely spread information retrieval method is WAP, the MExE CM 1 environment. Most devices ready for WAP are not able to identify their location and depend on the network to provide the location information. WAP applications are a set of services operated by a third-party information provider on a remote server, which means location-based data can only be accessed if the location information from the network and the user query of the mobile client are made available to the information provider at the same time.

Accessing Location Information Provided by the Network

To understand how location-based services are used in the WAP environment, the reader must understand how WAP works. In the case of user-initiated queries based on WAP, all queries are routed through a central server called the WAP gateway. Location-based

Figure 3. WAP location server

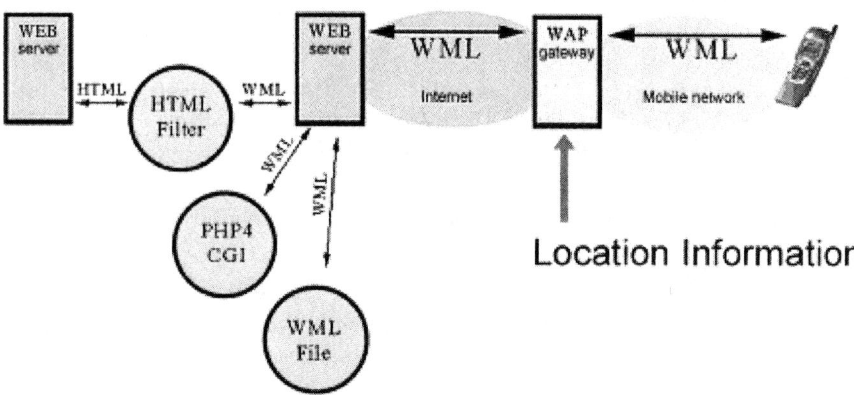

applications can use the WAP gateway as a location server. In this case, when a query issued by a mobile client passes through the gateway, the location information is appended to the query (Figure 3). This information is added to the header part of the request. The information provider that receives the query can process the header and is able to work with the location information. The results based on this query are returned using the WAP data format.

WAP uses the wireless markup language (WML) to format the information. WML is an Extensible Markup Language (XML)-based language and can be coded as text or in binary format. In both representations, "cards" that are organized into a deck carry the information. When a mobile terminal wants to receive information, it downloads a WML deck and displays the first card in the deck. The cards can contain formatted textual information and pictures. The most common picture format used on WAP pages is the wireless bitmap (WBMP) format. A WBMP picture is a small, black and white picture that is easy to process and display.

The upper layer communication protocol used in WAP is the Wireless Session Protocol (WSP). WSP can be used to download WML and WBMP data in a way very similar to the Hyptertext Transfer Protocol (HTTP) information-retrieval process used on the WWW. In HTTP, when an Internet browser would like to download some information, it issues a request to the appropriate Web server. The request contains an HTTP header and an HTTP body. The server, after receiving the request, does some server-side processing and then creates the response. The WAP browser and the WAP information server communicating over WSP behave exactly the same way. As a matter of fact, the WAP information server is a Web server, which is located on the Internet.

When a WAP browser running on a mobile device issues a request for location information, it sends a WSP header and a WSP body. The WAP gateway, which has a transparent role in the communication, appends the location information to the WSP header. The server receives the modified header and the body, and performs the server-side processing to send the response back.

Accessing Location Information in CM 2 and CM 3 Applications

Personal Java and J2ME applications work in a bit different way. When an application sitting on a wireless device wants to query location information, it needs to collect the location information first. Since these applications can interact with users locally and can communicate with servers over the network independently, they have many options. They can simply ask the user to enter the location information, or they can issue a request for location information to an appropriate server before the real location-based information request is composed.

Recent developments, such as the Java Location API, give significant advantages to these applications. They are not only able to find a way to find out their location (Figure 4), but can issue remote database requests and present results in a nonstandardized way. For example, an interactive map running on a mobile terminal can be developed, where the geographical coordinates are downloaded in textual format and the rendering of the image can be performed on the mobile client itself. Application developers have a lot more freedom when they use J2ME instead of WAP.

Location-Based Queries

As discussed in the previous section, wireless devices are able to access spatial information in several ways. They can download WML pages using WAP, they can retrieve XML data using HTTP, or they can issue direct database queries using a remote database access technology, such as JDBC. In one way, all of these data access methods are common. The information they retrieve is filtered and processed on the server side according to the remote query before it is downloaded. The reason for this is obvious:

Figure 4. CM 3 application architecture

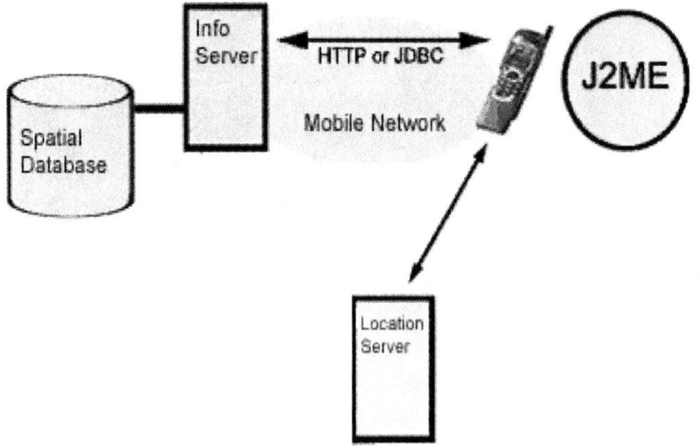

Wireless devices often communicate using low bandwidth, have limited storage and processing capabilities, and can consume only a limited amount of battery power, which means they cannot afford to download and work on large data sets. In order to deal with this issue, location information providers must have storage and processing facilities that can handle large amounts of data on the server side. The technology for data storage needs to be scalable and it must be able to adapt to different usage patterns. Existing database management system (DBMS) technology is not fit for this purpose. Traditional relational databases are not well equipped to work with large amounts of geographical data, to deal with spatial relations, or to handle continuously changing information, such as the location of moving objects.

Spatial Database Technology

Databases used for storing geographical information are called spatial databases. Spatial databases combine conventional and spatially related data. Handling such data requires significantly different access methods and storage architectures from conventional relational databases. Storing spatial information is not only used in geographic information systems (GISs; Frank, 1988), but is also important in image databases (Chang & Kunii, 1981) and remote sensing (Lohman et al., 1983). All application areas require a method for managing and manipulating large amounts of spatial data and for performing analysis. For location-based applications, the spatial database technology specialized for GIS is used. This specialization focuses on the vector model for storing location information. In the vector model, geographic regions and areas are represented using geographical coordinates and can be visualized on a map as points, lines, and polygons. The geographical regions can be related to each other and various operations can be performed on them. Spatial indexing techniques and proven data models can be used to handle geographical information.

Querying Location Information

Once geographical information is available for location-based applications, data associated with geographical regions must be added to the system. Once real-life data is added to a spatial database, the analysis based on the regions associated with a particular point or area of interest can be performed. Users of location-based services use various actions, such as filtering, classifying, and prioritizing of services. In the following section, a set of queries is discussed that needs to be handled by a spatial database system used for location-based services in order to satisfy the needs. These queries can be used as benchmark database queries to test the different DBMS systems (Theodoridis, 2003). The queries discussed can be used in various applications and are fit for different purposes. The query language is based on the structured query language (SQL).

SQL was extended to be able to handle spatial information in two ways (Egenhofer, 1994). New statements were added to be able to describe the information about to be retrieved, and the presentation language was improved to specify how to display query results. The extended language is often referred to as spatial SQL (SSQL). SSQL satisfies a set of crucial requirements which are not covered by conventional database systems (Egenhofer). Namely, a new data type for spatial data was introduced, a new set of operations was added to be able to work with this data type, and graphical presentation of query results was specified. Furthermore, an extension to automatically add complementary information to the result set to help interpret the results was also introduced. The ability to merge a combination of independent query results into one result set, with the option for the user to control presentation, makes these systems efficient.

The new data types in SSQL can be used to describe geographical areas of interest. They are used during the definition of the table structures in the database definition phase. Spatial data types have different dimensions. For example, a mobile phone in a user's bag can be represented as a point, or it can be represented as an area relevant for shopping information. In this case, an SSQL table is defined in the following way:

```
CREATE TABLE usersmobile (
        name varchar(100),
        location spatial_0,
        shoppingarea spatial_2 );
```

In order to improve information-description capabilities, during the extension of SQL, functions were added which determine the dimension of an object, its boundary, and its interior. The dimension is 0, 1, and 2 for points, lines, and areas, respectively. The boundary determines the bounding faces of an n-dimensional object and the interior calculates the interior faces. Unitary operations, such as length, area, extreme coordinates, complement, and convex hull, were also introduced. Binary topological relationships such as meet, disjoint, overlap, equal, inside, covers, cross, and neighbor (Figure 5) can be used to calculate the value between two regions.

Example Location Queries

Spatial queries can be categorized into classes. We talk about queries based on a stationary reference object, such as a point query, range query, or a distance query, or we can use queries based on moving reference objects. A point query can be used to ask questions like, "Are there any special offers in the store I am in now?"

Query 1: Point query

```
SELECT Offer.description
FROM Shopper, Shop, Offer
WHERE Offer.shop_id = Shop.id and Shopper.id = 12
AND Shopper.location INSIDE Shop.geometry;
```

Figure 5. Binary topological relationships between geographical areas

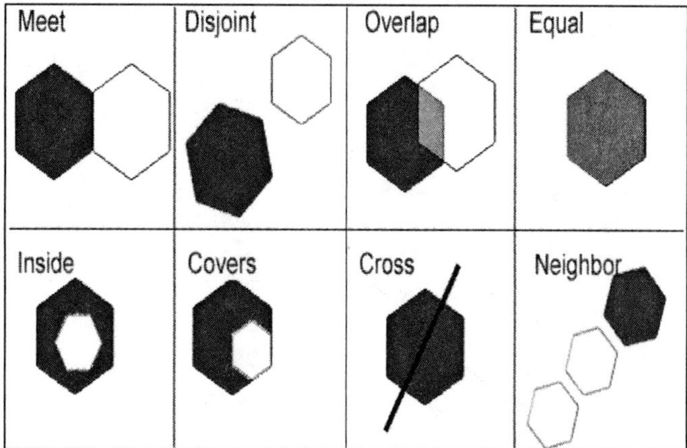

A large family of location applications does not rely on location information related to a certain wireless device, but works with a set of devices located in an area. These applications use a data-push technology, for example WAP push. An example for such an application could be an advertising service that sends information to every mobile user who travels to a selected area. This family of applications uses distance-based queries based on stationary reference objects. The example query gives the following command: "Find cars located in the eight-kilometer area of a certain gas station in Budapest!"

Query 2: Distance-based query

SELECT Car.phonenumber

FROM cars

WHERE car.route OVERLAP Circle((47.30145,19.0789),8000)

Spatial databases make it possible to combine time and location information together in a single query. The "life" operator (Theodoridis, 2003) is a good option to handle geographical information related to time. The life operator computes the temporal projection of a point onto the time axis, returning a set of coordinated intervals. This way, it is possible to do statistical analysis of moving objects within a geographical area in a period. For example, we can ask, "How many spectators visited the soccer match yesterday evening?"

Query 3: Range query restricted by a period

SELECT count (Human)

FROM Human

```
AND Human.location INSIDE Rectangle ((47.24578,19.17894),(47.24562,19.17178));
AND life (Human.route) RESTRICTED_BY Interval (2003-11-21 20:00, 2003-11-21 22:00)
TOGETHER;
```

Many interesting applications can be based on information served by the nearest-neighbor query. This query type can be used to isolate a set of objects that are located close to a stationary reference point in various directions. For example, a user could ask for the location of the nearest ATM machine. The query in this case would look like this.

```
Query 4: Nearest-neighbor query
SELECT atm.address
FROM atm
WHERE atm.status = "working"
ORDER BY Spatial_Neighbor (atm.location, (47.24578,19.17894)) ASC 2;
```

Binary and unitary spatial topological relationships can also be included in a query with the combination of time. For example, we can ask for a list of children crossing a street in the morning hours.

```
Query 5: Topological query
SELECT Children.name
FROM Children, Street
WHERE Street.id = "Kossuth St."
AND trajectory (Children.route) CROSS Street.shape
AND life (Children.route) RESTRICTED_BY Interval (2003-11-21 7:00, 2003-11-21 9:00)
TOGETHER;
```

Queries issued using moving reference objects can sound a little complicated. Handling continuously changing locations in a vector-based geographical model requires the introduction of a unitary operator called *trajectory*. The trajectory operator is required because continuous information is hard to work with, primarily because current DBMS technology is not able to store and manipulate infinite sets of data, and monitoring systems such as GPS are inherently discrete. The trajectory operator performs interpolation using the linear or some complex method. The following example shows how a distance-based query can be composed using the trajectory operator with a moving reference object. The query can be used to find people who have passed close to me (within 100 m) and can speak Hungarian.

Query 6: Simple distance-based query

```
SELECT Human.name
FROM Human X, Human Y
WHERE X.name=”Gyula Rábai” and Y.language = “Hungarian”
AND Y.route INSIDE Strip (trajectory(X.route), 100);
```

The operator *strip* used in Query 6 extends a geographical region by a number of units into each direction. Next-generation wireless applications that will be required to fulfill the requirements of emergencies will use the *distance* operator. The distance operator returns the distance of two areas in a two-dimensional geographical environment. This operator can be located after the SELECT and the WHERE keywords as well as other operators. As an example, we present a query returning the closest hospital to an address.

Query 7: Distance-based query using the distance operator

```
SELECT min (distance (building.geometry, hospital.geometry))
FROM building,building hospital
WHERE building.address=”13 Tailor Street” and hospital.type=”hospital”;
```

Spatial database technology can be used to perform pattern matching and similarity checks. The operator called *similarity* returns a numerical value describing the similarity of two areas. This value is calculated using the *average Euclidean distance* method. The similarity operator, most of the time, is used in the ORDER BY section of the SQL query. Query 8 and Query 9 are two examples of queries composed using the similarity operator. Query 8 shows how we can find out who was following us on our way to the office. Query 9 can be used to find the two most similar buildings in a city. Query 9 can be referred to as a similarity-join query because spatial information is used in the join statement.

Query 8: Similarity-based query

```
SELECT Y.numberplate
FROM Car X, Car Y
WHERE X.id = 20
AND life (X.route) RESTRICTED_BY Interval (2003-11-21 7:00, 2003-11-21 9:00)
AND life (Y.route) RESTRICTED_BY Interval (2003-11-21 7:00, 2003-11-21 9:00)
ORDER BY Similarity (trajectory (Y.route), trajectory (X.route), Distance()) DESC 3;
```

Query 9: Similarity-join query

```
SELECT A.address, B.address
FROM Building X, Building Y
```

```
WHERE NOT (X.address = Y.address)
ORDER BY Similarity (trajectory (X.route), trajectory (Y.route), Distance()) DESC 1;
```

Processing Results

Queries used in location-based applications, most of the time, return nongeographical information related to a geographical area. In some cases, the geographical data is also required to interpret the results. When nongeographical data is returned, result processing can be handled with the traditional methods. The most convenient one is often textual representation. If the geographical data is required and an appropriate spatial database technology is used, the geographical portion of the results can be returned in two ways. SSQL queries can produce geographical results as a set of coordinates represented in textual format, or they can produce an image containing the information as a graphical presentation. If query results are returned as graphics, a set of operators can be used to control the appearance. These operators allow the user to control colors, fill patterns, and line appearances. Information labeling and highlighting is also supported in some systems.

Query Validity

A single location application can serve many wireless clients in a particular geographical area. It might happen that spatial queries formed in SSQL produce similar results for most clients. If overlapping results of queries with stationary or moving reference objects are recalculated each time the information is needed, a large amount of valuable server resources and communication bandwidth can be wasted. Overlapping query results can be used more than one time at the server side to speed up response times. If overlapping results occur, the server can use a caching technique to use the result of previous queries partially or fully. This way, better performance can be reached. If a location-based application is developed that issues queries based on a moving reference point periodically, with careful programming, communication can be spared. All that should be done is to check the validity of previous queries. For example, a moving car can query a location server for available gas stations within a distance every five minutes. In this scenario, the results of consecutive queries could overlap if the car is moving slowly. If some intelligence is added to this system, the load on the infrastructure can be decreased. An intelligent terminal can determine the validity of previous queries based on its current location. Calculations are not the only options to perform validity checks in location-based applications. Spatial database technology can be used to set up a validity region around the returned result (Zhang, Zhu, Papadias, Tao, & Lee, 2003). Validity regions can be defined by nearest-neighbor and window queries.

Storing Location Information

The performance of a database system that supports location-based information retrieval depends on the data storage architecture, the access methods, and the query processing techniques. Query processing and information retrieval performance can be improved by using query validity, indexes, and metadata.

Speeding Up Query Processing

For geographical information, spatial indexing techniques can provide good performance. The two most common indexing solutions are R-tree and quadtree indexing. For example, they are implemented in the Oracle 9i DBMS. R-tree indexing approximates geographical areas with the smallest, single rectangle that encloses the appropriate geometry. Once the rectangles are identified, hierarchical layers are used to build a tree (Figure 6).

Quadtree indexing (Figure 7), on the other hand, uses a mosaic of covering tiles to represent an area. Location data is assigned to one or more tiles, which are represented in a binary format. The operation called *tessellation* is used to create the quadtree index. This operation defines exclusive and exhaustive tiles for every stored geometry, and it decomposes the coordinate space in a regular, hierarchical manner. Quadtree indexing has two subtypes: fixed indexing and hybrid indexing.

Location-based queries can be significantly faster if the storage of location information is well organized. If the SSQL queries that will be served are available when the database is defined, or if the database definition is subject to modification during run-time, adding appropriate indexes can optimize data storage.

Figure 6. R-tree indexing

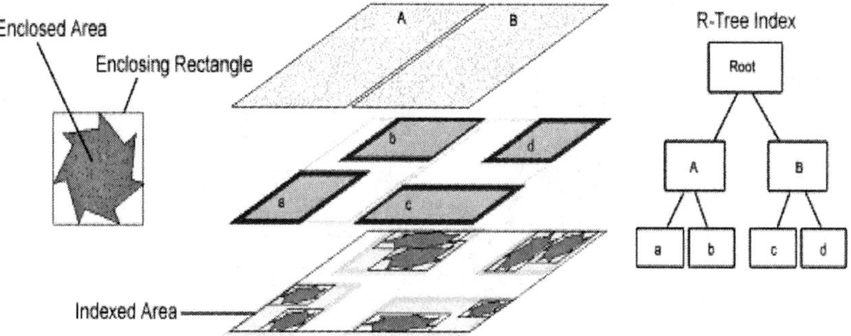

Figure 7. Quadtree decomposition

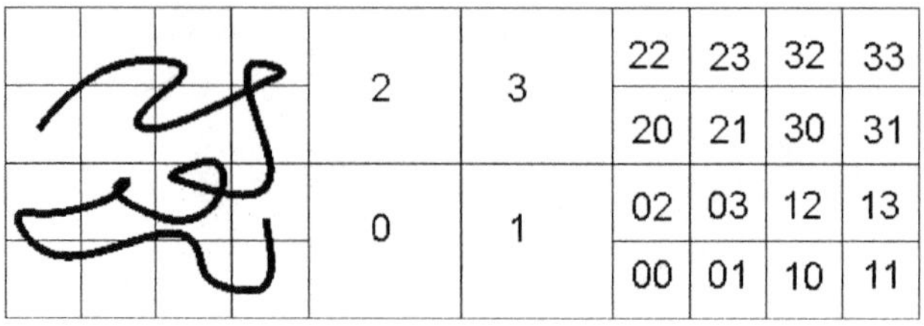

2	3
0	1

22	23	32	33
20	21	30	31
02	03	12	13
00	01	10	11

Geographically Distributed Data Access Methods

Although spatial indexing techniques and other clever storage technologies can improve query performance, it is often impractical or impossible to use a single location database server in a centralized location. A central architecture could create an intolerable performance problem if a large number of queries are issued simultaneously, or low bandwidth can be a limiting factor if information is required from a large distance. In such a scenario, it is necessary to use a geographically distributed database system.

In a geographically distributed environment, a partitioned data model is required to find out which data is stored at which location. For a purpose, a small reference database can be used that is replicated among storage and routing facilities. Along with distributing the data, responsibility needs to be delegated for maintenance purposes, and redundancy should be added to increase reliability. If the data is distributed among various geographical locations, the location-based queries need to be routed to the relevant serving nodes. Query routing can be done by query routers that analyze each query and forward them to the appropriate servers. Query routers highly depend on the replicated reference database.

The advantages we gain with geographical data partitioning are reduced response times for queries issued from a large distance. We can save communication bandwidth and can serve concurrent queries more efficiently. The overall throughput of a system can be increased.

Example Applications

Current and next-generation location-based data access solutions provide opportunities for businesses to create new services. In the next section, we are going to give a brief overview of some of the location-based data access solutions existing today and we are going to present interesting application areas. It is highly probable that the introduced applications will be even more successful in the near future.

Location-Based Applications Used in Present Wireless Networks

One of the most promising application areas in wireless networks is location-aware content delivery. Services based on this technology use user-location data to tailor the information sent to the user. Different information relevant to a certain location can be delivered using user-initiated queries and push technologies. Example applications in this area could be services that can be used to query the location of nearby hospitals, ATM machines, gas stations, or restaurants, or for offers of the nearest stores or instant coupons.

Other application areas include public, business, and government usage of location-based technology. Tourist services, mobile electronic commerce, mobile workforce management, automatic vehicle location, emergencies and requests, fleet management, logistics, transportation management and support, traffic control, and entertainment service are examples of this usage.

In Budapest, one of the most popular applications is hard to add to any application category. It is the friend finder. This application uses E-OTD in the Hungarian GSM network operated by Westel, and it makes it possible to find friends and relatives who are close to us. The user must register by filling out a WAP form and they can request information from the other registered parties using WAP or Short Message Service (SMS). In Rome, the tourist information city guide is the most attractive location-based service. If a tourist in Rome registers for this service, he or she receives information in SMS or MMS about historical places while walking through the city. The information is always sent at the relevant location.

Unfortunately, or fortunately, location-based data access methods can be used for military purposes. Research concerning military usage in the digital battlefield (Chamberlain, 1995) is currently taking place.

Configurable Radio with Advanced Software Technology Project

A good example of location-based data access in a wireless network was developed by the team working on the CAST project at the Mobile Communication and Computing Laboratory at the Budapest University of Technology and Economics. The CAST project was a European Fifth Framework project aimed at building a reconfigurable radio network using software radio technology. The project officially ended in the spring of 2003, but the research activities were continued and a software radio research group was set up as part of European Cooperation in the Field of Scientific and Technical Research COST 289 Spectrum and Power Efficient Broadband Communications (COST 289). The latest achievements in this work are focused on dynamic configuration update of mobile terminals using software radio technology.

According to the software radio concept, a universal radio terminal can be built using reconfigurable digital building blocks. This is possible if smart multiband antenna technology is used, a wideband analog-digital converter (ADC) is placed as close to the antenna as possible, and all the radio functionality is implemented in software, including the Intermediate Frequency (IF), base-band, and bit-stream functionality. Such a terminal could be used in present and future radio networks operating at different standards (GSM, Code-Division Multiple Access [CDMA], Universal Mobile Telecommunications Service [UMTS], etc.) by simply updating the software functions on the reconfigurable hardware. Technology was developed in the CAST project to dynamically update the software configuration in a software radio terminal. As an example, configuration update from GSM to UMTS was performed. One of the biggest problems with dynamic software updates was to find a way for the new software modules to get to the mobile terminal. In recent years, our research team has spent a lot of time in finding possible solutions for these problems. It seems the best option is to use a spatial database storing information about radio coverage of different radio systems, and to use location-dependent data access to figure out what services are available in a certain location.

We have compared two spatial database systems, Microsoft SQL Server and Oracle, and have decided to use Oracle 9i, which was used for other purposes in the CAST project as well. Location information was stored in the database using layers sharing a common coordinate system. The representation of an area included separate layers for radio coverage, radio standards, network operators, base stations, and so forth. (The same approach can be used in a general-purpose application, where layers with the locations of gas stations, restaurants, electrical lines, etc. can be added). The different options represented by the different layers were related by the geographical location.

Configuration updates in mobile terminals are managed by a resource controller (RSC). This module performs software installations, keeps track of available hardware capacity, and maintains a repository of downloaded software objects. Periodically and on certain asynchronous events, the RSC can initiate the download of new software functions and

Figure 8. Spatial database with layers

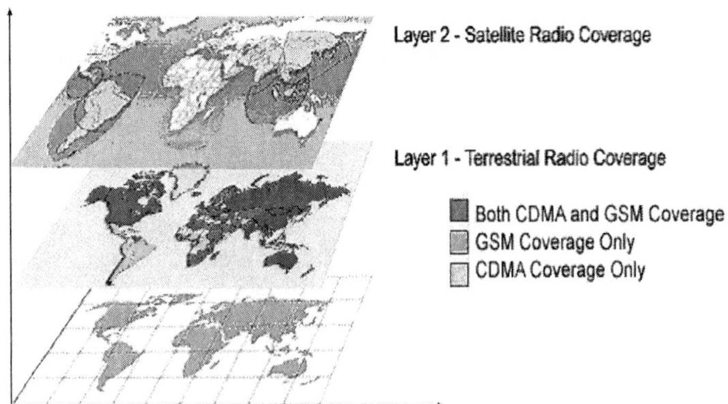

can perform dynamic software updates. If the user of a reconfigurable mobile terminal decides to travel to a new area, an event to make the mobile device ready for the new environment can be triggered. Once this event is triggered, a location-based query is issued to a central spatial database to find out what type of radio networks and services are available at the desired location. Once this remote discovery is made, the RSC downloads the necessary software to the terminal. The query used to find the necessary software functions is a point query using a stationary reference object.

Query A:

SELECT service.servicetype, service.operator

FROM service, location

WHERE location.address="2 Magyar T. krt., 1117 Budapest, Hungary" and location.geometry OVERLAP service.geometry

After the available service types are identified, another query is used for each service to find the software functions that should be downloaded to make the device ready to be used in the given location.

Query B":

SELECT downloadurl

FROM service

WHERE servicetype="UTRA"

ISSUEDFROM myLocation.geometry='[14.2342,24.343]'

Query B":

SELECT downloadurl

FROM service

WHERE servicetype="GSM"

ISSUEDFROM myLocation.geometry='[14.2342,24.343]'

One can ask why these queries are not written as one. The answer for this question is that if queries are issued independently with an appropriate query-routing technology, a more relevant result can be returned for each question. In the example, Query A would be routed to a central information database where available services are stored; meanwhile, Queries B1 through B*N* would be routed to a location which can provide better download capabilities.

The relative resource poverty of mobile elements, as well as their lower levels of security and robustness, argues for reliance on static servers. At the same time, the need to cope with unreliable, low-performance networks and be sensitive to power consumption argues for self-reliance.

The introduced system is using a distributed data model supporting location-based data access and query routing. The query routers are organized in a topology matching the topology of the wireless network, where nodes of the network connect up to host nodes and down to hosted nodes. This model is very efficient in a cellular network. Relaying of location-based queries is almost trivial. All queries are relayed up to the host node until the appropriate information becomes available. With this system, it is possible to use the same query on a mobile terminal in various locations. The result of the query depends on the location of the terminal and the moving characteristics of the terminal, such as its speed, path, and direction, since consequent queries can be served by different base stations serving the node. In the introduced system, we rely on the service to route the query to the appropriate data source, and to process the results before it is returned to the terminal.

Conclusion

The ability to obtain information on demand, wherever the user happens to be over a wireless link, opens up new opportunities. The introduction of positioning to wireless networks, using technologies like TA, E-OTD, and GPS, paves the way for location-based wireless services. These services rely highly on background information stored in spatial databases and need a method for querying this information. We have shown how traditional database technology can be extended to satisfy this need, and how location-based data can be accessed using different application architectures in existing and future wireless networks. A set of examples and hot research topics were also discussed as we expressed our point that location-based services will play a great role in future wireless information highways.

References

Bajada, J. (2003). Mobile positioning for location dependent services in GSM networks. *Computer Science Annual Workshop (CSAW).*

Chamberlain, S. (1995). Model-based battle command: A paradigm whose time has come. *Proceedings of the Symposium on Command & Control Research and Technology.*

Chang, S. K., & Kunii, T. (1981). Pictorial database systems. *IEEE Computer, 14*(11), 13-21.

Egenhofer, M. (1994). Spatial SQL: A query and presentation language. *IEEE Transactions on Knowledge and Data Engineering, 6*(1), 86-95.

Frank, A. (1988). Requirements for a database management system for a GIS. *Photogrammetric Engineering and Remote Sensing*, 54(11), 1557-1564.

Imielinski, T., & Navas, J. C. (1999). GPS-based geographic addressing, routing, and resource discovery. *Communications of the ACM, 42*(4), 86-92.

J2ME specifications. (n.d.). Retrieved April 19, 2004, from http://www.java.sun.com/products/cldc/

Java community process, JSR-179: Location API to J2ME. (n.d.). Retrieved April 19, 2004, from http://www.jcp.org/jsr/detail/179.jsp

Lohman, G., Stoltzfus, J., Benson, A., Martin, M., & Cardenas, A. (1983). Remotely sensed geophysical databases: Experience and implications for generalized DBMS. *Proceedings of the ACM International Conference on the Management of Data (SIGMOD)*, 146-160.

Madani, K., Bosch, B., Honary, B., Justo, G., Kovacs, J., Lohi, M., et al. (2000). Configurable radio with advanced software technology (CAST): Initial concepts. *IST Mobile Communications Summit 2000*, 139-144.

MExE: MS application execution environment. (n.d.). Retrieved April 19, 2004, from ftp://ftp.3gpp.org/Specs/2000-12/Rel-4/23_series/23057-400.zip

Theodoridis, Y. (2003). Ten benchmark database queries for location-based services. *The Computer Journal, 46*(5), 713-725.

WAE: Wireless Application Environment specification, version 2.0. (n.d.). *WAP Forum.* Retrieved April 19, 2004, from http://www.wapforum.org/

WAP: Wireless Application Protocol 2.0 specifications. (n.d.). *WAP Forum.* Retrieved April 19, 2004, from http://www.wapforum.org/what/technical.htm

WPA: WAP Push Architecture specification. (n.d.). *WAP Forum.* Retrieved April 19, 2004, from http://www.wapforum.org/what/technical.htm

Zhang, J., Zhu, M., Papadias, D., Tao, Y., & Lee, D. L. (2003). Location-based spatial queries. *Proceedings of the ACM International Conference on the Management of Data (SIGMOD)*, 443-454.

Section V

Advanced Topics

Chapter XV

Security in Pervasive Computing

Sajal K. Das, University of Texas at Arlington, USA

Afrand Agah, University of Texas at Arlington, USA

Mohan Kumar, University of Texas at Arlington, USA

Abstract

Security requirements for pervasive computing environments are different from those in fixed networks. This is due to the intensity and complexity of the communication between the user and the infrastructure, the mobility of the user, and dynamic sharing of limited resources. As pervasive computing makes information access and processing easily available for everyone from anywhere at anytime, the close relationship between distributed systems and mobile computing with a pervasive infrastructure leads us to take a closer look at different types of vulnerabilities and attacks in such environments. Pervasive computing includes numerous, often transparent, computing devices that are frequently mobile or embedded in the environment, and are connected to an increasingly ubiquitous network structure. For example, when an organization employs pervasive computing, the environment becomes more knowledgeable about the users' behavior and, hence, becomes more proactive with each individual user as time passes. Therefore, the user must be able to trust the environment and the environment must be confident of the user's identity. This implies security is an important concern in the success of pervasive computing environments. In this chapter we evaluate the suitability of existing security methods for pervasive environments.

Introduction

Advances in technology provide isolated means for detecting and perhaps preventing security violations reactively. However, there is a need to glue these disparate technologies together so as to provide proactive infrastructure support and services for managing security-related issues. It is an extremely challenging task to process the information collected from sensory devices, interpret them meaningfully in the context of ongoing events, and accordingly carry out automated security services. This requires continual and proactive, real-time collaborations among physical devices, software agents, and personnel in dynamic, heterogeneous, autonomous environments.

The fundamental principles that guide pervasive computing environment design evolved from distributed systems. So in order to understand the concept of pervasive computing, we first begin describing the two closely related fields: distributed systems and mobile computing. As described in Satayanarayanan (2001), the following five areas are fundamental to distributed systems: (a) remote communication, (b) fault tolerance, (c) high availability, (d) remote information access, and (e) security. High bandwidth and low error rate in distributed systems make it possible to break down centralized software systems into separate network-connected components.

Although many basic principles of distributed systems are common to mobile computing as well, the following additional key features are fundamental to mobile computing: (a) unpredictable variation in wireless network communication quality, (b) lowered trust, (c) limitations on local resources, and (d) battery power consumption.

The ability to communicate remotely, access distributed files, share resources, and roam are the underlying challenges of a pervasive computing environment. In such an environment, if the nodes of the network are like agents that are programmable, or are able to move with the program and execute in different locations of the network, we would be able to have the benefits of active networks, which are thus essential components of pervasive computing infrastructures.

Many difficult design and implementation problems must be solved to realize pervasive computing. The main challenge we try to address in this chapter is security. Any successful pervasive computing environment uses smart spaces effectively and masks uneven conditions. Usages of smart spaces and masking uneven conditions have some degree of conflict when we embed the security aspects into them. Eavesdropping on wireless links is very easy, so the security of wireless communications can be easily compromised, especially if transmissions happen in a large area or while users are allowed to cross security domains, like giving permission to use one service in one environment but prohibiting the use of another service in the same area. Protecting the identity of the user, securing the information flowing between a user and base station, and achieving security and authentication are very hard tasks to do in a wireless environment. We study the existing security methods and discuss how to enhance them for pervasive computing.

To have a better feeling of life in a pervasive computing world, let us consider an example (Satayanarayanan, 2001). Suppose "Fred" is in his office and is preparing for his presentation in a meeting room, which is about a 10-minute walk across the campus. As he is not completely done with his presentation, he grabs his handheld computer. His files

Figure 1. Medical scenario

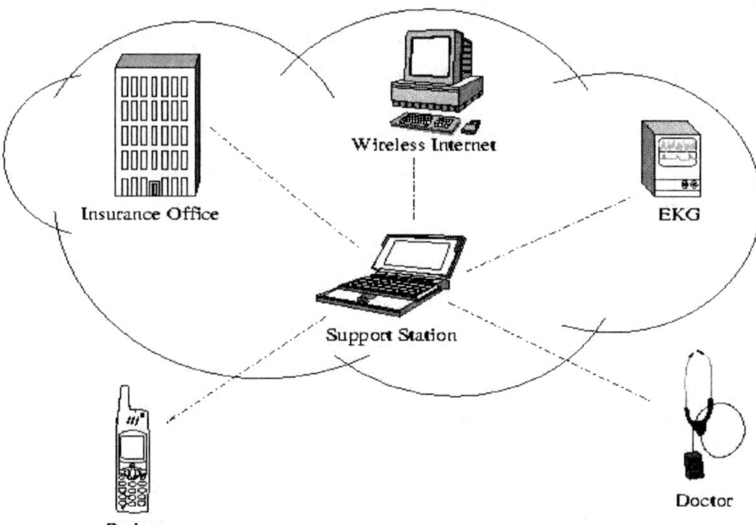

are transferred from his desktop to his handheld, so he edits the presentation while walking toward the meeting room. Then his talk is downloaded to the projection computer. In this scenario, security plays a major role. Fred must be sure that no one can fabricate or alter his files. It is to be ensured that the projection and handheld computers must get information from Fred and not someone impersonating him. While he is giving the talk, future presenters must be able to transparently download files on the projector computer. If Fred sends something to the projector, he should not be able to deny it later. While he is giving a talk, if there are some slides that some of the audiences should not see, the projector would not show them; this is a trust that Fred must have on the system.

Now let us consider two other scenarios, one from a medical application and another from a military application. "Jack" has a history of cardiac illness; his doctor wants to have the latest information about Jack's heart condition. Jack will use a heart monitor unit, which can be connected to his cell phone. Information will be transferred to his doctor's computer, and accordingly, Jack can get diagnostic notifications from his doctor. In this case, if someone alters Jack's information or impersonates the doctor, it may have fatal results. Jack's information is private, which he shares with his insurance company. Also, any unauthorized access to his information must be forbidden. Figure 1 illustrates this scenario.

In the military application, a soldier's personal digital assistant (PDA) or cell phone contains information about terrain, strategies, vital data, enemy positions, up-to-date commands from his or her commander, shared data with peers, and so forth. The PDA's or cell phone's connection to the wireless Internet is intermittent and noncontinuous. It is necessary to provide the cell phone or PDA with the most relevant data all the time from nearby support stations based on the soldier's current position. Leaders and commanders can constantly monitor the status of friendly troops and ammunition in a battlefield by use of some preattached sensors. Approach routes and paths can be covered with sensors and closely watched for battlefield surveillance from a nearby surface mobile station or from the sea. Alteration of any of this information can bring catastrophic damages to each party. No party should be able to eavesdrop on the other party's communications or be able to impersonate. Figure 2 depicts this scenario. Throughout this chapter, we refer to Figures 1 and 2 when we evaluate the security issues for pervasive computing. We talk about security in general, which includes security in physical and network layers, distributed systems, mobile computing environments, and active networks. We also discuss security in pervasive computing and present our current research on this topic.

Security

In this section, we will define security challenges and issues in mobile computing and other environments. Then as we proceed to the pervasive computing section, we shall show how hard it is to incorporate security aspects while satisfying pervasive computing capabilities. In order to have security in a wireless network, a range of issues needs to be taken into consideration. These are (a) identification of a mobile user, (b) anonymity of a mobile user, (c) authentication of a base station, (d) security of information flow, (e) prevention of attacks, (f) resource requirements (CPU, memory, bandwidth), (g) the cost

Figure 2. Military scenario

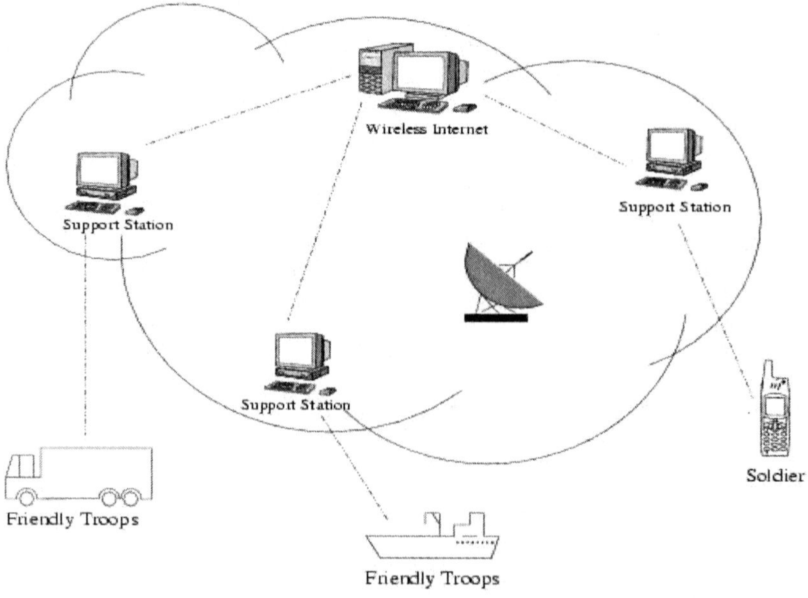

of establishing a session key between a user and a base station, and (h) the cost of communication between a user and a foreign domain.

Any security protocol in a mobile computing environment has the following goals (Bharghavan, 1994; Campbell, Qian, Liao, & Liu, 1996).

- *Authentication*: The portable computer used by a mobile user and the base station must be able to mutually authenticate each other. This authentication enables the base station to prevent unauthorized users from using its services, and enables the user to choose an authorized base station as a service provider.
- *Data privacy*: The portable computer used by a mobile user and the base station must be able to communicate in a secure way. Data privacy protects the data from being snooped, replayed, or forged.
- *Location privacy*: Location privacy is a unique goal of mobile communication. It is desirable or critical to not compromise the location and identity of the mobile user.
- *Accounting*: When a mobile user uses a portable computer and establishes communications with a base station, the service provider needs to charge the portable computer used by the mobile user for its service.

An attacker may try to steal information, contaminate messages, or impersonate a node. There are three main goals that must be achieved: (a) Nodes that are roaming in a hostile environment and have poor physical protection have a probability of being compromised, so attacks from inside the network must be considered as important as attacks from outside the network, (b) in a network with frequent changes in network topology, a trust relationship between nodes changes frequently, which is uncommon in static networks, and (c) a large network consists of hundreds of nodes, therefore security configuration must be scalable. Routing protocols must be able to cope with dynamically changing topology and malicious attacks. Many protocols have been proposed for mobile networks to cope with dynamic topology, but not much is being done to have a secure one. In most proposed routing protocols, routers exchange information to establish routes between different nodes. This information can be very inviting to attackers. Introducing excessive traffic load into the network or advertising incorrect routing information to other nodes are different kinds of attacks on routing protocols. The authors in Kagal, Undercoffer, Perich, Joshi, and Finin (2002) employ digital signatures to protect routing information and data traffic.

Now we briefly describe some of the techniques used in computer security, and will later consider their applicability to pervasive computing. Cryptographic primitives have two dual processes: encryption and decryption. The goal of encryption is to compute a special encoding of some secret information. Applying the encryption process to some text produces this encoding, and applying the decryption process can retrieve the original text. The intended recipient has a secret key, which he or she uses for the decryption process. The sender of a message, on the other hand, uses an encryption key to encrypt the original message. The following techniques are basic building blocks of security protocols (Kagal, Undercoffer, Perich, et al., 2002).

- *Symmetric Cryptography*: Secret-key cryptography (SKC) schemes rest on the basis that the key for the encryption and decryption of messages is the same. This is why such schemes are also referred to as *symmetric cryptography*, and they are usually used for confidentiality of stored data. Symmetric cryptography uses a single private key to both encrypt and decrypt data. Any party that has the key can encrypt and decrypt data. Symmetric cryptography algorithms are typically fast and suitable for processing large streams of data. The disadvantage is that it presumes two parties have agreed on a key and been able to exchange that key in a secure manner prior to communication. This is a significant challenge. Symmetric algorithms are usually mixed with public key algorithms to obtain a mixture of security and speed.

- *Asymmetric Cryptography*: This technique incorporates algorithms which use different keys for encryption and decryption. So, both parties that want to transfer data are able to exchange their encryption keys over even an insecure connection link. The general idea behind public-key cryptography is that each participant *A* in a communication network has two keys: one public key and one private key. Agent *A*'s public key is freely available or may be obtained from a certifying and trusted authority on demand. The agent's private key is meant to be a secret: only *A* knows this key, and letting others know it will allow them to assume *A*'s identity, as well as decrypt all messages that were addressed to *A*. Needless to say, public key and private key are different keys, so this is an example of an asymmetric scheme. It uses a secret key that must be kept from unauthorized users and a public key that can be made public to anyone. Both the public and private keys are mathematically linked; only the private key can decrypt data encrypted with the public key, and the data signed with the private key can only be verified with the use of the public key. The public key can be published to anyone. Both keys are unique to the communication session. Public-key cryptographic algorithms use a fixed buffer size, while private-key cryptographic algorithms use a variable-length buffer. With private-key algorithms, only a small block size can be processed, typically 8 or 16 bytes. There are many cryptographic toolkits to choose from. The choice may be dictated by one's development platform such as Java Cryptography Extensions (JCE) and the Java Secure Socket Extensions (JSSE; *Javasun*, n.d.). According to Javasoft, "JCE provides a framework and implementations for encryption, key generation, key agreement, and message authentication code algorithms. Support for encryption includes symmetric, asymmetric, block, and stream ciphers. The software also supports secure streams and sealed objects."
- *Certificate Authority*: When two parties are exchanging their public keys over an insecure channel, an attacker can substitute the public keys with fake ones. So, one way to verify that the public key really belongs to the sender is to use a certificate authority (CA), which signs the public key with some information about the owner. A CA is an authority in a network that issues and manages security credentials and public keys for message encryption. As part of a public-key infrastructure, a CA checks with a registration authority (RA) to verify information provided by the requester of a digital certificate. If the RA verifies the requester information, the CA can then issue a certificate. Depending on the public-key infrastructure implementation, the certificate includes the owner's public key, the expiration date of the certificate, the owner's name, and other information about the public-key owner. An RA is an authority in a network that verifies user requests for a digital certificate

and tells the CA to issue it. RAs are part of a public-key infrastructure, a networked system that enables users to exchange information safely and securely.

Now let us look at the security aspects on physical and network layers. Physical-layer security mainly prevents every unauthorized access to the devices. There are three aspects used to authenticate devices: (a) what the user knows (small devices, personal identification numbers [PINs], and passwords fall into this category), (b) what the user owns, such as key cards and electronic tokens to access control of some devices, and permanently installed devices, and (c) what the user is like, which is also known as biometric authentication, and is based on checking the user's physical characteristics. Usually a good authentication involves more than one of the above forms. As an example, automated teller machines (ATMs) require a physical card as well as a PIN.

By using wireless networks, we are able to use more flexible communication models than in traditional wireline networks. Security in mobile networks is difficult to achieve. In wired networks, an attacker must have physical access to the network or pass several firewalls and gateways, but attacks against a wireless network can come from all directions.

Security in Distributed Systems

A distributed system is always susceptible to threats both from legitimate users of the system and intruders. Two general security threats are *host compromise* and *communi-*

Figure 3. Attacks on a distributed system

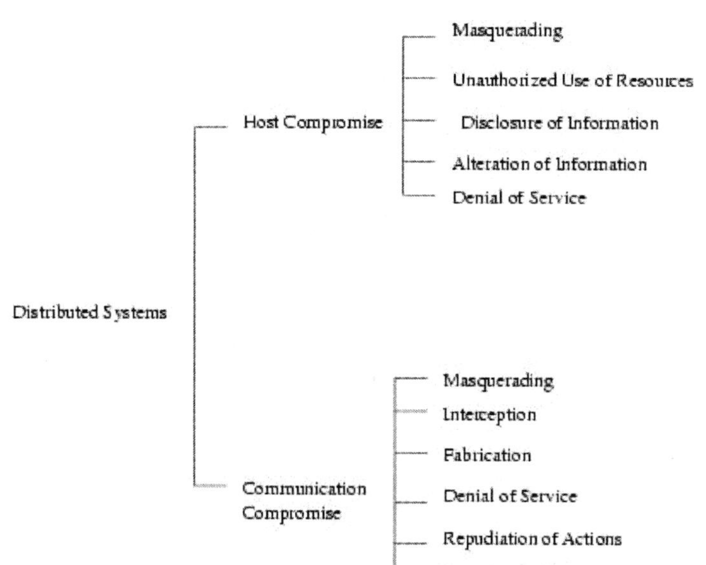

cation compromise. Figure 3 depicts possible attacks on a distributed system (Debar, Dacier, & Wespi, 1999).

Masquerading occurs when a user masquerades as another one to gain access to a system object which he or she is not authorized to use. Unauthorized use of resources occurs when a user accesses a system without proper authorization. Unauthorized reading of stored information is disclosure of information. Unauthorized editing of information is called alteration of information, and unauthorized insertion of information is called fabrication of information. By denying a resource to an entity, an attacker denies a service. Repudiation of actions is when the sender of a message denies sending it. Interception is when the opponent gains access to the data transmitted over the communication link so he or she can obtain the transmitted information, identities, and/or locations of the communication parties.

Security in the Mobile Computing Environment

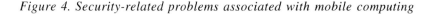

Mobility, portability, and wireless communication are essential properties of a mobile computing system. Each of these properties introduces a number of problems, which are illustrated in Figure 4.

In any wireless communication, depending on whether the system is plugged in or uses wireless access, bandwidth is a very critical component. Wireless communication suffers from frequent disconnections. To achieve wireless communication, a mobile host might get connected to different and heterogeneous networks. The general problem of heterogeneity can be addressed by emerging distributed-systems standards such as the Object Management Group's Common Object Request Broker Architecture (OMG-CORBA), or the Open Software Foundation's Distributed Computing Environment (OSF-DCE; Pangalos, 1997). Answering dynamic location queries requires knowing the

Figure 4. Security-related problems associated with mobile computing

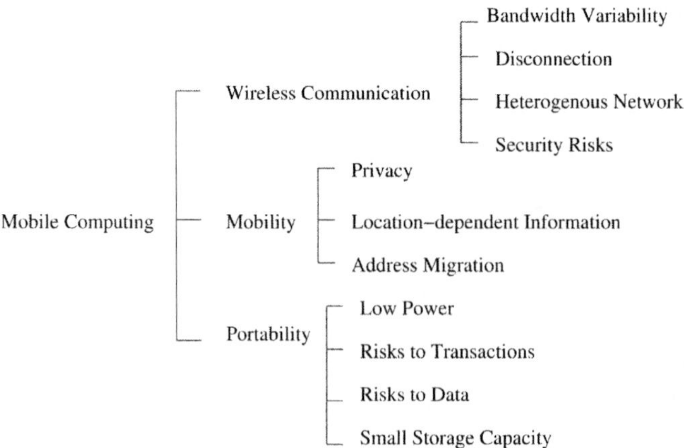

location of other mobile users, which introduces the privacy concept. Information needed to configure a computer, such as the local name server, available printers, and time zone, is location dependent. Address migration is a consequence of mobility, broadcasting, and home-based forwarding services (Pangalos). Portability introduces more problems. Low power can be due to the far distance of a source or being unable to recharge the temporary battery or any other source of energy. In order to make a device portable, weight plays a very important role, so portability comes with small storage capacity. In mobile computing, the risks are higher due to wireless communications; data, and even every transaction, is at the risk of eavesdropping.

The nature of the mobile computing environment makes it very vulnerable to attacks, ranging from eavesdropping to interfering. Eavesdropping is very easy; when one node sends a message over the radio path, everyone equipped with a suitable transceiver in the range of the transmission can eavesdrop on the message. In all wireless LAN standards, this is taken care by some kind of link-level ciphering. Damages to a wireless mobile network can include leaking secret information, message contamination, and node impersonation (Zhang, Lee, & Huang, 2002). So a wireless mobile network will not have a clear line of defense. Also, as nodes are autonomous, they can roam freely and this means nodes can be captured or compromised. Since tracking such nodes is hard, attacks from within the network by such compromised nodes is much harder to be detected. One other problem is decentralization of mobile networks, where lack of central authority invites more attacks toward the network. Applications and services in a wireless mobile network can be weak, too. In these networks, there are often proxies and software agents running in base stations and some nodes to achieve performance gains. Potential attacks may target these proxies or agents to gain information or launch denial of service attacks (Zhang et al., 2002). If the attacker has a powerful transceiver, he or she can generate such radio interference that the wireless LAN is unable to communicate using the radio path. Protection against this kind of attack is expensive and difficult. One other type of attack against mobile networks is fooling mobile nodes to trust the base station, which is controlled by the attacker. The attacker may let nodes log onto his or her network to find out passwords and secret keys. Or, he or she can just reject all log-on attempts, but record all the messages and find out authentication information. The distributed denial of service (DDoS) attack is the most advanced form of denial-of-service attacks (Geng, Huang, & Whiston, 2002). As the name suggests, this attack has the ability to deploy its weapons in a distributed way. DDoS attacks are distinct from prior denial-of-service attacks in that they never try to break into the victim's system, thus making any security defense irrelevant. The attacker first gets control of several master computers by hacking them. Then the master computers further get control of more daemon computers, often by using some automatic intrusion software. Finally, a command from the attacker synchronizes all daemons to send junk traffic to the victim to jam his or her entrance and block access by legitimate users.

Another significant difference is the communication pattern in a mobile computing environment. Mobile users tend to be stingy about communication and adopt new operations such as disconnected operations (Satayanarayanan, Kistler, Mummert, Ebling, Kumar, & Liu, 1993), which suggests anomaly models for wired networks cannot be used here. Mobile computing environments have inherent vulnerabilities that are not easily preventable.

Security in Active Networks

Active networking offers a technology where the application not only determines the communication control functions as necessary at the endpoints, but also injects new rules into the network. So, active networks can play a very important role in pervasive environments.

In active networks, routers are programmable. The programs executing at the router are either permanently installed or they exist for the duration of the session (Schwartz, Jackson, Strayer, Zhou, Rockwell, & Partridge, 2000). Nodes in an active network are called *active nodes* because they are programmable elements that allow applications to execute user-defined programs to implement new services. Active nodes perform the functions of receiving, scheduling, monitoring, and forwarding *smart packets*. In an active network, data packets are information entities. Smart packets (Schwartz et al., 2000) contain a destination address, user data, and methods that can be executed locally at any node. The code in a packet is executed at a node if the node has the correct processing environment. Firewalls implement filters to determine which packets must pass transparently and which must be stopped. Firewalls use applications and user-specific functions. By using active networks, these functions can be done automatically, just by allowing applications from approved vendors to authenticate themselves to the firewall and inject the appropriate modules into it (Tennenhouse & Wetherall, 1996).

It is also necessary to look at different types of heterogeneous migration or mobility. These types are (a) passive object, where the process contains only passive data, there is no executable code, and data can be converted from the source to the destination machine, (b) active object, which migrates when inactive, where the process has executable code and data, migration only occurs when the code is not active, and the executable code is on the destination machine, and (c) active object, an interpreted code, where the process executes the code by using an interpreter, and moving the process needs the translating of the state of the interpreter and all data values (Hut, 1996).

Since active networks are flexible, the numbers of safety and security issues that need to be addressed are tremendously increased. We should reduce the risk of mistakes or unintentional behaviors, as well as address data privacy, integrity, and availability issues. Moreover, a packet–carrying, executable code has the potential to change the state of a node. Nodes can be routers that are public resources and are very essential for proper and correct running of the system. Thus, active packets may misuse active nodes and resources. Denial of services can easily occur. Also, active nodes may misuse active packets. A node may erase an active packet before the completion of its job in the node. An active packet may destroy or change the resources or services of a node by modifying or erasing them from memory. Active packets may overload a resource, forcing it to not function properly. An active packet may access and steal private information. When active packets visit a node, they make that node vulnerable. A malicious user can send too many active packets toward one node and bring it down by consuming all the bandwidth (Tennenhouse & Wetherall, 1996).

So we have to consider two aspects: (a) protection of active nodes and (b) protection of active packets. Any active packet must have authenticating credentials produced by a public key signature; this will give assurance that someone else vouches for this packet. Time must be limited, like the amount of time that an active packet may be allowed to execute. The number of packets generated per unit of time from a single node must also be bounded. The packet forwarding cache must act as fast as possible. There must be a bound on the amount of processing time a thread can consume. As each thread takes up memory, there should be a bound on the number of threads a program can generate. Also, programs cannot generate arbitrary references to memory. We can have a monitoring system that restricts the information, services, and resources that an active packet is allowed to access. This monitor will check a database to see what access should be granted (Tennenhouse & Wetherall, 1996). When we use active networks in a pervasive computing environment, we can add distributed intrusion detection to the system. Each node of the active network can collect audit data and detect attacks based on some previously defined attacks. Smart packets can dynamically change the collected data, but they should have restricted access to the sensitive data that they carry.

In order to protect active packets, we can use encryption. We can also store packets for some short period of time to see if there is any node failure. Packets must be able to seek a new route for traversing. However, all these approaches consume memory and bandwidth. So the system must recognize the authenticity of the packets, and identify the network elements and users. Then it can authorize access and allow execution based on its policy. The system also monitors access to the system resources. One important application of active networks is time-sensitive resource access, where quality of service must be considered. However, much of the fundamental work on computer security has been based on a time-independent model of resource access. If the information does not get there in time, it is useless, so we have to push the security model to reflect time while using the active network.

The application of existing cryptographic techniques to the active networks environment presents certain challenges. Existing Internet interactions are typically between client and server, where the explicit individual identity of the client and server are important. The Internet community has begun to move away from explicit individual identities to attribute-based identities. Existing mechanisms for providing authentication protection of a packet are rooted in the existing Internet paradigm of client- and server-based communication. These will not be sufficient in an active network environment, where the packet needs to be authenticated at the source and destination and potentially every node in between. Some of the most challenging aspects of securing active networks concern the authentication support for authorization (Kagal, Undercoffer, Perich, et al., 2002).

In the military application, if we use active networks, by sending smart packets to soldiers' PDA's, some commands can be executed locally at any node. For example, if confidential information about the enemy is spreading into the network, then only those commanders who have the correct processing environment will be able to retrieve the information. If a soldier is last seen on a battlefield, then the monitoring part of active nodes will have this information; if commanders want to interact with the soldier, probably the best way is through his or her PDA and not a laptop. Also, as pervasive computing must have localized scalability, the number of active packets generated per unit of time from a single node must be bounded. This will allow everyone in the network to have a fair chunk of the resources.

Security in Pervasive Computing

The need to make mobile devices smaller and lighter means that the computing capabilities have to be compromised. Meeting the ever-growing expectations of mobile users requires computing and data-manipulation capabilities well beyond those of a lightweight, mobile computer with long battery life (Campbell et al., 1996).

A pervasive computing system must be aware of its user's state and surroundings, and must be able to modify its behavior based on this information. A key challenge is obtaining this information. In some cases, the information is part of a user's personal computing space. The thrust areas of pervasive computing are (Satayanarayanan, 2001) the effective use of smart spaces, invisibility, localized scalability, and masking uneven conditioning.

The smart space may be an enclosed area, like a meeting room or doctor's office, or it can be a well-defined open area, such as a campus or a battlefield. By embedding computing infrastructure, a smart space brings disjoint worlds together. It makes possible the adjustment of lighting levels in a room based on the occupant's profile, or the level of information that an ordinary soldier can receive on his or her PDA compared to the information that the commander will receive.

A smart space is the complete disappearance of devices from a user's consciousness. Consider the presenter example: When he is walking toward the meeting room on campus, the infrastructure must provide seamless connection and other services to the presenter despite mobility. According to Weiser (1991), "The most profound technologies are those that disappear. They weave themselves into the fabric of everyday life until they are indistinguishable from it." As smart spaces grow, the interactions between a user's personal computing space and his or her surroundings will grow, too. Consider the medical application: A doctor can access and monitor the information about one patient, but if there are tens of doctors in the clinic and they all want to access the information that they want, there should not be any shortages due to bandwidth or power. Or in the military application, when hundreds of soldiers from different positions in the battlefield want to send critical information to their commander, the quality of service should not be different from when only one soldier interacts with the commander. Having multiple users and assigning bandwidth, energy, and load between these users would be a very hard task to do. In the medical scenario, the delay caused by dividing the available bandwidth between many users could be fatal for one patient, whereas in the battlefield scenario, the delay could result in low performance by an order of magnitude and losing many types of ammunition.

The rate of penetration of pervasive computing into an infrastructure will depend on many nontechnical factors like economics and business models. When users get used to some level of smartness in one environment, the lack of smartness in another environment would be very visible and unwanted. A service providing critical information to a soldier should take into consideration the fact that the soldier may be with a laptop at his or her base, and a PDA on the battlefield; or, if they are under fire, then the soldier could be unable to even open the PDA, and information and commands must be given through a headset. So the information is the same and the receiver is the same, but delivery is based on the type of infrastructure that the receiver is using. Also, the level of security must be the same whether he or she uses a PDA or laptop.

The tasks of pervasive computing range from simple tasks, like switching on lights in a conference room or checking e-mail messages, to more complex ones, such as buying and selling stocks or booking airline tickets. Pervasive computing involves the interaction, cooperation, and coordination of several transparent computing devices. As computing is becoming more pervasive, "smart homes" and "smart offices" will become more common in future. In a smart space, having a central authority for a single building or even a group of rooms is infeasible because every possible access right will have to be specified for every user. Simple authentication and access control are only effective if the system knows in advance which users are going to access where, and what their access rights are. Traditional security systems often use an access matrix to model control policies. Each column of an access matrix represents a protected object and each row corresponds to a principal who wants to access that object. Each matrix entry defines access rights a potential principal has on that object. This model is applicable to only static access-control policies (Michalakis, 2002). Portable, handheld, and embedded devices have severely limited processing power, memory capacities, software support, and bandwidth characteristics. Also, security information in different domains is subject to inconsistent interpretations in such open distributed environments (Zhang, 2002).

One way to satisfy the requirements of pervasive computing is adding distributed trust to the security infrastructure (Cedilnik, Kagal, Perich, Undercoffer, & Joshi, 2002). This approach needs the (a) articulating of policies for user authentication and access control, (b) assigning of security credentials to individuals, (c) allowing of entities to modify access rights of other entities, and (d) providing of access control. Role-based control is probably one of the best-known methods for access control, where entities are assigned roles and there are rights associated with each role. This is difficult for systems where it is not possible to assign roles to all users, and foreign users are common. Trust models are very powerful but they do not meet requirements for trust management. Security systems should not only authenticate users, but also allow them to delegate their rights (Feamster, Balazinska, Harfst, Balakrishnan, & Karger, 2002).

Agents are authorized to access a certain service if they have the required credentials. In Bharghavan (1994), additional ontologies are used that include not just role hierarchies, but any properties expressed in a semantic language. For example, if an agent in a meeting room is using the projector, it is possible that the agent is a presenter and must be allowed to use the computer, too. So they assign roles dynamically without making new roles. They extend the Agent Management System (AMS) and the directory facilitator (DF) to manage security of the platform. The AMS, DF, and agents follow some security policies to give access rights. They use a handshaking protocol between the AMS and agents to verify both parties. Once an agent signs a request, it will be accountable for it. When an unknown agent tries to register with a platform, the platform checks the agent's credential and makes a decision.

In multiagent systems, as they are decentralized, it is impossible to have a central database of access rights. So here, the policy is being checked at two levels: controlling access to the AMS and DF, and specifying who can access the service. Achieving security can be classified into two levels: platforms or agents. In platform security, the

AMS and DF decide to allow an agent to register, search, or use other functions. An agent can send security information to the AMS and specifies its category as private, secure, or open. When an agent wants to register with the AMS, it makes a request and sends a digital identity certificate. If the request is valid, the AMS decides the access rights. If the agent does not have the right to register with the AMS, the AMS starts the handshaking protocol.

Security processing in the nodes would follow the following sequence of actions: (a) receive the packet, (b) verify hop-by-hop integrity, (c) assign packets to existing domains, (d) extract the credential list, (e) check the authenticity credential according to the authentication policy for the domain, (f) check the credentials against the access control policy for the domain, and (g) deliver the entire packet to the domain, including the credentials, authentication protection fields, and so forth.

Security processing in the execution environment would include (a) receiving a packet including credentials, (b) creating a subdomain providing the security context parameters to the domain, (c) modifying the access control policy of the domain, and (d) adding or removing cryptographic protections to user data.

The existing solutions can be used hop by hop in the path, but that provides little in the way of end-source authentication. When hop-by-hop protections do not provide sufficient end-to-end authentication of the principal associated with a packet, we can employ end-to-end protections. However, the use of end-to-end cryptographic techniques is also a challenge in active networks. Symmetric techniques could be installed at each node of the packet's path through the network.

The packet modifications at each node could be protected anew with the shared key. However, this has a similar trust drawback as using hop-by-hop protection: Every node on the path must be implicitly trusted. Also, the assurance of the authenticity of the principal, derived from the shared key, is diluted if the key is not unique to the principal and the path.

Asymmetric techniques can operate in a datagram model, but have difficulty protecting packets that change. Signing a packet with a digital signature provides a cryptographic association from the signer to every potential verifier of the future. Therefore, authentication by digital signature is suited for a datagram model of communication, where the packet may decide en route what nodes it will visit (Kagal, Finin, & Joshi, 2001).

This section presented security in pervasive computing. Our goals are describing the concept of security and showing that by incorporating security in a pervasive computing environment, we may lose some of the flexibilities of the network. In order to have security, we need to have authentication, data privacy, location privacy, and so forth. Let us consider the military application as a pervasive computing environment. Soldiers send up-to-date information to their commanders and in return, they receive up-to-date strategies and war tactics. The way a soldier receives information is very much dependent on his or her location. From the pervasive viewpoint, it is a masking of uneven conditioning, but from a security viewpoint, it is desirable to not compromise this location. Or, when a commander sends some confidential information to a soldier's PDA, is it invisibility of pervasive computing or is it a violation of data privacy? If a soldier dies, the PDA should not get more information, but how often should the commanders check out each soldier? Each check requires bandwidth and energy, and we need to have localized scalability, which requires enough resources for everyone and the ability to service everyone with the same quality as servicing only one.

Middleware of Seamless Links in Pervasive Computing

A novel middleware concept, called *community computing,* to provide seamless links among devices, software entities, and applications has been recently developed (Kumar, Shirazi, Das, Sung, Levine, & Singhal, 2003). It also proposes the Pervasive Information Community Organization (PICO) as a framework for creating mission-oriented, dynamic communities of autonomous software entities that perform tasks for users and devices. We provide an overview of the PICO framework. The main components of community computing are *camileuns*, *delegents*, and communities. A camileun represents an adaptable smart device, for example, a PDA, sensor, camera, metal detector, and so forth. Delegents perform goal-oriented tasks on behalf of camileuns, databases, or humans; for example, a delegent associated with a surveillance camera will continuously analyze, record, and transmit captured images. Community computing is the framework for collaboration and coordination among delegents to carry out application-specific services. For example, a delegent capturing the image of a suspected terrorist in a public place may form a community with delegents associated with the FBI and INS databases to exchange information and determine the next course of actions. A community consists of one or more delegents that share a common goal or mission. Delegents within a community closely interact and collaborate with each other in order to achieve the community's goal. Let us brightly illustrate PICO's applicability through an example. Suppose a car accident victim needs immediate attention by medical and other personnel who are in geographically distributed locations (Kumar et al., 2003). PICO dynamically creates mission-oriented communities of delegents representing doctors, hospitals, and ambulances. Then, it sets up effective collaboration to save the victim. Camileuns around the victim, such as a street camera, a cellular phone, and pocket personal computers (PCs) exchange sensory data. Delegents representing these devices create a community and contact an ambulance service. The ambulance community interfaces with the hospital to perform the required collaborative tasks. Pervasive computing challenges addressed by the community computing concept include (a) handling dynamically changing information, (b) adapting to changing situations, and (c) providing scalability in terms of number of users, devices, resources, and data sizes. Services and information available on and exchanged between delegents are vulnerable to security and privacy breaches that standard security approaches are not prepared to deal with. Mobile code executing on untrusted hosts is subject to corruption. It is important to develop such tools as security mediators and virtual firewalls to secure information in the PICO environment.

Incorporating Vigil System

Vigil is an open project based on role-based access control with trust management (Kagal, Undercoffer, Joshi, & Finin, 2002). It has five components: (a) service broker, (b) client, (c) certificate controller, (d) role-assignment manager, and (e) security agent. The Vigil system is divided into smart spaces and a broker controls each one. The broker finds a matching service for the user. All clients must register with the broker. Each client is only concerned with the trust relationship with its broker. Service brokers establish trust relationships with each other. A trust between clients is transitive through brokers. When a client registers with a broker, it transmits its digital certificate, a list of roles which can access it, and a flag indicating if the broker should publish a client's presence to other clients. After receiving the registration, the broker verifies the client's certificate in order to send the digital certificate. The certificate controller generates digital certificates. To get a certificate, an entity sends a request to the certificate controller, which responds with its own signed certificate. The role-assignment manager maintains the access control list for entities. This list is different for every space. An entity can have more than one role. When the role-assignment manager gets a query for a role, it compares the current time stamp on the capability file with the time stamp of the latest access time. If they are not the same, it rereads the access list file. A security agent manages trust in the space by enforcing the global policy and a local policy. Access rights can be delegated; they must be from an authorized entity to another entity. When a user needs to access a service that it does not have the right to use, it asks another user who has that right. If the requested entity has permission to delegate the access, it sends a message signed by its own private key with its certificate to the security agent. The security agent verifies that the delegator has the right to delegate, then sets a short period of validity for that permission. After this period expires, the delegation must be reprocessed; so if one entity in the chain loses the permission, it will propagate to the chain in a fast manner.

Now let us see how we can incorporate the existing security methods into two pervasive computing scenarios considered in the introduction section. As Figure 5 illustrates, we can have different smart spaces. Each client must register through a broker, which gets the digital certificates from the certificate controller, and has access to the list of roles provided by the role-assignment manager. Then the broker issues a certificate to the client. The broker also has access to a profile database. Each different client has a different profile based on his or her previous interaction with the environment. In the case of the presenter (Fred), using PowerPoint for providing slides or the time that it takes to walk from the office to the meeting room makes a profile for the user. In the case of the patient, the time of the day that he usually sends his information, or the fact that he usually has high blood pressure on weekends, are some of the information that we can consider as his profile. We can use the distributed trust concept as in Kagal, Undercoffer, Joshi, et al. (2002). So, entities can modify access rights of other entities by delegating their own rights. For example, Fred can delegate the right to his secretary to upload his files from his desktop to his PDA for a specific day. Also, it is possible to have a chain of delegations and then revoke them accordingly. It is worth mentioning that when Fred gives a right to the secretary, the secretary is able to access services without the system creating a new identity for it.

In the car accident framework described under the PICO framework, we have a certificate controller which issues digital certificates. Each client (hospital, medical staff, etc.) must register with the broker and get certified. The broker has a profile list of the patient's medical abnormalities. The ambulance driver will use this certification, for example, to show his or her true identification to the system. The broker has access to these digital certificates and to the list of roles, as it must decide which hospital is most suitable, and notifies a medical specialist, accordingly.

In the military application scenario, the smart space is the battlefield covered with 1,000s of sensors. Some sensors are even attached to critical ammunitions and tanks. The certificate controller issues digital certificates for commanders and their soldiers. Each client has a list of valid roles, and a profile of each person is accessible to the broker, which can be miles away from the battlefield. All the communication between broker and support terminals must be very secure. When sensor nodes collect information about the approach path that the enemy is taking, leaders will notify their commanders and unreachable soldiers on their PDAs about the series of actions that they should take.

Figure 5. A view of the Vigil system

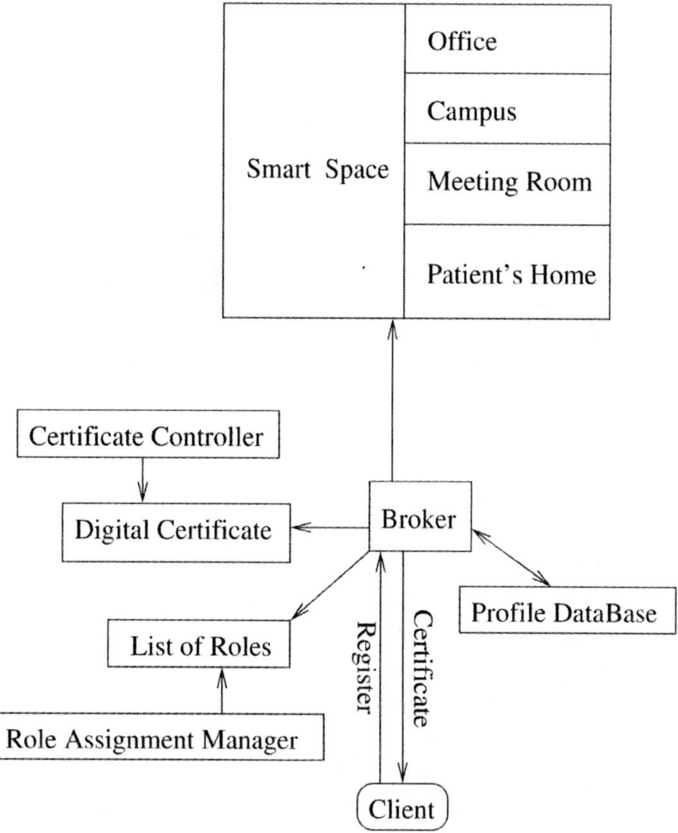

Conclusion

In this chapter we have described our vision of security in pervasive computing. To secure a pervasive computing environment, we need to consider security in each of the related areas such as distributed systems, mobile computing, active networks, and pervasive computing. Active networks challenge traditional thinking because computations performed within the network can be dynamically varied. Mobility and wireless communications introduce a number of problems with respect to security. We discussed that by incorporating the Vigil system, we can improve the level of security in a pervasive computing environment. Pervasive computing will be a source of challenging research problems in computer systems for many years in future. Privacy and trust are likely to be enduring problems in pervasive computing. Still, many questions need to be addressed, like what the best authentication technique is for pervasive computing, or how to express security constraints.

Acknowledgment

This work is supported by the NSF ITR grant under award number IIS-0326505.

References

Bacon, J. (2002). Toward pervasive computing. *IEEE Pervasive Computing*, *1*(2), 84.

Bharghavan, V. (1994). Secure wireless LANs. *Proceedings of the Second ACM Conference on Computers and Communications Security*, 10-17.

Campbell, R., Qian, T., Liao, W., & Liu, Z. (1996). *Active capability: A unified security model for supporting mobile, dynamic and application specific delegation.* Security White Paper. University of Illinois at Urbana-Champaign, System Software Research Group, Department of Computer Science.

Cedilnik, C., Kagal, L., Perich, F., Undercoffer, J., & Joshi, A. (2002). A secure infrastructure for service discovery and access in pervasive computing. *ACM MONET, Special Issues on Security in Mobile Computing Environments, 8*(2), 113-125.

Debar, H., Dacier, M., & Wespi, A. (1999). Towards a taxonomy of intrusion detection systems. *Computer Networks*, *31*, 805-822.

Feamster, N., Balazinska, M., Harfst, G., Balakrishnan, H., & Karger, D. (2002). Infranet: Circumventing Web censorship and surveillance. *Proceedings of the 11th USENIX Security Symposium*, (pp. 247-262).

Geng, X., Huang, Y., & Whiston, A. (2002). Defending wireless infrastructure against the challenge of DdoS attacks. *Mobile Networks and Applications*, *7*, 213-223.

Javasun. (n.d.). Retrieved May 11, 2004, from http://Java.sun.com/products

Kagal, L., Finin, T., & Joshi, A. (2001). Trust-based security in pervasive computing environments. *IEEE Computer, 34*(12), 154-157.

Kagal, L., Undercoffer, J., Joshi, A., & Finin, T. (2002). *Vigil: Enforcing security in ubiquitous environments.* Retrieved May 11, 2004, from http://www.csee.umbc.edu/~lkagal1/papers/vigil.pdf

Kagal, L., Undercoffer, J., Perich, F., Joshi, A., & Finin, T. (2002). A security architecture based on trust management for pervasive computing systems. *Proceedings of the Conference, Grace Hopper Celebration of Women in Computing.*

Kumar, M., Shirazi, B., Das, S. K., Sung, B., Levine, D., & Singhal, M. (2003). PICO: A middleware framework for pervasive computing. *IEEE Pervasive Computing, 2*(3), 72-79.

Michalakis, N. (2002). PAC: *Location aware access control for pervasive computing environments.* Retrieved May 11, 2004, from http://sow.csail.mit.edu/2002/proceedings/michalakis.pdf

Misra, A., Das, S., McAuley, A., & Das, S. K. (2001). Autoconfiguration, registration, and mobility management for pervasive computing. *IEEE Personal Communications, 8*(4), 24-31.

Pangalos, M. (1997). Security issues in a mobile computing paradigm. *Communications and Multimedia Security, 3*, 60-76.

Salo, T. (2001). *Security in pervasive computing.* Helsinki University of Technology. Retrieved May 11, 2004, from http://www.tml.hut.fi/Studies/Tik-111.590/001s/papers/tomi_salo.pdf

Satayanarayanan, M. (2000). Caching trust rather than content. *Operating System Review, 34*(4), 245-246.

Satayanarayanan, M. (2001). Pervasive computing: Vision and challenges. *IEEE Pervasive Computing, 1*(2), 10-17.

Satayanarayanan, M. (2002). Integrated pervasive computing environments. *IEEE Pervasive Computing, 1*(2), 2-3.

Satayarayananan, M., Kistler, J., Mummert, L., Ebling, M., Kumar, P., & Lu, Q. (1993). Experiences with disconnected operation in a mobile environment. *Proceedings of the USENIX Symposium on Mobile and Location Independent Computing*, (pp. 11-28).

Schwartz, B., Jackson, A., Strayer, W., Zhou, W., Rockwell, R., & Partridge, C. (2000). Smart packets: Applying active networks to network management. *ACM Transactions on Computer Systems, 18*(1), 67-88.

Stanford, V. (2002). Pervasive health care applications face tough security challenges. *IEEE Pervasive Computing, 1*(2), 8-12.

Tennenhouse, D. L., & Wetherall, D. J. (1996). Towards an active network architecture. *Multimedia Computing and Networking, 26*(2), 5-17.

Tock, T., & Sturman, D. (1994). *Security, delegation and extensibility* (Tech. Rep. No.). University of Illinois at Urbana-Champaign.

Weiser, M. (1991). The computer for twenty-first century. *Scientific American*, 94-100.

Zhang, T. (2002). *Security architectures for pervasive computing environments*. Retrieved May 11, 2004, from http://www.cc.gatech.edu/people/home/zhangtao/mini2-report.pdf

Zhang, Y., Lee, W., & Huang, Y. (2002). Intrusion detection techniques for mobile wireless networks. *Mobile Networks and Applications*, 1-16.

Chapter XVI

Transaction Processing in Broadcast Databases

Wai Gen Yee, Illinois Institute of Technology, USA

Abstract

Broadcasting is a popular way of disseminating data due to the scalability of its request performance with an increasing population, and its ability to "match" the unique characteristics of modern wireless communications. The necessary decoupling of the clients from the server in the broadcast architecture, however, complicates database-style transaction processing. In this chapter, we describe why broadcast transaction processing is complicated and offer solutions.

Introduction

A broadcast database consists of a single server that cyclically broadcasts a sequence of data items (the *broadcast program*) from the database to the client population. To satisfy a user request, the client scans the *broadcast channel* for the requested item (see Figure 1).

Figure 1. In the broadcast environment, the server cyclically broadcasts the program consisting of items A, B, and C (clients download desired items from the broadcast channel as they arrive)

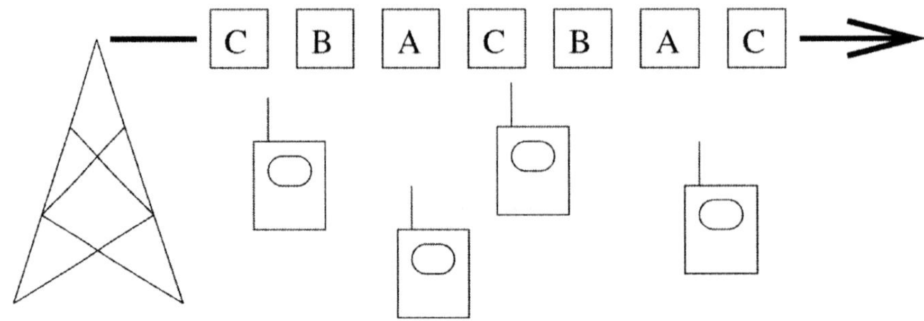

The main advantage of such databases is request performance that is scalable with the client population. The server achieves scalability because the single broadcast of an item potentially satisfies all outstanding requests for it. Broadcast transmission, however, complicates basic database functionality, such as *transaction processing*. The architecture of broadcast databases precludes conventional transaction-processing techniques. The goal of this chapter is to discuss what complications arise in broadcast transaction processing, and how they can be handled.

Motivation for Broadcast Data Transmission

Conventional database management systems (DBMSs) are unscalable with client populations because they serve each data request individually. For example, during the Clinton impeachment hearings, Web sites such as http://abcnews.com and http://cnn.com experienced severe performance degradation due to heavy request volume (Goodin, 1999). In fact, given a high client-request rate, the degradation of server performance is greater than the proportionate increase in the client-request rate (e.g., doubling the request rate more than doubles the time it takes the server to respond to a request). The single-server queuing theory predicts this phenomenon (Ross, 1997). In response to such phenomena, researchers have designed Web servers that push content to clients (Almeroth, Ammar, & Fei, 1998). These Web servers apply the same general principal as do broadcast databases: increase scalability by simultaneously satisfying multiple data requests.

Broadcast databases are also increasing in relevance due to technological trends. Because client populations are exploding with the spread of mobile and pervasive computing devices (e.g., cell phones, personal digital assistants [PDAs], and laptops), scalability of data services is of prime concern. Furthermore, in order to communicate, these devices are equipped with radio transceivers. Because battery life is at a premium in mobile devices, these transceivers are typically low-powered and more suited to receiving than transmitting data. Under such conditions, broadcast is the natural choice for communicating with clients.

The Challenge of Transaction Processing

Traditional DBMSs guarantee that clients always access a current and consistent set of data that reflects the effects of all successful transactions. These guarantees are made by regulating each transaction's data access, and by keeping a log of the updates made to the database state.

Such control is unavailable in broadcast databases. In fact, broadcast databases achieve their main benefit (i.e., scalability) because multiple clients can simultaneously and independently read a single data item. We highlight a problem caused by this unregulated data access in the following example.

Example: DBMSs guarantee that client transactions read a *consistent* (i.e., correct) set of data. Assume that at the beginning of each broadcast cycle, the server collects the prices of three stocks, *IBM*, *INT* and *MSF*, and then broadcasts them. The prices contained in each program are current and consistent with respect to the beginning of the cycle. The server considers these prices consistent for the duration of the cycle, regardless of whether the true prices might have changed.

Assume that a client has defined a transaction called GETPORTFOLIOVALUE that reads the program and returns the value of the stocks in his portfolio. This transaction has to be careful about how it performs reads. An unregulated read may yield an inconsistent view of the data. We give an example of this phenomenon for a stock portfolio containing *IBM* and *MSF* in Figure 2. In this case, the client reads a price for *MSF* that is current as of the beginning of cycle *i*, and a price for *IBM* that is current as of the beginning of cycle *i*+1, yielding a portfolio value of $24. Consistent reads of the program yield portfolio values of $27 and $25 in cycles *i* and *i*+1, respectively, but never $24. In fact, it is impossible to prove that the prices of *IBM* and *MSF* ever summed to $24 in the given time frame. The client's read, therefore, is considered inconsistent.

Figure 2. Consistent and inconsistent reads for the GETPORTFOLIOVALUE transaction

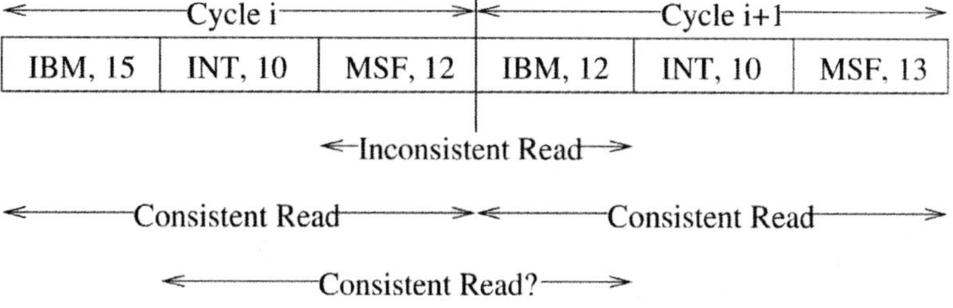

The cause of this inconsistency is an update to the stock prices that first appear in cycle $i+1$. Unfortunately, GETPORTFOLIOVALUE reads data that reflects only part of the update's effects (a price drop of \$2 for *IBM*). Conventional DBMSs avoid this anomaly by allowing update transactions to lock out all other transactions that access common data until the update is complete. As the broadcast channel is shared, however, the server cannot implement such a mechanism.

The simple solution of restricting a transaction's read operations to a single cycle does not work either. Such a restriction may be impossible to enforce and is overly strict. A transaction may read items in subsequent cycles based on values of the data in the current cycle. For example, a stock trader may pick the next stock price to read based on the current price of *IBM*, and the desired price may be broadcast in the next cycle. This restriction may also hurt performance. Assume that the trader's portfolio includes *IBM* and *INT* instead of *IBM* and *MSF*. Calling GETPORTFOLIOVALUE during cycle $i+1$ yields a value of \$22. However, reading *INT* in cycle i and *IBM* in cycle $i+1$ also yields a value of \$22 because *INT*'s price does not change between cycles. It is therefore possible that this transaction read a consistent view of the data even though the items were from different cycles. The benefit, in this case, is that a transaction that can read data items as soon as they are broadcast generally completes more quickly than those that restrict reads to a single cycle.

We have just given a rough example of the complexity of transaction processing in the broadcast environment. More precisely, broadcast databases must ensure that transactions have the *ACID properties* of *atomicity*, *consistency*, *isolation*, and *durability* (to be explained later in "Transaction Processing Primer"). As stated above, the goal of this chapter is to describe the new structures and protocols necessary to accomplish this.

Organization of the Chapter

In the next section ("Related Work"), we will put our work in perspective by mentioning some work related to broadcast databases. We will then define our model and state our problem in "Broadcast Model." In "Transaction Processing Primer," we will briefly review transaction-processing concepts. The transaction-processing review only includes those concepts that are relevant to this chapter. In "Issues with Transaction Processing in Broadcast Databases," we will describe a natural architecture for broadcast transaction processing. We will spend "Concurrency Control," "Recovery," and "Transaction Design" discussing concurrency control, recovery, and transaction design, respectively, in broadcast databases. In conclusion, we discuss outstanding work and make concluding remarks.

Related Work

One of the major shortcomings of data access via broadcast is its best-case performance. Items from the database are typically broadcast *flatly*, one after the other, in sequence.

That is, if there are N items in the program and each one requires a time unit to transmit, then the expected number of time units required to satisfy a data request is $N/2$, regardless of the request load. In contrast, in a lightly loaded *pull-based* system, the server can respond to a request almost instantaneously, regardless of database size.

If, however, the server has detailed information on data access patterns, it could tailor the program to reflect the particular needs of the client population. In particular, the server can transmit more popular data more frequently (Aksoy & Franklin, 1999; Bar-Noy, Dreizin, & Patt-Shamir, 2002; Vaidya & Hameed, 1999; Yee, Navathe, Omiecinski, & Jermaine, 2002). The algorithms described in the citations differ in their details and applications, but all yield near-optimal performance as defined in Ammar & Wong (1985), which is as much as 40% better than that of a flat schedule.

Researchers have also considered the problem of indexing programs. Indexes yield many advantages. First, they reduce a performance metric called *tuning time*. Tuning time is the amount of time a client must actively scan a broadcast channel in search of data, and has been linked to the amount of battery energy requests consume. By conveying when an item will be broadcast, the client can selectively tune in to the channel instead of naively scanning it. While not scanning, the client can conserve energy by "dozing." Indexes also allow clients to make other optimizations. For example, clients can decide which items to cache depending on when they are scheduled (Acharya, Alonso, Franklin, & Zdonik, 1995). Imielinski, Vishwanath, and Badrinath (1997) describe a way of allocating B+-tree buckets to a program. Datta, Celik, Kim, VanderMeer, and Kumar (1997) consider ways of indexing and broadcasting data sets dynamically determined by user interests. Yee and Navathe (2003) generalize the scheduling and indexing work to the case where data are broadcast on multiple channels. Although these works improve the performance of requests for individual items, they do not consider more complicated transaction processing.

Much other work has been done in the area of broadcast databases, but we limit our discussion of related work to scheduling and indexing. Our goal is only to give the reader a sense of other directions researchers have taken in order to improve the usability of broadcast databases. Furthermore, with some engineering, these works can be incorporated into broadcast transaction-processing systems.

Note that we are *not* covering an orthogonal topic known as *mobile transactions* (Chrysanthis, 1993; Dunham, Helal, & Balakrishnan, 1997). Mobile transactions deal with transactions performed by moving clients. The issues dealt with in mobile transactions include location dependence of data, request routing, resource discovery, and data replication over a network. Our work deals with transaction processing in an environment containing transmission-constrained clients in a given geographical area with a single server.

Broadcast Model

The database D consists of N_D items d_i, $1 \leq i \leq N_D$, each of which is identified by a key attribute. B, an N-sized subset of D, is the *broadcast set*. The broadcast set is organized into a broadcast program, which is broadcast on a channel. The span of time during which a program is broadcast is called a *cycle*, each of which is consecutively numbered. In this chapter, we assume that the program's organization is flat: Each item is broadcast once in a cycle.

The broadcast set and program may vary from cycle to cycle. The values of each item in a program are defined to be mutually consistent; that is, their collective values do not violate any logical application-level constraints and are assumed to be current as of the beginning of the cycle.

Each item is assumed to be unit-sized and fits into a unit of transmission called a *bucket*. The amount of time required to transmit a bucket is defined abstractly as a *tick*. Because the program is flat, the maximum number of ticks a client must wait for an item in the broadcast set is a function of the broadcast set's size.

In order to download items, the user supplies the client with the corresponding set of keys. The client then scans the broadcast channel until the bucket containing the key is broadcast. When the client finds the key, it downloads the contents of the corresponding bucket.

In this chapter, we consider transaction processing on the subset of the database contained in the broadcast program, not on the database as a whole. A query such as, "Select… from database," from a broadcast client will therefore be correct in returning the broadcast set *B*, not the database *D*.

In the example shown in Figure 2 ("Introduction"), *D* is a set of stock market symbols and their prices. The broadcast set *B* consists of $N = 3$ items: *IBM*, *INT*, and *MSF*. The program is organized in alphabetical order by symbol. Because a program contains three items, it takes three ticks to broadcast. Although the broadcast sets are the same in cycles i and $i + 1$, the programs are different because the prices are different. The user running GETPORTFOLIOVALUE tells the client to download items with the keys *MSF* and *INT*.

As described above, in the canonical broadcast environment, clients wirelessly communicate with the server. (The main concepts and arguments in this chapter also apply to the Datacycle and the Information Bus architectures, both of which, instead, use high-speed busses as their shared communications medium; Bowen, Gopal, Herman, Hickey, Lee, Mansfield, et al., 1992; Oki, Pflugel, Siegel, & Skeen, 1993) In this environment, clients are portable, battery-powered devices and communication with the server is kept at a minimum. There are three reasons for this:

1. **Performance.** Encoding and error-correcting delays, as well as bandwidth limitations, increase the communication latency in the wireless environment.
2. **Scalability.** The server cannot manage an unbounded number of client requests, especially using a high-latency medium.
3. **Energy conservation.** Wireless communications sap a client's battery power.

Furthermore, all communications are mediated by the server, and clients do not communicate directly with each other. In other words, the clients and server form a *star topology*, with the server at the center. This simple architectural model has some practical advantages. The server is assumed to be more reliable and available than clients and therefore constitutes a better repository for shared data. Furthermore, because data are centralized, the task of maintaining consistent, authoritative versions of shared data is simpler.

Transaction-Processing Primer

In this section, we give a brief overview of transaction processing. We offer this review in order to describe some concepts that will help motivate the rest of the chapter. Note that we are only considering the subset of transaction-processing concepts relevant to this chapter, leaving out many details. More comprehensive reviews on transaction processing are available in standard database texts, such as Elmasri and Navathe (2004). Readers already knowledgeable in transaction processing can skip this section without risk of confusion while reading later sections.

This section is divided into two parts. The first part ("Transaction-Processing Concepts") describes transaction-processing concepts. The second part ("Transaction Processing in Traditional Databases") explains how they are implemented in conventional DBMSs.

Transaction-Processing Concepts

A transaction is a set of logic and operations on a database that is supposed to perform some useful, well-defined task (e.g., compute the value of one's stock portfolio, as described in the introduction). In this section, the relevant operations of a transaction are its *read* and *write* operations since these directly affect the state of the database and the execution of the transaction. We will use the notations $r(d)$ and $w(d)$ to refer to the operations of reading and writing item d, respectively. For the sake of brevity, we generally exclude the value written to d by $w(d)$.

The goal of transaction processing is to ensure that transactions run correctly and efficiently by enforcing on them the properties of atomicity, consistency, isolation, and durability, collectively known as the ACID properties. We will now discuss each of these properties in turn.

Atomicity refers to the effect of a prematurely terminated transaction on the database. Premature termination occurs either because the transaction cannot find the necessary data in the database, or the system crashes. Such transaction effects should be undoneso that it appears to other transactions as if it never existed. The canonical example used to demonstrate the usefulness of atomicity is the BANKTRANSFER transaction. In this transaction, money is first withdrawn from one account, and then deposited into another. If a nonatomic transaction fails after debiting $100 from my savings, but before making the corresponding deposit into my checking account, then, according to the database, the bank customer just lost $100. An atomic transaction would rectify this problem by undoing the initial debit.

Consistency refers to the effect a transaction has on the correctness of the database. A database is consistent (i.e., correct) if all the data it contains satisfy high-level constraints defined by the database designer. For example, in a stock price database, no price should be negative. A transaction is consistent if its successful completion on an initially consistent database leaves it in a consistent state. Transaction designers are responsible for each transaction's consistency.

Isolation describes the perspective that each transaction has on the database. From the transaction's perspective, it has exclusive access to the database even though other transactions may be executed concurrently. The GETPORTFOLIOVALUE transaction described in "Introduction" did not have isolation because it was able to read the database in the middle of an update.

Durability is a property of *committed* transactions. Once the transaction successfully commits, the DBMS must ensure that its effects persist in the database, even in the event of system failure. Upon restarting the database after a failure, the effects of committed transactions are either already in the database, or can be reconstructed in a process known as *recovery* that uses a separate database *log*. Referring again to the BANKTRANSFER example, bank customers who are told by the database that their transfers succeeded do not have to redo them due to the transaction's property of durability.

Transaction Processing in Traditional Databases

We now consider how the ACID properties are ensured in traditional databases. The goal of this section is to motivate the explanation as to why traditional techniques that ensure the ACID properties are inapplicable in the broadcast environment. The discussion will also introduce concepts that will reoccur in our discussion of broadcast transaction processing.

Atomicity and Durability

Atomicity and durability are implemented using the *recovery subsystem* of the DBMS. The recovery subsystem restores the database's consistency in the event of a transaction or system failure. To preserve atomicity, a transaction that prematurely terminates must have its effects on the database undone. To preserve durability, a transaction that commits must have its effects appear subsequently in the database, even in the event of a system failure.

The recovery subsystem achieves these two goals by logging the intermediate effects of a transaction on the database as the transaction executes. The log contains records of all the updates made to the database by each transaction. A log record may either contain the value of an item right *before* it was updated (the item's *preimage*), the value of an item right *after* it was updated (the item's *postimage*), or both. While executing a transaction, therefore, the DBMS must maintain both the actual database, as well as the log. The log acts as a stable backup storage area that can be referred to in the event of a failure.

The exact way in which the database and log are maintained and used depends on the recovery protocol used. In one recovery protocol, the log records consist solely of postimages. In this case, the database must not be updated by a transaction until it commits. In addition, all postimages and a commit record must be written to the log at the time of the commit. If the transaction fails (does not commit), the recovery subsystem does nothing because the database has not been updated by the transaction. If the system fails after the transaction commits, then the recovery subsystem *replays* the log by writing postimages contained in the committed transactions' log records to the database. This is called *redo* recovery.

A corresponding *undo* recovery protocol also exists. In this case, the recovery subsystem records preimages in the log as the transaction updates the database. All updates made by a transaction must be in the database by the time it commits. If the system subsequently fails, nothing must be done because the effects of committed transactions have already been written to the database. If the transaction prematurely terminates, then the recovery subsystem must undo the effects of the transaction by replaying the preimages.

Both of these recovery protocols place severe restrictions on when data must be written to disk. In redo recovery, for example, updates must be kept in memory until a transaction commits or fails. A more practical alternative, therefore, is the *undo-redo* protocol, which requires log records to contain both preimages and postimages. By the time a transaction is ready to commit, all preimages and postimages must be in the log. In the event of a failure, the recovery subsystem undoes all transactions in the log that have not committed, and then redoes all the rest.

Note that the log is meant to be used for temporarily storing changes to the database state made by transactions. It is not meant to be a backup database. Therefore, it makes sense to occasionally prune it of old records by a process called *checkpointing*. Checkpointing is a process that makes sure that all committed updates up to a certain point (the checkpoint) are reflected in the database. All log records corresponding to these updates can be deleted. The details of checkpointing are beyond the scope of this chapter.

Consistency

Consistency refers to the way in which a transaction is designed, and is related to the transaction's intended goal. For a BANKTRANSFER transaction to be consistent, for example, the total balance of the bank accounts involved in the transfer must remain unchanged. Intuitively, a transaction must perform what is expected of it, so that the state of the database conforms to the state of the real world. If it cannot perform its expected task (e.g., complete a bank transfer), it must automatically abort. As consistency is application dependent, we will spend little time on it in this chapter.

Isolation

One way to guarantee isolation is to force transactions to execute *serially*. That is, all the read and write operations of one transaction must complete before those of another begin. Such a schedule guarantees consistency, but generally results in poor resource utilization. Consider two transactions, one that intensively uses the disk (i.e., input-output, or I/O intensive), and another that intensively uses the computer's central processing unit (i.e., CPU intensive). I/O and the CPU are separate computing components that can function concurrently, so executing these transactions serially leaves one component (i.e., the CPU or disk drive, respectively) underutilized.

Executing both concurrently by interleaving their operations may result in better resource utilization and potentially better throughput. The constraint is that the resultant schedule must affect the database in a way that is logically equivalent to some serial schedule. We call such schedules *serializable*, and the DBMS subsystem that guarantees serializability is the *concurrency control subsystem.*

The most common type of serializability is called *conflict serializability*. In conflict serializability, pairs of operations conflict if at least one of their effects is dependent on which one is executed first. Two read operations never conflict because they can be arbitrarily ordered without altering their effects. For the same reason, two write operations *on different items* also never conflict. Two operations on the same item, however, conflict if one of the operations is a write operation. Consider the two operations $r(INT)$ and $w(INT)$. The order in which $r(INT)$ executes relative to $w(INT)$ affects the value read, and therefore, these two operations conflict. Now consider two writes to INT, $w_1(INT)$ and $w_2(INT)$. The order of these writes affects the state of the database left by the schedule, and therefore, these two operations conflict as well.

We now define the notion of *conflict equivalence* and conflict serializability. Two schedules are conflict equivalent if the relative orders of all conflicting operations are the same in both. A schedule is conflict serializable if it is conflict equivalent to a serial schedule. A conflict serializable schedule leaves the database in a consistent state.

Consider the read and write operations for the following banking transactions:

BANKTRANSFER: $r_t(acct1)w_t(acct1)r_t(acct2)w_t(acct2)$, and

BANKBALANCES: $r_b(acct1)r_b(acct2)$.

The first two schedules below are conflict equivalent to the serial schedules BANKBALANCES and BANKTRANSFER, and BANKTRANSFER and BANKBALANCES, respectively, because of the conflict between the reads and writes of common bank account information. The third schedule is not serializable. The conflict involving $r_b(acct1)$ implies that BANKBALANCES must precede BANKTRANSFER, but the conflict involving $r_b(acct2)$ implies the reverse. The result of this schedule is giving the bank customer an incorrectly high value of his accounts.

Schedule 1: $r_t(acct1)r_b(acct1)r_b(acct2)w_t(acct1)r_t(acct2)w_t(acct2)$

Schedule 2: $r_t(acct1)w_t(acct1)r_t(acct2)r_b(acct1)w_t(acct2)r_b(acct2)$

Schedule 3: $r_b(acct1)r_t(acct1)w_t(acct1)r_t(acct2)w_t(acct2)r_b(acct2)$

A well-known way to test if a schedule is conflict serializable is to use a *serialization graph* (also known as a *precedence* or *dependency* graph). A serialization graph is a directed graph where the nodes represent transactions, and the edges represent conflicts in the schedule. An edge is drawn from transaction T_1 to transaction T_2 if they have conflicting operations and one of the conflicting operations in T_1 precedes the corresponding operation in T_2 in the schedule. The schedule is conflict serializable if, and only if, this graph does not contain a cycle. Depending on how the graph is implemented, detecting cycles may take up to V^2 time, where V is the number of transactions. In Figure 3, we show the serialization graphs for Schedules 1 and 3.

Remediation of cycles in serialization graphs involves aborting one of the transactions involved in the cycle. There are heuristics for picking the victim transaction, which consider costs such as how old the potential victim is or how many other transactions aborting it would affect. These heuristics are beyond the scope of this chapter.

Implementing serialization-graph testing may be complicated. For example, it may be difficult for the DBMS to keep track of the exact order of interleaved operations because they might be managed at the system level. Also, maintaining the graph for heavily loaded databases may be computationally costly. Therefore, database designers have come up with ways of allowing the database to only perform serializable schedules. These are called concurrency control techniques.

Among the most recognized concurrency control techniques are *two-phase locking* (2PL), *time stamp*, *multiversion*, and *validation* (or *optimistic*). We will restrict our descriptions of each of these techniques to their most basic forms, mentioning only concepts relevant to this chapter. Further details can be found in Bernstein and Goodman (1981), and Elmasri and Navathe (2004).

2PL is the most commonly used concurrency control technique in commercial database systems. In 2PL, transactions can only access items for which they have the locks. A transaction can only obtain a lock to an item if no other transaction holds it. In 2PL, a transaction accesses locks in two phases: the growing phase and the shrinking phase. The growing phase precedes the shrinking phase. During the growing phase, the

Figure 3. Serialization graphs for Schedules 1 and 3 (the labels on the edges refer to the items involved in the conflict)

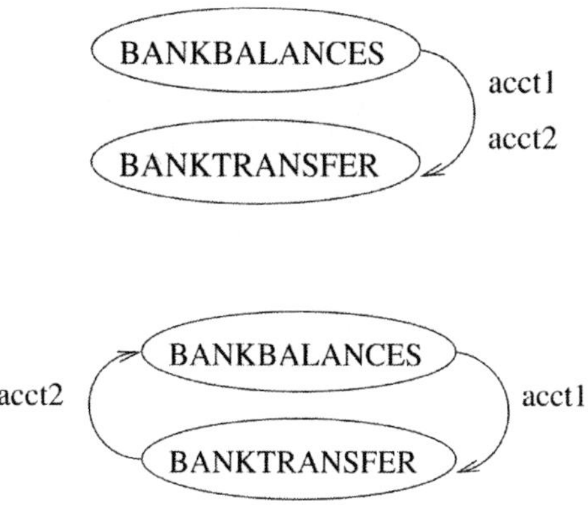

transaction can obtain locks to items but cannot release them. During the shrinking phase, the transaction can only release locks. By the end of the transaction, all locks are released. 2PL ensures conflict serializability of schedules.

In time-stamp-based protocols, each item is associated with two time stamps: a read time stamp and a write time stamp. In addition, a time stamp is associated with each transaction. The read and write time stamps of an item indicate the time stamps of the transactions that last successfully read or wrote them, respectively. These three time stamps govern legal access to each item. A transaction can read an item if the item is at least as old (does not have a higher write time stamp) as the transaction. In this case, the reading transaction must be serialized after the transaction that last wrote the item. If the transaction is older than the item, then it is trying to read an item that should not exist based on the state of the database at the time the transaction was initiated. In this case, the read operation fails.

A write operation can occur only if its transaction is younger than the last transactions that read *and* wrote the item (has a time stamp that is higher than the item's read and write time stamps). In this case, the writing transaction will be serialized after the transactions that last read and wrote the item. Violating either of these conditions means that the transaction is trying to create a database state that is older than what other successful transactions have already seen. In this case, the write operation fails. Note that if the read or write operations just described succeed and the transaction commits, then the respective time stamp is updated.

In multiversion protocols, the database maintains multiple versions of items, each of which has unique read and write time stamps. The time stamps indicate when the item was last successfully read and created, respectively. When multiple versions of items are available, a transaction reads the most recent version of each item that is older than itself. Transactions with updates also write to the most recent versions of items older than itself, but under the constraint that the read time stamp of the item is lower than the transaction's time stamp. (See the description of the time stamp protocol above for a justification of this heuristic.) If this is the case, then if the item's write time stamp equals that of the transaction, then the current value is overwritten. Otherwise, a new version of the item is created. The time stamps of a new version are equal to that of the transaction.

The validation or optimistic concurrency control (OCC) protocol is a transaction-processing architecture. With OCC, each transaction operates on local copies of data in three distinct phases. In the first phase, it acquires copies of items and executes writing updates locally. In the second phase, the client sends the partially committed transaction to a *certifier*, a server component, which checks if the transaction can be serialized in a process called validation. (Validation can use one of the concurrency control techniques described above.) In the final stage, the updates, if valid, are actually written to the database and the client is notified of the outcome.

Such a protocol works best when transactions either do not modify the database much, or modify disjoint parts of the database. In such cases, the increased transaction performance is not compromised by a high incidence of validation failures.

Issues with Transaction Processing in Broadcast Databases

The model given in "Broadcast Model" and the description of transaction processing given above in "Transaction Processing Primer" suggest the inherent difficulty of implementing transaction processing in the broadcast environment. Most significantly, broadcast clients access data as independent of the server as possible; this is precisely the source of a broadcast database's scalability. Unfortunately, conventional transaction-processing techniques typically require much interaction between client and server in order to ensure a degree of performance and robustness in execution. Consider 2PL: A transaction must lock items before they are read or written in order to exclude other transactions from concurrently accessing them. In a broadcast database, if one transaction can read an item, then all can, making locking impossible. Recovery protocols require that the log located at the server be regularly updated in the course of a transaction. As communications are limited in the broadcast environment, centralized log maintenance is generally impossible.

Implementing transaction atomicity, consistency, and durability is relatively straightforward; the authoritative database rests at the server, which can be implemented as a conventional DBMS. Supporting isolation with broadcast clients, however, introduces a new degree of complexity. Indeed, of the ACID properties, isolation is the most challenging to implement, as mentioned above. Consider the concurrency control protocols described above in the context of broadcast databases.

- **Locking-based techniques.** As mentioned, locks are not implementable in the broadcast environment.
- **Time-stamp-based techniques.** Time-stamp-based techniques require the ability of the clients to immediately update shared time stamps in order to work. Time stamps therefore suffer from the same problem as do locking-based techniques: Both require a high degree of interaction between client and server.
- **Multiversion techniques.** The server can broadcast multiple versions of data, increasing serializability as in conventional databases. There is a cost associated with using this technique that is unique to the broadcast environment, however. Maintaining multiple versions of data on disk consumes storage space, but does not necessarily hurt performance because disk access is random. Because broadcast access is linear, adding multiple versions of data to a program necessarily decreases the frequency of the most recent data.

None of these techniques can be directly applied to broadcast databases. We therefore spend most of this chapter studying concurrency control. We subsequently consider recovery and transaction design, but in less detail.

The Necessity of Optimistic Concurrency Control

Broadcast clients must naturally use OCC in executing transactions. As described in "Transaction Processing Primer," this protocol allows clients to execute transactions locally and send them to the server only for validation. The intended use of OCC, however, is to increase concurrency in an environment where transactions are known a priori to not overlap in their data requirements, where most transactions are read only, or where transactions occur at a low rate. In these cases, the probability of conflict is low, and the probability of successful validation is high. The motivation for using OCC in the broadcast environment is different; it is based primarily on its architectural restrictions, not on assumptions of clients' workloads.

The architecture of a broadcast database that uses OCC must then include a certifier to validate incoming transactions in order to ensure their serializability (see Figure 4). In addition, the broadcast program must be supplemented with information about the status of transactions that are pending validation.

The performance of broadcast databases complicates the implementation of OCC. OCC in a conventional DBMS has the advantage of low latency; data are randomly accessed and locks are not used. Broadcast clients generally must wait a while for data, depending on the organization of the program. The higher latency in the latter case increases the likelihood that some other transaction is updating the data, reducing the probability that another client's transaction will successfully validate. For example, consider a client that must wait some time for the arrival of an item (the current value). In that time, another client may have submitted a transaction that updates that item. If this update is validated, then the value ultimately read by the original client will be out of date. Consequently, subsequent updates based on this obsolete version of data are unlikely to be validated. On the other hand, in a conventional DBMS, updates are immediately reflected in the database, increasing the likelihood that reads are consistent and updates will be validated.

Another problem with traditional OCC is that *all* transactions are validated, including read-only transactions. This may be very costly for energy-limited clients because of the energy cost involved in transmitting a transaction. However, because serializability does not require the most current versions of data (consider multiversion concurrency

Figure 4. Architecture of a broadcast database that accepts updates using optimistic concurrency control (a certifier accepts only serializable updates from clients)

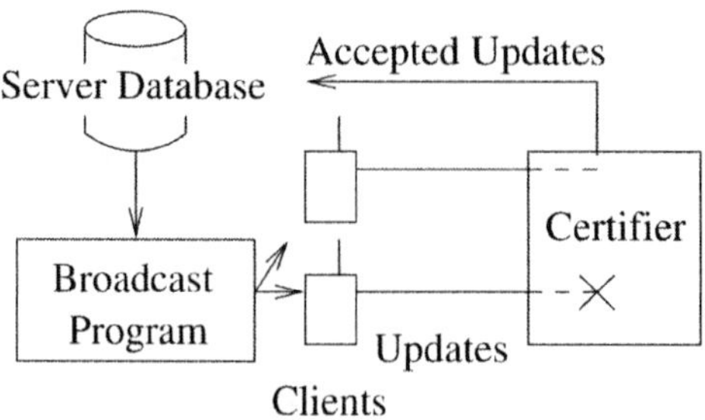

control), there is less of a need for the server to validate read-only transactions. Recall that a broadcast program is current as of the beginning of a cycle. Although it does not necessarily reflect the current state of the database, reads that are isolated to a single cycle are assumed to be consistent because the data within a program are consistent. Of course, some form of validation must still take place for reads that span multiple cycles, and we will consider this.

The challenge of implementing isolation in broadcast databases boils down to devising an OCC scheme that minimizes communication with the server, yet maximizes the possibility of successful validation. This scheme must consider the fact that data are linearly broadcast to clients and that these clients can access the broadcast medium at any time.

Concurrency Control

As mentioned above, a major challenge in transaction processing for broadcast databases is the implementation of concurrency control. Given the architectural constraints, OCC is the natural choice for handling transactions. Traditional OCC has limited applicability, though, because it still requires too much interaction with the server: All transactions must be validated by the certifier.

We will now consider ways of off-loading some of the work of the certifier to the clients. Of course, clients are already assumed to do some local validation (i.e., reads isolated to a single cycle are assumed to be consistent) and some validation must always be done at the server (e.g., updates). However, by off-loading validation:

1. the load on the server is reduced,
2. the amount of communication from the client to the server is reduced, and
3. the likelihood of success of transactions that must be validated at the server is increased.

In order to off-load work, the server must convey to the clients some information about the changes taking place at the server database. The simplest way of doing this is to add to a cycle a *certification report* containing the items that have changed since the last cycle. A transaction scans the broadcast channel for certification reports while active, and compares the items it reads with those in the certification report. If there is an overlap, then the transaction is aborted. The overlap implies that it must be serialized *before* the transaction that caused the update. A subsequent read by the active transaction, however, may force it to be serialized *after* the updating transaction. This causes unserializability in the schedule, resulting in the abort. Although such an interaction may be hard to detect with the given certification report, the transaction processor must be conservative and abort the transaction.

This type of certification, however, is sufficient to solve the problem we described in "Introduction" regarding the ambiguous reads of *IBM*'s and *INT*'s prices (which summed to $22). Although GETPORTFOLIOVALUE read data from two cycles, it seemed to have

Figure 5. Inserting certification information into a broadcast program (clients can use this information to validate reads that span multiple cycles)

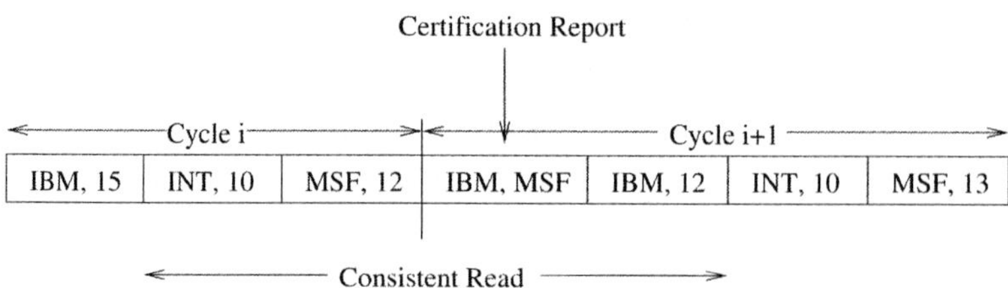

performed a consistent read. By inserting certification information into the program, we can tell whether or not this was truly the case. The example, shown in Figure 5, shows the case where the originally anomalous read should validate. The item *INT* is not in the certification report, meaning that its value did not change between cycles. According to the validation protocol described above, this execution of GETPORTFOLIOVALUE is valid. Specifically, because the value of *IBM* changed before the transaction read it, the transaction must be serialized after the update that occurred between the cycles.

The biggest limitation of the validation technique we just described is that it limits transactions to reading the most recent versions of data. Consider an alternative to the example shown in Figure 5 in which the certification report contains *INT* instead of *IBM* (and, consequently, the value of *IBM* does not change). GETPORTFOLIOVALUE would fail to validate because its read set is no longer current. However, if GETPORTFOLIOVALUE were allowed to continue to read *IBM*, inspection would reveal that it could be serialized before the update.

The most straightforward way of allowing clients to read more than the most recent data is to broadcast multiple versions of each item, including the cycles in which they were added to the broadcast, instead of a certification report. If data are updated during the lifetime of a transaction, then it is likely that the transaction can still read the most recent set of data older than itself, allowing it to be serialized immediately after the transactions that wrote the respective values.

The question is how many versions to support. Increasing the number of versions increases the likelihood of validation, but also increases the response time for each transaction. The answer lies in estimating the distribution of the lifetimes of transactions. Given this information, we can either be conservative, broadcasting *maxspan* versions (where *maxspan* is the maximum number of cycles a transaction can span), or we can be efficient, maximizing the average transaction rate by reducing the number of available versions.

In Pitoura and Chrysanthis (1999), the authors suggest broadcasting all the values of each item from the last *S* cycles, where *S* is a tunable parameter. In this case, transaction reads that span at most *S* cycles are guaranteed to be serializable, but transactions with longer spans fail. A variation of this is to broadcast the last *S* updates and their time stamps instead. Multiversion techniques in general, however, suffer from poor performance due to the high latencies they introduce.

The techniques described above require that read items be in the current broadcast. As described above (i.e., with the alternate GETPORTFOLIOVALUE transaction), this is often too strict of a requirement. Assume that T_i writes the values contained in cycle i, and T_{i+1} updates them for cycle i+1. We can then derive the serialization graph shown in Figure 6. The graph indicates that GETPORTFOLIOVALUE reads values (*IBM* and *INT*) from T_i, one of which (*INT*) is later overwritten by T_{i+1}. These conflicting operations result in the graph shown. Notice there are no cycles in the graph, implying that GETPORTFOLIOVALUE is serializable.

Pitoura and Chrysanthis (1999) suggest that each client executing a transaction maintain a serializability graph representing all the transactions committed on the server, as well as local read-only transactions. In order to do this, the certification report must include metadata on the most recent transactions and their conflicts with older transactions, as well as time stamps related to when data were updated. Local read-only transactions are valid as long as they do not form cycles in the local serialization graph.

A big limitation of Pitoura and Chrysanthis' technique is the client must constantly read certification reports in order to keep its graph up to date. Missing one, which is entirely possible in the wireless environment, invalidates the entire graph, forcing the transaction to abort. Another limitation is that the size of a certification report is a function of the update activity at the server. It can become very large depending on the workload. We now consider a validation technique that avoids these problems.

Shanmugasundaram, Nithrakashyap, Sivasankaran, and Ramamritham (1999) noticed an interesting aspect of using local serialization tests to validate read-only transactions, and took advantage of it in improving on the techniques described above. In particular, allowing clients to validate transactions based on only their local views allows global schedules (including all transactions occurring on all clients) that are not serializable.

This happens when read-only transactions induce different serial orders on a set of update transactions. Consider the following schedule based on server transactions S_1 and T_2 (based on the example in Shanmugasundaram et al., 1999):

$$w_{S1}(IBM)w_{S2}(INT). \quad (1)$$

Now consider the schedule with the following interleaved read operations from client transaction C_1:

Figure 6. Serialization graph of the alternate GETPORTFOLIOVALUE (Notice that there are no cycles in the graph.)

$$r_{C1}(IBM)w_{S1}(IBM)w_{S2}(INT)r_{C1}(INT). \tag{2}$$

The client that executed C_1 can only see this *projection* of the actual global schedule. This projection is conflict equivalent to $S_2 \rightarrow C_1 \rightarrow S_1$. The client can therefore validate this transaction using local serializability testing.

At the same time, assume that Client 2 performs C_2, resulting in the following schedule local to the client:

$$w_{S1}(IBM)r_{C2}(IBM)r_{C2}(INT)\,w_{S2}(INT)c_{S2}c_{S1}. \tag{3}$$

This projection is equivalent to the serial order $S_1 \rightarrow C_2 \rightarrow S_2$. This schedule is also valid according to local serializability testing. Notice that in this schedule, S_1 precedes S_2, whereas in the former schedule, S_2 precedes S_1, implying that the global schedule containing all transactions is not serializable. We will make this phenomenon clearer with an example.

The following global schedule consisting of S_1, S_2, C_1, and C_2 maintains the same conflicts as both client projections described above, but is not conflict serializable. Creating a serialization graph for this schedule reveals a cycle: $C_1 \rightarrow S_1 \rightarrow C_2 \rightarrow S_2 \rightarrow C_1$.

$$r_{C1}(IBM)w_{S1}(IBM)r_{C2}(IBM)r_{C2}(INT)w_{S2}(INT)r_{C1}(INT)c_{S2}c_{S1} \tag{4}$$

In general, a conservative client would assume such a situation and would have to abort its transaction.

In an extension to the example, C_1' contains all the operations of C_1 and additionally includes the operation $w_{C1}(MSF)$ at the end. Now assume that C_1' replaces C_1 in the examples above. Notice the effects of appending $w_{C1}(MSF)$ to the ends of Schedules 2 and 4. Schedule 2 is still serializable, and Schedule 4 is still not. C_1' is an update transaction that needs its update serialized with the rest of the schedule, even though the update operation does not conflict with any of the other transactions.

These examples demonstrate the difficulty of maintaining serializability in broadcast databases. With no access constraint on the program, different clients may induce different views of the evolution of the database. In particular, traditional consistency criteria are too strict for broadcast databases because they force:

1. all read-only transactions to induce the same global serializable schedule, and
2. all read-only transactions to be serializable with respect to update transactions, even if the updates do not conflict.

Update consistency is a correctness criterion that allows the above schedule (Bober & Carey, 1992). This correctness criterion is a relaxation of serializability, requiring:

1. read-only transactions to be serializable with respect to the set of (update) transactions from which they read, and
2. update transactions to be serializable with each other.

In Schedule 4, C_1 reads from S_2, and C_2 reads from S_1, and, as already mentioned above, each is independently serializable with the update transactions. This satisfies Condition 1. Furthermore, if we consider the same schedule with C_1' instead of C_1, we can serialize all update transactions, $S_2 \rightarrow C_1' \rightarrow S_1$. This satisfies Condition 2. The schedule is therefore update consistent.

One may be concerned that update consistency may cause some inconsistent behavior because it allows clients to see different serial orderings of update transactions. However, multiple read-only transactions can only see different serial orders of updates when the updates are unrelated. Thus, the clients are still guaranteed to see a consistent database state that will eventually converge to a single uniform state. Furthermore, if update and read-only transactions are interdependent (the reads depend on the values written by the updates), then Condition 1 for update consistency guarantees that they are serializable.

Implementing update consistency requires the use of a certification report. Because validation in this case can be done with a single certification report, the report must encode the update histories of all items, including the relationship each item's current value has with the values of other items.

We first define the set of transactions from which transaction T *directly* and *indirectly* reads. Assume that initially, all items in the database are written by transaction T_0. Consider all transactions T' that write a value that T reads, and all the transactions T'' that write a value that each T' reads, and so forth. This recursive closure of this set of transactions, which also includes T, is called the *live set* of T.

We can now define the *update log* of data item C_j. Assume that d_j was last updated by transaction T^j. The update log for d_j is a set of N integers, one for each item d_i in the broadcast set B. The ith integer in d_j's update log is the time stamp indicating when d_i was last updated by a transaction in T^j's live set. We combine the update logs for all N items in the broadcast set to create a certification report U. Let U_{ij} refer to the cycle of the last update of d_i by a transaction in T^j's live set.

The client uses the update log in the following way. Assume that it is performing transaction T with a read set R_T containing a set of (d_i, c_{di}) pairs, where c_{di} is the cycle during which T read d_i. If T attempts to read item d_j during cycle c, then it checks each (d_i, c_{di}) in R_T and compares c_{di} with the corresponding cycle number in d_j's update log. If c_{di} is greater than U_{ij}, then the read succeeds, and (d_j, c) is added to R_T. Otherwise, T fails. This is the *read condition*.

Basically, when T reads d_j, it must make sure that it can be serialized after T^j (the transaction that last wrote d_j), by transitivity, after all the transactions T^i in T^j's live set. U_{ij} is a time stamp indicating the last time d_i was written by a transaction T^i in T^j's live set. Now, assume d_i is already in T's read set R_T when it attempts to read d_j. Therefore, because $T^i \rightarrow T^j$, and $T^j \rightarrow T$, then $T^i \rightarrow T$. We ensure the last property by forcing T's read of d_i to have occurred in some cycle after U_{ij}.

The server can implement U_{ij} incrementally at the commit point of each update transaction T^U. U_{ij} is initially 0 for all i and j. Let U_{ij}' be the new U_{ij}, which incorporates T and is by default equal to U_{ij}. First, U_{ij}' equals the current cycle number if both d_i and d_j are written by T^U. In this case, T directly affects the values of both d_i and d_j.

If T^U writes d_j but not d_i, then U_{ij}' equals the maximum U_{ik} for some d_k in R_{TU}. That is, because T^U read d_k, then we must consider the version of d_i on which d_k might depend.

U consumes $\theta(N^2)$ space, which is significant if the relative size of each item is small. In order to reduce this space consumption, we may group together items so that column j of U^c, a compressed form of U, corresponds to $s_j < N$ items. The value of row i in U^c_{ij} therefore corresponds to the maximum value of the entries from the set of items that were grouped together. Checking the time stamp of an item d_i in R_T therefore requires checking the value in the group it belongs to in U^c. The price of reducing the space consumption of U is that some transactions may incorrectly fail the read condition and then abort. In the extreme case, all columns are grouped together, making U^c a 1xN matrix. In this case, when performing a read, T fails if any of the items in its read set R_T have changed since it read them.

Note that in some applications, clients do not need to read the most recent versions of data as long as they make consistent reads. Multiversion protocols operate on this principle. In such cases, the client may cache data which allows them to perform reads locally. In order to ensure consistency of these reads, the client must also cache relevant U matrices.

Dealing with Updates

In the discussion above, we tacitly assumed that client transactions were read only, even though the server could perform updates. In general, clients can also perform update transactions. Update transactions can write new values for items, insert items, and delete items. We will consider the details of such transactions in this section.

As we described in "Transaction Processing Primer," update transactions must be conflict serializable with other transactions. Because update transactions modify the state of the (primary) database, they cannot be fully validated by clients. All the client can do is to validate the update transaction's reads using the techniques described above before sending the transaction to the server for final validation of the updates. Our question is whether or not performing read validation is the best design when clients can perform updates.

One of the main characteristics of the read validation techniques described above is that only information or results that reflect committed transactions are sent to clients. Updates, however, may be queued at the server, awaiting validation. In all likelihood, some of these updates will be validated by the server, so clients must in fact validate their local update transactions with those that are queued up at the server.

Bowen et al. (1992) and Barbara (1997) address updates by sending information on updates whose effects may have not yet been reflected in the program. Barbara suggests creating certification reports that describe the update transactions that were submitted in the previous cycle and validated by the server. Each certification report is made up of two parts: The first lists the items read and the second lists the items written by the validated update transactions. The validated update transactions are therefore described to clients as one large (composite) transaction with given read and write sets. Organizing the certification report in this way is, ostensibly, an efficiency technique designed to reduce its size. In addition to the read and write sets, the server also broadcasts transaction identifiers, indicating those it has validated or rejected.

Clients use the certification report to detect read and write conflicts. If the client finds that its update transaction:

1. reads items that appear in the write set of the certification report, or
2. writes items that appear in either the read or write set of the certification report,

then the transaction aborts. The idea behind these conditions is to abort any transaction that conflicts with any that has been validated by the server during its lifetime. This technique was inspired by serialization graphs, which, as explained in "Isolation," indicate unserializability when they contain cycles. Conflicting operations create edges in the graph, which can lead to cycles. By disallowing conflicts, we disallow edges and, consequently, cycles.

This type of certification report requires, at most, on the order of log $(2N + N_T)$ space: N for each of the read and written items, and N_T transaction identifiers. The protocol described above is already very conservative, disallowing any concurrency allowable with conflicts. The organization of the certification report hides details of individual, validated update transactions that clients might use for validation. For example, consider three transactions:

T_1: $r_1(IBM)$,

T_2: $r_2(IBM)r_2(INT)w_2(IBM)$, and

T_3: $r_3(INT)w_3(INT)$.

Now, consider the schedule where c_i refers to T_i's commit point:

$$r_1(IBM)r_2(IBM)r_2(INT)w_2(IBM)r_3(INT)w_3(INT)c_1c_3c_2.$$

This schedule is obviously conflict equivalent to $T_1 \rightarrow T_2 \rightarrow T_3$. Notice, however, that T_1 and T_3 commit first. If T_2 commits in a cycle after these two commit, then the certification report will indicate that a transaction that performs $r(IBM)r(INT)w(INT)$ has committed. Due to the interaction between T_2 and the certification report's read and write sets, T_2 must abort. T_2 would in fact be serializable, using the same principle as does Thomas' write rule (Elmasri & Navathe, 2004), if T_1 and T_3 were kept distinct to the client.

Another problem with the work described so far is that it does not consider the effects of insertions and deletions of items. In fact, the broadcast set is assumed to be static.

Handling a delete operation is straightforward. A delete invalidates the value read by a transaction. It can therefore be treated as a write operation.

Handling an insertion operation is slightly more difficult. Insert and read operations conflict, but this conflict cannot be detected using the techniques described above. For example, a query T_1 may search for data using a *predicate*—all stocks priced over $12—returning *MSF*. T_2 then inserts "*LNX*, 14". T_2 must be serialized after T_1 because *LNX* would otherwise be in T_1's read set, but we cannot do this by comparing read and write sets. This type of conflict is known as the *phantom* phenomenon in transaction-processing theory.

The phantom phenomenon is often handled in conventional databases by locking the search structures used to access data (e.g., indexes). Inserts lock the parts of the index needed for the update. This action isolates the inserts and preserves consistent reading of the database.

Locks, as we have stated, cannot be implemented in broadcast databases. We therefore need another technique. Bowen et al. (1992) handle the phantom phenomenon by adding more information to the certification reports, and requiring clients to do more to validate their transactions. In particular, they include the time stamps and read and write sets of each of the update transactions that have been proposed but not validated, validated but not installed by the server, and validated and installed by the server. The server can afford to send so much information because data are assumed to be transmitted via a high-speed bus instead of a wireless medium.

In order to validate an active transaction, the client executes its predicate on the subset of the certification report that corresponds to the time after the transaction's initiation. Only if the predicate returns no tuples during its lifetime at the client does it qualify to be sent to the server for further validation. This test ensures that no phantoms exist. The client sends the transaction's time stamp, read set, write set, and predicate, and the server performs a similar test for final validation.

Limitations of Current Approaches and a Solution

One of the solutions we have considered above handles anomalies associated with read-only transactions, and another has considered the implementation of update-transaction validation. In this section, we consider the integration of these two techniques, yielding both of their benefits. Furthermore, we show how to handle phantoms.

Clients cannot detect conflicting insert operations that occur after transaction initiation. Returning to the example above, suppose a transaction *T* searches for stocks priced over $12. It finds *IBM* before another client inserts "*LNX*, 14". This insertion must therefore be serialized after *T*. *T* later reads *MSF*, whose price was affected by that of *LNX*. During the read, the client checks if its copy of *IBM* is valid with respect to that of *MSF*. Assuming that the price of *IBM* has not changed, the read of *MSF* succeeds, implying the *T* should be serialized after the insertion. The schedule is now unserializable.

The technique of Bowen et al. (1992) somewhat solves the phantom problem, but at the cost of increasing the amount of metadata contained in a certification report. Their technique also aborts any transaction that has read items whose values have changed since transaction initiation.

The obvious solution is to combine elements of the two approaches. We start by using the *U* matrix, described in "Concurrency Control," and assume that transactions span a maximum of *maxspan* cycles. When the server deletes item *d*, the item is removed from the next cycle's program, and the *U* matrix is updated as if a write were performed on *d*. Deleting *d* from the program guarantees that it cannot be added to any active or subsequent transactions. Updating the *U* matrix precludes the possibility of any active transactions that have already read *d* from reading an item that was involved in its deletion. After *maxspan* cycles, the server can delete the entries in *U* that correspond to *d* because by then, all transactions that have *d* in their read sets will have terminated.

When the server inserts an item *d*, it is added to the next cycle's program. The new row and column added to *U* for *d* are filled as if a write were just performed on *d*. The reason for this is obvious. Any unfilled *U* entries are filled with 0s. This implies that by reading *d*, a transaction is not precluding it from reading any other item that is unrelated to its insertion.

In order to handle phantoms, a certification report must also be part of the program and include recently inserted items. The transactions that should not see these items are those that would already have them in their read sets had the items been inserted earlier. The client must make sure that it does not subsequently read these items, which is what placing them in the certification report ensures. The client must also make sure that the transaction does not read any item that depends on the newly inserted ones. Suppose that d_i is a newly inserted item that the transaction should not see. Then, the transaction must not read d_j such that $U_{ij} > 0$.

The new items must be in the certification report for at least one cycle. In that cycle, active transactions must identify the new items to avoid by applying their (already run) predicates on the certification report. The item can be deleted subsequently as the active transactions have already had a chance to find insertions.

Recovery

The server uses traditional recovery techniques. Accepted transactions are installed into the database as log records appropriate to the recovery protocol are written to the log, as described in "Atomicity and Durability." The server can recover from a failure using these log records. The ability to use traditional recovery techniques stems from the fact that transaction processing at the server is independent of how the clients read data.

A client must implement recovery according to the fact that the partially committed transactions (transactions that have executed all their operations) they send to the server for validation may ultimately be aborted by the certifier. If a transaction is aborted, then its effects on the client's view of the database must be restored to its pretransaction state. If the client maintains local replicas of data, then restoring their state can be done by executing an undo protocol using a local log.

While waiting for a partially committed transaction to be certified, the client's transaction processor may allow the user to execute other transactions to possibly increase the user's productivity. If the transaction processor allows these transactions to read data written by partially committed transactions, then it runs the risk of *cascading aborts*. A cascading abort is the abort of a transaction that is necessary because of the abort of another transaction from which it read data. These aborts ensure database consistency as they keep transactions from reading uncommitted data.

The client-side techniques we described above ensure the atomicity of transactions. Durability can be also be handled by using a log. However, the client also has the option of referring to the data being broadcast by the server in order to see the effects of committed transactions. Both are options in maintaining transaction durability, and the difference between the two is that the broadcast data are more current, whereas local replicas are easier to access.

Note that a client can also suffer a *catastrophic failure* and lose the entire contents of its primary and secondary storage. Such a failure is more likely with mobile devices as they are often taken out of the controlled office environment. In the broadcast environment, recovery consists of requesting lost data from the server.

Transaction Design

Transaction design in traditional DBMSs must leave a database in a consistent state. If the state of the database does not allow the transaction to do this, it must automatically abort and may be restarted with different values. For example, a BANKTRANSFER transaction should require that the accounts exist and have enough money to transfer. If these conditions are not true, then the transaction should automatically abort, leaving the database in a consistent state.

A broadcast transaction must also leave the database in a consistent state. However, its design should include *compensating actions* in order to increase the likelihood that it is certified. For example, a TRADE transaction may have two outcomes depending on the value of a stock at the precise time of certification: "if IBM.price > 100 then sell 100 shares of IBM, else buy 100 shares of IBM." The performance improvement in the broadcast environment may be significant because of the latency involved in waiting for the news of whether or not a transaction has been certified. The design of the TRADE transaction guarantees its success (that it will commit, regardless of the value of IBM.price). This type of transaction has also been used in other mobile environments where synchronous communication with the server is impossible (Demers, Petersen, Spreitzer, Terry, Theimer, & Welch, 1994).

Conclusion and Future Work

The consequence of broadcast database scalability is complex transaction processing. There is no access control to the broadcast channel, disallowing most traditional concurrency control protocols. Seminal work done in the Datacycle system introduced the structures needed to support some form of transaction-processing concurrency control (Bowen et al., 1992). Unlike much other work on the subject, the Datacycle team implemented full transaction processing for clients, supporting both read-only and update transactions. Datacycle transaction processing, however, suffers from two problems that make it inapplicable for the broadcast environment. The certification reports it requires consume too much space, and its validation condition is too strict.

Shanmugasundaram et al. (1999) propose an alternative validation condition that is based on a relaxed consistency condition called update consistency. This type of validation decreases the time required for transaction completion by more than half by increasing the probability of certification.

This validation technique, however, does not allow updates. We show how to combine it with the Datacycle's validation technique to allow for updates, including insertions and deletions. Our technique takes care of the phantom phenomenon as well.

Note that complete transaction processing also requires transaction-design guidelines, as well as a recovery system to ensure the consistency of the database state, and the atomicity and durability of transactions. Because the broadcast server acts as a typical server (with a certifying component), we can simply borrow consistency and recovery techniques from traditional databases.

The Future of Broadcast Databases

The future of broadcast databases is unclear, but likely to gain significance. Common wireless communication is going digital. For example, the U.S. Federal Communications Commission has mandated that commercial radio and television stations eventually convert to digital transmission. Digital satellite radio is already in widespread use (Layer, 2001). This trend is being mirrored internationally. Furthermore, federal (United States) restrictions on owning multiple radio stations within a market have been relaxed. For example, the broadcast giant Clear Channel owns nine stations in the Atlanta area and six in Chicago. Finally, wireless networks are growing in popularity in local and larger areas. Taken as a whole, these trends suggest that the building blocks for large-area wireless data systems are being put in place.

Much more work must be done to improve the usability and general applicability of broadcast databases. In their current state, they are applicable for restricted applications requiring small data sets. One such application is in multiplayer gaming. Small data sets allow low latencies, and the techniques described in this chapter allow localized concurrency control, reducing the server load. Note that the Datacycle system, which was intended as a general-purpose DBMS, was a database machine. As of 1992, it consisted of many specialized parts, including a 52 MB/sec data channel. Such specialized architectures are not very popular.

Research in determining the set of items to broadcast must be done. An intuitive way to determine the broadcast set is to select the most popular items, as done in Almeroth et al.(1998) and Stathatos, Roussopoulos, and Baras (1997). However, as described in Lee,

Hu, and Lee (1999), items not in the program must be downloaded separately. So far, no studies have been done that consider the performance of transaction processing when data can come from data sources other than the broadcast channel.

References

Acharya, S., Alonso, R., Franklin, M., & Zdonik, S. (1995). Broadcast disks: Data management for asymmetric communication environments. *Proceedings of the ACM International Conference on Management of Data (SIGMOD)*, San Francisco, California, (pp. 1214-1221).

Aksoy, D., & Franklin, M. J. (1999). RxW: A scheduling approach for large-scale on-demand data broadcast. *IEEE/ACM Transactions on Networking*, *7*(6), 846-860.

Almeroth, K. C., Ammar, M. H., & Fei, Z. (1998). Scalable delivery of Web pages using cyclic best-effort (udp) multicast. *Proceedings of the IEEE International Conference on Computer Communications (INFOCOM)*, San Hose, California, (pp. 199-210).

Ammar, M. H., & Wong, J. W. (1985). The design of teletext broadcast cycles. *Performance Evaluation*, *5*(4), 235-242.

Barbara, D. (1997). Certification reports: Supporting transactions in wireless systems. *Proceedings of the IEEE International Conference on Distributed Computing Systems (ICDCS)*, 466-473.

Bar-Noy, A., Dreizin, V., & Patt-Shamir, B. (2002). Efficient periodic scheduling by trees. *Proceedings of the IEEE International Conference on Computer Communications (INFOCOM)*, New York, (pp. 791-800).

Bernstein, P. A., & Goodman, N. (1981). Concurrency control in distributed database systems. *ACM Computing Surveys*, *13*(2), 185-221.

Bober, P., & Carey, M. (1992). Multiversion query locking. *Proceedings of the International Conference on Very Large Databases (VLDB)*, Vancouver, BC, (pp. 497-510).

Bowen, T. F., Gopal, G., Herman, G., Hickey, T., Lee, K. C., Mansfield, W. H., et al. (1992). The datacycle architecture. *Communications of the ACM*, *35*(12), 71-80.

Chrysanthis, P. (1993). Transaction processing in a mobile computing environment. *Proceedings of the IEEE International Workshop on Advances in Parallel and Distributed Systems*, (pp. 77-82).

Datta, A., Celik, A., Kim, J., VanderMeer, D., & Kumar, V. (1997). Adaptive broadcast protocols to support power conservant retrieval by mobile users. *Proceedings of the IEEE International Conference on Data Engineering (ICDE)*, (pp. 124-133).

Demers, A., Petersen, K., Spreitzer, M., Terry, D., Theimer, M., & Welch, B. (1994). The Bayou architecture: Support for data sharing among mobile users. *Proceedings of the IEEE Workshop on Mobile Computing Systems and Applications (MCSA)*, Santa Cruz, California, (pp. 2-7).

Dunham, M. H., Helal, A., & Balakrishnan, S. (1997). A mobile transaction model that captures both the data and movement behavior. *ACM/Kluwer Mobile Networks and Applications*, *2*(2), 149-162.

Elmasri, R., & Navathe, S. B. (2004). *Fundamentals of database systems* (4th ed.). Reading, MA: Addison Wesley.

Goodin, D. (1999). Impeachment trial draws in netizens. *CNET News*. Retrieved August 16, 2004, from http://news.com.com/2100-1023-219861.html?legacy=cnet&tag=st.ne.1002.srch%res.ni

Imielinski, T., Vishwanath, S., & Badrinath, B. R. (1997). Data on air: Organization and access. *IEEE Transactions on Knowledge and Data Engineering*, *9*(3), 353-372.

Layer, D. H. (2001). Digital radio takes to the road. *IEEE Spectrum*, 38(7), 40-46.

Lee, W. C., Hu, Q., & Lee, D. L. (1999). A study on channel allocation for data dissemination in mobile computing environments. *ACM/Kluwer Mobile Networks and Applications*, *4*, 117-129.

Oki, B., Pfluegl, M., Siegel, A., & Skeen, D. (1993). The information bus: An architecture for extensible distributed systems. *Proceedings of the ACM International Symposium on Operating Systems Principles (SOSP)*, Asheville, North Carolina, (pp. 58-68).

Pitoura, E., & Chrysanthis, P. K. (1999). Exploiting versions for handling updates in broadcast disks. *Proceedings of the International Conference on Very Large Data Bases (VLDB)*, Edinburgh, Scotland (pp. 114-125).

Ross, S. M. (1997). *Introduction to probability models* (6th ed.). New York: Academic Press.

Shanmugasundaram, J., Nithrakashyap, A., Sivasankaran, R., & Ramamritham, K. (1999). Efficient concurrency control for broadcast environments. *Proceedings of the ACM International Conference on Management of Data (SIGMOD)*, (pp. 85-96).

Stathatos, K., Roussopoulos, N., & Baras, J. S. (1997). Adaptive data broadcast in hybrid networks. *Proceedings of the International Conference on Very Large Data Bases (VLDB)*, Athens, Greece (pp. 326-335).

Vaidya, N. H., & Hameed, S. (1999). Scheduling data broadcast in asymmetric communication environments. *ACM/Kluwer Mobile Networks and Applications*, *5*, 171-182.

Yee, W. G., & Navathe, S. B. (2003). Efficient data access to multi-channel broadcast programs. *Proceedings of the ACM International Conference on Information and Knowledge Management (CIKM)*, New Orleans, LA, (pp. 153-160).

Yee, W. G., Navathe, S. B., Omiecinski, E., & Jermaine, C. (2002). Bridging the gap between response time and energy efficiency in broadcast schedule design. *Proceedings of the International Conference on Extending Database Technology (EDBT)*, Prague, Czech Republic, (pp. 572-589).

About the Authors

Dimitrios Katsaros was born in Thetidio (Farsala), Greece, in 1974. He received a BS in informatics (March 1997) and a PhD in informatics (May 2004), both from the Aristotle University of Thessaloniki, Greece. He spent a year (July 1997-August 1998) as a visiting researcher in the Department of Pure and Applied Mathematics at the University of L'Aquila, Italy. His research interests include the Internet, particularly in caching, replication, prefetching, and content delivery, mobile computing, in particular, data management and delivery, and data mining, especially tree and graph mining. He is the author or coauthor of more than 20 papers in the above areas.

Alexandros Nanopoulos was born in Craiova, Romania, in 1974. He graduated from the Department of Informatics (November 1996) at the Aristotle University of Thessaloniki, and obtained a PhD from the same institute (February 2003). The subject of his dissertation was "Techniques for Nonrelational Data Mining." He is a coauthor of 20 articles in international journals and conferences, and also a coauthor of the monograph *Advanced Signature Techniques for Multimedia and Web Applications*. His research interests include spatial and Web mining, integration of data mining with DBMSs, and spatial database indexing.

Yannis Manolopoulos received a BEngr (1981) in electrical engineering and a PhD (1986) in computer engineering, both from the Aristotle University of Thessaloniki. Currently, he is a professor there with the Department of Informatics. He has been with the Department of Computer Science of the University of Toronto, Department of Computer Science of the University of Maryland at College Park, and the University of Cyprus. He has published more than 140 papers in refereed scientific journals and conference proceedings. He is coauthor of the books titled *Advanced Database Indexing* and *Advanced Signature Indexing for Multimedia and Web Applications* (Kluwer). He is also author of two textbooks on data structures and file structures, which are recommended by the vast majority of the computer science and engineering departments in Greece. He served as PC cochair of the Eighth Panhellenic Conference in Informatics (2001), the Sixth ADBIS Conference (2002), the Fifth WDAS Workshop (2003), the Eighth SSTD Symposium (2003), and the First Balkan Conference in Informatics (2003). Currently, he is vice-chairman of the Greek Computer Society. His research interests include access methods and query processing for databases, data mining, and performance evaluation of storage subsystems.

* * *

Afrand Agah is currently pursuing a PhD in computer science at the University of Texas at Arlington, USA. She received an MS in computer science from Kansas State University (2001). Her work involves research in security of sensor networks, mobile ad hoc networks, and intrusion detection techniques.

Hitha Alex is an assistant instructor and PhD student of computer science and engineering at the University of Texas at Arlington, USA. Her research interests are in the areas of pervasive and mobile computing, service discovery, and information fusion. She obtained her MS in computer science and engineering from the University of Texas at Arlington (May 2003).

Angelos Batistakis received his BS in informatics and telecommunications from the University of Athens, Greece, and his MS in distributed systems from the University of Kent, UK (1998 and 1999, respectively). Since then, he has worked in the IT industry in the field of B2B e-Business and in the banking sector as an IT architect. Currently, he is a research associate in the Communication Networks Laboratory of the University of Athens (UoA-CNL), involved in the Information Society Technologies (IST) project PoLoS (integrated platform for location-based services). His research interests are in the area of mobile and distributed computing.

Panayiotis Bozanis received his five-year diploma and his PhD from the Department of Computer Engineering & Informatics at the University of Patras, Greece (1993 and 1997, respectively). He is currently a lecturer with the computer and communication engineering department at the University of Thessaly, Greece. His main research interests include data structures, computational geometry, information retrieval, storage and indexing techniques, databases for large sets, and mobile data. His publications comprise several journal and conference papers concerning problems of data structuring, computational geometry, information retrieval, and spatio-temporal indexing, and two books in Greek about data structures and design, and analysis of algorithms. He is an EATCS member.

Erdal Cayirci graduated from the Turkish Army Academy (1986) and from the Royal Military Academy, Sandhurst (1989). He received his MS from the Middle East Technical

University (METU), Ankara, Turkey, and a PhD from Bogazici University in computer engineering (1995 and 2000, respectively). He was a visiting researcher with the Broadband and Wireless Networking Laboratory and a visiting lecturer with the School of Electrical and Computer Engineering at the Georgia Institute of Technology, Atlanta, Georgia (USA), in 2001. He is the director of the combat models operations department in the Turkish War College's Wargaming and Simulation Center, and a faculty member with the computer engineering department of the Istanbul Technical University. His research interests include sensor networks, mobile communications, tactical communications, and military constructive simulation. He is an editor for *IEEE Transactions on Mobile Computing, AdHoc Networks* (Elsevier Science), *ACM/Kluwer Wireless Networks,* and *ASP Sensor Letters.* He received the 2002 IEEE Communications Society Best Tutorial Paper Award for his paper titled "A Survey on Sensor Networks" published in the *IEEE Communications Magazine* in August 2002.

Yon Dohn Chung received a BS in computer science from Korea University, Seoul, Korea (1994), and an MS and PhD in computer engineering from Korea Advanced Institute of Science and Technology (KAIST), Daejon, Korea (1996 and 2000, respectively). In 2003, he joined the faculty of the Department of Computer Engineering at Dongguk University, Seoul, Korea, where currently he is an assistant professor. Before joining the department, he worked in the Department of Computer Science of KAIST as a postdoctoral research associate and a research professor. His research interests include mobile databases, spatio-temporal databases, XML databases, data stream processing, and database systems.

Sajal K. Das is currently a professor of computer science and engineering and founding director of the Center for Research in Wireless Mobility and Networking (CReWMaN) at the University of Texas at Arlington (USA). His current research interests include resource and mobility management in wireless networks, mobile and pervasive computing, sensor networks, mobile Internet, parallel processing, and grid computing. He has published more than 250 research papers and holds four U.S. patents in wireless mobile networks. He received the Best Paper Awards in ACM MobiCom '99, ICOIN-16, ACM, MSWiM '00, and ACM/IEEE PADS '97. Dr. Das serves on the editorial boards of *IEEE Transactions on Mobile Computing*, *ACM/Kluwer Wireless Networks*, *Parallel Processing Letters*, and *Journal of Parallel Algorithms and Applications*. He served as general chair of IEEE PerCom '04, IWDC '04, MASCOTS '02, and ACM WoWMoM '00-'02; general vice chair of IEEE PerCom '03, ACM MobiCom '00, and IEEE HiPC '00-'01; program chair of IWDC '02 and WoWMoM '98-'99; TPC vice-chair of ICPADS '02; and as TPC member of numerous IEEE and ACM conferences.

Margaret H. Dunham (formerly Eich) received a BA and MS in mathematics from Miami University, Oxford, Ohio, and a PhD in computer science from Southern Methodist University in Dallas, USA (1970, 1972, and 1984, respectively). From August 1984 to the present, she has been first an assistant professor, associate professor, and now professor in the Department of Computer Science and Engineering at Southern Methodist University. In addition to her academic experience, professor Dunham has nine years of industry experience. Dr. Dunham's current research interests are in mobile computing and data mining. Dr. Dunham served as editor of the *ACM SIGMOD Record* (1986-1988). She has served on the program and organizing committees for several ACM and IEEE conferences. She served as guest editor for a special section of *IEEE Transactions on Knowledge and Data Engineering* devoted to main memory databases as well as a special issue of the *ACM SIGMOD Record* devoted to mobile computing in databases. She was general chair of the ACM SIGMOD conference held in Dallas in May 2000. She is currently an associate editor for *IEEE Transactions on Knowledge Engineering* and is author of a recently published book, *Data Mining: Introductory and Advanced Topics*, available from Prentice Hall. She has published over 70 technical papers on various database areas. Dr. Dunham lives in Dallas with husband Jim, daughters Stephanie and Kristina, and cat Missy.

Stathes Hadjiefthymiades earned a BS in informatics from the Department of Informatics at the University of Athens (1993), and an MS in advanced information systems from the same department (1996). In 1999, he earned a PhD from the University of Athens (Department of Informatics and Telecommunications). In 2002 he received a joint engineering-economics MS from the National Technical University of Athens, Greece. In 1992 he joined the Greek consulting firm Advanced Services Group, Ltd., where he was extensively involved in the analysis and specification of information systems and the design implementation of telematic applications. In 1995 he became a member of the UoA-CNL. From September 2001 to July 2002, he served as a visiting assistant professor with the University of Aegean's Department of Information and Communication Systems Engineering. In July 2002 he joined the faculty of the Hellenic Open University (Department of Informatics), Patras, Greece, as an assistant professor in telecommunications and computer networks. Since December 2003, he has belonged to the faculty of the University of Athens' Department of Informatics and Telecommunications. He has participated in numerous projects realized in the context of EU programs (Advanced Communication Technologies and Services [ACTS], ORA, TAP, and IST), EURESCOM projects, as well as national initiatives. His research interests are in the areas of wireless/mobile computing, Web engineering, and networked multimedia applications. He is the author of more than 70 publications in these areas.

Haibo Hu is currently a PhD candidate in the Department of Computer Science at Hong Kong University of Science and Technology. He received his BEngr in computer science and engineering from Shanghai Jiao Tong University, Shanghai, China. His research interests include wireless data-dissemination techniques, location-based services, location modeling, and mobile spatial databases.

A.R. Hurson is a computer science and engineering faculty-member at The Pennsylvania State University (USA). His research for the past 20 years has been directed toward the design and analysis of general as well as special-purpose computer architectures. His research has been supported by NSF, the Office of Naval Research, DARPA, NCR Corp., IBM, Lockheed Martin, and The Pennsylvania State University. He has published over 200 technical papers in areas including database systems, heterogeneous (multi-)

databases, mobile computing environments, mobile databases, object-oriented databases, computer architecture and cache memory, parallel and distributed processing, dataflow architectures, and VLSI algorithms.

Sandor Imre received MS and PhD degrees in electrical engineering from Budapest University of Technology and Economics, Hungary (BUTE; 1993 and 1999, respectively). He is currently an associate professor at the Department of Telecommunications at BUTE and is working in the field of mobile communications.

Anastasios Ioannidis received a BS in computer science from the Department of Informatics at the University of Athens (1997), and an MS (Honours) in information security with the Royal Holloway and Bedford New College at University of London (1998). Before joining the University of Athens, he won two prizes, one at the Greek National Olympiad of Informatics and another at the Greek National Olympiad of Mathematics. He has worked at Nortel Networks UK as a GSM services developer and, since 2000, he has been a member of the UoA-CNL. He is currently a PhD candidate in the area of network security and he has been involved in the development of the EURO-CITI (European cities platform for online transaction services) project implemented in the context of IST (Fifth Framework Program), and is currently working on the PoLoS IST project. His research interests include security protocols, peer-to-peer networks, as well as wireless communications.

Y. Jiao is a PhD candidate in the computer science and engineering department at The Pennsylvania State University (USA). She received a BS from the Civil Aviation Institute of China, Tianjin, China, and an MS from The Pennsylvania State University (1997 and 2002, respectively), both in computer science. Her main research interests include mobile data-access-system security, mobile agent security, and energy-efficient information system design.

Myoung Ho Kim received a BS and MS in computer engineering from Seoul National University, Seoul, Korea (1982 and 1984, respectively), and a PhD in computer science from Michigan State University, East Lansing, MI (1989). In 1989 he joined the faculty of the Department of Computer Science at the Korea Advanced Institute of Science and Technology (KAIST) where currently he is a professor. His research interests include database systems, OLAP, XML, mobile computing, transaction management, information retrieval, workflow, and distributed processing. He is a member of the ACM and IEEE Computer Society.

Mohan Kumar is an associate professor of computer science and engineering at the University of Texas at Arlington (USA). His current research interests are in pervasive computing, wireless networks and mobility, active networks, mobile agents, and distributed computing. He has published over 95 articles in refereed journals and conference proceedings, and supervised master's and doctoral theses in the above areas. Kumar is on the editorial board of *The Computer Journal*. He is a cofounder of the IEEE

International Conference on Pervasive Computing and Communications (PerCom), has served as the program chair for PerCom 2003, and is the general chair for PerCom 2005. Mohan Kumar obtained his PhD (1992) and MTech (1985) degrees from the Indian Institute of Science and his BE (1982) from Bangalore University in India. Prior to joining the University of Texas at Arlington in 2001, he held faculty positions at the Curtin University of Technology, Perth, Australia (1992-2000), the Indian Institute of Science (1986-1992), and Bangalore University (1985-1986).

Lazaros Merakos received a diploma in electrical and mechanical engineering from the National Technical University of Athens (1978) and an MS and PhD in electrical engineering from the State University of New York, Buffalo (1981 and 1984, respectively). From 1983 to 1986, he was on the faculty of electrical engineering and computer science at the University of Connecticut, Storrs. From 1986 to 1994, he was on the faculty of the Electrical and Computer Engineering Department at Northeastern University, Boston, Massachusetts. During the period from 1993 to 1994, he served as director of the Communications and Digital Processing Research Center at Northeastern University. During the summers of 1990 and 1991, he was a visiting scientist at the IBM T. J. Watson Research Center, Yorktown Heights, New York. In 1994 he joined the faculty of the University of Athens, where he is presently a professor in the Department of Informatics and Telecommunications and director of the UoA-CNL and Networks Operations and Management Center. His research interests are in the design and performance analysis of broadband networks and wireless/mobile communication systems and services. He has authored more than 150 papers in the above areas. Since 1995 he has led the research activities of UoA-CNL in the area of mobile communications in the framework of the ACTS and IST programs funded by the European Union (projects RAINBOW [Radio Access Independent Broadband on Wireless], Magic WAND, WINE, MOBIVAS, PoLoS, and ANWIRE). He is chairman of the board of the Greek Universities Network and the Greek Schools Network, and member of the board of the Greek Research Network. In 1994 he received the Guanella Award for the best paper presented at the International Zurich Seminar on Mobile Communications.

Petros Nicopolitidis received a BS and PhD in computer science from the Department of Informatics at the Aristotle University of Thessaloniki (1998 and 2002, respectively). Since 2004 he has been a lecturer at the same department. His research interests are in the areas of wireless networks and mobile communications. He is coauthor of the book *Wireless Networks* (Wiley, 2003).

Georgios I. Papadimitriou received a diploma and PhD degree in computer engineering from the University of Patras (1989 and 1994, respectively). From 1989 to 1994 he was a teaching assistant at the Department of Computer Engineering of the University of Patras and a research scientist at the Computer Technology Institute, Patras, Greece. From 1994 to 1996 he was a postdoctoral research associate at the Computer Technology Institute. From 1997 to 2001, he was a lecturer at the Department of Informatics of Aristotle University of Thessaloniki. Since 2001 he has been an assistant professor at the Department of Informatics of Aristotle University of Thessaloniki. His research interests include optical networks, wireless networks, high speed LANs, and learning automata. Professor Papadimitriou is associate editor of five scholarly journals, including the *IEEE Transactions on Systems, Man and Cybernetics-Part C*, the *IEEE Transactions on Broadcasting*, and the *IEEE Communications Magazine*. He is coauthor of the books *Multiwavelength Optical LANs* (Wiley, 2003) and *Wireless Networks* (Wiley, 2003), and coeditor of the book *Applied System Simulation* (Kluwer, 2003). He is the author of more than 100 refereed journal and conference papers. He is a senior member of IEEE.

Samuel Pierre received a BEngr in civil engineering from Ecole Polytechnique de Montreal, Québec, Canada (1981), a BS and MS in mathematics and computer science (1984 and 1985, respectively) from the UQAM, Montréal, an MS in economics in 1987 from the Université de Montréal, and a PhD degree in electrical engineering in 1991 from École Polytechnique de Montréal. He is currently a professor of computer engineering at École Polytechnique de Montréal, where he is director of the Mobile Computing and Networking Research Laboratory (LARIM) and NSERC/Ericsson Industrial Research chair in next-generation mobile networking systems. Dr. Pierre is the author of four books, and coauthor of two books and six book chapters, as well as over 200 other technical publications including journal and proceedings papers. He received the Best Paper Award of the Ninth International Workshop in Expert Systems and Their Applications (France, 1989) and a Distinguished Paper Award from OPNETWORK2003 (Washington, USA). One of these coauthored books, *Télécommunications et Transmission de données* (Eyrolles, 1992), received special mention from *Telecoms Magazine* (France, 1994). His research interests include wireline and wireless networks, mobile computing, performance evaluation, artificial intelligence, and electronic learning. He is a fellow of the Engineering Institute of Canada, senior member of IEEE, and a member of the ACM and IEEE Communications Society. He is an associate editor of *IEEE Communications Letters* and *IEEE Canadian Review*. He also serves on the editorial board of *Telematics and Informatics* published by Elsevier Science.

Andreas S. Pomportsis received a BS in physics and an MS in electronics and communications from the University of Thessaloniki, Greece, and a diploma in electrical engineering from the Technical University of Thessaloniki, Greece. In 1987, he received a PhD in computer science from the University of Thessaloniki. Currently, he is a professor at the Department of Informatics of Aristotle University of Thessaloniki. He is coauthor of the books *Wireless Networks* (Wiley, 2003) and *Multiwavelength Optical LANs* (Wiley, 2003). His research interests include computer networks, learning automata, computer architecture, parallel and distributed computer systems, and multimedia systems.

Ioannis Priggouris received a BS in informatics from the Department of Informatics and Telecommunications at the University of Athens (1997), and an MS in communication systems and data networks from the same department (2000). Over the last year he has been a PhD candidate in the department. Since 1999, he has been a member of the UoA-CNL. He has participated in the RAINBOW and the EURO-CITI projects implemented in the contexts of ACTS and IST, correspondingly. Currently, he is extensively involved in the development of the PoLoS project implemented in the context of IST. His research interests are in the areas of mobile computing, quality of service (QoS), and mobility support for IP networks. He is the author of many publications on the above areas.

Gyula Rabai received an MS in technical informatics from Budapest University of Technology and Economics (BUTE), Hungary (2001). Currently he is completing his PhD at the Department of Telecommunications at BUTE. He has taken part in the work carried out in an EU-sponsored Fifth Framework research project, CAST (Configurable Radio with Advanced Software Technology). He is currently working for the Computer and Automation Research Institute of the Hungarian Academy of Sciences.

Ayse Yasemin Seydim received a BS and MS in computer engineering from METU (1986 and 1989, respectively). She received a PhD degree in computer science from Southern Methodist University in 2003. She has been working as a system specialist and researcher in Central Bank of the Republic of Turkey since 1989. She is also teaching in METU as a part-time faculty member. She is interested in database management, mobile computing, wireless communications, and data mining. She has also been externally reviewing for several journals and conferences such as IEEE TKDE (2002), ACM SIGKDD (2002), SSTD (2001), and JSS (2004). Her thesis, "Location-Dependent Query Processing in Mobile Environments," has an architectural model for the location-dependent services management and location leveling approaches and benchmarking in this architecture. Information retrieval subjects, such as precision and recall, are first used in the evaluation of the quality of query results for such location-dependent ones.

Behrooz A. Shirazi is a professor and chair of the Department of Computer Science and Engineering at the University of Texas at Arlington (USA). He has conducted research in the areas of pervasive computing, software tools, and distributed real-time systems. He has served on the editorial boards of *IEEE Transactions on Computers* and *Journal of Parallel and Distributed Computing*. He is the principal founder of the IEEE Symposium on Parallel and Distributed Processing and a cofounder of the IEEE International Conference on Pervasive Computing and Communications. He has served on the program committee of many international conferences as an IEEE Distinguished Visitor (1993-1996) and as an ACM lecturer (1993-1997).

Manos Spanoudakis received a BS in computer science and telecommunications from the Department of Informatics and Telecommunication at the University of Athens in 2000 and his MS (Honours) in computer systems technology from the same department in 2003. He is currently a PhD candidate at the same department, specializing in content distribution networks. He has performed research on World Wide Web architectures for wireless/mobile systems and caching. Since 2000, he has been a member of the UoA-CNL. He is involved in the IST projects EURO-CITI and PoLoS. His research interests are in the area of wireless/mobile computing, caching technologies, and CDNs.

Xueyan Tang is an assistant professor in the School of Computer Engineering at Nanyang Technological University, Singapore. He received a BEngr in computer science and engineering from Shanghai Jiao Tong University (1998), and a PhD in computer science from the Hong Kong University of Science and Technology (2003). His research interests include the Web and Internet (particularly caching, replication, and content delivery), mobile and pervasive computing (especially data management), peer-to-peer and sensor networks, and distributed systems. He has served as a program committee member of the IEEE INFOCOM '04 conference.

Tuna Tugcu is a visiting assistant professor in Georgia Tech Savannah at Georgia Institute of Technology (USA). He received his BS (1993) and PhD (2001) degrees in computer engineering from Bogazici University, and an MS (1994) in computer and information science from New Jersey Institute of Technology. He worked as a postdoctorate in the Broadband and Wireless Networks Laboratory at Georgia Institute of Technology until August 2002. Dr. Tugcu's research interests include real-time systems, computer networks, and wireless communication systems. Currently, he is focused on location and resource management in next-generation wireless systems. Dr. Tugcu will start working as an assistant professor at Bogazici University as of September 2004.

Wenye Wang (M '98/ACM '99) received a BS and MS from Beijing University of Posts and Telecommunications, Beijing, China, in 1986 and 1991, respectively. She also received MSEE and PhD degrees from Georgia Institute of Technology in 1999 and 2002, respectively. She is now an assistant professor with the Department of Electrical and Computer Engineering at North Carolina State University. Her research interests are in mobile and secure computing, QoS-sensitive networking protocols, mobility, security, and resource management in single- and multihop networks. Dr. Wang has served on program committees for IEEE INFOCOM, ICC, and ICCCN. She is also serving on the editorial board of *Computer Networks* (COMNET; Elsevier).

Jianliang Xu is an assistant professor in the Department of Computer Science at Hong Kong Baptist University. He received a BEngr degree in computer science and engineering from Zhejiang University, Hangzhou, China, in 1998, and a PhD degree in computer science from Hong Kong University of Science and Technology in 2002. His research interests include mobile and pervasive computing, Web content delivery, and wireless networks. He has served as a program committee member and an executive committee member for several international conferences including IEEE INFOCOM '04, IEEE MDM '04, and ACM SAC '04. He is serving as an executive committee member of the ACM Hong Kong chapter.

Wai Gen Yee's area of specialization is in distributed database systems. His PhD dissertation (Georgia Institute of Technology, 2003) focused on the performance of synchronization of mobile databases in intermittently connected environments. In addition to distributed databases, Wai Gen is also studying peer-to-peer systems and information retrieval. In the fall of 2003, Wai Gen joined the faculty of the computer science department at the Illinois Institute of Technology (USA)as an assistant professor.

Baihua Zheng is an assistant professor in the School of Information Systems at Singapore Management University. She received a BEngr degree in computer science and engineering from Zhejiang University in 1999 and a PhD degree in computer science from Hong Kong University of Science and Technology in 2003. Her research interests include mobile/pervasive computing and spatial databases. She is a member of ACM and IEEE, and has served as a program committee member for ACM SAC '04.

Index

A

B

C

D

E

F

G

T

U

V

W

X